Crystal Dislocations: Their Impact on Physical Properties of Crystals

Crystal Dislocations: Their Impact on Physical Properties of Crystals

Special Issue Editor
Peter Lagerlof

MDPI • Basel • Beijing • Wuhan • Barcelona • Belgrade

MDPI

Special Issue Editor
Peter Lagerlof
Case Western Reserve University
USA

Editorial Office
MDPI
St. Alban-Anlage 66
4052 Basel, Switzerland

This is a reprint of articles from the Special Issue published online in the open access journal *Crystals* (ISSN 2073-4352) from 2017 to 2018 (available at: https://www.mdpi.com/journal/crystals/special_issues/Crystal_Dislocations)

For citation purposes, cite each article independently as indicated on the article page online and as indicated below:

LastName, A.A.; LastName, B.B.; LastName, C.C. Article Title. *Journal Name* **Year**, *Article Number, Page Range.*

ISBN 978-3-03897-465-9 (Pbk)
ISBN 978-3-03897-466-6 (PDF)

Contents

About the Special Issue Editor

K. Peter D. Lagerlof received his Ph.D. from Case Western Reserve University in 1984. Following a post-doctoral fellowship at CWRU, he worked at Chalmers University of Technology and the Swedish Ceramic Institute in Gothenburg, Sweden, and in 1987 joined the faculty at CWRU as an Assistant Professor; he was promoted to Associate Professor in 1994. His major interest is in the mechanical properties of ceramic materials, particularly how low temperature deformation twinning is related to deformation via dislocation slip at elevated temperatures. This relationship was studied both theoretically and experimentally in sapphire (α-Al$_2$O$_3$), and was generalized to other materials systems, including bcc metals and other ceramics. He has over 80 publications in peer-reviewed journals.

crystals

MDPI

Editorial

Crystal Dislocations: Their Impact on Physical Properties of Crystals

Peter Lagerlof[iD]

Department of Materials Science and Engineering, Case Western Reserve University, 10900 Euclid Avenue, Cleveland, OH 44106-7204, USA; peter.lagerlof@case.edu; Tel.: +1-216-368-6488

Received: 1 November 2018; Accepted: 1 November 2018; Published: 3 November 2018

It is rare to find technical applications involving a material of any crystal structure that is not impacted by dislocations—which affect the material's mechanical properties, interfaces, martensitic phase transformations, crystal growth, and electronic properties, to name a few. In many systems the properties are controlled by the formation of partial dislocations separated by a stacking fault; for example, plastic deformation via dislocation slip and plastic deformation via deformation twinning. In other systems the electronic properties are affected by acceptor or donor states associated with changes of the electronic state of atoms corresponding to the dislocation core of perfect or partial dislocations. Crystal growth often occurs at growth ledges, which can be associated with dislocations. This Special Issue on "Crystal Dislocations: Their Impact on Physical Properties of Crystals" covers a broad range of physical properties involving dislocations and their impact on crystal properties, and contains a mixture of review articles and original contributions.

The influence of dislocations on properties in transition metal oxides was reviewed by Szot et al. [1]. Their review focuses on the important role of dislocations in the insulator-to-metal transition and for redox processes in prototypic binary and ternary oxides (such as TiO_2 and $SrTiO_3$) examined using transition electron microscopy (TEM) and scanning probe microscopy (SPM) techniques combined with classical etch pits methods. Since dislocations play a critical role in the plastic deformation of materials, several contributions deal with a detailed understanding of dislocations and their influence on the mechanical properties of materials. The plastic deformation in magnesium oxide crystals, MgO, being an archetypal ionic ceramic with refractory properties of interest in several fields of applications such as ceramic materials fabrication, nanoscale engineering, and earth sciences, was reviewed by Amodeo et al. [2]. Their review describes how a combined approach of macro-mechanical tests, multi-scale modeling, nano-mechanical tests, and high-pressure experiments and simulations have helped to improve the understanding of the mechanical behavior of MgO and elementary dislocation-based processes. The structure of the basal edge dislocations in ZnO was examined by Nakamura et al. [3]. The dislocation core structure was observed using scanning transmission electron microscopy (STEM) at atomic resolution, and it was found that a basal edge dislocation dissociated into $1/3\langle 1\bar{1}00\rangle$ and $1/3\langle 10\bar{1}0\rangle$ partial dislocations on the (0001) plane, separated by a stacking fault with a stacking fault energy of 0.14 J/m^2. The importance of stoichiometry on the mechanical properties of $SrTiO_3$ was also examined by Nakamura et al. [4] through studies of the room-temperature plasticity of strontium titanate crystals grown from source materials having varying Sr/Ti ratios. It was found that the flow stresses of $SrTiO_3$ crystals grown from a powder with a Sr/Ti ratio of 1.04 were almost independent of the strain rate which, in turn, is believed to be due to the high dislocation mobility in such crystals.

Several contributions examined the impact of dislocations on the mechanical properties in metallic systems. Keshavarz et al. [5] developed a framework to obtain the flow stress of nickel-based super alloys as a function of $\gamma - \gamma'$ morphology, as the yield strength is a major factor in the design of such alloys. In order to obtain the flow stress, non-Schmid crystal plasticity constitutive models at two length scales were employed and bridged through a homogenized multi-scale framework. The importance of stacking faults in *Al7075* formed through precipitate and dislocation interactions was examined by Li et al. [6] using high-resolution electron microscopy. Stacking faults due to Frank partial dislocations were found following deformation using low strain and strain-rates, and extrinsic stacking faults were found to be surrounded by dislocations and precipitates whereas an intrinsic stacking fault was found between two Guinier-Preston II (GP II) zones when the distance of the two GP II zones was 2 nm. The anisotropic plastic behavior in two metallic systems was examined; Wei et al. [7] reported on the anisotropic plastic behavior in aluminum single crystals by crystal plasticity finite element methods and Bian et al. [8] examined the anisotropic plastic deformation in the compression of single crystalline copper nanoparticles. The dislocation-dopant ions interaction in ionic crystals by strain-rate cycling tests with the Blaha effect measurement was reviewed by Kohzuki [9]. The strain-rate cycling test during Blaha effect measurement has successively provided information on the dislocation motion breaking away from the strain fields around dopant ions with the help of thermal activation, and seems to separate the contributions arising from the interaction between dislocation and the point defects and those from dislocations themselves during the plastic deformation of ionic crystals.

The importance of dislocations on phase transformation was examined by Hu et al. [10], in which they studied phase transformation and hydrogen storage properties of an $La_{7.0}Mg_{75.5}Ni_{17.5}$ hydrogen storage alloy. Differential thermal analysis showed that the initial hydrogen desorption temperature of its hydride was 531 K and, compared to Mg and Mg_2Ni, $La_{7.0}Mg_{75.5}Ni_{17.5}$ was found to be a promising hydrogen storage material that demonstrates fast adsorption/desorption kinetics as a result of the formation of an $La - H$ compound. Since some phase transformations, e.g., martensitic phase transformations, involve the motion of interface dislocations, it is important to understand the properties of such dislocations. In addition, the details of partial dislocation making up perfect dislocations are important for both the plastic deformation of materials via the dislocation motion of dissociated perfect dislocations and deformation twinning involving the motion of partial dislocations. One way of producing such dislocations is to fuse crystals with surfaces having controlled crystallographic planes to make low-angle grain boundaries for subsequent characterization. Tochigi et al. [11] reviewed the dislocation structures in $\alpha - Al_2O_3$, obtained using systematically fabricated alumina bi-crystals with low-angle grain boundaries and characterized using transmission electron microscopy (TEM). Wang et al. [12] examined the interface effects on screw dislocations in *Al/TiC* hetero-structures, which was used as a model interface to study the unstable stacking fault energies and dislocation properties of interfaces. It was found that the mismatch of lattice constants and shear modulus at the interface resulted in changes of the stacking fault. However, in many cases it is desirable to eliminate interface dislocations altogether, and Montalenti et al. [13] reviewed dislocation-free *SiGe/Si* hetero-structures grown on deeply patterned $Si(001)$, providing possibilities of growing micron-sized *Ge* crystals largely free of thermal stress and hosting dislocations only in a small fraction of their volume. They also analyzed the role played by the shape of the pre-patterned substrate in directly influencing the dislocation distribution.

Finally, the effect of dislocations on peak broadening anisotropy and the contrast factor in metal alloys characterization using X-ray diffraction was reviewed by Simm [14]. Peak broadening anisotropy, in which the broadening of a diffraction peak does not change smoothly with d-spacing, is an important aspect of diffraction peak profile analysis (DPPA) and is a valuable method to understand the microstructure and defects present in the material examined. There are numerous approaches to deal with this anisotropy in metal alloys, which can be used to gain information about the dislocation types present in a sample and the amount of planar faults. However, there are problems in determining which method to use and the potential errors that can result, in particular for hexagonal close-packed (*hcp*) alloys. There is, however, a distinct advantage of broadening anisotropy in that it provides a unique and potentially valuable way to develop crystal plasticity and work-hardening models.

The present Special Issue on "Crystal Dislocations: Their Impact on Physical Properties of Crystals" can be considered as a status report reviewing the progress that has been achieved over the past several years in several subject areas affected by crystal dislocations.

References

1. Szot, K.; Rodenbücher, C.; Bihlmayer, G.; Speier, W.; Ishikawa, R.; Shibata, N.; Ikuhara, Y. Influence of Dislocations in Transition Metal Oxides on Selected Physical and Chemical Properties. *Crystals* **2018**, *8*, 241. [CrossRef]
2. Amodeo, J.; Merkel, S.; Tromas, C.; Carrez, P.; Korte-Kerzel, S.; Cordier, P.; Chevalier, J. Dislocations and Plastic Deformation in MgO Crystals: A Review. *Crystals* **2018**, *8*, 240. [CrossRef]
3. Nakamura, A.; Tochigi, E.; Nagahara, R.; Furushima, Y.; Oshima, Y.; Ikuhara, Y.; Yokoi, T.; Matsunaga, K. Structure of the Basal Edge Dislocation in ZnO. *Crystals* **2018**, *8*, 127. [CrossRef]
4. Nakamura, A.; Yasufuku, K.; Furushima, Y.; Toyoura, K.; Lagerlöf, K.P.D.; Matsunaga, K. Room-Temperature Plastic Deformation of Strontium Titanate Crystals Grown from Different Chemical Compositions. *Crystals* **2017**, *7*, 351. [CrossRef]
5. Keshavarz, S.; Molaeinia, Z.; Reid, A.C.E.; Langer, S.A. Morphology Dependent Flow Stress in Nickel-Based Superalloys in the Multi-Scale Crystal Plasticity Framework. *Crystals* **2017**, *7*, 334. [CrossRef]
6. Li, S.; Luo, H.; Wang, H.; Xu, P.; Luo, J.; Liu, C.; Zhang, T. Stable Stacking Faults Bounded by Frank Partial Dislocations in Al7075 Formed through Precipitate and Dislocation Interactions. *Crystals* **2017**, *7*, 375. [CrossRef]
7. Wei, P.; Lu, C.; Liu, H.; Su, L.; Deng, G.; Tieu, K. Study of Anisotropic Plastic Behavior in High Pressure Torsion of Aluminum Single Crystal by Crystal Plasticity Finite Element Method. *Crystals* **2017**, *7*, 362. [CrossRef]
8. Bian, J.; Zhang, H.; Niu, X.; Wang, G. Anisotropic Deformation in the Compressions of Single Crystalline Copper Nanoparticles. *Crystals* **2018**, *8*, 116. [CrossRef]
9. Kohzuki, Y. Study on Dislocation-Dopant Ions Interaction in Ionic Crystals by the Strain-Rate Cycling Test during the Blaha Effect. *Crystals* **2018**, *8*, 31. [CrossRef]
10. Hu, L.; Nan, Ru.; Li, Ji.; Gao, L.; Wang, Yu. Phase Transformation and Hydrogen Storage Properties of an $La_{7.0}Mg_{75.5}Ni_{17.5}$ Hydrogen Storage Alloy. *Crystals* **2017**, *7*, 316. [CrossRef]
11. Tochigi, E.; Nakamura, A.; Shibata, N.; Ikuhara, Y. Dislocation Structures in Low-Angle Grain Boundaries of α-Al_2O_3. *Crystals* **2018**, *8*, 133. [CrossRef]
12. Wang, J.; Sun, T.; Xu, W.; Wu, X.; Wang, R. Interface Effects on Screw Dislocations in Heterostructures. *Crystals* **2018**, *8*, 28. [CrossRef]
13. Montalenti, F.; Rovaris, F.; Bergamaschini, R.; Migli, L.; Salvalaglio, M.; Isella, G.; Isa, F.; von Känel, H. Dislocation-Free SiGe/Si Heterostructures. *Crystals* **2018**, *8*, 257. [CrossRef]
14. Simm, T.H. Peak Broadening Anisotropy and the Contrast Factor in Metal Alloys. *Crystals* **2018**, *8*, 212. [CrossRef]

crystals

MDPI

Article

Structure of the Basal Edge Dislocation in ZnO

Atsutomo Nakamura [1,*] , Eita Tochigi [2] , Ryota Nagahara [1], Yuho Furushima [1], Yu Oshima [1] ,
Yuichi Ikuhara [2,3], Tatsuya Yokoi [1] and Katsuyuki Matsunaga [1,3]

[1]   Department of Materials Physics, Nagoya University, Furo-cho, Chikusa-ku, Nagoya 464-8603, Japan;
      nagahara.ryouta@e.mbox.nagoya-u.ac.jp (R.N.); furushima.yuuho@b.mbox.nagoya-u.ac.jp (Y.F.);
      ooshima.yuu@f.mbox.nagoya-u.ac.jp (Y.O.); yokoi@mp.pse.nagoya-u.ac.jp (T.Y.);
      kmatsunaga@nagoya-u.jp (K.M.)
[2]   Institute of Engineering Innovation, University of Tokyo, 2-11-16 Yayoi, Bunkyo-ku, Tokyo 113-8686, Japan;
      tochigi@sigma.t.u-tokyo.ac.jp (E.T.); ikuhara@sigma.t.u-tokyo.ac.jp (Y.I.)
[3]   Nanostructures Research Laboratory, Japan Fine Ceramics Center, 2-4-1, Mutsuno, Atsuta-ku,
      Nagoya 456-8587, Japan
*   Correspondence: anaka@nagoya-u.jp; Tel.: +81-52-789-3366

Received: 19 January 2018; Accepted: 6 March 2018; Published: 8 March 2018

Abstract: Basal dislocations having a Burgers vector of $1/3<2\bar{1}\bar{1}0>$ in zinc oxide (ZnO) with the
wurtzite structure are known to strongly affect physical properties in bulk. However, the core
structure of the basal dislocation remains unclear. In the present study, ZnO bicrystals with a
$\{2\bar{1}\bar{1}0\}/<01\bar{1}0>$ 2° low-angle tilt grain boundary were fabricated by diffusion bonding. The resultant
dislocation core structure was observed by using scanning transmission electron microscopy (STEM)
at an atomic resolution. It was found that a basal edge dislocation in α-type is dissociated into two
partial dislocations on the (0001) plane with a separation distance of 1.5 nm, indicating the glide
dissociation. The Burgers vectors of the two partial dislocations were $1/3<1\bar{1}00>$ and $1/3<10\bar{1}0>$, and
the stacking fault between the two partials on the (0001) plane has a formation energy of 0.14 J/m².
Although the bicrystals have a boundary plane of $\{2\bar{1}\bar{1}0\}$, the boundary basal dislocations do not
exhibit dissociation along the boundary plane, but along the (0001) plane perpendicular to the
boundary plane. From DFT calculations, the stacking fault on the (0001) plane was found to be much
more stable than that on $\{2\bar{1}\bar{1}0\}$. Such an extremely low energy of the (0001) stacking fault can realize
transverse dissociation of the basal dislocation of ZnO.

Keywords: zinc oxides (ZnO); wurtzite structure; line defects; low-angle grain boundaries; scanning
transmission electron microscopy (STEM); generalized stacking fault (GSF) energy

---

1. Introduction

The wurtzite structure is a stable crystal structure for binary compound semiconducting materials
such as GaN, AlN, CdS, ZnO, and so on. In these materials, ZnO is a representative wide and direct
band gap semiconductor and has long been used as a main constituent material of varistors because
of its highly nonlinear current–voltage characteristics [1–4]. In addition, ZnO has received broad
interest due to its high electron mobility, high thermal conductivity and large exciton binding energy
suitable for a variety of applications [1–3]. Although a large number of studies have been performed
so far, most of them have aimed to control point defects and grain boundaries, both of which play
a critical role in functional properties [1–6]. On the other hand, it is likely that atomic structures of
dislocations, which dominate mechanical properties and also affect functional properties, have been
poorly understood. This situation is also applied to the other wurtzite crystals.

Since the Burgers vector of the $1/3<2\bar{1}\bar{1}0>$ dislocation on the (0001) basal plane corresponds to
the minimum translation of the wurtzite structure, the $<2\bar{1}\bar{1}0>$ slip on (0001) can work as an easy slip

system in ZnO [7,8]. It is interesting that such a basal dislocation in ZnO brings about localized energy levels within the band-gap, which may influence local optical properties [8]. It is also remarkable that hardness and flow deformation stresses of ZnO can be affected by light exposure [9]. However, atomic structures of basal dislocations in ZnO still remain unclear, as they have not been investigated using scanning transmission electron microscopy (STEM) with an atomic resolution.

There exist four candidates for core structures of the basal edge dislocation in ZnO. This is because in the wurtzite structure the (0001) slip can occur on either of two inequivalent atomic planes [10–13], which are shown in Figure 1 as "glide type" and "shuffle type", and moreover, the [0001] and [000$\bar{1}$] directions are not equivalent due to lack of inversion symmetry. In the case of the glide type, the basal edge dislocation will have its slip plane between the narrowly spaced Zn and O planes along [0001]. In this case, a basal dislocation has the potential to dissociate into two partial dislocations according to the Shockley partial reaction [10,13]. On the other hand, the shuffle type dislocation has its slip plane between the widely spaced Zn and O planes, where the basal dislocation cannot dissociate into partials because of the absence of stable stacking faults resulting from the crystal structure. Thus, two candidates of core structures are present due to their slip planes. Additionally, since there is no symmetry along [0001], two kinds of cores can be formed for both of the glide and shuffle type dislocations depending on whether an extra half plane of the dislocations is inserted toward [0001] or [000$\bar{1}$] [13]. Although there are four candidates, the core structure of the basal edge dislocation has not, thus far, been experimentally characterized. Thus, it is of great importance to reveal a core structure of the basal edge dislocation for understanding an effect of dislocations on material properties in ZnO.

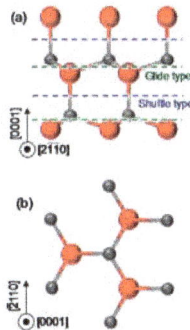

Figure 1. Schematic illustration showing the crystal structure of ZnO with the wurtzite structure. The arrangement of ions along [2$\bar{1}\bar{1}$0] and [0001] are represented in (**a**,**b**), respectively. Large red circles correspond to oxygen ions while small gray circles do zinc ions. Locations of slip planes for glide type and shuffle type dislocations are indicated in (**a**) by dotted lines.

In the present study, therefore, we focus on the core structure of the basal edge dislocation in wurtzite ZnO. Bicrystal experiments with artificial fabrication of a low-angle grain boundary are performed. The bicrystal experiment has proven to be an efficient method for evaluating atomic structures and properties of dislocations [14–17]. This is because a low-angle tilt grain boundary consists of periodically arranged edge dislocations, which have a Burgers vector perpendicular to the boundary plane. Accordingly, by fabricating a bicrystal with a controlled crystallographic relationship, periodical dislocations with a desirable Burgers vector can be produced at the boundary. A ZnO bicrystal with a {2$\bar{1}\bar{1}$0}/<01$\bar{1}$0> low-angle tilt grain boundary was fabricated by diffusion bonding of two single crystals. In this case, basal edge dislocations that have a Burgers vector of 1/3<2$\bar{1}\bar{1}$0> should be periodically formed at the boundary. Resultant dislocation structures were characterized by transmission electron microscopy (TEM) and STEM. Moreover, formation energies of stacking faults in

ZnO were evaluated by density functional theory (DFT) calculations to understand the origin of the observed atomic structure of the basal edge dislocation.

2. Experimental Procedure

Wurtzite ZnO single crystals grown by the hydrothermal synthesis were used to fabricate a bicrystal with a {$2\bar{1}\bar{1}0$}/<$01\bar{1}0$> low-angle tilt grain boundary. Figure 2a shows schematic illustrations of two pieces of the single-crystal plates used for fabricating the bicrystal. Each single crystal plate was inclined at +1° or −1° from the ($2\bar{1}\bar{1}0$) plane around the common [$01\bar{1}0$] axis. The size of the plates was set as $10 \times 5 \times 1$ mm³, and their surfaces were polished by a diamond slurry and a colloidal silica to achieve a mirror finish. The two single-crystal plates were then joined by diffusion bonding at 1120 °C in air for 10 h under a uniaxial load of 25 N. Owing to the 1° inclination of the crystallographic orientation of the individual grains, the tilt misorientation angle of 2° was introduced at the bonding interface. As a result, a ZnO bicrystal with a {$2\bar{1}\bar{1}0$}/<$01\bar{1}0$> 2° tilt grain boundary was fabricated as shown in Figure 2b. Here, the relation between the spacing between periodic basal dislocations d and the misorientation angle θ in a low-angle tilt grain boundary is given by the equation of $\theta = b/d$, where b is the magnitude of the edge component of the Burgers vector of boundary dislocations, according to the Frank's formula [18]. Figure 2c shows the d, b and θ at the boundary. Therefore, if $\theta = 2°$ and $b = 0.325$ nm ($|1/3<2\bar{1}\bar{1}0>|$) [1] are substituted in the above equation, the spacing d is obtained as 9.3 nm.

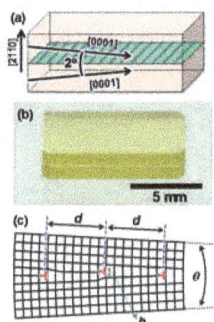

Figure 2. Schematics and optical micrograph of the fabricated ZnO bicrystal with a ($2\bar{1}\bar{1}0$)/[$01\bar{1}0$] 2° tilt grain boundary. (**a**) Schematic illustration showing the crystallographic orientations of two pieces of used ZnO single-crystal plates. (**b**) A fabricated ZnO bicrystal. (**c**) Schematic illustration showing periodic dislocations at the boundary.

Since ZnO has a polar crystal structure along the [0001] direction according to the wurtzite structure, the crystal lattice structure along [0001] differs from the one along the opposite direction of [000$\bar{1}$]. It should be mentioned that the [0001] directions in the fabricated bicrystals were set up to face into the bonding interface as shown in Figure 2a. In this case, an extra half plane of dislocations introduced at the boundary is inserted toward [000$\bar{1}$] when viewed from the dislocation cores. This type of dislocation is called "α-dislocation", while the other type of dislocation with the extra half plane toward [0001] is called "β-dislocation" [13]. There is a difference between α-dislocation and β-dislocation in terms of atomic species at the edge of extra half plane; Zn or O.

The grain boundary of bicrystals thus fabricated was observed by TEM and STEM. Specimens for the observations were prepared using a standard technique involving mechanical grinding to a thickness of 60 μm, dimpling to a thickness of about 30 μm and ion beam milling to electron transparency. Observations were conducted by a conventional TEM (Hitachi H-800, 200 kV, Japan) and an atomic resolution STEM (JEOL JEM-ARM200F, 200 kV, Japan).

3. DFT Calculations

Stacking fault energies in ZnO were calculated using DFT calculations based on the projector augmented wave (PAW) method as implemented in VASP code [19,20]. In the PAW potentials, Zn $3d4s$ and O $2s2p$ electrons were treated as valence electrons. The generalized gradient approximation (GGA) parameterized by Perdew, Burke and Ernzerhof was used for the exchange-correlation term [21]. To correct for the on-site Coulomb interaction of the $3d$ orbitals of Zn atoms, the rotationally invariant $+U$ method [22] was applied with $U = 8$ eV [23]. Wavefunctions were expanded by plane waves with a cut-off energy of 600 eV.

Since the glide type dislocation dissociates into two partial dislocations according to the Shockley reaction of 1/3<2$\overline{1}\overline{1}$0> → 1/3<10$\overline{1}$0> + 1/3<1$\overline{1}$00>, the sheared structure along <10$\overline{1}$0> should be considered for the stacking faults on (0001). Here, the (0001) atomic layers consist of like ions, and accordingly, each (0001) plane becomes a polar plane. Moreover, the <10$\overline{1}$0> direction does not have mirror symmetry. In order to calculate stacking fault energies on (0001), therefore, 72-atom supercells containing three slabs with twelve {0001} atomic layers each were employed. In this case, no vacuum layer was involved in the supercells, because the polar surfaces of the atomic slabs may induce spurious electric-dipole interactions normal to the slab surfaces. It is noted that, when three slabs are relatively displaced by a same amount toward three different <10$\overline{1}$0> directions, three stacking faults then produced in the supercells are equivalent to one another [24,25]. Brillion zone integration was performed with an 8 × 8 × 1 k-point mesh for the stacking faults on (0001). In contrast, supercells containing two atomic slabs with a vacuum layer of 1.6 nm were employed to calculate stacking fault energies on (2$\overline{1}$10) as the (2$\overline{1}$10) planes are not polar. In this case, the supercell for the stacking fault on (2$\overline{1}$10) contains 40 atoms (corresponding to 10 atomic layers) and a stacking fault. Brillion zone integration was performed with a 6 × 6 × 1 k-point mesh for the stacking faults on (2$\overline{1}$10). Structure optimizations for all the calculations were conducted until residual forces of atoms reached to less than 0.01 eV/Å.

4. Results and Discussion

Figure 3 shows a typical bright field TEM image of the (2$\overline{1}\overline{1}$0)/[01$\overline{1}$0] 2° tilt grain boundary taken along [01$\overline{1}$0]. As can be seen from the image, the dot-like contrasts were distinctly observed at the boundary, which should be caused by boundary dislocations. A spacing between the contrasts was estimated to be about 9 nm, which is in good agreement with the expected spacing between dislocations of b = 1/3[2$\overline{1}\overline{1}$0] at the boundary (9.3 nm, see Section 2). Diffraction spots originating from both of the two grains around the boundary were separated from each other by approximately 2°, which coincides with the designed tilt angle. Thus, the (2$\overline{1}\overline{1}$0)/[01$\overline{1}$0] 2° tilt grain boundary was successfully fabricated, and it was suggested that 1/3[2$\overline{1}\overline{1}$0] basal dislocations were periodically introduced at the boundary.

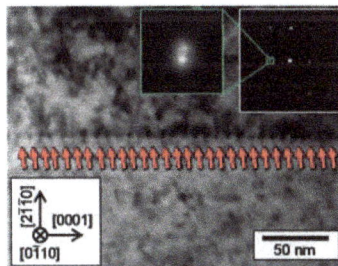

Figure 3. A typical bright field TEM image of the (2$\overline{1}\overline{1}$0)/[01$\overline{1}$0] 2° tilt grain boundary taken along [01$\overline{1}$0]. A corresponding selected-area diffraction pattern is inset right above.

Figure 4a shows a typical high-angle annular dark field (HAADF) STEM image taken from one of the dislocations at the grain boundary. In this image, the bright spots correspond to Zn columns. It can be seen that two lattice discontinuities clearly appear in the Burgers circuits on the image. This means that a dislocation dissociates into two partial dislocations. Here, the spacings of the bright points on the HAADF-STEM image correspond to the magnitude of $1/2[0001]$ along $[0001]$ and the magnitude of $1/6[2\bar{1}\bar{1}0]$ along $[2\bar{1}\bar{1}0]$. Accordingly, the Burgers circuits on the image show that the each partial dislocation has the same edge component of $1/6[2\bar{1}\bar{1}0]$, and thus the total edge component is $1/3[2\bar{1}\bar{1}0]$. It was confirmed that the periodic dislocations in Figure 3 are characterized as basal dislocations represented by the same Burgers circuits in Figure 4a. Figure 4b shows an inverse Fast-Fourier-Transformed (FFT) image reconstructed from a mask-applied FFT image of the area shown in Figure 4a. The separation distance between the two partial dislocations is 1.5 nm along $[2\bar{1}\bar{1}0]$, while the two seem to be adjacently located along $[0001]$. It can thus be said that the basal edge dislocation with $\boldsymbol{b} = 1/3[2\bar{1}\bar{1}0]$ dissociates into two partial dislocations with the edge component of $1/6[2\bar{1}\bar{1}0]$ along (0001). Such a feature of the basal dislocation dissociating into two partials ensures that the basal dislocations at the fabricated boundary have the core structure of the glide-type dislocation. In this case, each partial dislocation should have the Burgers vector of $\boldsymbol{b_1} = 1/3[10\bar{1}0]$ or $\boldsymbol{b_2} = 1/3[1\bar{1}00]$, according to the Shockley partial reaction of the glide type basal dislocation. In fact, the edge component of $1/6[2\bar{1}\bar{1}0]$ corresponds to the projection vector of the partial dislocations with $\boldsymbol{b_1}$ and $\boldsymbol{b_2}$ onto the $(01\bar{1}0)$ plane. This also supports that the basal dislocations dissociate into two partials according to the Shockley partial reaction.

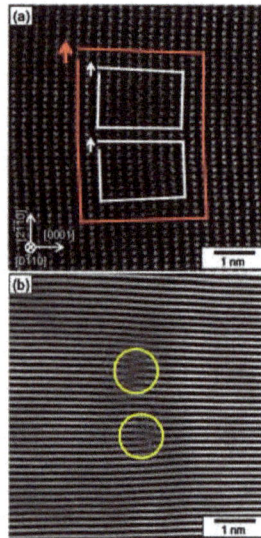

Figure 4. (a) A typical HAADF-STEM image taken from one of the contrasts on the image of Figure 3. Burgers circuit for a basal edge dislocation is drawn by red color line while Burgers circuits for two partial dislocations are drawn by white color lines. (b) The inverse Fast-Fourier-Transformed (FFT) image reconstructed from a mask-applied FFT image of (a). Yellow circles indicate the positions of cores of partial dislocations.

When a dislocation dissociates into two partial dislocations, a stacking fault is formed between the partials. In the present case, the stacking fault is formed along the (0001) plane perpendicularly to the $(2\bar{1}\bar{1}0)$ boundary plane. The fault vector of the stacking fault corresponds to the Burgers vector of the partial dislocation of $1/3[10\bar{1}0]$ (or $1/3[1\bar{1}00]$). Here, the separation distance between partials is determined by a balance of the two forces acting in the dissociated dislocation, that is, the repulsive elastic force between partial dislocations and the attractive force due to the stacking fault energy. According to the Peach-Koehler equation [10], the balance in the present dissociated basal dislocation can be expressed by the following equation,

$$\gamma = \frac{\mu b_p{}^2 (2 + \nu)}{8 \pi r\,(1 - \nu)} \qquad (1)$$

where γ is the stacking fault energy, μ is the shear modulus (44.3 GPa [26]), ν is the Poisson' ratio (0.3177 [26]), b_p is the magnitude of Burgers vectors of the partial dislocations and r is the separation distance. See Supplementary Materials for the derivation of Equation (1). By substituting 1.5 nm for r in Equation (1), the formation energy of the (0001) stacking fault with the fault vector of $1/3[10\bar{1}0]$ was estimated to be about 0.14 J/m^2. It should be mentioned that the equation is based on a conventional elastic theory for an isotropic elastic medium. Accordingly, the estimation may slightly lose accuracy due to the anisotropic elasticity of ZnO. Here, the ratio of c_{44} to $(c_{11}-c_{12})/2$ in ZnO was calculated to be 0.96 according to the elastic constants in the former report [26]. Since the value is close to 1, ZnO seemingly exhibits almost isotropic elasticity.

Figure 5a,b shows schematic illustrations of a dissociated dislocation and dissociated dislocations array at the boundary, respectively. The stacking sequence of the $(2\bar{1}\bar{1}0)$ plane along $[2\bar{1}\bar{1}0]$ in ZnO corresponds to $ABAB\dots$, as shown in (a). Figure 5b explains the periodic formation of dissociated basal dislocations and stacking faults between the two partial dislocations. In general, dislocations at symmetrical tilt grain boundaries tend to dissociate on the boundary plane [16,17] as shown in Figure 5c. This is believed to be due to the fact that total elastic energy derived from dislocations array is minimized when the dislocations are located in such a linear arrangement according to the elastic theory [10]. Accordingly, it is noteworthy that the dislocations dissociated perpendicular to the boundary plane as shown in Figure 5b. Here, a lower stacking fault energy makes a longer separation distance at a dissociated dislocation according to the Peach-Koehler equation [10]. In addition, two partial dislocations with a longer separation distance do not suffer excess elastic energies so much as those with a shorter separation distance since their excess elastic energies are accommodated with increasing separation distance. It is thus suggested that stacking fault energy in ZnO should be relatively lower on the (0001) basal plane than the $(2\bar{1}\bar{1}0)$ boundary plane, resulting in the dissociation on (0001).

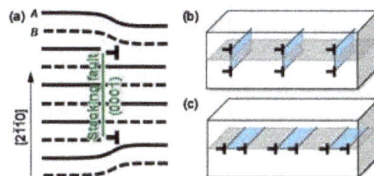

Figure 5. (**a**) Schematic illustration showing the observed structure of a dissociated basal dislocation. (**b**) Schematic illustration showing an array of the dissociated basal dislocations at the boundary. (**c**) Schematic illustration of usual structure of boundary dislocations, where dislocations dissociate on the boundary plane.

In order to confirm such a difference in stacking fault energy, DFT calculations were made for stacking faults on different planes of (0001) and $(2\bar{1}\bar{1}0)$. Figure 6a,b shows a schematic of the $[10\bar{1}0]$ vector on the (0001) plane and the energy curve of the stacking faults along $[10\bar{1}0]$ on the (0001) plane in ZnO. The $u/\,|b_m|$ is employed as the horizontal axis in the energy curve, where b_m is $[10\bar{1}0]$ and u is the displacement. The energies gradually increase up to the local maximum at $u/\,|b_m| = 1/6$ and inversely decrease to the local minimum at $u/\,|b_m| = 1/3$. Then, they again rapidly increase up to $u/\,|b_m| = 2/3$ and inversely decrease to the initial state at $u/\,|b_m| = 1$. The value at the local minimum of $u/|b_m| = 1/3$ represents the energy of the stacking fault formed with the $1/3[10\bar{1}0]$ partial dislocation, which is the Shockley partial of the glide type basal dislocation. It can be said that it is energetically favorable for a glide-type basal dislocation to dissociate into two Shockley partials along (0001).

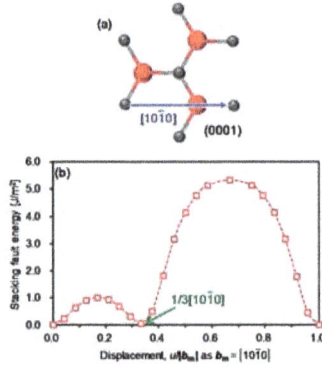

Figure 6. (a) Schematic illustration showing the $[10\bar{1}0]$ vector on (0001). (b) Stacking fault energies along $[10\bar{1}0]$ on (0001) in ZnO. It can be seen that the energy indicates the local minimum at the displacement of $1/3[10\bar{1}0]$.

Figure 7a,b shows the perfect crystal structure and the relaxed atomic structure of the stacking fault at $u/|b_m| = 1/3$. It was found that the relaxed stacking fault structure in Figure 7b is very close to the rigidly sheared atomic structure and individual atoms did not shift distinctly from the original positions before structure optimization. This also ensures the rather low energy of the stacking fault. The calculated stacking fault energy of 73 mJ/m^2 is in good agreement with that of 56 mJ/m^2 in the previous study [27]. It should be mentioned that the calculated value of the stacking fault energy at $u/|b_m| = 1/3$ are 67 mJ/m^2 lower than the experimental value of 0.14 J/m^2, which was estimated from the separation distance between partial dislocations. Although the difference between the theoretical and experimental stacking fault energies is not distinct in terms of the absolute value, such a difference was also reported in Al$_2$O$_3$ [28,29]. This difference may be brought about by temperature-dependent effects such as vibrational entropy.

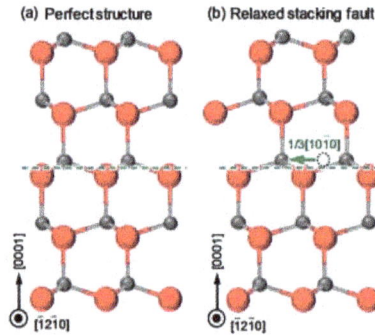

Figure 7. Schematic illustrations showing atomic structure of a stacking fault on (0001). (a) Perfect structure of ZnO along $[\bar{1}2\bar{1}0]$. (b) A relaxed stacking fault structure with the fault vector of $1/3[10\bar{1}0]$ after structure optimization.

Figure 8a,b shows a schematic of the [0001] and [01$\bar{1}$0] vectors on (2$\bar{1}$$\bar{1}$0) and stacking fault energies on (2$\bar{1}$$\bar{1}$0) as functions of $u/|\boldsymbol{b}_{m'}|$ and $u/|\boldsymbol{b}_c|$, where $\boldsymbol{b}_{m'}$ = [01$\bar{1}$0] and $\boldsymbol{b}_c$ = [0001], respectively. It should be noticed that the stacking faults on (2$\bar{1}$$\bar{1}$0) take the minimum energy of 0.40 J/m^2 at the center of the energy map in (b), where the fault vector is 1/2[01$\bar{1}$1]. Here, the 1/2[01$\bar{1}$1] dislocations cannot be induced by dissociation of a basal dislocation because the size of 1/2[01$\bar{1}$1] is larger than that of 1/3<2$\bar{1}$$\bar{1}$0>. In addition, the stacking fault on (2$\bar{1}$$\bar{1}$0) with the fault vector of 1/3[10$\bar{1}$0] due to the Shockley partial reaction has an energy of over 2.4 J/m^2, as shown in Figure 8c. Such a stacking fault energy is close to twice the (2$\bar{1}$$\bar{1}$0) surface energy (about 2.7 J/m^2). Moreover, the value of 2.4 J/m^2 is about 33 times higher than the stacking fault energy on (0001), which was estimated to be 73 mJ/m^2 in the present calculation. These indicate that such a stacking fault cannot be stably formed on the (2$\bar{1}$$\bar{1}$0) plane, and thus the basal dislocation does not show the Shockley partial reaction along the (2$\bar{1}$$\bar{1}$0) boundary plane. Instead, the basal dislocation can be dissociated into two partials with the rather stable stacking fault on the (0001) plane, as observed experimentally.

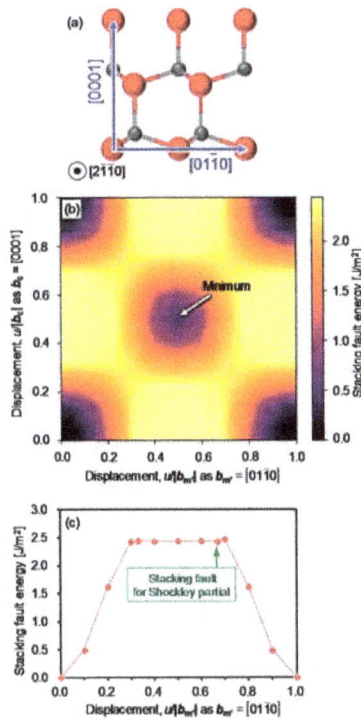

Figure 8. (**a**) Schematic illustration showing the [0001] and [01$\bar{1}$0] vectors on (2$\bar{1}$$\bar{1}$0). (**b**) A map of stacking fault energies on (2$\bar{1}$$\bar{1}$0) in ZnO. This map indicates the minimum value of 0.40 J/m^2 at the center, where the fault vector is 1/2[01$\bar{1}$1]. (**c**) Stacking fault energies along [01$\bar{1}$0] on (2$\bar{1}$$\bar{1}$0) in ZnO, where the fault vector does not have the component along [0001]. Here, the (2$\bar{1}$$\bar{1}$0) stacking fault with the 2/3[01$\bar{1}$0] fault vector indicated by the green arrow in (**c**) is the same as that with the 1/3[0$\bar{1}$10] fault vector. In addition, owing to the symmetry of the wurtzite crystal structure, the (2$\bar{1}$$\bar{1}$0) stacking fault with the 1/3[0$\bar{1}$10] fault vector is equivalent to the (2$\bar{1}$$\bar{1}$0) stacking fault with the 1/3[10$\bar{1}$0] fault vector, which corresponds to a Burgers vector of the Shockley partials in ZnO. Eventually, the energy of the stacking fault with the 2/3[01$\bar{1}$0] fault vector in (**c**) represents the energy of a stacking fault structure that would arise by the hypothetical Shockley dissociation on (2$\bar{1}$$\bar{1}$0).

5. Conclusions

The $(2\bar{1}\bar{1}0)/[01\bar{1}0]$ 2° tilt grain boundary in wurtzite ZnO was observed by using atomic-resolution STEM in order to investigate the core structure of the $1/3<2\bar{1}\bar{1}0>$ basal edge dislocation in α-type. It was found that the basal edge dislocation in ZnO dissociates into two partial dislocations along (0001) with a separation distance of 1.5 nm. This indicates that the basal edge dislocation in ZnO has a core structure of the glide-type dislocation, where the Burgers vector of the two partial dislocations should be $1/3<10\bar{1}0>$ and $1/3<1\bar{1}00>$. Applying the separation distance of 1.5 nm to the equation based on an elastic theory, the stacking fault energy on (0001) was experimentally estimated to be 0.14 J/m^2. Although the bicrystals have the $\{2\bar{1}\bar{1}0\}$ boundary plane, the basal dislocations do not exhibit dissociation along the boundary plane but dissociate along the (0001) basal plane perpendicular to the boundary plane. From DFT calculations, the stacking fault on the (0001) plane was found to be much more stable than that on $\{2\bar{1}\bar{1}0\}$. It is suggested that such an extremely low energy of the (0001) stacking fault causes dissociation of the basal dislocation on the (0001) plane.

Supplementary Materials: Supplementary texts and figures on the derivation of the Equation (1) and the initial and relaxed structures of supercells for the DFT calculations are available online at www.mdpi.com/2073-4352/8/3/127/s1.

Acknowledgments: The authors acknowledge I. Matsukura for technical assistance with TEM samples preparation. This work was mainly supported by a Grant-in-Aid for Scientific Research on Innovative Areas "Nano Informatics" (grant numbers JP25106002 and JP25106003) from Japan Society for the Promotion of Science (JSPS). A part of this study was supported by JSPS KAKENHI Grant Numbers JP15H04145, JP15K20959, JP17H06094, JP17K18982 and JP17K18983. Additionally, this work was partly supported by Nanotechnology Platform Program (Advanced Characterization Nanotechnology Platforms of Nagoya Univ. and Univ. of Tokyo) of the Ministry of Education, Culture, Sports, Science and Technology (MEXT), Japan.

Author Contributions: Atsutomo Nakamura and Katsuyuki Matsunaga conceived the research idea, and Atsutomo Nakamura designed the experiments; Atsutomo Nakamura and Eita Tochigi performed the experiments; Yuho Furushima, Yu Oshima and Atsutomo Nakamura analyzed the data; Ryota Nagahara and Atsutomo Nakamura performed the theoretical calculations; Katsuyuki Matsunaga, Tatsuya Yokoi and Yuichi Ikuhara advised the experiments and calculations; All the authors discussed the results and wrote the paper.

Conflicts of Interest: The authors declare no conflict of interest.

References

1. Pearton, S.J.; Norton, D.P.; Ip, K.; Heo, Y.W.; Steiner, T. Recent progress in processing and properties of ZnO. *Prog. Mater. Sci.* **2005**, *50*, 293–340. [CrossRef]
2. Janotti, A.; Van de Walle, C.G. Fundamentals of zinc oxide as a semiconductor. *Rep. Prog. Phys.* **2009**, *72*, 126501. [CrossRef]
3. Klingshirn, C.F.; Meyer, B.K.; Waag, A.; Hoffmann, A.; Geurts, J. *Zinc Oxide: From Fundamental Properties Towards Novel Applications*, 2010th ed.; Springer: Berlin/Heidelberg, Germany, 2010.
4. Sato, Y.; Yamamoto, T.; Ikuhara, Y. Atomic Structures and Electrical Properties of ZnO Grain Boundaries. *J. Am. Ceram. Soc.* **2007**, *90*, 337–357. [CrossRef]
5. Oba, F.; Nishitani, S.R.; Isotani, S.; Adachi, H.; Tanaka, I. Energetics of native defects in ZnO. *J. Appl. Phys.* **2001**, *90*, 824–828. [CrossRef]
6. Oba, F.; Ohta, H.; Sato, Y.; Hosono, H.; Yamamoto, T.; Ikuhara, Y. Atomic structure of [0001]-tilt grain boundaries in ZnO: A high-resolution TEM study of fiber-textured thin films. *Phys. Rev. B* **2004**, *70*, 125415. [CrossRef]
7. Osipiyan, Y.A.; Smirnova, I.S. Perfect dislocations in the wurtzite lattice. *Phys. Status Solidi* **1968**, *30*, 19–29. [CrossRef]
8. Ohno, Y.; Koizumi, H.; Taishi, T.; Yonenaga, I.; Fujii, K.; Goto, H.; Yao, T. Optical properties of dislocations in wurtzite ZnO single crystals introduced at elevated temperatures. *J. Appl. Phys.* **2008**, *104*, 073515. [CrossRef]
9. Carlsson, L.; Svensson, C. Photoplastic effect in ZnO. *J. Appl. Phys.* **1970**, *41*, 1652–1656. [CrossRef]
10. Anderson, P.M.; Hirth, J.P.; Lothe, J. *Theory of Dislocations*, 3rd ed.; Cambridge University Press: New York, NY, USA, 2017.

11. Osipiyan, Y.A.; Smirnova, I.S. Partial dislocations in the wurtzite lattice. *J. Phys. Chem. Solids* **1971**, *32*, 1521–1530. [CrossRef]
12. Hirsch, P.B. The structure and electrical properties of dislocations in semiconductors. *J. Microsc.* **1980**, *118*, 3–12. [CrossRef]
13. Holt, D.B.; Yacobi, B.G. *Extended Defects in Semiconductors*; Cambridge University Press: Cambridge, UK, 2007.
14. Nakamura, A.; Mizoguchi, T.; Matsunaga, K.; Yamamoto, T.; Shibata, N.; Ikuhara, Y. Periodic nanowire array at the crystal interface. *ACS Nano* **2013**, *7*, 6297–6302. [CrossRef] [PubMed]
15. Furushima, Y.; Nakamura, A.; Tochigi, E.; Ikuhara, Y.; Toyoura, K.; Matsunaga, K. Dislocation structures and electrical conduction properties of low angle tilt grain boundaries in LiNbO$_3$. *J. Appl. Phys.* **2016**, *120*, 142107. [CrossRef]
16. Furushima, Y.; Arakawa, Y.; Nakamura, A.; Tochigi, E.; Matsunaga, K. Nonstoichiometric [012] dislocation in strontium titanate. *Acta Mater.* **2017**, *135*, 103–111. [CrossRef]
17. Tochigi, E.; Kezuka, Y.; Nakamura, A.; Nakamura, A.; Shibata, N.; Ikuhara, Y. Direct observation of impurity segregation at dislocation cores in an ionic crystal. *Nano Lett.* **2017**, *17*, 2908–2912. [CrossRef] [PubMed]
18. Frank, F.C. Crystal dislocations. Elementary concepts and definitions. *Philos. Mag.* **1951**, *42*, 809–819. [CrossRef]
19. Kresse, G.; Furthmüller, J. Efficient iterative schemes for ab initio total-energy calculations using a plane-wave basis set. *Phys. Rev. B* **1996**, *54*, 11169. [CrossRef]
20. Blöchl, P.E. Projector augmented-wave method. *Phys. Rev. B* **1994**, *50*, 17953. [CrossRef]
21. Perdew, J.P.; Burke, K.; Ernzerhof, M. Generalized gradient approximation made simple. *Phys. Rev. Lett.* **1996**, *77*, 3865–3868. [CrossRef] [PubMed]
22. Liechtenstein, A.I.; Anisimov, V.I.; Zaanen, J. Density-functional theory and strong interactions: Orbital ordering in Mott-Hubbard insulators. *Phys. Rev. B* **1995**, *52*, R5467–R5470. [CrossRef]
23. Karazhanov, S.Z.; Ravindran, P.; Kjekshus, A.; Fjellvåg, H.; Grossner, U.; Svensson, B.G. Coulomb correlation effects in zinc monochalcogenides. *J. Appl. Phys.* **2006**, *100*, 043709. [CrossRef]
24. Nakamura, A.; Ukita, M.; Shimoda, N.; Furushima, Y.; Toyoura, K.; Matsunaga, K. First-principles calculations on slip system activation in the rock salt structure: Electronic origin of ductility in silver chloride. *Philos. Mag.* **2017**, *97*, 1281–1310. [CrossRef]
25. Oshima, Y.; Nakamura, A.; Matsunaga, K. Extraordinary plasticity of an inorganic semiconductor. submitted for publication.
26. Bateman, T.B. Elastic Moduli of Single-Crystal Zinc Oxide. *J. Appl. Phys.* **1962**, *33*, 3309–3312. [CrossRef]
27. Yan, Y.; Dalpian, G.M.; Al-Jassim, M.M.; Wei, Su-H. Energetics and electronic structure of stacking faults in ZnO. *Phys. Rev. B* **2004**, *70*, 193206. [CrossRef]
28. Jhon, M.H.; Glaeser, A.M.; Chrzan, D.C. Computational study of stacking faults in sapphire using total energy methods. *Phys. Rev. B* **2005**, *71*, 214101. [CrossRef]
29. Tochigi, E.; Mizoguchi, T.; Okunishi, E.; Nakamura, A.; Shibata, N.; Ikuhara, Y. Dissociation reaction of the 1/3<$\bar{1}$01> edge dislocation in α-Al$_2$O$_3$. *J. Mater. Sci.* **2018**, in press. [CrossRef]

![crystals logo] *crystals*

MDPI

Article

Anisotropic Deformation in the Compressions of Single Crystalline Copper Nanoparticles

Jianjun Bian [1,2], **Hao Zhang** [3], **Xinrui Niu** [2] **and Gangfeng Wang** [1,*] [iD]

[1] Department of Engineering Mechanics, SVL, Xi'an Jiaotong University, Xi'an 710049, China; jianjun@stu.xjtu.edu.cn
[2] CASM and Department of Mechanical and Biomedical Engineering, City University of Hong Kong, Hong Kong SAR, China; xinrui.niu@cityu.edu.hk
[3] Department of Chemical and Materials Engineering, University of Alberta, Edmonton, AB T6G 1H9, Canada; hao.zhang@ualberta.ca
* Correspondence: wanggf@mail.xjtu.edu.cn

Received: 25 December 2017; Accepted: 4 February 2018; Published: 1 March 2018

Abstract: Atomistic simulations are performed to probe the anisotropic deformation in the compressions of face-centred-cubic metallic nanoparticles. In the elastic regime, the compressive load-depth behaviors can be characterized by the classical Hertzian model or flat punch model, depending on the surface configuration beneath indenter. On the onset of plasticity, atomic-scale surface steps serve as the source of heterogeneous dislocation in nanoparticle, which is distinct from indenting bulk materials. Under [111] compression, the gliding of jogged dislocation takes over the dominant plastic deformation. The plasticity is governed by nucleation and exhaustion of extended dislocation ribbons in [110] compression. Twin boundary migration mainly sustain the plastic deformation under [112] compression. This study is helpful to extract the mechanical properties of metallic nanoparticles and understand their anisotropic deformation behaviors.

Keywords: nanoparticle; compression; anisotropy; dislocation

---

1. Introduction

Single crystal nanoparticles play increasing important roles in a wide variety of fields such as fuel cells, energetic materials and high performance catalyst. Among these applications, some crucial functions are closely in coordination with the mechanical properties of nanoparticles. It is well known that the mechanical properties of such surface-confined materials are drastically different from their macro-scale counterparts. For example, both the Young's modulus and hardness of single crystal nanoparticles are a bit larger than those of their bulk counterparts [1,2]. Extensive efforts have been devoted to investigate the unique properties of nanoparticles.

Anisotropy is one inherent characteristic of crystalline materials, and strongly affects their elastic and plastic deformation. For bulk single crystal materials, the reduced Young's modulus, the location of nucleation sites and the nucleation stress vary with different lattice orientations under indentation [3,4]. In incipient plasticity, dislocation nucleation depends largely on the available slip systems [5]. Anisotropic effects become even more prominent in low dimensional nanostructures such as nanowires and nanopillars. For example, lattice orientation in axial direction significantly affects the yield stress of gold nanowires [6]. The exhibiting tension-compression asymmetry in nanopillars depends on crystallographic orientation [7]. The plastic deformation of nanowires under torsion can be either homogeneous or heterogeneous, depending on the wire orientation [8]. In the case of silicon nanowires under uniaxial tension, the fracture mechanism would switch from brittle to ductile for varying axial crystallographic orientation [9]. The prominent anisotropy in low dimension materials is closely related to the intrinsic crystal structure and the extrinsic surface morphology. Free surfaces usually serve as preferential dislocation nucleation sites. Changing surface structure by removing weakly bound atoms produces a striking rise in yield strength [10].

For nanoparticles, such factors as surface facets, geometric profile and internal twin boundaries contribute to their unique properties [11,12]. Extensive experiments and simulations were performed to investigate the deformation mechanisms in nanoparticles. For example, phase transition was observed in single-crystal silicon nanoparticles during uniaxial compression [13]. The in situ TEM indentation of silver nanoparticle revealed reversible dislocation plasticity in nanoparticles [14]. During uniaxial compression, lateral free surface strengthened defect-free gold nanoparticles by draining dislocations from particles [15]. For uniaxial [001] compression of spherical copper nanoparticles, deformation twinning dominated the severe plastic deformation [16]. Microstructural evolution of tin dioxide nanoparticles under compression exhibited the formation of shear bands, twinning and stacking faults [17].

Up to now, there is still a lack of systematic investigation of anisotropic behaviors of face-centred-cubic (fcc) metallic nanoparticles. In the present study, we conduct molecular dynamics (MD) simulations to illuminate both the elastic characterization and plastic deformation mechanisms of copper nanoparticles, aiming to present a landscape of anisotropic deformation of low dimensional nanostructures.

2. Materials and Methods

MD simulations are conducted using the open-source simulator, LAMMPS, developed by Sandia National Laboratories. The embedded atom method (EAM) model is utilized to describe the atomic interaction of copper atoms. According to the EAM model, the total energy U of a system containing N atoms is expressed as

$$U = \sum_{i=1}^{N} \left(F_i(\overline{\rho}_i) + \frac{1}{2} \sum_{i \neq j}^{N} \phi_{ij}(r_{ij}) \right), \tag{1}$$

where $F_i(\overline{\rho}_i)$ is the embedding energy depending on atomic electron density $\overline{\rho}_i$ at the position of the i-th atom, $\phi_{ij}(r_{ij})$ is the pair-wise interaction energy related to the interatomic distance r_{ij} between atoms pairs i and j. In the present study, we use an EAM potential for copper parameterized by Mishin et al. [18], which was constructed by fitting both experimental and ab-initio computational data, for instance, cohesive energy, bulk modulus, elastic constants, intrinsic stacking fault energy, and vacancy formation and migration energies. This potential has been widely used to investigate mechanical properties and deformation mechanisms in different nanostructures [19,20].

The simulation model of uniaxially compressing a nanoparticle is depicted in Figure 1. One single crystal copper nanoparticle with radius of 10 nm is carved out of bulk defect-free single crystal copper, and contains about ~0.35 million atoms. Then the particle is placed between two rigid planar indenters. A repulsive potential is utilized to model the frictionless compression as

$$U_i(z_i) = \begin{cases} K(z_i - h_T)^3 & z_i \geq h_T \\ 0 & h_b < z_i < h_T , \\ K(h_B - z_i)^3 & z_i \leq h_B \end{cases} \tag{2}$$

where K is a specified constant representing the rigidity of the planar indenter. Compression direction is parallel to the z-axis, and z_i, h_T and h_B represent the positions of the i-th atom, the top indenter, and the bottom indenter, respectively. Based on the previous studies [16], we chose three typical orientations with lower Miller indices, [111], [110] and [112], as the uniaxial compression directions in the present study.

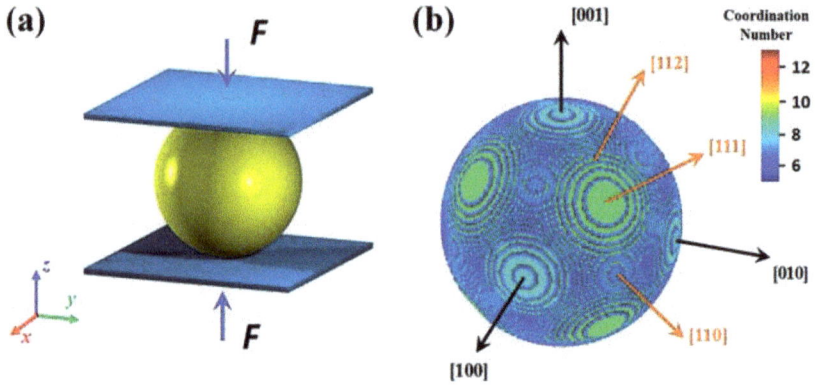

Figure 1. (a) Schematic of the uniaxial compression of nanoparticle; and (b) initial configuration of a copper nanoparticle (surface atoms are colored by their coordination number).

Loading procedure is implemented within the framework of canonical (NVT) ensembles. Before compression, the as-carved spherical nanoparticle firstly is performed structure relaxation using the conjugate gradient method, and then is equilibrated at 10 K for about 20 ps to relief the internal stress. When compression is conducted, the top and bottom planar indenters are simultaneously move towards the center of nanoparticle with a speed of ~0.1 Å/ps, and the compression depth δ is denoted by the displacement of one indenter. During the loading process, the temperature is controlled at 10 K using a Nosé-Hoover thermostat, and the time step of velocity-Verlet integration is chosen as 2.0 fs. Simulations at other temperatures, e.g., 1 K, 100 K and 300 K have also been conducted. Elevating temperature leads to larger fluctuations on the loading curves, but similar mechanical responses and dislocation activities have been observed.

To identify the characteristics of nucleated defects inside the particles and visualize their evolution processes, the local crystal structure of each atom is computed based on the common neighbor analysis (CNA) [21]. Atomic configurations and defect structures are visualized using Ovito [22] and Paraview [23]. To reveal the mechanical response of nanoparticles, the variations of contact area and averaged contact stress are examined. Contact area is determined by the Delaunay triangulation algorithm, and the averaged contact stress is defined as the loading force divided by contact area.

3. Results and Discussions

Compression of nanoparticle clearly demonstrates that the elastic response, initial dislocation nucleation and the following defects evolution vary with different loading directions. These orientation dependent features originate from the intrinsic crystallographic structure, surface configurations and the activated slip systems. In this section, we will analyze these behaviors in details.

3.1. [111] Compression

Figure 2a shows the loading response under [111] compression. The overall load response can be clearly subdivided into elastic and plastic stages, connected by the critical yielding point. In the elastic stage, compression load linearly increases with the accumulation of compression depth. After reaching the yielding point, the load is abruptly punctuated by an evident drop, indicating the initiation of plastic deformation. In the plastic stage, load fluctuates serratedly with further compression, as a reflection of inside atomic defect evolutions. Figure 2b demonstrates the variations of the averaged contact stress and contact area with respect to compression depth. In the elastic stage, the contact stress obtains its maximum value as high as ~26.0 GPa, followed by a sudden drop with the occurrence of yielding. Under further compression, contact stress fluctuates and maintains at a low level. Contact area is constant in elastic stage, and exhibits a stepwise increasing at yielding point, and then keeps increasing over the rest loading stage.

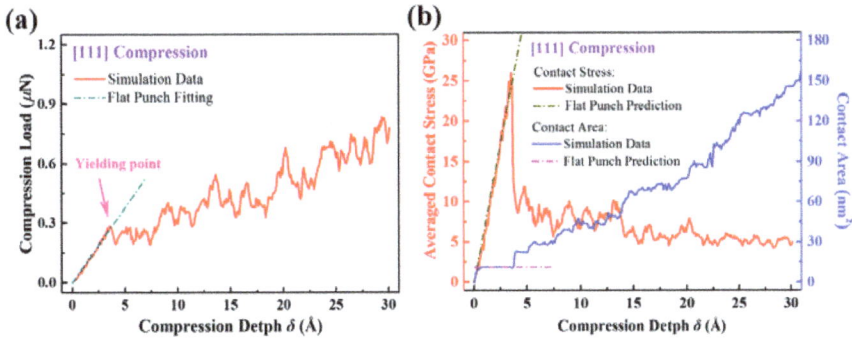

Figure 2. (**a**) load response under [111] compression; and (**b**) variation of averaged contact stress and contact area with respect to compression depth.

Atomic surface structures govern the elastic response of nanoparticles. Along [111] direction, atomic-scale facets are the most prominent characteristics of spherical surface. When compression begins, planar indenters firstly touch the outmost facet, rather than an ideal smooth spherical surface. It is manifested in recent studies that the flat punch model should be utilized when surface step is compressed [24,25]. In this model, compressive load F is expressed as a function of compression depth δ by

$$F = 2E^*a\delta, \tag{3}$$

where a and E^* are the contact radius and the reduced modulus, respectively. In Figure 2a, we use Equation (3) to fit the loading curve in the elastic stage, and the fitted reduced modulus is ~201.2 GPa. In the [111] nanoindentation of bulk copper, the reduced modulus is extracted as ~193 GPa [5]. It is noted that nanoparticle have a slightly larger elastic modulus than that of bulk material, which may be attributed to surface effects. Based on this model, the theoretical predictions of contact stress and contact area are plotted in Figure 2b, which coincide well with the MD computational results in elastic stage.

From yielding point, dislocation initiates inside the nanoparticle. The slip systems available for [111] compression is depicted in Figure 3a, and initial surface morphology and surface steps are displayed in Figure 3b. The outmost surface facet has a hexagonal shape, whose edges are preferential sites for dislocation nucleation. When yielding occurs, heterogeneous dislocations nucleate around surface steps as depicted in Figure 3c. Embryo size is indicative of the sequence of dislocation nucleation, and three larger embryos as one set (marked by blue arrows) nucleate earlier than the other three (marked by black arrows). When the outmost {111} facets are flattened, the embryos quickly grow up, turn into full dislocations, and then expand toward central region. Dislocations on different {111} slip planes react and are joined by pinning points, as marked in Figure 3d. Typical dislocation structures in the following stage are shown in Figure 3e,f. Slipping of the jogged dislocations dominates the plastic deformation. There are two events giving birth to dislocation jogs, i.e., one is the intersection of dislocations at different slip planes, and the other is the directly cross-slipping of dislocation loops nucleated from contact surface. During dislocation evolution, extended dislocation segments shrink or expand, and change the dislocation morphology intensively. In bulk materials, dislocation jogs and junctions decrease the mobility of dislocations, whereas in nanoparticles they can only exist for a short time before terminating at surface, similar to the case in metallic thin film [26].

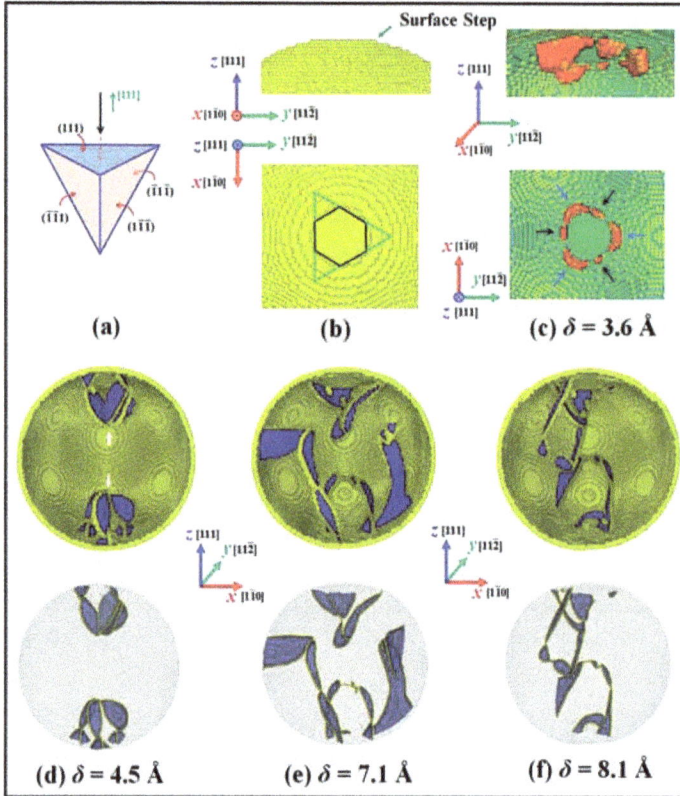

Figure 3. Atomic structure and defect evolution under [111] compression. (Panel (**a**) gives the available slip system, and (**b**) depicts initial surface configuration. In panel (**c**), atoms are colored by coordination number, and those with 12 are not shown for clarify. In panel (**d–f**), top parts are cross section views, atoms are colored by CNA parameter, and those in preface lattice are not shown. Atoms in yellow represent surface and dislocation cores, atoms in blue are in hcp lattice. Lower parts are the whole views of nanoparticle with only dislocation, stacking fault and surface are extracted and shown).

In addition to the surface, single-arm source is another important dislocation source in confined volume structures [27,28]. Figure 4 depicts typical single-arm dislocation sources inside the deformed nanoparticle. In Figure 4a–c, two spiraling arms of a source are connected by a short {001}<110> Lomer dislocation, and their other ends terminate at free surface. The dislocation arms glide on two parallel {111} slip planes, while the Lomer dislocation glides along {001} plane. In Figure 4d–f, the Lomer dislocation of another source has larger line length, and two arms are jogged extended dislocations. During compression, the dislocation arms cross slip and revolve around the Lomer dislocations. With the evolution of deformation, both the extended dislocation and Lomer dislocation finally annihilate at surface.

In the nanoindentation of bulk copper, initial dislocation homogeneously nucleated under indenter [4,29]. While in the compression of nanoparticle, initial dislocations are always heterogeneously intrigued under surface steps. Besides the nucleation events, dislocation morphologies are also different in these two cases. In nanoparticle under [111] compression, jogged dislocations with short extended segments dominate the defects structures. While in [111] indentation of bulk material, extended dislocations are not distorted heavily [5], and prismatic loops usually emanate from contact zone [30].

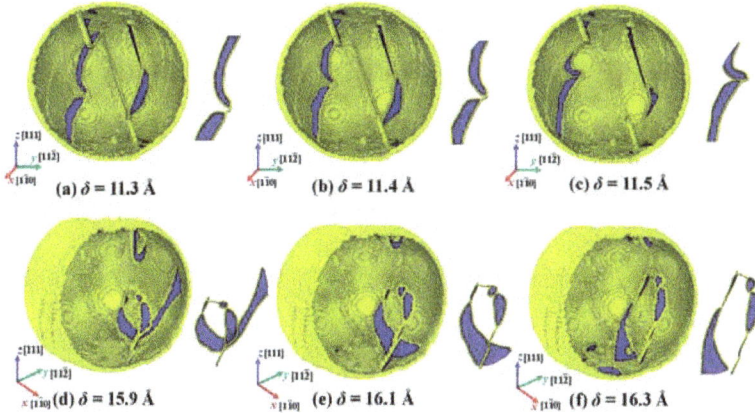

Figure 4. Single-arm dislocation source inside nanoparticle under [111] compression (Atoms are colored by CNA parameter, and color scheme is the same as in Figure 3).

3.2. [110] Compression

Under [110] compression as shown in Figure 5a, the loading curve increases continuously up to the yielding point, following a power-law function rather than the linear relation of [111] compression. Figure 5b demonstrates the evolution of contact stress and contact area with respect to compression depth. Before yielding point, three local peaks appear on the contact stress curve, and meanwhile stepwise increase emerges on the contact area curve. In this case, three [110] atom layers are flattened before yielding. Each time when one atom layer is flattened, contact area increases suddenly, leading to the drop of contact stress. When multiple atom layers are involved in contact, the classical Hertzian model can be used to capture the elastic response [24]. In this model, the load F as a function of compression depth δ is given as

$$F = \frac{4}{3}E^* R^{1/2}\delta^{3/2},\tag{4}$$

where E^* is the reduced modulus of the nanoparticle, R is the radius of nanoparticle. The fitted modulus in [110] direction is ~176.1 GPa. For nanoindentation of bulk copper, the modulus in [110] is ~163 GPa [5]. Nanoparticle has a larger modulus than that of bulk material, similar to [111] compression. Figure 5b illustrates the comparisons between the simulation data and the Hertzian prediction. It is seen that the overall elastic behaviors can be approximately characterized by the Hertzian model.

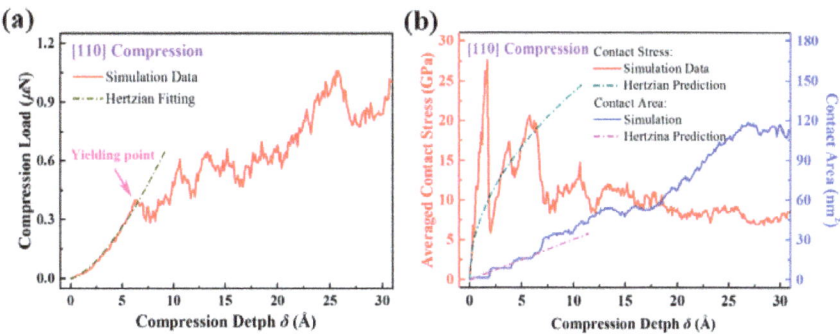

Figure 5. (a) load response under [110] compression; and (b) variation of averaged contact stress and contact area with respect to compression depth.

In the case of [110] compression, Figure 6a depicts two types of <111>{110} available slip systems. Figure 6b shows different views of the initial surface configuration under indenter. Initial dislocation firstly nucleates beneath surface steps, as shown in Figure 6c. In this stage, slip system of type I in Figure 6a is activated. Initial dislocations nucleate and glide in the {111} slip planes parallel to the compressive direction. Following the nucleation and propagation of leading partial dislocations, trailing partial dislocations also nucleate from contact surface, forming extended dislocation ribbons. These extended dislocation segments on four adjacent slip planes connect with each other, and compose a prismatic loop, which moves along its glide prism and then emanates from contact region, as shown in Figure 6d–h. The prismatic loop is highlighted in Figure 6i, and the intersection between neighboring extended ribbons is stair-rod dislocation.

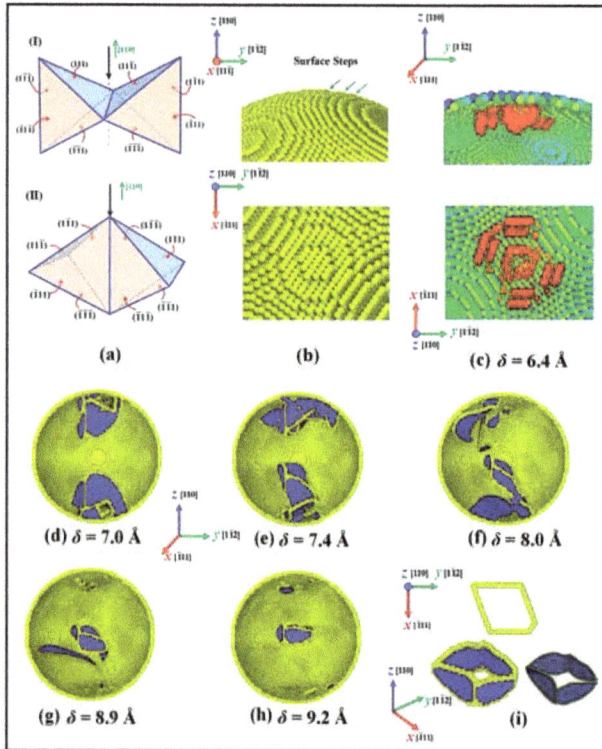

Figure 6. Atomic structures and evolution of prismatic dislocation loop in nanoparticles under [110] compression (Panel (**a**) gives the available slip systems, and (**b**) depicts initial surface configuration. In panel (**c**), atoms are colored by coordination number. In panel (**d–h**), atoms are colored by CNA parameter. Color scheme is the same as in Figure 3).

After releasing the prismatic loop, dislocations are prone to nucleate on slip system II, shown in Figure 6a. Extended dislocation ribbons nucleate from surface, glide across the center region, and finally exhaust at the lateral surface. After being impacted by these extended dislocations, the prismatic loop in center region collapses (Figure 7a), and finally is expelled out of the nanoparticle, as shown in Figure 7b,c. In further compression, only the nucleation and gliding of extended dislocations repeat to sustain plastic deformation, and Figure 7d–i depict this atomic process. Surface ledges are formed at the position where dislocation exhausts, as marked in Figure 7i. The annihilation of dislocations leaves nanoparticle starved of dislocations, which is different from indenting bulk materials in [110] orientation [5,31].

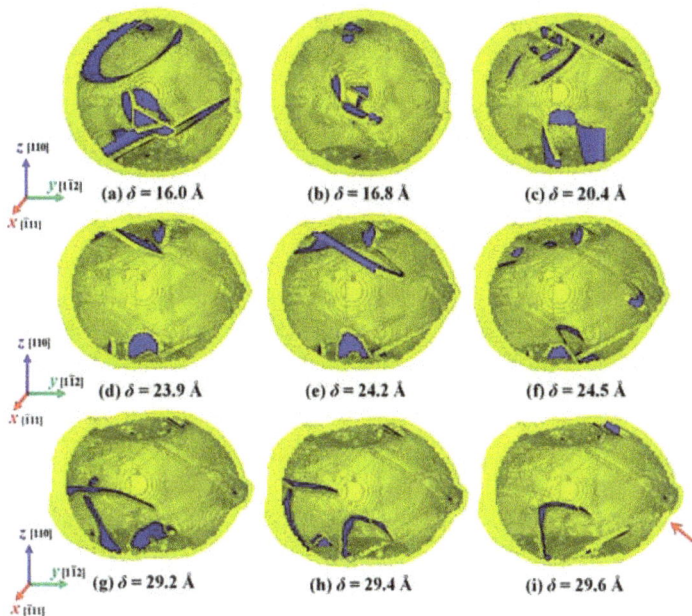

Figure 7. Nucleation and exhaustion of extended dislocation ribbons in nanoparticle under [110] compression (Atoms are colored by CNA parameter, and color scheme is the same as in Figure 3).

3.3. [112] Compression

Figure 8a illustrates the loading response under [112] compression. Similar to [110] compression, load increase in elastic stage follows a power-law function. However, the transition from elastic stage to plastic stage is more smooth, and no evident load-drop can be observed. Figure 8b gives the variations of averaged contact stress and contact area. When compression depth is smaller than ~4.7 Å, contact stress fluctuates and the curve exhibits several apparent peaks. Over the rest of compression, the magnitude of stress fluctuation decreases, and the stress keeps around ~11.0 GPa. Owing to the small crystal plane spacing in this orientation, contact area is raised stepwise for compression depth smaller than ~5.0 Å. Contact area keeps increasing until compression depth approaches ~9.5 Å, then it maintains almost at a constant value. Since multiple (112) atom layers are involved in elastic stage, Hertzian contact model is utilized to fit the elastic regime, as shown in Figure 8a. The fitted modulus is ~155.9 GPa.

For [112] compression, Figure 9 depicts the available slip system and the initial surface configuration. Since (112) planes are not close-packed, the initial (112) surface facet is rough and composed of $[\bar{1}10]$ surface steps. As the compression progresses, initial dislocations nucleate beneath these surface steps, and grow on the vertical $(\bar{1}\bar{1}1)$ slip planes. Four dislocation embryos are observed, as shown in Figure 9c. Since there is no resolved shear stress on the vertical {111}-type slip planes, these dislocation embryos are suppressed to expand. It is noted that nucleation of these small embryonic defects does not induce a significant load drop. In subsequent deformation, new nucleated dislocations prefer to glide and expand on the inclined (111) slip planes, which intrigue the yielding of nanoparticle, as shown in Figure 9d. Upon further compression, partials generate at the intersections between stacking fault and surface, then glide on adjacent slip plane, and change the existing stacking fault to extrinsic fault. The migration of subsequent dislocations gradually broadens these nano-twinned structures, as shown in Figure 9e,f. After the twin boundary (TB) departs from the contact region, perfect partial dislocation loops directly nucleate on the twin plane, expand and annihilate at surface, which is depicted in Figure 9g–i. This results in the migration of TBs toward center. When two TBs from the top and bottom regions get close to each other, dislocations begin to nucleate at contact surface, and glide on slip planes inclined to TBs, as show in Figure 9j–l. The existing TBs serve as barriers to these new dislocations. A few dislocations may slip across the TBs, and make TBs defective [32]. Under [112] compression, owing to the orientation of the activated <110>{111} slip system, migration of TBs mainly sustains the plastic deformation.

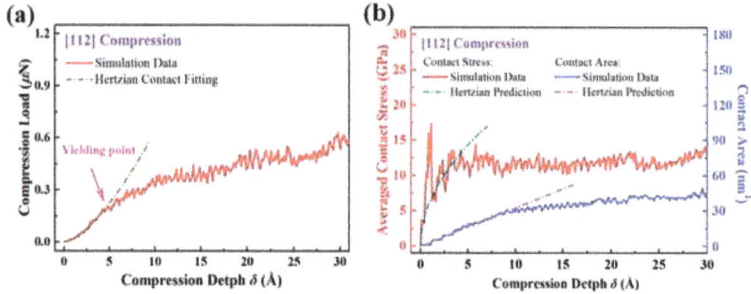

Figure 8. (**a**) Load response under [112] compression; and (**b**) variation of averaged contact stress and contact area with respect to compression depth.

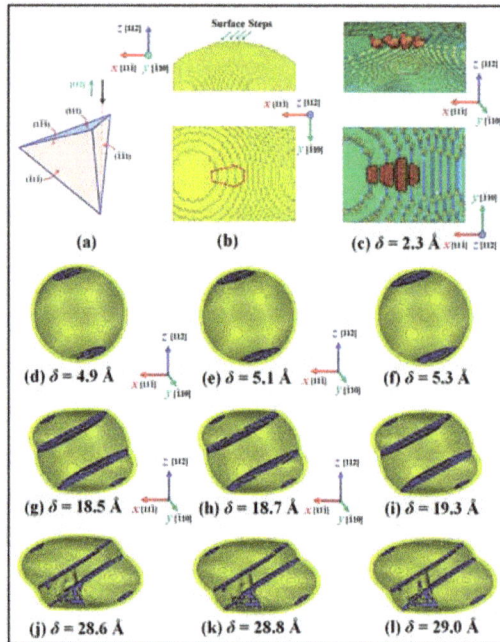

Figure 9. Atomic structures and twin boundary migration in nanoparticle under [112] compression (Panel (**a**) gives the available slip system, and (**b**) depicts initial surface configuration. In panel (**c**), atoms are colored by coordination number. In panel (**d–l**), atoms are colored by CNA parameter. Color scheme is the same as in Figure 3).

4. Conclusions

We perform MD simulations to investigate the compression of fcc copper nanoparticles in different directions. The results demonstrate that both the elastic and plastic behavior vary with compression orientation. Discrete atomic-scale surface steps play crucial roles in the mechanical response of nanoparticles. Under [111] compression, the elastic behavior follows the flat punch contact model. Under [110] and [112] compression, the classical Hertzian contact model is applicable in elastic regime. In nanoparticles, heterogeneous dislocations nucleate around surface steps and initiate the yield, different from the homogeneous dislocations under indenter tip for bulk materials. In severe compression, jogged dislocations dominate the plastic deformation in nanoparticles under [111] compression. For [110] compression, the plasticity is governed by extended dislocation ribbons. The annihilation of dislocations at surface leaves nanoparticle starved of dislocations. In [112] compression, deformation twinning is the major plastic deformation mechanism. This work reveals new atomic-scale features of mechanical response of metallic fcc nanoparticles.

Acknowledgments: Supports from the National Natural Science Foundation of China (Grant No. 11525209) are acknowledged.

Author Contributions: G.W. conceived and designed the project; J.B. performed the MD simulations; J.B. and G.W. drafted the manuscript; H.Z. and X.N. revised the paper. All authors read and approved the final manuscript.

Conflicts of Interest: The authors declare no conflict of interest.

References

1. Kim, T.; Myung, S.; Kim, T.H.; Hong, S. Robust single-nanoparticle probe for contact-mode analysis and dip-pen nanolithography. *Small* **2008**, *4*, 1072–1075. [CrossRef] [PubMed]
2. Gerberich, W.W.; Mook, W.M.; Perrey, C.R.; Carter, C.B.; Baskes, M.I.; Mukherjee, R.; Gidwani, A.; Heberlein, J.; McMurry, P.H.; Girshick, S.L. Superhard silicon nanospheres. *J. Mech. Phys. Solids* **2003**, *51*, 979–992. [CrossRef]
3. Zhong, Y.; Zhu, T. Simulating nanoindentation and predicting dislocation nucleation using interatomic potential finite element method. *Comput. Methods Appl. Mech. Eng.* **2008**, *197*, 3174–3181. [CrossRef]
4. Zhu, T.; Li, J.; Van Vliet, K.J.; Ogata, S.; Yip, S.; Suresh, S. Predictive modeling of nanoindentation-induced homogeneous dislocation nucleation in copper. *J. Mech. Phys. Solids* **2004**, *52*, 691–724. [CrossRef]
5. Tsuru, T.; Shibutani, Y. Anisotropic effects in elastic and incipient plastic deformation under (001), (110), and (111) nanoindentation of Al and Cu. *Phys. Rev. B* **2007**, *75*, 035415. [CrossRef]
6. Hyde, B.; Espinosa, H.D.; Farkas, D. An atomistic investigation of elastic and plastic properties of Au nanowires. *JOM* **2005**, *57*, 62–66. [CrossRef]
7. Kim, J.; Jang, D.; Greer, J.R. Crystallographic orientation and size dependence of tension-compression asymmetry in molybdenum nano-pillars. *Int. J. Plast.* **2012**, *28*, 46–52. [CrossRef]
8. Weinberger, C.R.; Cai, W. Orientation-dependent plasticity in metal nanowires under torsion: Twist boundary formation and Eshelby twist. *Nano Lett.* **2010**, *10*, 139–142. [CrossRef] [PubMed]
9. Liu, Q.; Shen, S. On the large-strain plasticity of silicon nanowires: Effects of axial orientation and surface. *Int. J. Plast.* **2012**, *38*, 146–158. [CrossRef]
10. Rabkin, E.; Srolovitz, D.J. Onset of plasticity in gold nanopillar compression. *Nano Lett.* **2007**, *7*, 101–107. [CrossRef] [PubMed]
11. Johnsion, C.; Snoeck, E.; Ezcurdia, M.; Rodriguez-Gonzalez, B.; Pastoriza-Santos, I.; Liz-Marzan, L.M.; Hytch, M.J. Effects of elastic anisotropy on strain distributions in decahedral gold nanoparticles. *Nat. Mater.* **2008**, *7*, 120–124. [CrossRef] [PubMed]
12. Casillas, G.; Palomares-Baez, J.P.; Rodriguez-Lopez, J.L.; Luo, J.; Ponce, A.; Esparza, R.; Velazquez-Salazar, J.J.; Hurtado-Macias, A.; Gonzalez-Hernandez, J.; Jose-Yacaman, M. In situ TEM study of mechanical behavior of twinned nanoparticles. *Philos. Mag.* **2012**, *92*, 4437–4453. [CrossRef]
13. Valentini, P.; Gerberich, W.W.; Dumitrica, T. Phase-transition plasticity response in uniaxially compressed silicon nanospheres. *Phys. Rev. Lett.* **2007**, *99*, 175701. [CrossRef] [PubMed]

14. Carlton, C.E.; Ferreira, P.J. In situ TEM nanoindentation of nanoparticles. *Micron* **2012**, *43*, 1134–1139. [CrossRef] [PubMed]

15. Mordehai, D.; Kazakevich, M.; Srolovitz, D.J.; Rabkin, E. Nanoindentation size effect in single-crystal nanoparticles and thin films: A comparative experimental and simulation study. *Acta Mater.* **2011**, *59*, 2309–2321. [CrossRef]

16. Bian, J.J.; Wang, G.F. Atomistic deformation mechanisms in copper nanoparticles. *J. Comput. Theor. Nano* **2013**, *10*, 2299–2303. [CrossRef]

17. Armstrong, P.; Knieke, C.; Mackovic, M.; Frank, G.; Hartmaier, A.; Goken, M.; Peukert, W. Microstructural evolution during deformation of tin dioxide nanoparticles in a comminution process. *Acta Mater.* **2009**, *57*, 3060–3071. [CrossRef]

18. Mishin, Y.; Mehl, M.J.; Papaconstantopoulos, D.A.; Voter, A.F.; Kress, J.D. Structural stability and lattice defects in copper: Ab initio, tight-binding, and embedded-atom calculations. *Phys. Rev. B* **2001**, *63*, 224106. [CrossRef]

19. Jang, D.; Li, X.; Gao, H.; Greer, J.R. Deformation mechanisms in nanotwinned metal nanopillars. *Nat. Nanotechnol.* **2012**, *7*, 594–601. [CrossRef] [PubMed]

20. Zhang, Y.; Zhou, L.; Huang, H. Size dependence of twin formation energy of metallic nanowires. *Int. J. Smart Nano Mater.* **2013**, *2*, 112–118. [CrossRef]

21. Tsuzuki, H.; Branicio, P.S.; Rino, J.P. Structural characterization of deformed crystals by analysis of common atomic neighborhood. *Comput. Phys. Commun.* **2007**, *177*, 518–523. [CrossRef]

22. Stukowski, A. Visulization and analysis of atomistic simulation data with OVITO-the Open Visualization Tool. *Model. Simul. Mater. Sci. Eng.* **2010**, *18*, 015012. [CrossRef]

23. Ahrens, J.; Geveci, B.; Law, C. ParaView: An end-user tool for large data visualization. In *Visualization Handbook*; Elsevier: Amsterdam, The Netherlands, 2005; ISBN 978-0123875822.

24. Wang, G.F.; Bian, J.J.; Feng, J.; Feng, X.Q. Compressive behavior of crystalline nanoparticles with atomic-scale surface steps. *Mater. Res. Express* **2015**, *2*, 015006. [CrossRef]

25. Yang, L.; Bian, J.J.; Wang, G.F. Impact of atomic-scale surface morphology on the size-dependent yield stress of gold nanoparticles. *J. Phys. D Appl. Phys.* **2017**, *50*, 245302. [CrossRef]

26. Lee, S.; Aubry, S.; Nix, W.D.; Cai, W. Dislocation junctions and jogs in a free-standing fcc thin film. *Model. Simul. Mater. Sci. Eng.* **2011**, *19*, 025002. [CrossRef]

27. Weinberger, C.R.; Cai, W. The stability of Lomer-Cottrell jogs in nanopillars. *Scr. Mater.* **2011**, *64*, 529–532. [CrossRef]

28. Tucker, G.J.; Aitken, Z.H.; Greer, J.R.; Weinberger, C.R. The mechanical behavior and deformation of bicrystalline nanowires. *Model. Simul. Mater. Sci. Eng.* **2013**, *21*, 015004. [CrossRef]

29. Wang, W.; Zhong, Y.; Lu, K.; Lu, L.; McDowell, D.L.; Zhu, T. Size effects and strength fluctuation in nanoscale plasticity. *Acta Mater.* **2012**, *60*, 3302–3309. [CrossRef]

30. Shin, C.; Osetsky, Y.N.; Stoller, R.E. Dislocation nucleation and defect formation in copper by stepped spherical indeneter. *Philos. Mag.* **2012**, *92*, 3158–3171. [CrossRef]

31. Liang, H.; Woo, C.H.; Huang, H.; Ngan, A.H.W.; Yu, T.X. Crystalline plasticity on Copper (001), (110), and (111) surface during nanoindentation. *Comput. Model. Eng. Sci.* **2014**, *6*, 105–114.

32. Wang, Y.M.; Sansoz, F.; LaGrange, T.; Ott, R.T.; Marian, J.; Barbee, T.W., Jr.; Hamza, A.V. Defective twin boundaries in nanotwinned metals. *Nat. Mater.* **2013**, *12*, 697–702. [CrossRef] [PubMed]

crystals

MDPI

Article

Interface Effects on Screw Dislocations in Heterostructures

Jianwei Wang [1], Ting Sun [2], Weiwei Xu [2], Xiaozhi Wu [2,3,*] and Rui Wang [2]

[1]   Microsystem and Terahertz Research Center, CAEP, Chengdu 610200, China; wangjianwei@mtrc.ac.cn
[2]   Institute for Structure and Function, Chongqing University, Chongqing 401331, China;
      sunting@cqu.edu.cn (T.S.); xuweiwei0308@cqu.edu.cn (W.X.); rcwang@cqu.edu.cn (R.W.)
[3]   College of Materials Science and Engineering, Chongqing University, Chongqing 400044, China
*   Correspondence: xiaozhiwu@cqu.edu.cn

Received: 14 November 2017; Accepted: 8 January 2018; Published: 10 January 2018

Abstract: The governing equation of screw dislocations in heterostructures is constructed using image method. The interface type ($-1 \leq \gamma \leq 1$) and distance between dislocation and interface h are considered in the new equation. The Peierls–Nabarro equations for screw dislocations in bulk and semi-infinite materials can be recovered when $\gamma = 0$ and $\gamma = -1$. The soft ($\gamma < 0$) and hard ($\gamma > 0$) interfaces can enhance and reduce the Peierls stress of screw dislocations near the interface, respectively. The interface effects on dislocations decrease with the increasing of distance h. The Al/TiC heterostructure is investigated as a model interface to study the unstable stacking fault energy and dislocation properties of the interface. The mismatch of lattice constants and shear modulus at the interface results in changes of the unstable stacking fault energy. Then, the changes of the unstable stacking fault energy also have an important effect on dislocation properties, comparing with γ and h.

Keywords: interface effects; screw dislocation; image method; Peierls–Nabarro model

1. Introduction

Dislocations play an important role in understanding the mechanical properties of crystalline solids [1]. The continuum elasticity theory can describe the long-range elastic field of dislocations well, but it fails in dislocation core region due to the singularity. The influence of the lattice periodicity of real crystals on dislocation properties should be taken into account. The classical Peierls–Nabarro (P-N) model provides a useful approach to determining the core properties of dislocations [2,3]. Based on the finite scale of the dislocation core, we can estimate the minimum external stress required to move a dislocation in a perfect crystal—the so-called Peierls stress. The predictions can be improved when the sinusoidal force law is replaced by the generalized stacking fault (GSF) energy [4–6]. Then, the P-N model combined GSF energy is extensively and successfully used to predict the dislocation properties in bulk materials [7,8].

It is widely accepted that the dislocation core and mobility can be influenced by various inhomogeneities, such as surface, interfaces, and cavities in materials [9]. Lee and Li yielded a half-space P-N model for dislocations near a free surface in semi-infinite materials with single surface [10,11]. Cheng et al. constructed the semidiscrete P-N model to characterize the surface on dislocations using the image dislocation method [12]. The Peierls stress is increased when dislocations are located near the free surface. The results are in agreement with the molecular dynamics simulation. The heterostructures and laminated structures are very common in modern science for the design of strong structural materials. Within continuum elasticity theory, Head investigated the interaction of screw dislocations and interfaces in heterostructures using the image method [13]. When the screw dislocation is located in a rigid (soft) material, the dislocations are attracted (repelled) by the interface.

The attraction and repulsion between dislocation and interface are only affected by the mismatch in the shear modulus of two semi-infinite materials. However, two other sources also arise in the interaction of dislocations and interface. One is the mis-fit of lattice constants at the interface. This will change the atomic arrangement near the interface. Then, the GSF energy near the interface will be altered. The other is the distance between the interface and dislocations. When the distance is large enough, the atomic arrangement on GSF energy is small due to the locality of GSF energy. Only the elastic interaction is expected.

In this paper, the interface effects on screw dislocation are investigated using the image method in heterostructures. The image dislocations are introduced to satisfy the interface boundary condition. The stress and displacement should be continuous across the interface. The governing equation for screw dislocations is presented according to the procedure for the P-N equation in an elegant manner. The mismatch of shear modulus, mis-fit of lattice constants, and the distance between interface and dislocations on the dislocation core and Peierls stress are considered. Furthermore, the interface effects on screw dislocations in Al/TiC heterostructure are also presented.

2. Screw Dislocation Equation with Interfacial Effects

In order to take the interfacial effects on screw dislocations into account, an original right-handed screw dislocation parallel to the interface is represented by O (see Figure 1). In a cartesian set of coordinates xyz, the original dislocation line is set to along the z axis, the xoz plane is the slip plane, and y is normal to the slip plane. The $y = h$ plane is the interface of material-I and material-II. Material-I and material-II are placed at $y < h$ and $y > h$. The shear moduli of material-I and material-II are μ_I and μ_{II}, respectively. Therefore, h is the distance between interface and dislocation. Due to the equivalence of dislocations and point charges in elastic strain field and electrostatics, the image method is used to describe the interface effects on the dislocation elastic strain field. The interface effects on dislocations are represented by image dislocations. The image dislocation of O is at A. As we know, there is no strain field in material-II while the interface is not taken into account. The interface between material-I and material-II will introduce a strain field in the material. Accordingly, the interface effects on dislocation O is represented by image dislocation A . To account for finite strain in material-II, an additional image is placed at B to describe the interface effects. The dislocations O, A, and B have Burgers vectors of magnitudes b, γb, and βb, respectively. According to the required boundary conditions that the displacements and stress must be continuous across the interface, the three dislocations must lie in the plane $x = 0$, and in fact, the results verify that such an arrangement fulfills the boundary conditions. Additionally, in order for the displacements at the interface to vary in the same way with varying x, it is necessary that $L_O = L_A = L_B$. The displacement and stress of the original screw dislocation are

$$u_z = \frac{b}{2\pi} \arctan \frac{y}{x}, \tag{1}$$

$$\sigma_{yz} = \frac{\mu b}{2\pi} \frac{x}{x^2 + y^2}. \tag{2}$$

It is important to elucidate that both image dislocations A and B are the images of the original screw dislocation O. In other words, the strain field in material-I is the same as that which would be produced by dislocation O and the image dislocation A, while the strain field in material-II is the same as that produced by dislocation B. Therefore, the displacement in material-I and material-II can be written as

$$u_{z_I} = \frac{b}{2\pi} \left(\arctan \frac{y}{x} + \gamma \arctan \frac{y - 2h}{x} \right), \tag{3}$$

$$u_{z_{II}} = \frac{\beta b}{2\pi} \arctan \frac{y}{x}, \tag{4}$$

and the stress in material-I and material-II can be written as

$$\sigma_{yz_I} = \frac{\mu_I bx}{2\pi(x^2 + y^2)} + \frac{\mu_I \gamma bx}{2\pi[(x^2 + (y - 2h)^2]}, \tag{5}$$

$$\sigma_{yz_{II}} = \frac{\mu_{II} \beta b}{2\pi(x^2 + y^2)}. \tag{6}$$

At the interface ($y = h$), the displacement and stress must be continuous; namely,

$$u_{z_I} = u_{z_{II}}, \tag{7}$$

$$\sigma_{yz_I} = \sigma_{yz_{II}}. \tag{8}$$

One finds

$$\beta = 1 - \gamma, \tag{9}$$

$$\mu_I(1 + \gamma) = \mu_{II}\beta. \tag{10}$$

It is easy to yield

$$\gamma = \frac{\mu_{II} - \mu_I}{\mu_I + \mu_{II}}, \tag{11}$$

$$\beta = \frac{2\mu_I}{\mu_I + \mu_{II}}. \tag{12}$$

γ can be used to characterize the interface type. Dislocations are in the soft material for $\mu_{II} > \mu_I$ ($1 \geq \gamma > 0$) and the interface can be named as a rigid interface. When $\mu_{II} < \mu_I$ ($-1 < \gamma < 0$), dislocations are in a rigid material and the interface can be named as a soft interface. When $\gamma = -1$, the interface is reduced to surface.

Now, the stress field on the slip plane ($y = 0$) with the contribution of image dislocation can be written as

$$\sigma_{yz} = \frac{\mu_I b}{2\pi} \left[\frac{1}{x} + \gamma \frac{x}{x^2 + (2h)^2} \right]. \tag{13}$$

The above discussions are focused on the properties of dislocations in the continuum theory. A dislocation is a singular line in continuum theory, and can be displaced without any application of force. However, the influence of the lattice periodicity on dislocation properties should be taken into account. The P-N model is the first to determine the dislocation width and Peierls stress. In the P-N model, the crystal is set to be two semi-infinite crystals. The dislocation is produced from the nontrivial contact of two semi-infinite crystals. The dislocation core is assumed to spread, and the displacements are assumed to be produced by a continuous distribution of infinitesimal dislocations at every point x' of Burgers vectors $\rho(x') = \frac{du(x')}{dx'}$. The total Burgers vector $b = \int_{-\infty}^{+\infty} \rho(x') dx'$. The infinitesimal dislocations combined with the stress field from elastic theory are used to derive equations for dislocations in bulk and misfit dislocations at the interface [5,14]. Accordingly, the governing equation for screw dislocation in heterostructure with the interfacial effects can easily yield

$$\frac{\mu_I}{2\pi} \int_{-\infty}^{+\infty} \left[\frac{1}{x - x'} + \gamma \frac{x - x'}{(x - x')^2 + (2h)^2} \right] \frac{du(x')}{dx'} dx' = f(u). \tag{14}$$

The restoring force $f(u)$ can be obtained from the gradient of GSF energy $\gamma(u)$

$$\gamma(u) = \frac{\gamma_{us}}{2} \left(1 - \cos \frac{2\pi u}{b} \right), \tag{15}$$

with γ_{us} the unstable stacking fault energy [15]. γ_{us} is the maximum energy barrier for a moving dislocation.

Then, we have the restoring force

$$f(u) = \sigma_{max} \sin \frac{2\pi u}{b} , \tag{16}$$

with the maximum restoring stress $\sigma_{max} = \frac{\pi \gamma_{us}}{b}$.

It is interesting to discuss the dislocation equation for two limitations:

(i) When $\mu_2 = \mu_1$ ($\gamma = 0$), material-I and material-II are the same material. Both the interface and image dislocation do not exist. One obtains the classical P-N equation for an isolate screw dislocation in bulk:

$$\frac{\mu}{2\pi} \int_{-\infty}^{+\infty} \frac{1}{x - x'} \frac{du(x')}{dx'} dx' = f(u) \tag{17}$$

(ii) When $\mu_2 = 0$ ($\gamma = -1$), material-II is vacuum. The interface reduces to the surface of material-I. One obtains

$$\frac{\mu}{2\pi} \int_{-\infty}^{+\infty} \left[\frac{1}{x - x'} - \frac{x - x'}{(x - x')^2 + (2h)^2} \right] \frac{du(x')}{dx'} dx' = f(u) \tag{18}$$

This equation is the same as the previous dislocation equation obtained by Cheng et al. for a semi-infinite material with seldom free surface [8].

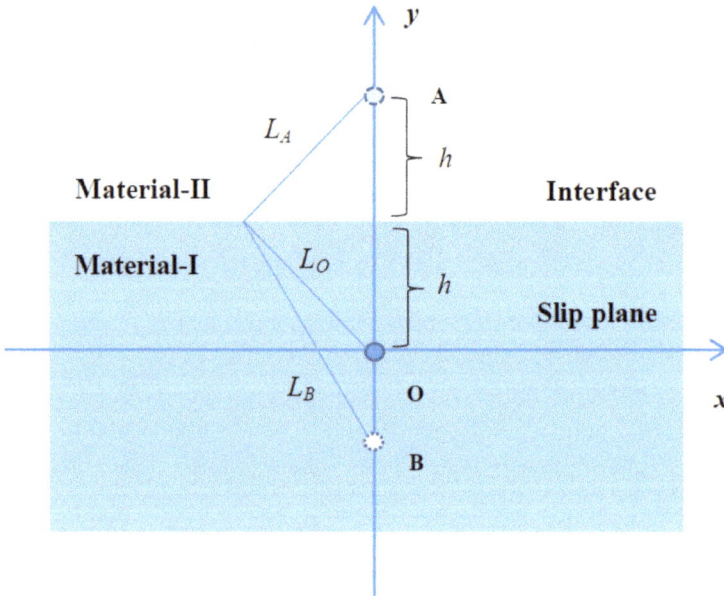

Figure 1. The original dislocation and image dislocations in heterostructure. h is the distance between dislocation and interface. The slip plane is parallel to the interface.

3. Interfacial Effects on Dislocation Core and Peierls Stress

Foreman introduced an effective method (Foreman method) to solve the original P-N dislocation equation with nonsinusoidial restoring force [16]. In this paper, we take the Foreman method to solve the screw dislocation with the interfacial effects. The typical Foreman solution for the screw dislocation is given as follows:

$$u(x) = \frac{b}{\pi}\left[1 - (\zeta - 1)\frac{\partial}{\partial\zeta}\right]\arctan\frac{p}{\zeta}, \tag{19}$$

with $p = 2x/d$ and d the spacing distance of the planes parallel to the slip plane. The core structure parameter ζ is crucially determined by the maximum restoring force. $u(x)$ ($\zeta = 1$) reduces to the solution of the original P-N equation with $\sigma_{max} = \frac{\mu b}{2\pi d}$ and the core width $\zeta = d/2$. The core width is determined by $\rho(\zeta) = \frac{1}{2}\rho(0)$. However, the core width will be changed, while the unstable stacking fault energy and interface effects are taken into account. The Foreman method is a modification of the classical P-N solution for the classical P-N dislocation equation. The core structure parameter ζ is introduced into the classical P-N solution to obtain the Foreman solution, Equation (19). It is clear that the classical P-N solution can be obtained from Equation (19) if the core structure parameter $\zeta = 1$. Furthermore, the Foreman solution is introduced a parametric derivative form. This is helpful in solving the governing dislocation equation, Equation (14).

Substituting the Foreman solution to the dislocation equation and letting $p = \cot\frac{\theta}{2}$, the stress and displacement can be expressed as the parametric forms

$$\sigma(\theta) = \frac{\mu\kappa b}{2\pi}\left\{\frac{\cot\frac{\theta}{2}\left[1 + \zeta^2(-1 + 2\zeta)\cot^2\frac{\theta}{2}\right]}{\left(1 + \zeta^2\cot^2\frac{\theta}{2}\right)^2} + \gamma\frac{\cot\frac{\theta}{2}\left[(1 + 2\kappa h)^2 + \zeta^2(-1 + 2\zeta)\cot^2\frac{\theta}{2}\right]}{\left[(1 + 2\kappa h)^2 + \zeta^2\cot^2\frac{\theta}{2}\right]^2}\right\} \tag{20}$$

$$u(\theta) = \frac{b}{\pi}\left(\frac{\pi}{2} - \frac{\theta}{2} + \frac{\zeta - 1}{2\zeta}\sin\theta\right) \tag{21}$$

with $\kappa = \frac{2}{\zeta d}$ and θ in the range from 0 to 2π. The core structure parameter ζ can be determined by fitting Equation (20) for the maximum restoring force. It is interesting to find that $\sigma(\theta) = \frac{\mu b}{2\pi d}\sin\theta$ when $\zeta = 1$ and $\gamma = 0$.

The variations of ζ with respect to distance h for different interface types $\gamma = -1, -0.5, 0, 0.5, 1$ are shown in Figure 2. For comparison, the results of $\sigma_{max} = \mu b/8\pi d$ and $\mu b/2\pi d$ are displayed. It worth noting that ζ decreases with h for $\gamma > 0$ and increases with h for $\gamma < 0$ when $\sigma_{max} = \mu b/8\pi d$. The larger the value of h, the fewer influence the interface has on the dislocation core. ζ tends towards the results of $\gamma = 0$. The surface and interface effects on core structure become small for large h. For the same h, $\gamma > 0$ leads to a larger ζ (wider dislocation core) than $\gamma < 0$. Namely, a rigid material will extend the dislocation core in a soft material. However, a soft material will compress the dislocation core in a rigid material. In other words, the image dislocation has the same (different) chirality with the original dislocation for $\gamma > 0$ ($\gamma < 0$). This means that dislocations at the parallel slip planes with same chirality lead to a wide core, and different chirality lead to a narrow core. It should be kept in mind that the above discussions are based on a fixed maximum restoring force for different h. The core structure parameter ζ is near the same for $\sigma_{max} = \mu b/2\pi d$ for different h and γ (see Figure 2). Figure 3a illustrates that the core structure parameter ζ decreases with increasing maximum restoring force σ_{max} for $h = 4d$. It is clear that $\zeta > 1$ and $\zeta < 1$ for $\sigma_{max} < \mu b/2\pi d$ and $\sigma_{max} > \mu b/2\pi d$, respectively. The difference between ζ and γ is very small for higher σ_{max}.

Generally, the mobility is controlled by the dislocation core width ζ and the unstable stacking fault energy γ_{us}. The following formula can be used to predict the Peierls stress of dislocations:

$$\sigma_P = \frac{2\pi^2\gamma_{us}\zeta}{b^2}\exp^{-\frac{2\pi\zeta}{b}}. \tag{22}$$

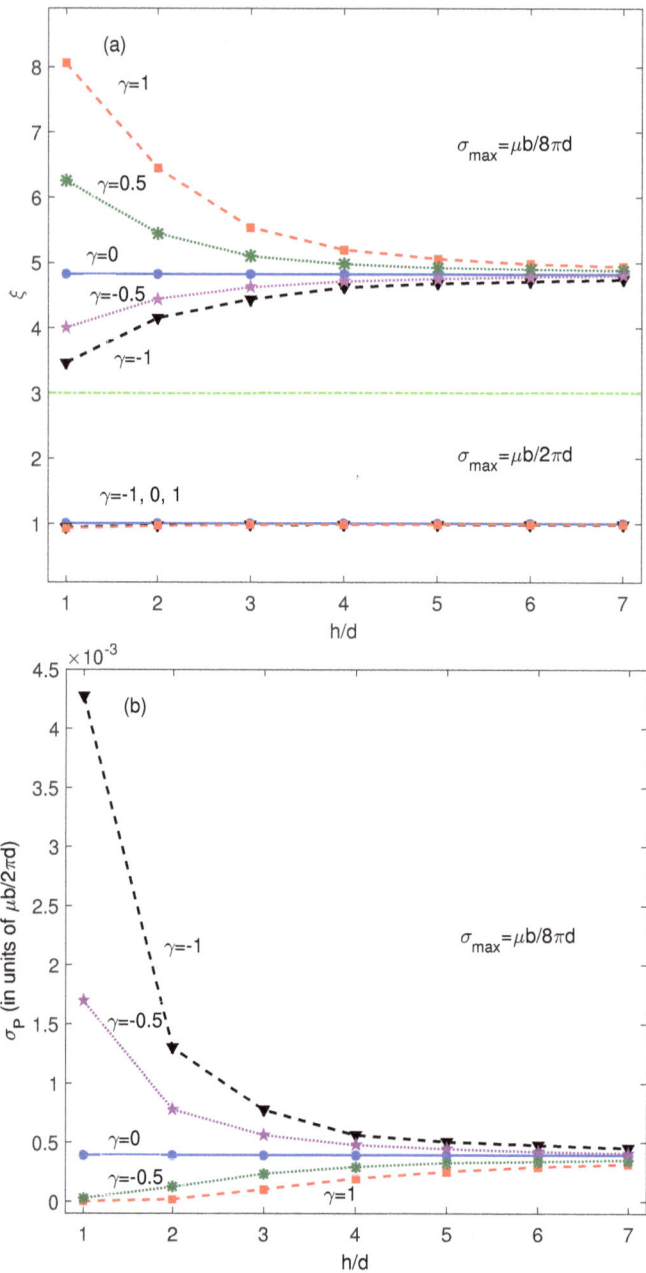

Figure 2. The core structure parameters ξ vary with the distance between dislocation (**a**) and interface h (**b**).

Peierls stress for $\sigma_{max} = \frac{\mu b}{8\pi d}$ and different interface types are illustrated in Figure 2b. It is easy to see that Peierls stress can be enhanced by the interface for $\gamma < 0$ ($\mu_{II} < \mu_I$), and reduced by the interface for $\gamma > 0$ ($\mu_{II} > \mu_I$). The Peierls stress of screw dislocations in bulk material-I is $3.91 \times 10^{-4} \sigma_{max}$. We take $h = 2d$ as an example. When $\gamma = -1$ ($\mu_{II} = 0$), the Peierls stress of screw dislocations is $1.3 \times 10^{-3}\sigma_{max}$. The Peierls stress is enhanced by the surface by about three times. When $\gamma = -0.5$ ($\mu_{II} = \frac{1}{3}\mu_I$), the Peierls stress of screw dislocations is $7.76 \times 10^{-4}\sigma_{max}$. The Peierls stress decreases with the increasing γ. The Peierls stress of screw dislocations is $1.24 \times 10^{-4}\sigma_{max}$ for $\gamma = 0.5$ ($\mu_{II} = 3\mu_I$). The Peierls stress is reduced by the interface. Furthermore, the interface effect decreases with increasing h and σ_{max} for both $\gamma > 0$ and $\gamma < 0$. This can be easily understood using Figure 4. When $h < \zeta$, the original core region will superpose with the image core region. The original dislocation core will be affected by the image core. The surface and interface effects cannot be ignored. The original core width decreases with increasing σ_{max}. The distance between the original core region and the image core region become large, and the superposition of core regions becomes weak. The surface and interface effects on the original core are also reduced. Figure 4 also shows that the superposition decreases with increasing h.

Based on the above discussions, the conclusion can be made that the interface effects on the core width and Peierls stress are determined by the maximum restoring force σ_{max} (the unstable stacking fault energy γ_{us}), the distance to interface h, and the interface type γ. In order to elucidate their contributions, we take $h = 4d$ as an example. The Peierls stress has the same order for $\gamma = -1$, $\gamma = 0$, and $\gamma = 1$. This is also true for $h = d$, although the Peierls stress has the biggest difference for $\gamma = -1$, $\gamma = 0$, and $\gamma = 1$. The Peierls stress also has the same order when h in the range of $1d \sim 7d$. However, the Peierls stress for $\sigma_{max} = \frac{\mu b}{2\pi d}$ is 1000 times that for $\sigma_{max} = \frac{\mu b}{8\pi d}$ (see Figure 3). However, the Peierls stress for $\sigma_{max} = \frac{\mu b}{2\pi d}$ is only 1.3 times of that for $\sigma_{max} = \frac{3\mu b}{8\pi d}$. The Peierls stress for $\sigma_{max} = \frac{3\mu b}{8\pi d}$ is about 21 times that for $\sigma_{max} = \frac{\mu b}{8\pi d}$. Therefore, the interface effect on dislocation will be mainly controlled by the difference between the unstable stacking fault energy (the maximum restoring force) with and without interface. Furthermore, the interface effect is clearer when the unstable stacking energy in bulk material-I is smaller than $\frac{\mu b}{2\pi d}$.

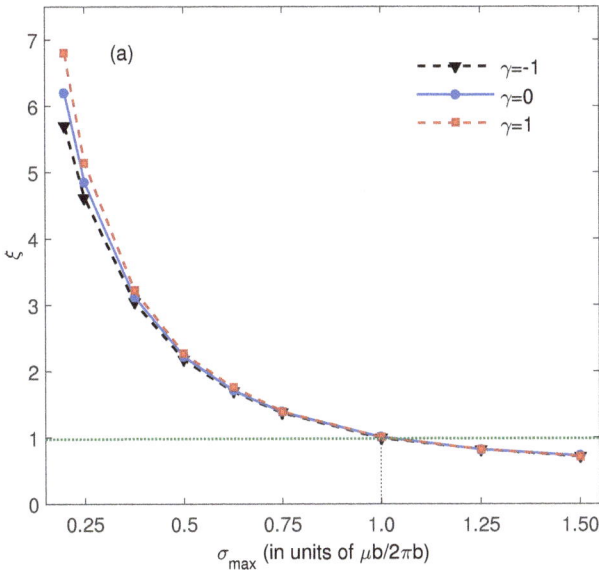

Figure 3. *Cont.*

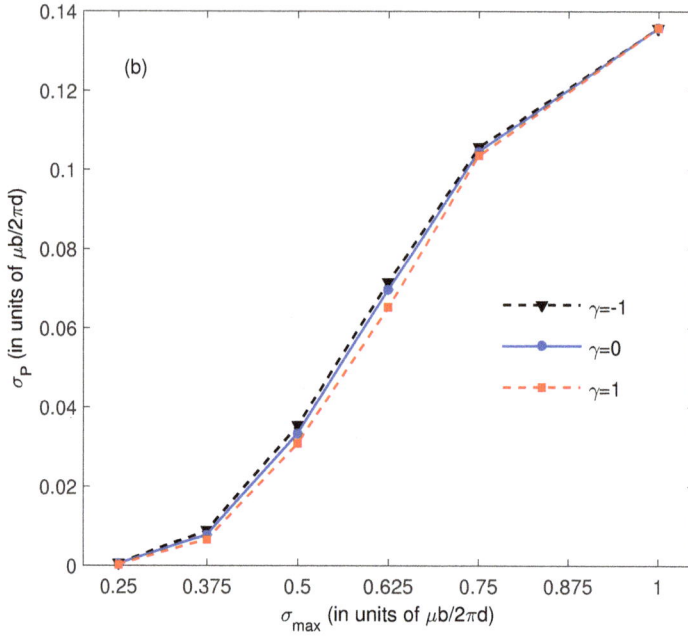

Figure 3. The maximum restoring force σ_{max} on the core structure parameter ξ (**a**) and Peierls stress σ_p (**b**) (take $h = 4d$ as an example).

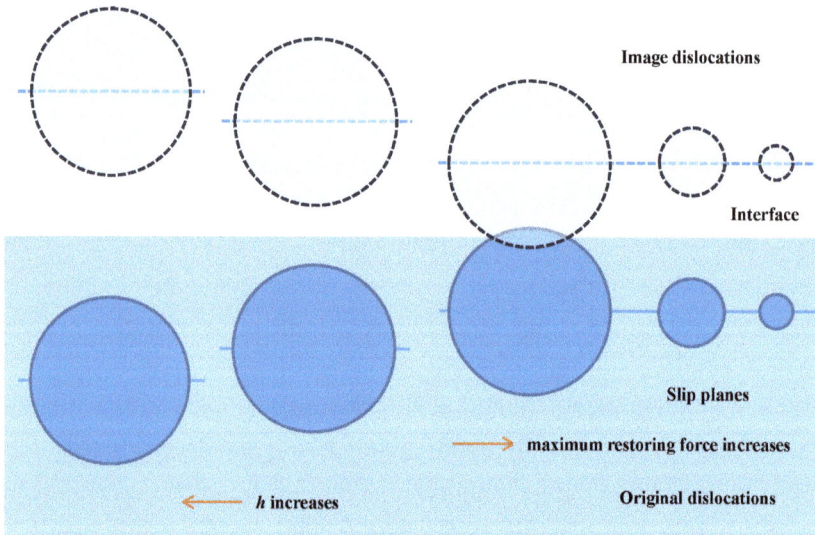

Figure 4. Image dislocations core structures for different maximum restoring force σ_{max} and the distance between dislocation and interface h. Larger σ_{max} leads to narrower dislocation core, and the superposition between original core and image core will become smaller. h has the same effects.

4. Screw Dislocations in Al/TiC Heterostructure

Generally, material-I and material-II have different lattice constants and shear moduli. The atoms near the interface adjust their position to reduce the elastic strain energy. These will change the unstable stacking fault energy in material-I and material-II. Compared to interfacial type (γ) and distance (h), the contributions of maximum restoring force (unstable stacking fault energy) needs to be discussed further. In order to investigate the direct effect of interface, we calculated the unstable stacking fault energy for Al/TiC heterostructures as a model interface using first principles methods. In order to make the paper easy to read, the detailed computation procedures are not listed (for detailed computation procedures, refer to [17]). The interface is (100) plane, and the Burgers vector of screw dislocation is along $\langle 110 \rangle$ directions. The calculated lattice constants of Al and TiC are 4.04 and 4.33, respectively. The shear modulus is 26.0 GPa for Al and 164.5 GPa for TiC. According to the large difference in shear modulus between Al and TiC, we can use the coherent interface approximation. The softer Al is stretched to match the dimension of TiC. The axis normal to the interface is still the dimension of Al. If the difference of shear modulus is small, the material with larger constants will be stretched, and the other material will be compressed.

The unstable stacking fault energy is listed in Table 1. The unstable stacking fault energy for bulk Al is 0.837 J/m^2. When Al is stretched, the unstable stacking fault (USF) energy is 0.527 J/m^2. The stretch reduces the USF energy by 0.31 J/m^2. When the Al/TiC interface is formed, the unstable stacking fault energies in Al are 0.396, 0.540, 0.545, and 0.475 J/m^2 for distances h = 1d, 2d, 3d, and 4d, respectively. The changes of energy between stretched Al and Al in the interface are 0.131, -0.013, -0.018, and 0.052 J/m^2, respectively. The USF energy for bulk TiC is 3.582 J/m^2. In the coherent interface approximation, TiC is not stretched. After the Al/TiC interface is formed, the unstable stacking fault energies in TiC are 3.650, 3.568, 3.576, and 3.582 J/m^2. The energy differences are -0.068, 0.014, 0.006, and 0 J/m^2. The interface formation can be postulated including two steps : (1) stretch soft Al, (2) form Al/TiC interface by TiC and stretched Al. The above calculation results show that step (1) provides the main contribution to the unstable stacking fault energy in comparison with step (2). Therefore, the interface effects on USF stacking are originated from the stretching of Al.

The change of energy between $h = 1d$ and stretched Al is larger than $h = 2d$, 3d, and 4d. The calculated interfacial USF energy is 1.397 J/m^2, which is larger than bulk Al and stretched Al. This means that Al and C atoms at the interface have strong covalent bonding, resulting in a smaller unstable stacking fault energy for $h = 1d$. However, the interfacial bonds have little impact on $h = 2d$, 3d, and 4d, due to the locality of stacking fault energy.

For screw dislocations in Al and TiC, the interface types (γ) are 0.727 (rigid interface) and -0.727 (soft interface), respectively. The core width and Peierls stress is listed in Table 1. The Peierls stress of screw dislocations in interface-Al is smaller than that in bulk Al due to the stretching when forming the Al/TiC interface. The lattice constants are unaltered when forming the Al/TiC interface due to large shear modulus. The Peierls stress is nearly the same in bulk TiC and interface-TiC.

Table 1. The unstable stacking fault energy γ_{us} (in units of J/m^2), the maximum restoring force σ_{max} (in units of GPa), the half width of dislocation ζ (in units of Burgers vector), and the Peierls stress σ_p (in units of GPa).

	Al				TiC			
	γ_{us}	σ_{max}	ζ	σ_p	γ_{us}	σ_{max}	ζ	σ_p
Bulk	0.837	9.20	0.225	3.16	3.582	36.75	0.356	8.77
Bulk$_s$	0.527	5.80	0.357	1.38	–	–	–	–
1d	0.396	4.35	0.476	0.65	3.650	37.45	0.350	9.15
2d	0.540	5.94	0.348	1.46	3.568	36.61	0.358	8.70
3d	0.545	5.99	0.345	1.48	3.576	36.69	0.357	8.74
4d	0.475	5.22	0.396	1.08	3.582	36.75	0.356	8.77

5. Conclusions

Due to the equivalence of dislocations and point charges in elastic strain field and electrostatics, the image method is used to describe the interface effects on the dislocation elastic strain field. The interface effects on dislocations are represented by image dislocations. The governing equation for screw dislocation in heterostructure is constructed using the image dislocation method. The interface type (γ) and distance (h) effects are contained in the equation. The P-N dislocation equations in bulk and in semi-infinite materials can be recovered when $\gamma = 0$ and $\gamma = -1$, respectively. The Peierls stress can be enhanced by a soft interface and decreased by a hard interface. The hard and soft interfaces can be described by $\gamma < 0$ and $\gamma > 0$. With the increasing of h, the effects on dislocations of the interface become small. In order to investigate the interface on the unstable stacking fault energy and dislocation properties, the Al/TiC heterostructures were studied as a model interface. It was found that the changes of the unstable stacking fault energy mainly originated from the mismatch of lattice constants and shear modulus at the interface. Then, the changes of the unstable stacking fault energy also have important effects on dislocation properties comparing with γ and h.

Acknowledgments: Project Supported by Science Challenge No.TZ2016003 and Project Supported by the Fundamental Research Funds for the Central Universities No.2017CDJQJ308822, Project Supported by the Natural Science Foundation of China (91634106).

Author Contributions: J.W. and X.W. conceived the paper; T.S. and W.X. performed the first principle calculations; X.W. and R.W. analyzed the results; J.W. wrote the paper.

Conflicts of Interest: We declare that we have no conflict of interest.

References

1. Hirth, J.P.; Lothe, J. *Theory of Dislocations*, 2nd ed.; Wiley: New York, NY, USA, 1982.
2. Peierls, R. The size of a dislocation. *Proc. Phys. Soc.* **1940**, *52*, 34.
3. Nabarro, F.R.N. Dislocations in a simple cubic lattice. *Proc. Phys. Soc.* **1947**, *59*, 256.
4. Vitek, V. Intrinsic stacking faults in body-centred cubic crystals. *Philos. Mag.* **1968**, *18*, 773–786.
5. Joos, B.; Ren, Q.; Duesbery, M.S. Peierls-Nabarro model of dislocations in silicon with generalized stacking-fault restoring forces. *Phys. Rev. B* **1994**, *50*, 5890.
6. Pei, Z.R.; Ma, D.C.; Friak, M.; Svendsen, B.; Raabe, D.; Neugebouer, J. From generalized stacking fault energies to dislocation properties: Five-energy-point approach and solid solution effects in magnesium. *Phys. Rev. B* **2015**, *92*, 064107.
7. Gouriet, K.; Carrez, P.; Cordier, P.; Guitton, A.; Joulain, A.; Thilly, L.; Tromas, C. Dislocation modelling in Ti2AlN MAX phase based on the Peierls–Nabarro model. *Philos. Mag.* **2015**, *95*, 2539–2552.
8. Chen, C.; Meng, F.C.; Sun, J. Effects of Mg and Al doping on dislocation slips in GaN. *J. Appl. Phys.* **2016**, *119*, 064302.
9. Xiao, Z.M.; Chen, B.J. A screw dislocation interaction with a coated fiber. *Mech. Mater.* **2000**, *32*, 485–494.
10. Lee, C.L.; Li, S. The size effect of thin films on the Peierls stress of edge dislocations. *Math. Mech. Solids* **2008**, *13*, 316–335.
11. Gars, B.; Markenscoff, X. The Peierls stress for coupled dislocation partials near a free surface. *Philos. Mag.* **2012**, *92*, 1390–1421.
12. Cheng, X.; Shen, Y.; Zhang, L.; Liu, X.H. Surface effect on the screw dislocation mobility over the Peierls barrier. *Philos. Mag. Lett.* **2012**, *92*, 270–277.
13. Head, A.K. The Interaction of Dislocations and Boundaries. *Philos. Mag.* **1953**, *44*, 92–94.
14. Yao, Y.; Wang, T.C.; Wang, C.Y. Peierls-Nabarro model of interfacial misfit dislocation: An analytic solution. *Phys. Rev. B* **1999**, *59*, 8232.
15. Rice, J.R. Dislocation nucleation from a crack tip: An analysis based on the Peierls concept. *J. Mech. Phys. Solids* **1992**, *40*, 239–271.

16. Foreman, A.J.; Jaswon, M.A.; Wood, J.K. Factors Controlling Dislocation Widths. *Proc. Phys. Soc. A* **1951**, *64*, 156.
17. Wu, X.Z.; Sun, T.; Wang, R.; Liu, L.L.; Liu, Q. Energy investigations on the adhesive properties of Al/TiC interfaces: First-principles study. *Physica B* **2014**, *449*, 269–273.

crystals

MDPI

Article

Stable Stacking Faults Bounded by Frank Partial Dislocations in Al7075 Formed through Precipitate and Dislocation Interactions

Sijie Li [1,2,3], Hongyun Luo [1,2,3,*], Hui Wang [1,2,3], Pingwei Xu [1,2,3] [iD], Jun Luo [1,2,3], Chu Liu [1,2,3] and Tao Zhang [1,2,3]

[1] Key Laboratory of Aerospace Materials and Performance (Ministry of Education), School of Materials Science and Engineering, Beijing University of Aeronautics and Astronautics, Beijing 100191, China; 18014128516@163.com (S.L.); huiwang@buaa.edu.cn(H.W); xpw2012buaa@gmail.com (P.X.); luojunxt@126.com (J.L.); hustliuchu@163.com (C.L.); ztwuai@126.com (T.Z.)

[2] The Collaborative Innovation Center for Advanced Aero-Engine (CICAAE), Beijing University of Aeronautics and Astronautics, Beijing 100191, China

[3] Beijing Key Laboratory of Advanced Nuclear Materials and Physics, Beijing University of Aeronautics and Astronautics, Beijing 100191, China

* Correspondence: luo7128@163.com; Tel.: +86-010-8233-9905; Fax: +86-010-8231-7108

Academic Editor: Peter Lagerlof
Received: 9 November 2017; Accepted: 11 December 2017; Published: 13 December 2017

Abstract: Through high-resolution electron microscopy, stacking faults (SFs) due to Frank partial dislocations were found in an aluminum alloy following deformation with low strain and strain rate, while also remaining stable during artificial aging. Extrinsic stacking faults were found surrounded by dislocation areas and precipitates. An intrinsic stacking fault was found between two Guinier-Preston II (GP II) zones when the distance of the two GP II zones was 2 nm. Defects (precipitates and dislocations) are considered to have an influence on the formation of the SFs, as their appearance may cause local strain and promote the gathering of vacancies to lower the energy.

Keywords: stacking faults; frank partial dislocation; precipitates; vacancies

---

1. Introduction

As is well known, with high stacking-fault energy (SFE) and twin-boundary energy, aluminum and its alloys rarely produce stacking faults (SFs) or microtwins. Producing SFs in high-SFE metals, which could potentially improve the ductility of the metals, has raised considerable interest among scholars [1–7]. For example, SFs and twins were found in nanocrystal aluminum powders produced by ball milling [5]. Twins were produced in Al-Mg alloy [3] at a high strain rate (10^2–10^6 s^{-1}). Most of these findings were made under very extreme conditions, such as high strain and high strain rate.

The SFs and microtwins mentioned above are commonly explained by the generation of Shockley partial dislocation [8,9] (a perfect dislocation divided into two Shockley partial dislocations), while SFs caused by Frank partial dislocations are seldom investigated. Although some SFs led by Frank loop have been found [10–12], unfortunately, these SFs were unstable during heating or irradiation.

Alternatively, another effective and widely researched way of improving the performance of aluminum alloys is to produce precipitates [13–15]. However, the correlation between the precipitates and SFs has also rarely been investigated.

In our former investigation, we discovered SFs in aluminum alloys; however, these were not discussed in detail [16]. This paper shall focus on stable SFs with Frank partial dislocations that were found, as well as their formation process.

2. Experimental Details

The experiment used an Al7075 alloy, with a chemical composition of (wt %): Zn-5.35, Mg-2.3, Cu-1.41. A bulk sample was solution-treated at 500 °C for 2 h, and immediately quenched in water. Subsequently, it was rolled from 15 mm to 7.5 mm (50% reduction, with a strain rate <1 s^{-1}). The rolling direction–transverse direction plane was observed, showing {111} <112> orientation, as determined in our former research [16], which is in line with findings in [6]. Artificial aging was carried out at 70 °C for up to 75 and 175 h after rolling. Figure 1 shows the XRD patterns of the aluminum alloy before (SST) and after deformation treatment (50%), as well as after aging for 175 h (50%-175 h). The sample maintains its (111) preferred orientation during the rolling and aging process. Microstructures were characterized using a JEOL 2100 F transmission electron microscope (TEM), and were observed from the <110> axis. The samples for TEM characterization were prepared by mechanically grinding to 75 μm–95 μm, before punching them into discs. Subsequently, the 3 mm diameter discs were thinned using an ion beam thinner (Gatan691, V = 2.5–3.5 KV, Time = 10–30 min).

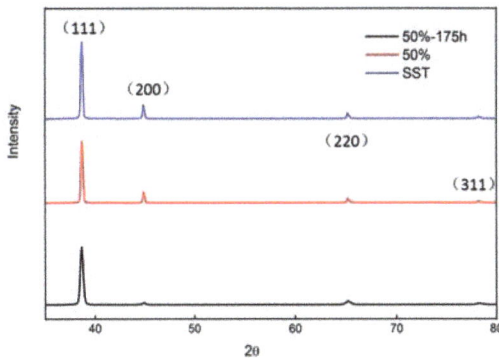

Figure 1. XRD patterns of the aluminum alloy before (SST) and after deformation treatment (50%), as well as the one after aging for 175 h (50%-175 h).

3. Results and Discussion

Figure 2(a$_1$,a$_2$) shows the high-resolution electron microscopy (HREM) images of rolled 7075 Al alloys after artificial aging at 70 °C for 75 h. In Figure 2(a$_1$), the dislocations, marked with 'T', form a circle, as shown by the white dotted line. Interestingly, a stacking fault is found in the center of the circle, highlighted by the yellow frame. An enlarged image of the yellow square of Figure 2 (a$_1$) is shown in Figure 2 (b$_1$), and the insert image shows the Fast Fourier Transformation (FFT) of Figure 2 (b$_1$). An inserted atomic plane along the [−112] direction in the (111) plane is observed at the red spots. The insert plane (red spots), along with the two rows around it (green lines), forms a stacking fault with three layers of atoms. Moreover, emanative lines in the insert FFT spectrum could indicate the emergence of SF. Figure 2(c$_1$) is the Inverse Fast Fourier Transformation (IFFT) image of the (111) reflections. Every stripe represents a (111) plane, and there is an inserted (111) plane at the red dotted line, further confirming our viewpoint. Each end of the inserted plane is a Frank partial dislocation.

Similarly, in Figure 2(a$_2$), a stacking fault with several rows of atoms emerges in the center of a circle, which has a radius of about 9–10 nm formed by precipitates and dislocations. Figure 2(b$_2$) is the magnified picture of Figure 2(a$_2$), in which two inserted atomic planes marked with red spots can be observed. Through Burgers circuits (made up of the yellow lines), it is confirmed that the partial dislocations in this paper are Frank partial dislocations with **b** = a/3<111>. Figure 2(c$_2$), the IFFT of the (111) reflections, also clearly shows two inserted atomic planes. As the two partial dislocations are not on a straight line, the stacking fault in Figure 2(b$_1$) is regarded as precursory, before one or both

Frank partial climbs along the direction perpendicular to the (111) plane. As a result, the SF broadens to eight layers.

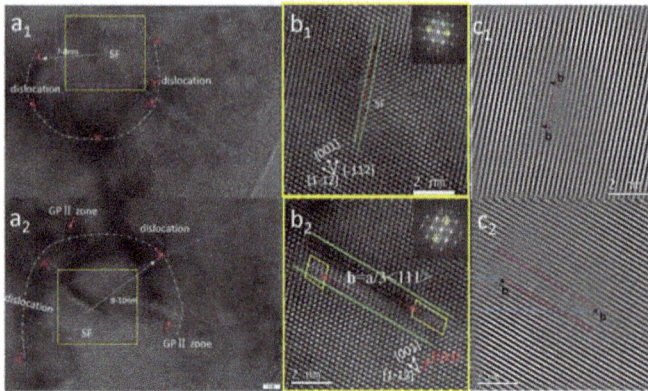

Figure 2. HREM images of rolled 7075 Al alloys after aging at 70 °C for 75 h; (a_1,a_2) HREM images of stacking faults surrounded by dislocations and GP II zones; (b_1,b_2) the magnified pictures of the square in (a_1,a_2) show inserted atomic planes along [−112] direction; (c_1,c_2) IFFT images of (111) patterns show inserted (111) planes.

Figure 3a shows an HREM image with longer artificial aging (175 h). Figure 3b shows the magnified image of Figure 3a, with the insert picture as the corresponding FFT. The morphology and the diffraction patterns of the two GP II zones can be clearly observed (the arrows' points). Interestingly, there is an intrinsic stacking fault between the two GP II zones. The atoms by the lines are well arranged, except for a little distortion (the green curve), while a layer of atoms seems to be drawn out between the lines. Similarly, the IFFT image shows that there is an absence of a (111) plane, which, in this case, is an intrinsic stacking fault.

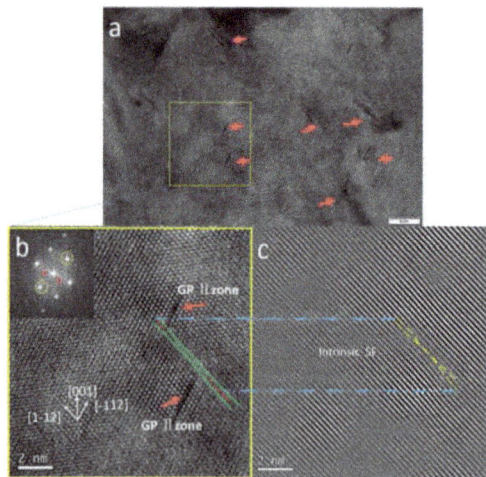

Figure 3. HREM images of rolled 7075 Al alloys after aging at 70 °C for 175 h: (**a**) HREM image; (**b**) the magnified picture of the yellow square in (**a**) shows an intrinsic stacking fault between two GP II zones; (**c**) IFFT shows one drawn-out (111) plane.

A phenomenon worth mentioning is that the SFs nucleate around dislocations or precipitates (Figure 2 shows 7–10 nm and Figure 3 shows 2 nm). Hence, the precipitates and dislocations here are speculated to have effects on the formation of SFs.

The formation process of stacking faults will be analyzed using, as an example, an intrinsic stacking fault, the formation model of which is shown in Figure 4. The smaller red spheres are Zn atoms, while the larger blue spheres represent Al atoms. To simplify the model, only Zn atoms are taken into consideration, as the GP II zone is the aggregation of Zn atoms [15] on {111}$_{Al}$. Moreover, the motion of the atoms during deformation is not considered in this paper. Figure 4a is the atomic arrangement of solid solution state along a <110> axis, and Zn distributes uniformly. Vacancies and dislocations are induced after deformation, as shown in Figure 4b. Dislocation cores act as fast paths for diffusing atoms, since the disorder in the core region effectively lowers the activation energy for diffusion [14]. Hence, Zn atoms diffuse rapidly through defects, and accumulate near the dislocation. During the aging process, vacancies migrate towards the dislocations to release the stress caused by dislocations [17]. Once the vacancies approach, Zn-vacancy clusters form and grow into GP II zones on {111}$_{Al}$, and the vacancies left are usually absorbed by dislocations or grain boundaries [18]. However, in this paper, two GP II zones formed very closely (2 nm) in Figure 4c and, as a result, high strain may be produced between them. In this case, the vacancies left tended to gather together (an intrinsic stacking fault formed) to release the strain and lower the energy. In other words, two GP II zones and the SF in the middle formed a stable GP II-SF group. On one hand, the emergence of SF reduces the free energy of the system. On the other hand, the two GP regions play a role in the pinning of the SF formed by the migration and coalescence of the vacancies. This is why, in this paper, it was able to stably exist during heating, while some other SFs with Frank partial dislocations have tended to disappear during heating [10] or irradiation [11].

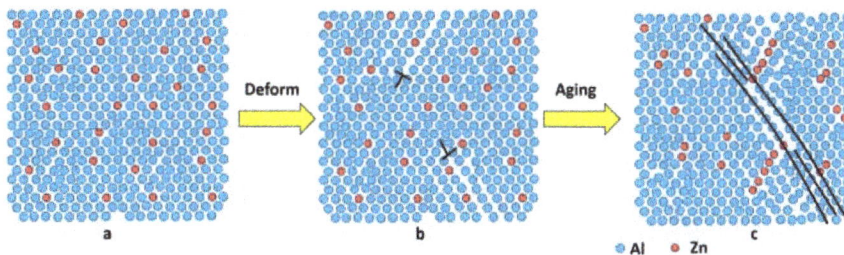

Figure 4. A schematic image showing the formation process of an intrinsic stacking fault: (**a**) atomic arrangement of solid solution state; (**b**) vacancies and dislocations induced by deformation; (**c**) a stacking fault emerges between two precipitates during aging through vacancies gathering to lower the energy.

A stacking fault is the destruction of the normal stacking sequence of close-packed planes. For aluminum, the plane is {111}, indicating that the stacking faults in aluminum always occur in {111}. GP II zones also generate in this plane, and influence atomic arrangement, which may increase the possibility of stacking faults. This is another important reason for the GP II zones in this paper promoting SFs to emerge.

The formation of extrinsic SFs in this paper is considered to have a similar mechanism to that of intrinsic SFs. As shown in Figure 2 (a$_1$), a large number of dislocations are produced after deformation. Interestingly, these dislocations form a radius of about 9 nm, as shown in the white line of the figure. In this case, vacancies migrate towards dislocations, since the stress field caused by dislocations readily affects vacancies [17], and stacking faults are formed as the vacancies gather. However, this kind of stacking fault is not stable, as it is not observed after aging for 175 h. This is predominantly because, under the influence of heat, the Frank partial dislocation at both ends of the fault continuously absorbs more vacancies [10], making it climb along the direction perpendicular to the (111) plane (as shown in Figure 2(b_1,b_2)), thus widening the stacking fault.

4. Conclusions

In summary, stabled SFs with Frank partial dislocations are found in rolled (50% reduction, with a strain rate <1 s^{-1}) aluminum alloy after aging. Precipitates and dislocations have a great effect on the formation of the SFs, which offers a new perspective for SFs in high stacking-fault energy alloys. However, the critical conditions necessary for precipitates to influence SFs, such as how close the precipitates should be or how much strain they will produce, need to be further investigated, as not all precipitates produce SFs.

Acknowledgments: This work was financially supported by National Key Technology R&D Program of China (2015BAG20B04 and 2015BAF06B01-3), the National Key Research and Development Program of China (2016YFC0801903 and 2016YFF0203301) and National Natural Science Foundation of China (No. 51175023 and No. U1537212).

Author Contributions: Sijie Li and Hongyun Luo conceived and designed the experiments; Sijie Li performed the experiments; Sijie Li, Hongyun Luo and Hui Wang analyzed the data; Pingwei Xu, Jun Luo, Chu Liu and Tao Zhang contributed materials and analysis tools; Sijie Li wrote the paper.

Conflicts of Interest: The authors declare no conflict of interest.

References

1. Liao, X.Z.; Zhou, F.; Lavernia, E.J.; He, D.W.; Zhu, Y.T. Deformation twins in nanocrystalline Al. *Appl. Phys. Lett.* **2003**, *83*, 5062–5064. [CrossRef]
2. Yamakov, V.; Wolf, D.; Phillpot, S.R.; Gleiter, H. Deformation twinning in nanocrystalline Al by molecular dynamics simulation. *Acta Mater.* **2002**, *50*, 5005–5020. [CrossRef]
3. Jin, S.B.; Zhang, K.; Bjorge, R.; Tao, N.R.; Marthinsen, K.; Lu, K.; Li, Y.J. Formation of incoherent deformation twin boundaries in a coarse-grained Al-7Mg alloy. *Appl. Phys. Lett.* **2015**, *107*, 091901. [CrossRef]
4. Cao, B.; Daphalapurkar, N.P.; Ramesh, K.T. Ultra-high-strain-rate shearing and deformation twinning in nanocrystalline aluminum. *Meccanica* **2015**, *50*, 561–574. [CrossRef]
5. Li, B.Q.; Sui, M.L.; Li, B.; Ma, E.; Mao, S.X. Reversible Twinning in Pure Aluminum. *Phys. Rev. Lett.* **2009**, *102*, 205504. [CrossRef] [PubMed]
6. Han, W.Z.; Cheng, G.M.; Li, S.X.; Wu, S.D.; Zhang, Z.F. Deformation induced microtwins and stacking faults in aluminum single crystal. *Phys. Rev. Lett.* **2008**, *101*, 115505. [CrossRef] [PubMed]
7. Zhao, F.; Wang, L.; Fan, D.; Bie, B.X.; Zhou, X.M.; Suo, T.; Li, Y.L.; Chen, M.W.; Liu, C.L.; Qi, M.L.; et al. Macrodeformation Twins in Single-Crystal Aluminum. *Phys. Rev. Lett.* **2016**, *116*. [CrossRef] [PubMed]
8. Wang, Y.B.; Sui, M.L. Atomic-scale in situ observation of lattice dislocations passing through twin boundaries. *Appl. Phys. Lett.* **2009**, *94*, 021909. [CrossRef]
9. Liao, X.Z.; Zhou, F.; Lavernia, E.J.; Srinivasan, S.G.; Baskes, M.I.; He, D.W.; Zhu, Y.T. Deformation mechanism in nanocrystalline Al: Partial dislocation slip. *Appl. Phys. Lett.* **2003**, *83*, 632–634. [CrossRef]
10. Wu, X.L.; Li, B.; Ma, E. Vacancy clusters in ultrafine grained Al by severe plastic deformation. *Appl. Phys. Lett.* **2007**, *91*, 141908. [CrossRef]
11. Yang, W.; Dong, R.; Jiang, L.; Wu, G.; Hussain, M. Unstable stacking faults in submicron/micron Al grammins in multi-SiCp/multi-Al nanocomposite. *Vacuum* **2015**, *122*, 1–5. [CrossRef]
12. Monnet, G. New insights into radiation hardening in face-centered cubic alloys. *Scr. Mater.* **2015**, *100*, 24–27. [CrossRef]
13. Zheng, Y.; Luo, B.; Bai, Z.; Wang, J.; Yin, Y. Study of the Precipitation Hardening Behaviour and Intergranular Corrosion of Al-Mg-Si Alloys with Differing Si Contents. *Metals* **2017**, *7*, 387. [CrossRef]
14. Hu, T.; Ma, K.; Topping, T.D.; Schoenung, J.M.; Lavernia, E.J. Precipitation phenomena in an ultrafine-grained Al alloy. *Acta Mater.* **2013**, *61*, 2163–2178. [CrossRef]
15. Berg, L.K.; Gjonnes, J.; Hansen, V.; Li, X.Z.; Knutson-Wedel, M.; Waterloo, G.; Schryvers, D.; Wallenberg, L.R. GP-zones in Al-Zn-Mg alloys and their role in artificial aging. *Acta Mater.* **2001**, *49*, 3443–3451. [CrossRef]

16. Xu, P.; Luo, H. Improving the ductility of nanostructured Al alloy using strongly textured nano-laminated structure combined with nano-precipitates. *Mater. Sci. Eng. A Struct.* **2016**, *675*, 323–337. [CrossRef]
17. Sha, G.; Cerezo, A. Early-stage precipitation in Al-Zn-Mg-Cu alloy (7050). *Acta Mater.* **2004**, *52*, 4503–4516. [CrossRef]
18. Kelly, A.; Nicholson, R.B. Precipitation hardening. *Prog. Mater. Sci.* **1963**, *10*, 151–391.

crystals

MDPI

Article

Study of Anisotropic Plastic Behavior in High Pressure Torsion of Aluminum Single Crystal by Crystal Plasticity Finite Element Method

Peitang Wei [1,*], Cheng Lu [2], Huaiju Liu [1], Lihong Su [2], Guanyu Deng [2,3] and Kiet Tieu [2]

[1] State Key Laboratory of Mechanical Transmissions, Chongqing University, Chongqing 400044, China; huaijuliu@cqu.edu.cn
[2] School of Mechanical, Materials and Mechatronic Engineering, University of Wollongong, Wollongong, NSW 2522, Australia; chenglu@uow.edu.au (C.L.); lihongsu@uow.edu.au (L.S.); gd577@uowmail.edu.au (G.D.); ktieu@uow.edu.au (K.T.)
[3] Department of Materials Science and Engineering, Kyoto University, Sakyo-ku 606-8501, Japan
* Correspondence: peitangwei@cqu.edu.cn

Academic Editor: Peter Lagerlof
Received: 28 October 2017; Accepted: 4 December 2017; Published: 6 December 2017

Abstract: In this study, a crystal plasticity finite element method (CPFEM) model has been developed to investigate the anisotropic plastic behavior of (001) aluminum single crystal during high-pressure torsion (HPT). The distributions of equivalent plastic strain and Mises stress recorded on the sample surface are presented. The directional variations of plastic strain and Mises stress with the development of four-fold symmetry pattern are observed along the sample circumference. The crystallographic orientation evolution along the tangential direction is studied, and the corresponding lattice rotation and slip trace are predicted, respectively. The plastic anisotropy mechanism is discussed in detail based on the theory of crystal plasticity. The simulation results reveal that the differences in slip systems activation (dominant slip and multiple slips) are responsible for the anisotropic plastic deformation in HPT.

Keywords: high pressure torsion (HPT); crystal plasticity; anisotropic; crystallographic orientation; slip trace; lattice rotation

---

1. Introduction

Interest in the use of severe plastic deformation (SPD) techniques to produce ultra fine-grained (UFG) or nanocrystalline structures of different materials has developed a great deal over the last two decades [1,2]. High-pressure torsion (HPT) is a SPD technique in which a sample is subjected to torsion straining under a high hydrostatic pressure [3]. Compared with the other SPD methods, such as equal-channel angular pressing (ECAP) [4] and accumulative roll bonding (ARB) [5], the most important advantage of the HPT process is that extremely high shear stain can be continuously achieved via simple means [6,7]. Due to its incomparable straining capacity, HPT has attracted extensive research interest for fabrication of UFG materials of pure metals and alloys, where polycrystals are typically used as starting materials [8,9].

Up until now, a number of studies have been reported on HPT processing of single crystals. In the case of single crystals, it is convenient to use a single-crystal structural state to exclude the influence of initial grain boundaries and overall initial orientations, aiding in the understanding of deformation behaviors during HPT. Astafurova et al. [10] and Pilyugin et al. [11,12] investigated the microstructure evolution of different single-crystalline materials during the HPT process. Kashihara and his co-workers [13] examined the texture evolution in {112} <111> aluminum single crystal processed by severe plastic deformation. Smirnova et al. [14,15] studied the developments of structure and orientational instability in face-centered cubic (FCC) single crystals (copper, nickel and alloys KhN77TYuR) deformed by torsion at quasi-hydrosatic pressure. Their results showed the differences in deformation behavior of different initially orientated single crystals. Hafok and Pippan [16,17] experimentally examined the texture evolution of nickel and copper single crystals subjected to HPT deformation. They reported that the texture of the (001)-oriented single crystal developed a four-fold symmetry along the sample tangential direction. Méric and Qailletaud [18] undertook conventional FEM simulation and anisotropic elasto-viscoplastic calculation derived from the slip theory for FCC single crystals loaded in torsion, respectively. The FEM method gave a homogeneous strain distribution along the circumference of the initial (001) single crystal specimen, while the micro-macro approach presented four more strained regions around <110> areas separated by four less-strained regions located in <100> areas. Nouailhas et al. [19] reported similar observations of strong anisotropic features using a crystallographic model based on the Schmid theory. Kaluza and Le [20] and Le and Piao [21] have quantitatively calculated the dislocation distribution inside a single crystal rod loaded in torsion within the framework of continuum dislocation theory. The non-uniform distribution of dislocations in equilibrium were found and a wavy deformation pattern comprising of four periods was observed during torsion. Even though research has already been done in the past, the reason underlying such anisotropic plastic bahavior during HPT processing of single-crystal metals is still unclear.

In our previous work [22], a crystal plasticity finite element method (CPFEM) model was developed to simulate the HPT processing of single crystals. It has been widely accepted that the CPFEM model is one of the best models for simulating the plastic deformation of crystalline materials [23,24], The predicted texture results were in good coherence with the experimental measurements, and considerable attention was devoted to revealing the texture evolution mechanism during the HPT process. In addition, the anisotropic deformation pattern on the surface of the HPT-processed sample could also be clearly observed. The purpose of this study is to further investigate such plastic anisotropy phenomena in HPT-deformed single crystals, adopting the CPFEM model. The underlying mechanism is discussed in detail based on the theory of crystal plasticity.

2. CPFEM Simulation Procedure

2.1. Crystal Plasticity Constitutive Model

The wide variety of currently available crystal plasticity models can be classified into two major types: the orientation gradient crystal plasticity method, which follows the crystal plasticity constitutive relations described by Asaro [25]; and the strain gradient crystal plasticity method, which is based on a scalar dislocation density-based constitutive frame proposed by Ashby [26]. The first type of model was employed in the simulation, as reported by Asaro [25]; under load, the crystalline material undergoes crystallographic slip, due to dislocation motion on the active slip systems and elastic deformation, which includes stretching and rotating of the crystal. The total deformation gradient (F) can be decomposed into two components:

$$F = F^* \cdot F^P \tag{1}$$

The velocity gradient (L) is evaluated from the deformation gradient by

$$L = \dot{F}F^{-1} = L^* + L^P \tag{2}$$

where

$$L^* = \dot{F}^* \cdot F^{*-1}; \; L^P = F^* \cdot \dot{F}^P \cdot F^{P-1} \cdot F^{*-1} \tag{3}$$

Taking the symmetric and asymmetric parts of the above relations leads to the elastic and plastic strain rates D^* and D^P, and the plastic spins Ω^P and Ω^* induced by the lattice rotation and stretching:

$$D = D^* + D^P; \ \Omega = \Omega^* + \Omega^P \tag{4}$$

By introducing the following symmetric and asymmetric tensors, respectively, for each slip system α,

$$P^{(\alpha)} = \frac{1}{2}(s^{(\alpha)} \otimes m^{(\alpha)} + m^{(\alpha)} \otimes s^{(\alpha)}); \ W^{(\alpha)} = \frac{1}{2}(s^{(\alpha)} \otimes m^{(\alpha)} - m^{(\alpha)} \otimes s^{(\alpha)}) \tag{5}$$

the plastic strain rate and spin for the crystal can be written as

$$D^P = \sum_{\alpha=1}^{N} P^{(\alpha)}\dot{\gamma}^{(\alpha)}; \ \Omega^P = \sum_{\alpha=1}^{N} W\dot{\gamma}^{(\alpha)} \tag{6}$$

where $\dot{\gamma}^{(\alpha)}$ is the resolved shear strain rate of the slip system α, $s^{(\alpha)}$ and $m^{(\alpha)}$ are the slip direction vector and the normal vector, respectively, in the current configuration.

The crystalline slip is assumed to follow the power law, which states that slip begins when the resolved shear stress reaches a critical value,

$$\dot{\gamma}^{(\alpha)} = \dot{\gamma}_0^{(\alpha)} sgn\left(\tau^{(\alpha)}\right)\left|\frac{\tau^{(\alpha)}}{\tau_c^{(\alpha)}}\right|^n \ \text{for} \ \left|\tau^{(\alpha)}\right| \gg \tau_c^{(\alpha)} \tag{7a}$$

$$\dot{\gamma}^{(\alpha)} = 0 \ \text{for} \ \left|\tau^{(\alpha)}\right| < \tau_c^{(\alpha)} \tag{7b}$$

and

$$sgn(x) = \begin{cases} -1, & x < 0 \\ 1, & x \geq 0 \end{cases} \tag{7c}$$

where $\dot{\gamma}_0^{(\alpha)}$ and n are the material parameters, $\tau_c^{(\alpha)}$ is the critical resolved shear stress (CRSS) of the slip system α. The rate of change of the CRSS is expressed as

$$\dot{\tau}_c^{(\alpha)} = \sum_{\beta=1}^{N} h_{\alpha\beta}\dot{\gamma}^{(\beta)} \tag{8}$$

The hardening models have been extensively studied by Taylor [27,28], Hutchinson [29], Peirce et al. [30], and Bassani and Wu [31,32]. Many comparative studies of the aforementioned hardening models have been carried out to simulate the finite- and large-strain deformations [33–36]. It has been found that the Bassani and Wu model was able to reflect the hardening of face-centered cubic (FCC) crystals more exactly, and was an accurate predictor of texture. Recently, the Bassani and Wu hardening model has been successfully employed to predict deformation behavior and texture evolution during ECAP [37,38] and HPT [22]. Therefore, in this study, the hardening model of Bassani and Wu was adopted. Their expressions for self and latent hardening are expressed as:

$$h_{\alpha\alpha} = \left[(h_0 - h_s)\sec h^2\left(\frac{(h_0 - h_s)\gamma^{(\alpha)}}{\tau_1 - \tau_0}\right) + h_s\right]\left[1 + \sum_{\substack{\beta=1 \\ \beta \neq \alpha}}^{N} f_{\alpha\beta} \tan h\left(\frac{\gamma^{(\beta)}}{\gamma_0}\right)\right] \ \text{for} \ \alpha = \beta \tag{9a}$$

$$h_{\alpha\beta} = qh_{\alpha\alpha} \ \text{for} \ \alpha \neq \beta \tag{9b}$$

where $h_{\alpha\alpha}$ is the self-hardening modulus and $h_{\alpha\beta}$ denotes the latent hardening modulus, q is a latent hardening parameter, h_0, h_s, τ_0 and τ_s are hardening moduli and shear stresses, γ is the shear strain, $f_{\alpha\beta}$ is the interaction parameter between the two slip systems α and β. The factors $f_{\alpha\beta}$ depend on the geometric relation between two slip systems. There are five constants for $f_{\alpha\beta}$; namely, α_1 (no junction), α_2 (Hirth lock), α_3 (coplanar junction), α_4 (glissile junction) and α_5 (sessile junction).

2.2. Finite Element Implementation

The commercial software ABAQUS 6.9-1 was used to simulate the deformation process of HPT, as shown in Figure 1. The disk-shaped sample with 10 mm of diameter and 0.8 mm of thickness was assumed to be a deformable body, while the upper and lower HPT anvils were set as rigid bodies. The starting material was oriented with the (001) crystallographic plane normal parallel to the Z axis, while the [100] crystallographic direction lying with the X axis; also known as the cube orientation. Because HPT deformation is applied on the thin disk-shaped samples, a cylindrical-polar coordinate system, C_S, was established with the associated orthonormal base vector (e^r, e^θ, e^z), where R, θ and Z denote the radial, tangential and axial directions, respectively. The angular value φ stands for the extent that material flows away from the X direction around the axial direction on the shear plane (the counterclockwise direction is positive, while the clockwise direction is negative), as illustrated in Figure 1a. The sample was meshed into 23,600 elements and 26,895 nodes in total, and C3D8R elements were applied, as shown in Figure 1b. During the simulation, the lower anvil was fixed. The rotation boundary condition along the Z axis was applied to the upper anvil, while the other freedoms of the upper anvil were set to be constrained.

The simulated material was aluminum single crystal, the elastic moduli of which were: $C_{11} = 112,000$ MPa, $C_{12} = 66,000$ MPa and $C_{44} = 28,000$ MPa. The parameters in the constitutive equations employed in this study are listed in Table 1. Franciosi et al. obtained the slip system interaction parameters α_1-α_5 by conducting the latent hardening experiment using aluminum single crystal [39]. Other parameters were identified by fitting the simulated stress-strain curve with the experimental results of single crystal aluminum [40,41]. These parameters have been validated in the CPFEM simulations of nano-indentation [42], cold rolling [43], and equal-channel angular pressing [37]. In the deformed aluminum material with FCC structure, it is assumed that slips occur on the {111} slip planes along the <110> slip directions. Their combination defines 12 different slip systems as indicated in Table 2.

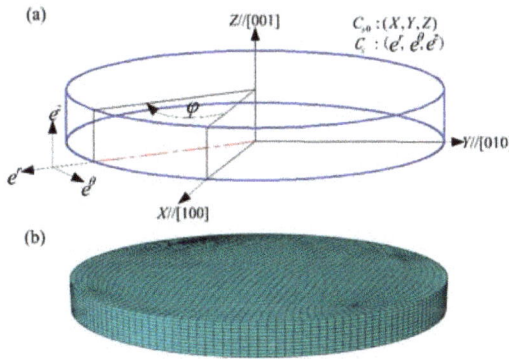

Figure 1. Three-dimensional HPT model: (**a**) configuration of the sample and coordinate systems, (**b**) meshes of the sample.

Table 1. Parameters in the constitutive model.

n	$\dot{f}l_0$ (1/s)	h_0 (MPa)	h_s (MPa)	$\varnothing_1$ (MPa)	$\varnothing_0$ (MPa)	fl_0	ff_1	ff_2	ff_3	ff_4	ff_5	q
300	0.0001	100	0.01	6.3	6	0.001	1.75	1.75	1.75	2	2.25	1

Table 2. Notations of slip systems for FCC crystals.

System	a1	a2	a3	b1	b2	b3	c1	c2	c3	d1	d2	d3
Slip plane	(111)	(111)	(111)	($\bar{1}$11)	($\bar{1}$11)	($\bar{1}$11)	(1$\bar{1}$1)	(1$\bar{1}$1)	(1$\bar{1}$1)	(11$\bar{1}$)	(11$\bar{1}$)	(11$\bar{1}$)
Slip direction	[0$\bar{1}$1]	[10$\bar{1}$]	[$\bar{1}$10]	[101]	[110]	[0$\bar{1}$1]	[011]	[110]	[10$\bar{1}$]	[011]	[101]	[$\bar{1}$10]

3. Results

3.1. Distributions of Equivalent Plastic Strain and Mises Stress

Figure 2a,b shows the contour plots of equivalent plastic strain (variable PEEQ used in ABAQUS software 6.9-1) and Mises stress recorded on the top surfaces of (001) aluminum single crystal after $N = 1/12$ turn of HPT deformation, respectively. A pronounced strain gradient exists on the sample section, with the values increasing gradually from the center of the sample to the edge along the radial direction, as shown in Figure 2a. Moreover, Figure 2a reveals that the strain is non-homogeneously distributed along the circumference of the sample, especially close to the periphery. Four more-strained regions, separated by four-less strained ones with interval angles of 90°, can be observed. The similar strain distribution on the section of the sample has been predicted by simulation and observed experimentally in [18,19,44], which reported that the torsion deformation of a cylinder sample with respect to a [001] axis gave rise to the formation of four zones of intense deformation. As can be seen from Figure 2b, the directional variations of Mises stress along the circumferential direction are significant, and a four-fold symmetry distribution pattern develops. Within one section of the four-fold symmetry, the deformation localization can be obviously observed. These results differ greatly from those obtained through classic finite element simulation, as described in [45–47], which gave uniform strain and stress distributions along the circumferential direction of the sample deformed by HPT.

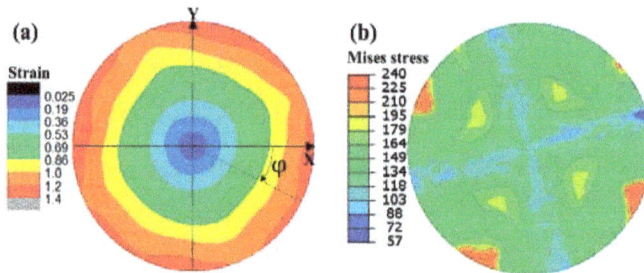

Figure 2. The distributions of (a) equivalent plastic strain and (b) Mises stress on the surfaces of (001) aluminum single crystal after $N = 1/12$ turn of HPT deformation.

3.2. Texture Development Along the Tangential Direction

Figure 3 shows the texture development in the tangential direction of the initial (001) aluminum single crystal after $N = 1/12$ turn of HPT deformation in terms of {111} pole figures. The simulated pole figures were obtained from positions on the top surface of the disk sample, which have the same radius (r_0/r ~=1), but different values of φ. All the pole figures are recorded on the $\theta - Z$ planes, and the locations of the main components of ideal torsion texture are also visualized by different symbols (after [48]). The black dots in the pole figures stand for the initial crystallographic orientations under un-deformed states, while the red dots denote the predicted orientations after deformation. Since HPT deformation is exerted on the small metallic discs, the initial crystallographic orientation changes between individual selected positions; it is close to the cube orientation (Figure 3a) for position $\varphi = -30°$, gradually approaches the ideal C component as the sample position changes from $\varphi = -15°$, 0° to $\varphi = 18°$ (Figure 3b–d), then rotates gradually away from the ideal C (Figure 3e,f).

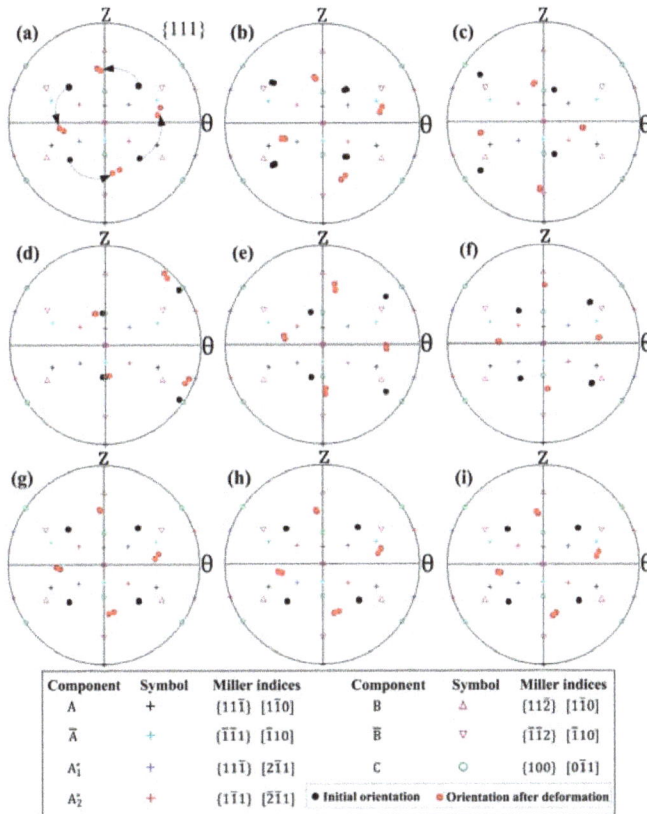

Component	Symbol	Miller indices	Component	Symbol	Miller indices
A	+	{11$\bar{1}$} [1$\bar{1}$0]	B	△	{11$\bar{2}$} [1$\bar{1}$0]
$\bar{A}$	+	($\bar{1}\bar{1}$1) [$\bar{1}$10]	$\bar{B}$	▽	($\bar{1}\bar{1}$2) [$\bar{1}$10]
A$'_1$	+	{11$\bar{1}$} [2$\bar{1}$1]	C	○	{100} [0$\bar{1}$1]
A$'_2$	+	(1$\bar{1}$1) [$\bar{2}\bar{1}$1]	● Initial orientation		○ Orientation after deformation

Figure 3. The predicted {111} pole figures of the (001) aluminum single crystal in the tangential direction after $N = 1/12$ turn of HPT deformation. The pole figure were recorded at radius ($r_0/r \sim =1$) with $\varphi =$ (a) $-30°$, (b) $-15°$, (c) $0°$, (d) $18°$, (e) $30°$, (f) $45°$, (g) $60°$, (h) $150°$ and (i) $-120°$.

It can be seen from Figure 3a–f that the resultant orientations occupy positions deviating significantly from each other in the corresponding pole figures. For instance, the developed orientations have rotated to the rotated cube orientation around the radial axis in an anti-clockwise direction in Figure 3a, while the resulting orientations illustrate only limited divergence from the ideal C component in Figure 3d. The local crystal orientation results obtained in the present work are in good agreement with the previously published reports. Tóth et al. [49,50] reported that in simple shear of cube orientation the orientation change took place solely around the sample radial axis, leading to the rotated cube component. Hafok and Pippan [16] and Arzaghi et al. [51] have examined the texture evolution during HPT, respectively, and they reported that the C orientation could be maintained over a wide range of straining in HPT. Further examination of the pole figure in Figure 3g reveals that the simulated orientation of $\varphi = 60°$ position is very similar with that of position $\varphi = -30°$. Due to the similarity, additional pole figures of $\varphi = 150°$ and $-120°$ were plotted, as shown in Figure 3h,i, respectively. It can be readily seen that the developed orientations recorded at these four different circumferential positions, with an interval of 90° about the torsion axis, occupy almost the same positions in the corresponding pole figures. From these observations, it can be concluded that the texture also develops a four-fold symmetry along the sample circumference. Hafok and Pippan [16] experimentally measured the texture evolution in the tangential direction of (001) nickel single crystal deformed by HPT. They also reported the phenomenon of repetition of the texture development and the evolution of a four-fold symmetry at the early stage of HPT deformation, which helps to validate the established CPFEM model of the HPT process.

3.3. Lattice Rotation and Slip Trace along the Tangential Direction

Figure 4a–c show the contour maps of lattice rotation angles around the R, θ and Z axes, respectively, recorded on the top surfaces of the initial (001) aluminum single crystal after $N = 1/12$ turn of HPT deformation. The calculation of the lattice rotation was based on the method proposed by Wert et al. [52]. In all maps, positive values mean anti-clockwise lattice rotations, while negative angles mean clockwise rotations. It can be seen from Figure 4 that the lattice rotates along all three directions. The rotation angles around the R axis are much larger than the other axes. Tóth et al. reported the same finding of predominant single-sense rotations around the sample radial direction in [53,54]. When considering the lattice rotation distribution in Figure 4a, it is apparent that the R-axis lattice rotation also exhibits four-fold symmetry. Within one section of the four-fold symmetry, a careful inspection reveals that the R-axis lattice rotation varies significantly. Sample position $\varphi = 18°$ has the smallest R-axis rotation, at less than $10°$. The initial orientation of this particular position is close to the ideal C component, and the resulting orientation in Figure 2d therefore illustrates limited divergence. By contrast, the sample position close to $\varphi = -30°$ has the largest R-axis rotation, at around $65°$. This sample position has an initial orientation close to the cube orientation, and HPT deformation has caused the poles to rotate significantly away in Figure 2a. These coincide with the observations that the lattice rotation vectors converge for the components of ideal torsion texture, while for other orientations the opposite is true and divergence occurs [53–55].

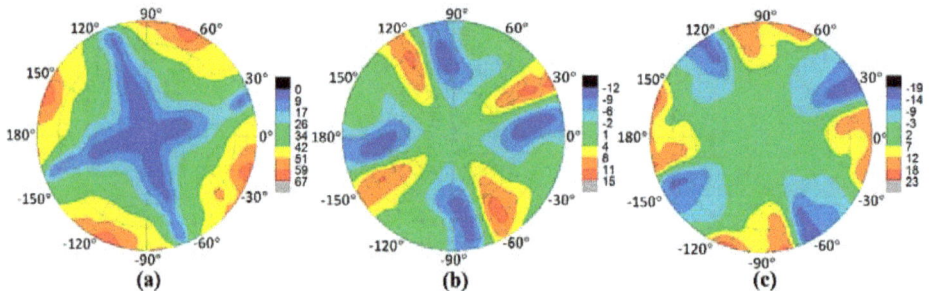

Figure 4. Lattice rotation angles recorded on the surfaces of the initial (001) aluminum single crystal after $N = 1/12$ turn of HPT deformation: (**a**) around R axis, (**b**) around θ axis and (**c**) around Z axis.

Figure 5 displays the slip traces at various tangential positions (marked out by red dots, $r_0/r \sim = 1$) on the top surface of the initial (001) aluminum single crystal after $N = 1/12$ turn of HPT deformation. To save space, only the right-half surface is shown here. For each sample position, the six slip systems with the largest magnitudes of accumulative shear strain are presented. The direction of each straight line infers the orientation of the slip trace, while the length of the straight line represents the relative magnitude of the accumulative shear strain. The black, blue, red, cyan, green and magenta colors refer to the slip systems with the first largest magnitude to the sixth largest magnitude of the accumulative shear strain, respectively. All slip trace lines were plotted on the θ–Z planes, as illustrated in the bottom right-hand corner of Figure 5.

As can be seen from Figure 5, the simulated slip traces indicate different activation behaviors of slip systems along the circumferential direction. At $\varphi = -30°$ position, the slips occur simultaneously on different systems, and the accumulative shear strains of the slip systems involved are approximately equal. The six slip systems can be treated as three sets that are all differently oriented. From position $\varphi = -30°$, through $-15°$ and $0°$, to $18°$, the magnitudes of the accumulative shear strain corresponding to the first and second largest slip systems tend gradually larger. The slip trace developed at the

$\varphi = 18°$ position indicates that the deformation is governed by two dominant slip systems, namely a3 and d3, as defined in Table 2. The accumulative shear strains on the primary two slip systems are significantly large compared with the other slip systems. Furthermore, the slip directions of the two dominant slip systems are parallel with each other, and almost align with the shear direction. When further examining the sample positions from $\varphi = 18°$ to $60°$, it can be seen that the activated slip develops gradually from a dominant slip to multiple slips. The slip trace obtained at $\varphi = 60°$ position possesses almost the same characteristics as those of $\varphi = -30°$, indicating the repeatability of the slip trace pattern along the sample circumference.

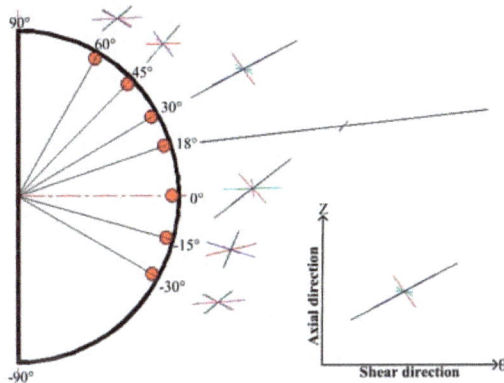

Figure 5. Slip traces of various tangential positions on the surface of the initial (001) aluminum single crystal after $N = 1/12$ turn of HPT deformation.

4. Discussion

In FCC crystals, the four slip planes are arranged symmetrically about the axes for the (001) aligned orientation. In HPT, a thin, disk-shaped sample is commonly adopted. When the sample is cylindrical, the (001) single crystal responds somewhat like a cube, and causes four-fold symmetrical deformation behaviors around the circumference. Within one section of the symmetry, pronounced local deformation variations are revealed, which is a consequence of the relative orientation of the shear applied by HPT, and of the local crystallographic axes. To achieve a deeper understanding of such deformation behaviors, the underlying mechanisms are discussed based on the theory of crystal plasticity.

During the HPT process, the rigid body rotation, Ω, is usually larger than for other SPD techniques due to severe shear deformation. Equation (4) suggests that Ω consists of two components: Ω^P, caused by the slip, and Ω^*, caused by the lattice rotation. The Ω requirement in HPT for different material points along the tangential direction may be met by Ω^P and Ω^*. Moreover, the torsional strain introduced in HPT activates different sets of slip systems for different circumferential positions. The slip mode with two dominant slip systems, as well as the multiple slips mode, in which many systems function together, are revealed in Figure 5.

At some sample positions—for example, $\varphi = 18°$—the slip mode is dominant slip. The accumulated shear strains are concentrated on the primary two slip systems, and the slip directions coincide with the macroscopic shear direction (see Figure 5). Under such circumstances, they are capable of generating a sufficiently large overall plastic spin (Ω^P) to almost fully satisfy the requirement for the whole rigid body rotation (Ω). Therefore, the lattice rotation (Ω^*) is nearly zero at the $\varphi = 18°$ position (see Figure 4a), leading to the near-initial orientation after deformation (see the pole figure in Figure 3d). The dominant slip is basically not influenced by other slip systems, and the material flows easily on the shear plane. Thus, relatively limited macroscopic strain hardening occurs at the $\varphi = 18°$ position in Figure 2b.

For certain sample positions, such as $\varphi = -30°$, a multiple-slips mode can also be activated. The activated systems possess almost the same amount of accumulated shear strain, but orient along different directions (see Figure 5). Each of the slip systems of interest produces a rotation component, and as a result, the overall glide rotation (Ω^P) is relatively small. The lattice must rotate to generate a large Ω^* to compensate the difference between Ω and Ω^P, as shown at $\varphi = -30°$ in Figure 4a, which leads to a considerable change in crystallographic orientation (see the pole figure in Figure 3a. The simultaneously activated slip systems interact with each other, and hinder the flow of the material. Therefore, strong macroscopic strain-hardening takes place at the $\varphi = -30°$ position in Figure 2b.

At other sample positions along the circumferential direction, the slip conditions are somewhere in between these two slip modes of dominant slip and multiple slips. Accordingly, there are moderate increases in Mises stress, as shown in Figure 2b.

5. Conclusions

A CPFEM model integrated with Bassani and Wu hardening modulus has been successfully implemented to investigate the anisotropic plastic behavior of aluminum single crystal in HPT. The macroscopic equivalent plastic strain and Mises stress, crystallographic orientation evolution, lattice rotation and slip activities at the crystal level of various tangential positions have been predicted and analyzed. The main conclusions are summarized as follows.

1. The predicted equivalent plastic strain and Mises stress are directionally dependent along the sample circumference, and the development of a four-fold symmetry pattern is found on the sample surface.
2. The R-axis lattice rotations, slip traces, and the resulting crystallographic orientations change significantly along the circumferential direction, and also develop the feature of four-fold symmetry.
3. The dominant slip and multi-slip conditions are revealed along the sample circumference. It has been found that the differences in slip system activation contribute significantly to the anisotropic plastic behavior of single crystal during the HPT process.

Acknowledgments: Peitang Wei would like to acknowledge the financial support from China Scholarship Council. Guanyu Deng acknowledges Australian Academy of Science (AAS) and Japan Society for the Promotion of Science (JSPS) for awarding him an international fellowship. The authors acknowledge the fundamental research funds supported by Chongqing Research Program of Basic Research and Frontier Technology (No. cstc2017jcyjAX0101) and the Central Universities Project (No. 106112016CDJXY110005, 0903005203365, and 106112017CDJQJ118847).

Author Contributions: P.W. proposed the idea, performed the simulations, and analyzed the data; C.L. and H.L. analyzed the data and discussed the results; L.S., G.D. and K.T. contributed to the results discussion; All authors contributed to the document written and preparation of publication material.

Conflicts of Interest: The authors declare no conflict of interest.

References

1. Langdon, T.G. Twenty-five years of ultrafine-grained materials: Achieving exceptional properties through grain refinement. *Acta Mater.* **2013**, *61*, 7035–7059. [CrossRef]
2. Estrin, Y.; Vinogradov, A. Extreme grain refinement by severe plastic deformation: A wealth of challenging science. *Acta Mater.* **2013**, *61*, 782–817. [CrossRef]
3. Bridgman, P.W. On torsion combined with compression. *J. Appl. Phys.* **1943**, *14*, 273. [CrossRef]
4. Deng, G.; Lu, C.; Su, L.; Tieu, A.K.; Li, J.; Liu, M.; Zhu, H.; Liu, X. Influence of outer corner angle (OCA) on the plastic deformation and texture evolution in equal channel angular pressing. *Comput. Mater. Sci.* **2014**, *81*, 79–88. [CrossRef]
5. Su, L.; Lu, C.; Deng, G.; Tieu, A.K.; Sun, X. Microstructure and mechanical properties of 1050/6061 laminated composite processed by accumulative roll bonding. *Rev. Adv. Mater. Sci.* **2013**, *33*, 33–37.
6. Zhilyaev, A.; Langdon, T. Using high-pressure torsion for metal processing: Fundamentals and applications. *Prog. Mater. Sci.* **2008**, *53*, 893–979. [CrossRef]

7. Edalati, K.; Horita, Z. A review on high-pressure torsion (HPT) from 1935 to 1988. *Mater. Sci. Eng. A* **2015**, *652*, 325–352. [CrossRef]
8. Edalati, K.; Horita, Z. High-pressure torsion of pure metals: Influence of atomic bond parameters and stacking fault energy on grain size and correlation with hardness. *Acta Mater.* **2011**, *59*, 6831–6836. [CrossRef]
9. Valiev, R.Z.; Zhilyaev, A.P.; Langdon, T.G. Hpt processing of metals, alloys, and composites. In *Bulk Nanostructured Materials*; John Wiley & Sons, Inc.: Hoboken, NJ, USA, 2013; pp. 88–151.
10. Astafurova, E.G.; Tukeeva, M.S.; Maier, G.G.; Melnikov, E.V.; Maier, H.J. Microstructure and mechanical response of single-crystalline high-manganese austenitic steels under high-pressure torsion: The effect of stacking-fault energy. *Mater. Sci. Eng. A* **2014**, *604*, 166–175. [CrossRef]
11. Khlebnikova, Y.V.; Egorova, L.Y.; Pilyugin, V.P.; Suaridze, T.R.; Patselov, A.M. Evolution of the structure of an α-titanium single crystal during high-pressure torsion. *Tech. Phys.* **2015**, *60*, 1005–1013. [CrossRef]
12. Solov'eva, Y.V.; Pilyugin, V.P.; Starenchenko, S.V.; Tolmachev, T.P.; Starenchenko, V.A. Structure and mechanical properties of ni3ge single crystals under severe plastic deformation and heating. *Bull. Russ. Acad. Sci. Phys.* **2017**, *81*, 311–314. [CrossRef]
13. Kashihara, K.; Tsujimoto, Y.; Terada, D.; Tsuji, N. Texture evolution in {112} <111> aluminum single crystals processed by severe plastic deformation. *Mater. Charact.* **2013**, *75*, 129–137. [CrossRef]
14. Smirnova, N.A.; Levit, V.I.; Pilyugin, V.I.; Kuznetsov, R.I.; Davydova, L.S.; Sazonova, V.A. Evolution of structure of FCC Single crystals during strong plastic deformation. *Phys. Met. Metallogr.* **1986**, *61*, 127–134.
15. Smirnova, N.A.; Levit, V.I.; Degtyarev, M.V.; Gundyrev, V.M.; Pilyugin, V.P.; Davydova, L.S. Development of orientational instability in FCC Single crystals at high degrees of plastic deformation. *Phys. Met. Metallogr.* **1988**, *65*, 141–151.
16. Hafok, M.; Pippan, R. High-pressure torsion applied to nickel single crystals. *Philos. Mag.* **2008**, *88*, 1857–1877. [CrossRef]
17. Pippan, R.; Scheriau, S.; Taylor, A.; Hafok, M.; Hohenwarter, A.; Bachmaier, A. Saturation of fragmentation during severe plastic deformation. *Annu. Rev. Mater. Res.* **2010**, *40*, 319–343. [CrossRef]
18. Méric, L.; Cailletaud, G. Single crystal modeling for structural calculations: Part 2—Finite element implementation. *Appl. Mech. Mater.* **1991**, *113*, 171–182. [CrossRef]
19. Nouailhas, D.; Cailletaud, G. Tension-torsion behavior of single-crystal superalloys: Experiment and finite element analysis. *Int. J. Plast.* **1995**, *11*, 451–470. [CrossRef]
20. Kaluza, M.; Le, K.C. On torsion of a single crystal rod. *Int. J. Plast.* **2011**, *27*, 460–469. [CrossRef]
21. Le, K.C.; Piao, Y. Distribution of dislocations in twisted bars. *Int. J. Plast.* **2016**, *83*, 110–125. [CrossRef]
22. Wei, P.; Lu, C.; Tieu, K.; Su, L.; Deng, G.; Huang, W. A study on the texture evolution mechanism of nickel single crystal deformed by high pressure torsion. *Mater. Sci. Eng. A* **2017**, *684*, 239–248. [CrossRef]
23. Wenk, H.R.; Van Houtte, P. Texture and anisotropy. *Rep. Prog. Phys.* **2004**, *67*, 1367–1428. [CrossRef]
24. Roters, F.; Eisenlohr, P.; Hantcherli, L.; Tjahjanto, D.D.; Bieler, T.R.; Raabe, D. Overview of constitutive laws, kinematics, homogenization and multiscale methods in crystal plasticity finite-element modeling: Theory, experiments, applications. *Acta Mater.* **2010**, *58*, 1152–1211. [CrossRef]
25. Asaro, R.J. Crystal plasticity. *J. Appl. Mech.* **1983**, *50*, 921–934. [CrossRef]
26. Ashby, M.F. The deformation of plastically non-homogeneous materials. *Philos. Mag.* **1970**, *21*, 399–424. [CrossRef]
27. Taylor, G.I. The mechanism of plastic deformation of crystals. Part I. *Theoretical. Proc. R. Soc. Lond. Ser. A Math. Phys. Eng. Sci.* **1934**, *145*, 362–387. [CrossRef]
28. Taylor, G.I. The mechanism of plastic deformation of crystals. Part II. Comparison with observations. *Proc. R. Soc. A Math. Phys. Eng. Sci.* **1934**, *145*, 388–404. [CrossRef]
29. Hutchinson, J.W. Elastic-plastic behaviour of polycrystalline metals and composites. *Proc. R. Soc. A Math. Phys. Eng. Sci.* **1970**, *319*, 247–272. [CrossRef]
30. Peirce, D.; Asaro, R.J.; Needleman, A. An analysis of nonuniform and localized deformation in ductile single crystals. *Acta Metall.* **1982**, *30*, 1087–1119. [CrossRef]
31. Wu, T.Y.; Bassani, J.L.; Laird, C. Latent hardening in single crystals I. Theory and experiments. *Proc. R. Soc. A Math. Phys. Eng. Sci.* **1991**, *435*, 1–19. [CrossRef]
32. Bassani, J.L.; Wu, T.Y. Latent hardening in single crystals II. Analytical characterization and predictions. *Proc. R. Soc. A Math. Phys. Eng. Sci.* **1991**, *435*, 21–41. [CrossRef]

33. Wu, P.D.; Neale, K.W.; Van, D.G.E. Simulation of the behaviour of fcc polycrystals during reversed torsion. *Int. J. Plast.* **1996**, *12*, 1199–1219. [CrossRef]
34. Lin, G.; Havner, K.S. A Comparative study of hardening theories in torsion using the taylor polycrystal model. *Int. J. Plast.* **1996**, *12*, 695–718. [CrossRef]
35. Kumar, A.V.; Yang, C. Study of work hardening models for single crystals using three dimensional finite element analysis. *Int. J. Plast.* **1999**, *15*, 737–754. [CrossRef]
36. Siddiq, A.; Schmauder, S. Crystal plasticity parameter identification procedure for single crystalline material during deformation. *J. Comput. Appl. Mech.* **2006**, *7*, 1–15.
37. Lu, C.; Deng, G.Y.; Tieu, A.K.; Su, L.H.; Zhu, H.T.; Liu, X.H. Crystal plasticity modeling of texture evolution and heterogeneity in equal channel angular pressing of aluminum single crystal. *Acta Mater.* **2011**, *59*, 3581–3592. [CrossRef]
38. Deng, G.Y.; Lu, C.; Su, L.H.; Liu, X.H.; Tieu, A.K. Modeling texture evolution during ECAP of copper single crystal by crystal plasticity FEM. *Mater. Sci. Eng. A* **2012**, *534*, 68–74. [CrossRef]
39. Franciosi, P.; Berveiller, M.; Zaoui, A. Latent hardening in copper and aluminium single crystals. *Acta Metall.* **1980**, *28*, 273–283. [CrossRef]
40. Akef, A.; Driver, J.H. Orientation splitting of cube-oriented face-centred cubic crystals in plane strain compression. *Mater. Sci. Eng. A* **1991**, *132*, 245–255. [CrossRef]
41. Liu, Q.; Hansen, N.; Maurice, C.; Driver, J. Heterogeneous microstructures and microtextures in cube-oriented al crystals after channel die compression. *Metall. Mater. Trans. A* **1998**, *29*, 2333–2344. [CrossRef]
42. Liu, M.; Lu, C.; Tieu, K.A.; Peng, C.T.; Kong, C. A combined experimental-numerical approach for determining mechanical properties of aluminum subjects to nanoindentation. *Sci. Rep.* **2015**, *5*, 15072. [CrossRef] [PubMed]
43. Si, L.Y.; Lu, C.; Huynh, N.N.; Tieu, A.K.; Liu, X.H. Simulation of rolling behaviour of cubic oriented Al single crystal with crystal plasticity fem. *J. Mater. Process. Technol.* **2008**, *201*, 79–84. [CrossRef]
44. Quilici, S.; Forest, S.; Cailletaud, G. On size effects in torsion of multi- and polycrystalline specimens. *J. Phys. IV* **1998**, *8*, Pr8-325–Pr8-332.
45. Yoon, S.C.; Horita, Z.; Kim, H.S. Finite element analysis of plastic deformation behavior during high pressure torsion processing. *J. Mater. Process. Technol.* **2008**, *201*, 32–36. [CrossRef]
46. Figueiredo, R.B.; Cetlin, P.R.; Langdon, T.G. Using finite element modeling to examine the flow processes in quasi-constrained high-pressure torsion. *Mater. Sci. Eng. A* **2011**, *528*, 8198–8204. [CrossRef]
47. Wei, P.; Lu, C.; Tieu, K.; Deng, G.; Wang, H.; Kong, N. Finite element analysis of high pressure torsion. *Steel Res. Int.* **2013**, *84*, 1246–1251. [CrossRef]
48. Montheillet, F.; Cohen, M.; Jonas, J.J. Axial stresses and texture development during the torsion testing of Al, Cu and α-Fe. *Acta Metall.* **1984**, *32*, 2077–2089. [CrossRef]
49. Tóth, L.S.; Jonas, J.J. Analytic prediction of texture and length changes during free-end torsion. *Texture Stress Microstruct.* **1989**, *10*, 195–209. [CrossRef]
50. Skrotzki, W.; Tóth, L.S.; Klöden, B.; Brokmeier, H.G.; Arruffat-Massion, R. Texture after ecap of a cube-oriented ni single crystal. *Acta Mater.* **2008**, *56*, 3439–3449. [CrossRef]
51. Arzaghi, M.; Fundenberger, J.J.; Toth, L.S.; Arruffat, R.; Faure, L.; Beausir, B.; Sauvage, X. Microstructure, texture and mechanical properties of aluminum processed by high-pressure tube twisting. *Acta Mater.* **2012**, *60*, 4393–4408. [CrossRef]
52. Wert, J.A.; Liu, Q.; Hansen, N. Dislocation boundary formation in a cold-rolled cube-oriented al single crystal. *Acta Mater.* **1997**, *45*, 2565–2576. [CrossRef]
53. Toth, L.S.; Gilormini, P.; Jonas, J.J. Effect of rate sensitivity on the stability of torsion textures. *Acta Metall.* **1988**, *36*, 3077–3091. [CrossRef]
54. Tóth, L.S.; Neale, K.W.; Jonas, J.J. Stress response and persistence characteristics of the ideal orientations of shear textures. *Acta Metall.* **1989**, *37*, 2197–2210. [CrossRef]
55. Barnett, M.R.; Montheillet, F. The generation of new high-angle boundaries in aluminium during hot torsion. *Acta Mater.* **2002**, *50*, 2285–2296. [CrossRef]

crystals

MDPI

Article

Room-Temperature Plastic Deformation of Strontium Titanate Crystals Grown from Different Chemical Compositions

Atsutomo Nakamura [1],* (iD)**, Kensuke Yasufuku [1], Yuho Furushima [1], Kazuaki Toyoura [2],**
K. Peter D. Lagerlöf [3] and Katsuyuki Matsunaga [1,4]

[1] Department of Materials Physics, Nagoya University, Furo-cho, Chikusa-ku, Nagoya 464-8603, Japan;
   keeenb9lox@gmail.com (K.Y.); furushima.yuuho@b.mbox.nagoya-u.ac.jp (Y.F.);
   kmatsunaga@nagoya-u.jp (K.M.)
[2] Department of Materials Science and Engineering, Kyoto University, Yoshida Honmachi, Sakyo-ku,
   Kyoto 606-8501, Japan; toyoura.kazuaki.5r@kyoto-u.ac.jp
[3] Department of Materials Science and Engineering, Case Western Reserve University, Cleveland, OH 44106,
   USA; peter.lagerlof@case.edu
[4] Nanostructures Research Laboratory, Japan Fine Ceramics Center, 2-4-1, Mutsuno, Atsuta-ku,
   Nagoya 456-8587, Japan
* Correspondence: anaka@nagoya-u.jp; Tel.: +81-52-789-3366

Academic Editor: Helmut Cölfen
Received: 11 October 2017; Accepted: 21 November 2017; Published: 22 November 2017

Abstract: Oxide materials have the potential to exhibit superior mechanical properties in terms of high yield point, high melting point, and high chemical stability. Despite this, they are not widely used as a structural material due to their brittle nature. However, this study shows enhanced room-temperature plasticity of strontium titanate ($SrTiO_3$) crystals through the control of the chemical composition. It is shown that the deformation behavior of $SrTiO_3$ crystals at room temperature depends on the Sr/Ti ratio. It was found that flow stresses in deforming $SrTiO_3$ crystals grown from a powder with the particular ratio of $Sr/Ti = 1.04$ are almost independent of the strain rate because of the high mobility of dislocations in such crystals. As a result, the $SrTiO_3$ crystals can deform by dislocation slip up to a strain of more than 10%, even at a very high strain rate of 10% per second. It is thus demonstrated that $SrTiO_3$ crystals can exhibit excellent plasticity when chemical composition in the crystal is properly controlled.

Keywords: strontium titanate (STO); dislocations; transmission electron microscopy (TEM); single crystals; oxides

1. Introduction

Oxide crystals are typically brittle at low temperatures because of their predominant ionic and/or covalent bonding. It is known that slip systems that can be activated in ionic crystals are limited by electrostatic interactions between ions [1], whereas slip system activation in covalent crystals is affected by strongly localized and directional bonds that share electrons [2]. In addition, oxide crystals have large unit cells, leading to large lattice constants, since an oxide is a compound between one or more metallic elements and oxygen atoms. As a result, oxide crystals have large Burgers vectors, which represent the magnitude and direction of dislocations that produce plastic deformation. The larger Burgers vectors make it more difficult to move dislocations. Since the stresses required for dislocation motion are often larger than stresses required for crack propagation, oxide crystals are typically brittle by nature.

The mechanical properties of strontium titanate (SrTiO$_3$) crystals have received growing interest due to the fact that they can exhibit a large degree of plasticity even at low temperatures [3–11]. SrTiO$_3$ is an oxide with a cubic perovskite structure belonging to the $Pm\bar{3}m$ space group. It has been reported that SrTiO$_3$ crystals can plastically deform in temperature ranges from 78 K to 1050 K [3,4]. In addition, SrTiO$_3$ crystals can deform up to and over 10% if they deform at a slow strain rate of 10^{-4} (s^{-1}) [6,7]. Recently, the deformation behavior and dislocation structure have been reported to be temperature dependent [9]. In addition, Kondo et al. demonstrated using in-situ observation in transmission electron microscopy (TEM) that dislocation glide can take place in SrTiO$_3$ crystals even at room temperature [10,11]. Thus, SrTiO$_3$ crystals are believed to exhibit plasticity even at low temperatures due to mobility of dislocations belonging to the {110}<110> slip systems. However, there is still little known about any key factor providing the high dislocation mobility of SrTiO$_3$.

In the present study, SrTiO$_3$ single crystals grown from powders with varying Sr/Ti ratios were deformed at room temperature. This is believed to be due to the fact that a variety of point defects can be formed in SrTiO$_3$ crystals depending on the specific chemical composition and/or chemical environments around the crystals [12,13]. It is well known that point defects in a crystal can interact strongly with dislocations. In general, point defects act as barriers to dislocation motion and result in work hardening of materials followed by nucleation and growth of cracks, leading to brittle fracture. In fact, both Sr and O vacancies have been reported to be present around dislocations in SrTiO$_3$ [14–16], forming a Cottrell atmosphere [1]. However, the influence of point defects on the dislocation mobility in SrTiO$_3$ has not been examined.

2. Experimental

SrTiO$_3$ single crystals, grown by the Verneuil method from high purity SrTiO$_3$ powder (99.9 wt % and Sr/Ti = 1.002) and high purity SrCO$_3$ powder (99.99 wt %), were used for the present deformation experiments. Typically, Sr deficiency is induced during crystal growth of SrTiO$_3$ crystals due to the evaporation of Sr ahead of the solidification front [17]. In addition, Sr vacancies are easily formed in SrTiO$_3$ because of the low formation energy [12,13]. For this reason, single crystals grown with a non-stoichiometric starting powder having excess Sr (Sr/Ti > 1) would result in fewer Sr vacancies in the resultant SrTiO$_3$ crystal. Thus, single crystals grown using starting powders with two composition ratios, Sr/Ti = 1.00 and Sr/Ti = 1.04, respectively, were selected to be used in this study. As the starting powder of Sr/Ti = 1.00, 1000.00 g of the high purity SrTiO$_3$ powder was employed as it was. On the other hand, a mixed powder of 1000.00 g high purity SrTiO$_3$ powder and 32.18 g high purity SrCO$_3$ powder was employed as the starting powder of Sr/Ti = 1.04. The growing of all single crystals was performed using a Verneuil furnace at Shinkosha Co., Ltd. (Yokohama, Japan). Here, we note that, considering the evaporation loss of Sr, excess Sr does not necessarily exist even in the crystals grown from the powder with Sr/Ti = 1.04. Meanwhile, it is difficult to display the real atomic ratios of Sr/Ti in the grown crystals because a slight change in the ratios is buried in a mass of Sr and Ti atoms. The grown crystals from Sr/Ti = 1.04 should have fewer Sr vacancies than those from Sr/Ti = 1.00.

Figure 1 shows the shapes and crystallographic orientation of the specimens for deformation tests. Specimens were deformed by compression along [001] at strain rates of $\dot{\varepsilon}$ = 1.0 × 10^{-5} (s^{-1}) and $\dot{\varepsilon}$ = 1.0 × 10^{-1} (s^{-1}) in air at 297 K. Note that $\dot{\varepsilon}$ = 1.0 × 10^{-1} (s^{-1}) is a high strain rate because specimens must change their length by 10% every second. In the case of the deformation along [001], four equivalent slip systems ((011)[0$\bar{1}$1], (0$\bar{1}$1)[011], (101)[$\bar{1}$01], and ($\bar{1}$01)[101], respectively, which are illustrated in Figure 1) are expected to be primary slip systems because they all have a Schmid factor equal to 0.5. During plastic deformation, the specimen's length and shape were recorded using a high-speed optical camera in order to confirm slip lines formed on their polished surfaces. The deformed crystals were examined by TEM (Hitachi H-800, 200 kV) to identify the density and type of dislocations that were introduced during the plastic deformation. Thin foils for TEM observations were prepared by a standard technique. That is, the specimens for deformation were mechanically sliced and ground to a thickness of about 60 μm and then attached with a stainless steel single hole mesh for reinforcement. The center of the foils was milled by an Ar ion beam using an ion polishing system (Gatan PIPS II) to obtain electron transparency.

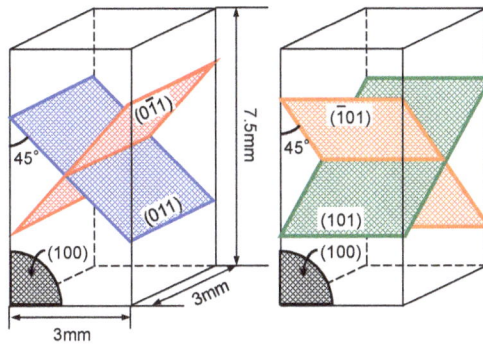

Figure 1. Schematic illustrations showing specimen shapes and available slip systems in the present deformation tests.

3. Results and Discussion

Figure 2a,b show the stress–strain curves that were obtained for the specimens with the two different chemical compositions of $Sr/Ti = 1.00$ (Figure 2a) and $Sr/Ti = 1.04$ (Figure 2b) as starting powders, respectively. It can be seen that stresses in all the curves gradually increase with increasing strain after yielding, and the deformed specimens give rise to brittle failure after a significant amount of work hardening. It was found that the $Sr/Ti = 1.04$ specimens exhibit lower yield and flow stresses as well as larger fracture strain, compared with the $Sr/Ti = 1.00$ specimens. The yield stresses, flow stresses (at a plastic strain of $\varepsilon_p = 3\%$), and fracture strains of each specimen are shown in Table 1. As can be seen from the flow stresses at $\varepsilon_p = 3\%$, the stress for $Sr/Ti = 1.04$ at $\dot{\varepsilon} = 1.0 \times 10^{-1}$ (s^{-1}) is higher by only 20% compared with the specimen deformed at $\dot{\varepsilon} = 1.0 \times 10^{-5}$ (s^{-1}), in spite of the fact that the strain rate is 10^4 times higher. In contrast, the stress for $Sr/Ti = 1.00$ increases by 31% with respect to the higher strain rate. The fracture stresses for the $Sr/Ti = 1.04$ specimens are significantly higher than those for the $Sr/Ti = 1.00$ specimens.

Table 1. Deformation stresses and fracture strains for SrTiO$_3$ single crystals.

Sr/Ti	Strain Rate (s^{-1})	Yield Stress (MPa)	Flow Stress (at $\varepsilon_p = 3\%$) (MPa)	Fracture Strain (%)
1.04	1.0×10^{-5}	112	122	13.6
	1.0×10^{-1}	135	147	13.1
1.00	1.0×10^{-5}	119	138	7.4
	1.0×10^{-1}	160	180	8.5

Figure 3a,b show shapes and surface morphologies of the $Sr/Ti = 1.00$ and $Sr/Ti = 1.04$ specimens deformed at $\dot{\varepsilon} = 1.0 \times 10^{-1}$ (s^{-1}), respectively. In both specimen types, slip lines become evident at around $\varepsilon_p = 3\%$ and continue to be generated until failure. Thus, it is evident that slip lines are formed even at the high strain rate of $\dot{\varepsilon} = 1.0 \times 10^{-1}$ (s^{-1}), indicating prominent multiplication and glide motion of dislocations, although the $Sr/Ti = 1.00$ specimen fails at an earlier stage of deformation. These results are similar to those observed in ordinary single crystals deformed at a slower strain rate. It should be mentioned that the formed slip lines originate from two particular primary slip systems, namely, the two systems of (011)[0$\bar{1}$1] and (0$\bar{1}$1)[011] or the two systems of (101)[$\bar{1}$01] and ($\bar{1}$01)[101], although the Schmid factors for all four of the slip systems are equal. This means that either the two systems of (011)[0$\bar{1}$1] and (0$\bar{1}$1)[011] or the two systems of (101)[$\bar{1}$01] and ($\bar{1}$01)[101] are preferentially activated during plastic deformation. As a result, only the horizontal slip lines were observed on a side surface of deformed specimens, as shown in Figure 3. This tendency is consistent with the previous reports [6,9] of plastic deformation of SrTiO$_3$ crystals.

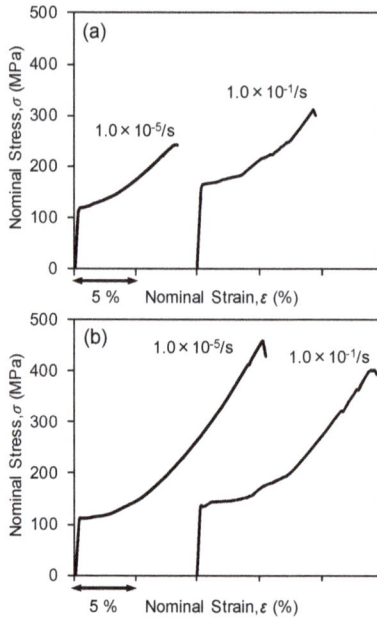

Figure 2. Stress–strain curves of SrTiO$_3$ single crystals with different chemical compositions, deformed at constant strain rates of $\dot{\varepsilon} = 1.0 \times 10^{-5}$ (s^{-1}) and $\dot{\varepsilon} = 1.0 \times 10^{-1}$ (s^{-1}). (a) $Sr/Ti = 1.00$, (b) $Sr/Ti = 1.04$ as the starting powders.

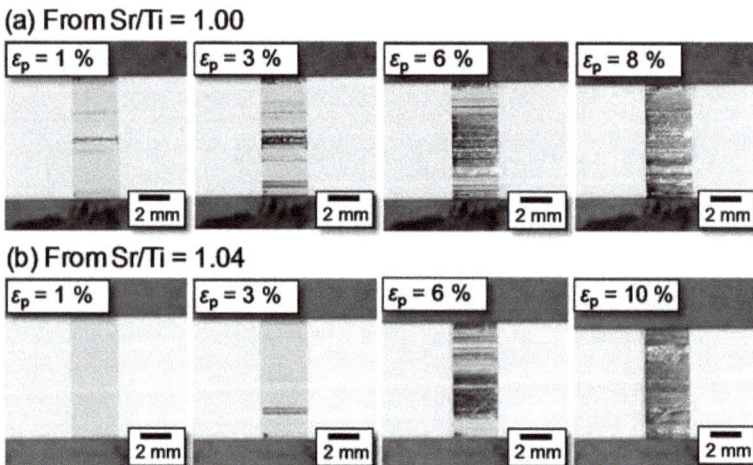

Figure 3. Shapes and surface morphologies of the deformed specimens. (a) The $Sr/Ti = 1.00$ specimen deformed at $\dot{\varepsilon} = 1.0 \times 10^{-1}$ (s^{-1}). (b) The $Sr/Ti = 1.04$ specimen deformed at $\dot{\varepsilon} = 1.0 \times 10^{-1}$ (s^{-1}). Here, each ε_p in the figures indicates the plastic strain.

Figure 4a,b show typical TEM bright field images taken from a $Sr/Ti = 1.04$ specimen deformed at $\dot{\varepsilon} = 1.0 \times 10^{-1}$ (s^{-1}) up to $\varepsilon_p = 3\%$. The incident beam direction is normal to the (011) primary slip plane, where the observed foil for TEM was prepared parallel to the slip lines formed by a primary slip system of (011)[0$\bar{1}$1]. It can be seen that many dislocations are introduced by the deformation, and most of the dislocations are present parallel to the (011) primary slip plane. Since glide dislocations in SrTiO$_3$ have Burgers vectors of the 1/2<011> type, the Burgers vectors of the dislocations in deformed specimens should be ±1/2[0$\bar{1}$1] according to activation of the (011)[0$\bar{1}$1] slip system. The image in Figure 4b is obtained from the same field as Figure 4a by using a different diffraction vector of $g = 100$. It can be seen that contrasts from almost all dislocations disappear under the diffraction vector of $g = 100$. This is because the ±1/2[0$\bar{1}$1] dislocations originating from the (011)[0$\bar{1}$1] slip system should lose contrasts according to the $g \cdot b = 0$ condition [18]. Thus, the dislocation substructure formed by a deformation test is believed to be mainly formed due to the activation of a primary slip system.

Figure 5 represents the dislocation substructure of a $Sr/Ti = 1.04$ specimen deformed at $\dot{\varepsilon} = 1.0 \times 10^{-5}$ (s^{-1}), which is also observed normally to the (011) primary slip plane. It can be also seen from the images that the density and type of introduced dislocations are almost same as those in the specimen deformed at $\dot{\varepsilon} = 1.0 \times 10^{-1}$ (s^{-1}). It should be noticed from comparison of the dislocation substructures that the density and type of introduced dislocations are very similar in spite of the great difference in the given strain rates. Additionally, dislocation densities of the specimens deformed up to $\varepsilon_p = 3\%$ were about 1×10^{13}/m^2, estimated using the Keh method [19] from the TEM images, regardless of given strain rates. The obtained dislocation densities are consistent with the values in earlier research [9], where the densities of dislocations in the un-deformed specimen and the specimen deformed up to 1% plastic strain at 25 °C were reported to be 3.7×10^{12}/m^2 and 7.9×10^{12}/m^2, respectively. In fact, since the difference in flow stresses is not great, as can be seen from the stresses at $\varepsilon_p = 3\%$ in Table 1, then it would be reasonable to conclude that dislocation densities were almost independent of given strain rates. Meanwhile, the image in Figure 6 shows the dislocation substructure of a $Sr/Ti = 1.00$ specimen deformed at $\dot{\varepsilon} = 1.0 \times 10^{-5}$ (s^{-1}). The density and type of introduced dislocations are also similar, regardless of the difference in chemical compositions as starting powders. Consequently, it can be said that the ratio of Sr/Ti does not strongly affect the dislocation substructure.

From the above discussion, it has been shown that the yield and flow stresses become lower in the $Sr/Ti = 1.04$ specimens without changing the character of the dislocation substructures. The $Sr/Ti = 1.00$ specimens are believed to have more point defects, such as Sr vacancies and Ti anti-sites that substitute for Sr, since Sr deficiency should be induced through Sr evaporation during crystal growth. Conversely, it is plausible that the $Sr/Ti = 1.04$ specimens have a composition that is closer to the stoichiometric composition in a grown crystal. A more stoichiometric composition in the grown crystal will give rise to fewer point defects, resulting in less interaction with moving dislocations. This may be the reason why the $Sr/Ti = 1.04$ specimens exhibit lower yield and flow stresses.

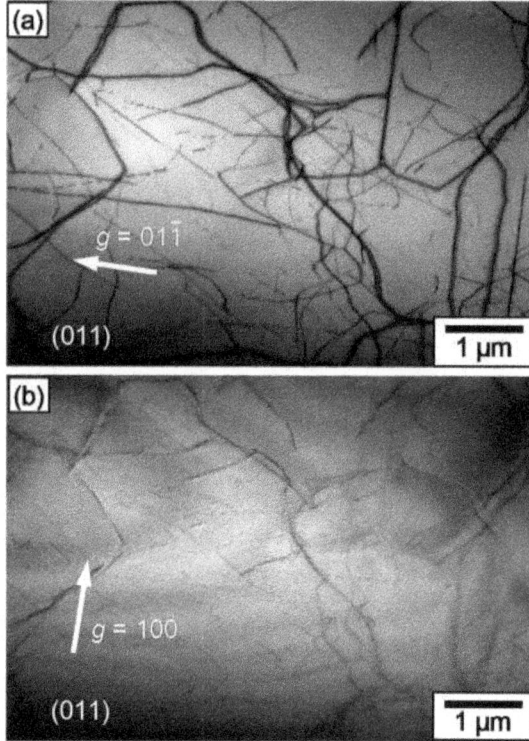

Figure 4. Typical TEM bright field images taken from the $Sr/Ti = 1.04$ specimen deformed at $\dot{\varepsilon} = 1.0 \times 10^{-1}$ (s^{-1}) up to $\varepsilon_p = 3\%$. The image in (**b**) was obtained from the same field as in (**a**) using a different diffraction vector.

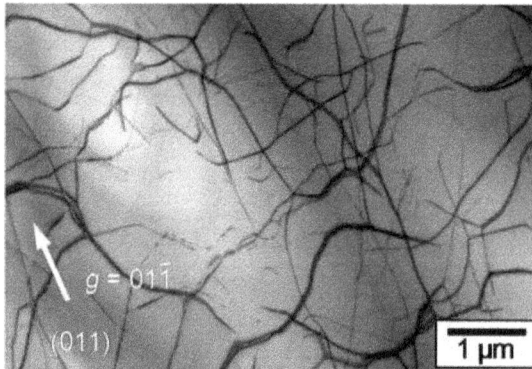

Figure 5. A typical TEM bright field image taken from the $Sr/Ti = 1.04$ specimen deformed at $\dot{\varepsilon} = 1.0 \times 10^{-5}$ (s^{-1}) up to $\varepsilon_p = 3\%$.

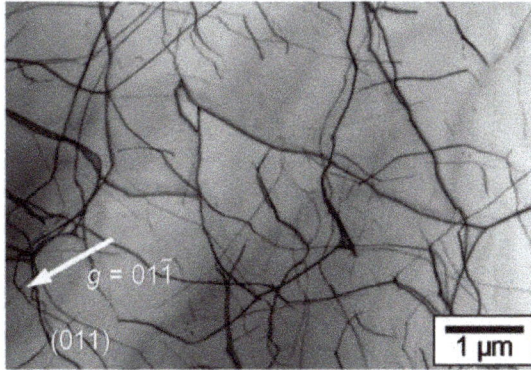

Figure 6. A typical TEM bright field image taken from the Sr/Ti = 1.00 specimen deformed at $\dot{\varepsilon} = 1.0 \times 10^{-5}$ (s^{-1}) up to ε_p = 3%.

It is interesting to note that strain-rate dependency of the deformation stress is small for both the Sr/Ti = 1.00 and Sr/Ti = 1.04 specimens, although the degree is more prominent in the case of Sr/Ti = 1.04. In order to evaluate the strain-rate dependency with respect to dislocation velocity in these specimens, an analysis will be done based on the Orowan relationship [1],

$$\dot{\gamma} = \rho b \bar{v} \tag{1}$$

where $\dot{\gamma}$ is the shear strain rate, ρ is the density of mobile dislocations, b is the magnitude of the Burgers vector of the mobile dislocations, and $\bar{v}$ is the average motion speed of the mobile dislocations. In addition, the relation between $\bar{v}$ and deformation stress is given by

$$\bar{v} \approx \tau^m \tag{2}$$

where τ is the shear stress and m is the stress exponent [20,21]. As can be seen from TEM images at ε_p = 3% in Figures 4 and 5, the dislocation densities of deformed specimens are similar even after increasing the strain rate by a factor of 10^4. Here, we note that dislocations that are observable in the TEM include both mobile and immobile dislocations, and most of the observable dislocations in some cases can be immobile [22]. Accordingly, the densities of mobile dislocations in deforming specimens are still unknown. First of all, we assume that the density of mobile dislocations should be independent of the strain rate as the dislocation substructures introduced by the deformation tests are very similar in spite of great difference in the given strain rates. In this assumption, the dislocation velocity of the mobile dislocations—$\bar{v}$ in Equation (1)—must increase by a factor of 10^4 in the case of the high strain rate of $\dot{\varepsilon} = 1.0 \times 10^{-1}$ (s^{-1}) since the strain rate, $\dot{\gamma}$, should be proportional to the dislocation velocity, $\bar{v}$. Moreover, since the flow stresses at ε_p = 3% have been determined (see Figure 1 and Table 1) for both cases of $\dot{\varepsilon} = 1.0 \times 10^{-5}$ (s^{-1}) and $\dot{\varepsilon} = 1.0 \times 10^{-1}$ (s^{-1}), the stress exponent m for SrTiO$_3$ crystals with a particular chemical composition can be estimated by using Equation (2). The values of m in this assumption are calculated to be 35 and 49 for the Sr/Ti = 1.00 and Sr/Ti = 1.04 specimens, respectively. Next, we assume that the density of mobile dislocations depends on the strain rate. Two hypotheses were considered: (i) there are ten times as many mobile dislocations, and (ii) there are one-tenth as many mobile dislocations in the case of the high strain rate of 1.0×10^{-1} (s^{-1}). Even in these hypothetical situations, the value of m can be estimated by using Equations (1) and (2). The values of m derived from (i) i.e., the hypothesis of ten times as many mobile dislocations, can be calculated to be 26 and 37 for the Sr/Ti = 1.00 and Sr/Ti = 1.04 specimens, respectively. Meanwhile, the values of m derived from (ii) i.e., the hypothesis of one-tenth as many mobile dislocations can be calculated to be 43 and 62 for the $Sr/Ti = 1.00$ and $Sr/Ti = 1.04$ specimens, respectively. Although the m in Equation (2) mainly depends on deformation flow stresses, the obtained values of m still have a dependency on the densities of mobile dislocations for each of the strain rates. Nevertheless, it is worth pointing out the fact that the stress exponent m for most crystalline materials, including metals, is less than 10 [21]. The m for SrTiO$_3$ would have definitely high values although there is still room for argument about the densities of mobile dislocations. Regardless, it can be said that the dislocation dynamics in SrTiO$_3$ single crystals are remarkable and interesting. We note that a higher value of stress exponent corresponds to smaller strain-rate dependency of deformation flow stresses. In addition, it can be seen from Equation (2) that a larger value of the stress exponent, m, results in a higher value of the dislocation velocity of the mobile dislocations, $\bar{v}$, for the same stress level, τ. That is, the dislocations in the $Sr/Ti = 1.04$ specimens would have higher mobility as compared with the dislocations in the $Sr/Ti = 1.00$ specimens. This is believed to be due to fewer point defects for a SrTiO$_3$ crystal made from a starting powder with $Sr/Ti = 1.04$.

Finally, the larger fracture stresses for crystals made from a starting powder with $Sr/Ti = 1.04$ will be considered. As seen from Figure 2, fracture stresses for the $Sr/Ti = 1.04$ specimens are significantly larger than those for the $Sr/Ti = 1.00$ specimens. As mentioned previously, the mobility of dislocations for the $Sr/Ti = 1.04$ specimens is significantly higher than that for the $Sr/Ti = 1.00$ specimens. The higher dislocation mobility can reduce stress concentrations during plastic deforming since the dislocations can accommodate the applied stress. The larger fracture stresses for the $Sr/Ti = 1.04$ specimens would arise from the higher dislocation mobility. In contrast, the $Sr/Ti = 1.00$ specimens fail at lower stresses. This can be explained by the relatively lower dislocation mobility, which has a potential to induce higher stress concentrations during plastic deformation.

4. Conclusions

From deformation tests of SrTiO$_3$ single crystals, it was found that a crystal grown with the ratio of $Sr/Ti = 1.04$ as the starting powder can deform with a high degree of plasticity up to a strain of more than 10% at room temperature, even at a very high strain rate of $\dot{\varepsilon} = 0.1$ (s^{-1}). It is remarkable that the strain-rate dependency of deformation flow stress in SrTiO$_3$ is rather small, in particular for crystals made from powders with $Sr/Ti = 1.04$, meaning that dislocations exhibit a relatively high mobility in these crystals. As a result, larger fracture stresses and strains could be attained for the $Sr/Ti = 1.04$ specimens. Here, it should be also mentioned that it is still unclear whether the crystals grown with the ratio of $Sr/Ti = 1.04$ exhibit the highest plasticity or not. Namely, there is a possibility that a better composition of powders for higher plasticity of SrTiO$_3$ crystals will be found in the future. It is likely that the crystals grown from $Sr/Ti = 1.04$ should have fewer Sr vacancies than those from $Sr/Ti = 1.00$, since additional Sr elements are mixed with the starting powder. Moreover, the crystals from $Sr/Ti = 1.04$ exhibited higher plasticity as an actual fact. It can be said that the plasticity of SrTiO$_3$ crystals could be enhanced if the chemical composition is properly controlled so as to decrease point defects in a crystal. This suggests that it may be possible to enhance the plasticity of other oxide materials by controlling the chemical composition and resulting point defect populations.

Acknowledgments: The authors gratefully acknowledge the financial support by a Grant-in-Aid for Scientific Research on Innovative Areas "Nano Informatics" (JP25106002) from Japan Society for the Promotion of Science (JSPS). A part of this study was supported by JSPS KAKENHI Grant Numbers JP15H04145, JP16K14414, JP17H06094 and JP17K18983. A.N. also thanks Iketani Science and Technology Foundation for the financial support. Additionally, this work was partly supported by Nanotechnology Platform Program (Advanced Characterization Nanotechnology Platforms of Nagoya Univ.) of the Ministry of Education, Culture, Sports, Science and Technology (MEXT), Japan.

Author Contributions: Atsutomo Nakamura and Katsuyuki Matsunaga conceived the research idea, and Atsutomo Nakamura designed the experiments; Kensuke Yasufuku, Yuho Furushima, and Atsutomo Nakamura performed the experiments and analyzed the data; K. Peter D. Lagerlöf, Kazuaki Toyoura, and Katsuyuki Matsunaga advised the experiments; all authors discussed the results and wrote the paper.

Conflicts of Interest: The authors declare no competing financial interests.

References

1. Hirth, J.P.; Lothe, L. *Theory of Dislocations*, 2nd ed.; Wiley: New York, NY, USA, 1982.
2. Hull, D.; Bacon, D.J. *Introduction to Dislocations*; Elsevier: Oxford, UK, 2011; pp. 124–133.
3. Brunner, D.; Taeri-Baghbadrani, S.; Sigle, W.; Ruhle, M. Surprising Results of a Study on the Plasticity in Strontium Titanate. *J. Am. Ceram. Soc.* **2001**, *84*, 1161–1163. [CrossRef]
4. Gumbsch, P.; Taeri-Baghbadrani, S.; Brunner, D.; Sigle, W.; Ruhle, M. Plasticity and an inverse brittle-to-ductile transition in strontium titanate. *Phys. Rev. Lett.* **2001**, *87*, 085505. [CrossRef] [PubMed]
5. Taeri, S.; Brunner, D.; Sigle, W.; Rühle, M. Deformation behaviour of strontium titanate between room temperature and 1800 K under ambient pressure. *Z. Metall.* **2004**, *95*, 433–446. [CrossRef]
6. Brunner, D. Low-temperature plasticity and flow-stress behaviour of strontium titanate single crystals. *Acta Mater.* **2006**, *54*, 4999–5011. [CrossRef]
7. Sigle, W.; Sarbu, C.; Brunner, D.; Ruhle, M. Dislocations in plastically deformed SrTiO₃. *Philos. Mag.* **2006**, *86*, 4809–4821. [CrossRef]
8. Yang, K.-H.; Ho, N.-J.; Lu, H.-Y. Plastic Deformation of <001> Single-Crystal SrTiO₃ by Compression at Room Temperature. *J. Am. Ceram. Soc.* **2011**, *94*, 3104–3111. [CrossRef]
9. Patterson, E.A.; Major, M.; Donner, W.; Durst, K.; Webber, K.G.; Rodel, J. Temperature-Dependent deformation and dislocation density in SrTiO₃ (001) single crystals. *J. Am. Ceram. Soc.* **2016**, *99*, 3411–3420. [CrossRef]
10. Kondo, S.; Shibata, N.; Mitsuma, T.; Tochigi, E.; Ikuhara, Y. Dynamic observations of dislocation behavior in SrTiO₃ by in situ nanoindentation in a transmission electron microscope. *Appl. Phys. Lett.* **2012**, *100*, 181906. [CrossRef]
11. Kondo, S.; Mitsuma, T.; Shibata, N.; Ikuhara, Y. Direct observation of individual dislocation interaction processes with grain boundaries. *Sci. Adv.* **2016**, *2*, e1501926. [CrossRef] [PubMed]
12. Tanaka, T.; Matsunaga, K.; Ikuhara, Y.; Yamamoto, T. First-principles study on structures and energetics of intrinsic vacancies in SrTiO₃. *Phys. Rev. B* **2003**, *68*, 205213. [CrossRef]
13. Liu, B.; Cooper, V.; Xu, H.; Xiao, H.; Zhang, Y.; Weber, W. Composition dependent intrinsic defect structures in SrTiO₃. *Phys. Chem. Chem. Phys.* **2014**, *16*, 15590–15596. [CrossRef] [PubMed]
14. Takehara, K.; Sato, Y.; Tohei, T.; Shibata, N.; Ikuhara, Y. Titanium enrichment and strontium depletion near edge dislocation in strontium titanate [001]/(110) low-angle tilt grain boundary. *J. Mater. Sci.* **2014**, *49*, 3962–3969. [CrossRef]
15. Choi, S.Y.; Kim, S.D.; Choi, M.; Lee, H.S.; Ryu, J.; Shibata, N.; Mizoguchi, T.; Tochigi, E.; Yamamoto, T.; Kang, S.J.L. Assessment of strain-generated oxygen vacancies using SrTiO₃ bicrystals. *Nano Lett.* **2015**, *15*, 4129–4134. [CrossRef] [PubMed]
16. Furushima, Y.; Arakawa, Y.; Nakamura, A.; Tochigi, E.; Matsunaga, K. Nonstoichiometric [012] dislocation in strontium titanate. *Acta Mater.* **2017**, *135*, 103–111. [CrossRef]
17. Scheel, H.J.; Bednorz, J.G.; Dill, P. Crystal growth of strontium titanate. *Ferroelectrics* **1976**, *13*, 507–509. [CrossRef]
18. Williams, D.B.; Carter, C.B. *Transmission Electron Microscopy: A Textbook for Materials Science*, 2nd ed.; Springer: New York, NY, USA, 2009.
19. Keh, A.S. Dislocation Arrangement in Alpha Iron during Deformation and Recovery. In *Direct Observation of Imperfections in Crystal*; Interscience Publishers (Wiley): New York, NY, USA, 1961.
20. Messerschmidt, U. *Dislocation Dynamics during Plastic Deformation*; Springer: Berlin, Germany, 2010; pp. 73–280.
21. Nadgornyi, E. *Dislocation Dynamics and Mechanical Properties of Crystals*; Pergamon: Oxford, UK, 1988; pp. 1–536.
22. Bergsröm, Y. A dislocation model for the stress-strain behaviour of polycrystalline α-Fe with special emphasis on the variation of the densities of mobile and immobile dislocations. *Mater. Sci. Eng.* **1970**, *5*, 193–200. [CrossRef]

crystals

MDPI

Article

Morphology Dependent Flow Stress in Nickel-Based Superalloys in the Multi-Scale Crystal Plasticity Framework

Shahriyar Keshavarz [1,*], Zara Molaeinia [2], Andrew C. E. Reid [3] and Stephen A. Langer [3]

[1] National Institute of Standards and Technology/Theiss Research, 7411 Eads Ave, La Jolla, CA 92037, USA
[2] Department of Materials Engineering, Purdue University, 701 West Stadium Avenue, West Lafayette, IN 47907, USA; zmolaein@purdue.edu
[3] National Institute of Standards and Technology, 100 Bureau Dr, Gaithersburg, MD 20899, USA; andrew.reid@nist.gov (A.C.E.R.); stephen.langer@nist.gov (S.A.L.)
*   Correspondence: shahriyar.keshavarzhadad@nist.gov

Academic Editor: Peter Lagerlof
Received: 18 September 2017; Accepted: 26 October 2017; Published: 2 November 2017

Abstract: This paper develops a framework to obtain the flow stress of nickel-based superalloys as a function of γ-γ' morphology. The yield strength is a major factor in the design of these alloys. This work provides additional effects of γ' morphology in the design scope that has been adopted for the model developed by authors. In general, the two-phase γ-γ' morphology in nickel-based superalloys can be divided into three variables including γ' shape, γ' volume fraction and γ' size in the sub-grain microstructure. In order to obtain the flow stress, non-Schmid crystal plasticity constitutive models at two length scales are employed and bridged through a homogenized multi-scale framework. The multi-scale framework includes two sub-grain and homogenized grain scales. For the sub-grain scale, a size-dependent, dislocation-density-based finite element model (FEM) of the representative volume element (RVE) with explicit depiction of the γ-γ' morphology is developed as a building block for the homogenization. For the next scale, an activation-energy-based crystal plasticity model is developed for the homogenized single crystal of Ni-based superalloys. The constitutive models address the thermo-mechanical behavior of nickel-based superalloys for a large temperature range and include orientation dependencies and tension-compression asymmetry. This homogenized model is used to obtain the morphology dependence on the flow stress in nickel-based superalloys and can significantly expedite crystal plasticity FE simulations in polycrystalline microstructures, as well as higher scale FE models in order to cast and design superalloys.

Keywords: flow stress; morphology; Ni-based superalloys; homogenization; crystal plasticity

1. Introduction

Some of the major materials used in turbine engines are nickel-based superalloys. There has been enormous investments in improving nickel superalloys in order to reach the desired mechanical properties. To make this process efficient, computational tools are needed to predict the mechanical behaviors of these alloys with consequences for the microstructural design. Flow stress is one of the main aspects of designing these alloys, which has impacts on the other mechanical properties of these alloys, such as fatigue and creep responses. The study is mainly focused on the behavior of single crystals of these alloys in service duties where the morphology of the microstructures can significantly change the mechanical behavior of these materials [1,2]. The morphology of the two-phase nickel superalloys is directly connected to the heat treatment processes. Different heat treatments [3] result in different γ-γ' arrangements including different shapes, volume fractions and sizes of precipitates. In general, these alloys have a two-phase γ-γ' microstructure as shown in Figure 1a. The channel γ phase (white) is mainly nickel, while the precipitate γ-γ' phase (black) is an ordered $L1_2$ crystal structure. The ordered structure γ' phase is a strengthening constituent with special thermo-mechanical properties in the overall nickel superalloy. The crystal structures of γ-γ' are shown in Figure 1b,c. The material in the channel has a regular FCC crystal structure, while in the Ni$_3$Al crystal, the corner sites are occupied by the minority (Al) atoms, and the face-centered sites are occupied by the majority (Ni) atoms. The precipitates act as obstacles to the motion of dislocations, which either loop around or shear the precipitates depending on the temperature and stress level. A full dislocation or super-dislocation in $L1_2$ crystal is $\langle 110 \rangle$, as opposed to $\frac{1}{2}\langle 110 \rangle$ in regular FCC crystals for a full dislocation. There is a big difference in the micro-mechanical deformation mechanisms of Ni$_3$Al ordered structure from those of a regular FCC structure. The length of the Burgers vector for a full dislocation in the Ni$_3$Al ordered structure is different from a regular FCC structure.

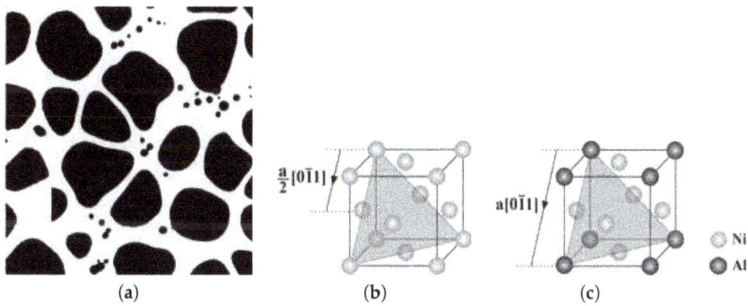

Figure 1. Two-phase nickel superalloys: (**a**) morphology of the two-phase γ-γ' sub-grain microstructure of Rene 88-DT [4]; (**b**) crystal structure of the γ phase; and (**c**) crystal structure of the γ' phase.

The mechanical properties, including the dislocation mechanisms under various loading and temperature conditions, have been studied extensively both for single-crystal [5] and polycrystalline [6] Ni-based superalloys. At lower temperatures, octahedral slip systems are mainly active, and slip occurs on these slip systems in both phases. According to the experimental reports, the flow stress in nickel-based superalloys increases when the temperature is elevated up to 1000 K. In this temperature range, most of the dislocations in the intermetallic γ' phase become immobile screw dislocations that are locked in a Kear–Wilsdorf (or KW) configuration due to cross-slip [7]. Therefore, they act as barriers for further dislocation motions and result in increasing the flow stress. A 3D configuration of the cross-slip and lock formation is shown in Figure 2. Above 1000 K, cube planes, which are not primary slip systems in FCC materials, can be activated, which has negative impacts on the flow stress; therefore, the flow stress begins to decrease above 1000 K. Above this temperature, edge and screw dislocations on cube planes occur without any cross-slip [8].

The mechanical behavior including creep and fatigue responses must be improved for the next generation of these alloys. This requires studying the lower scales of these materials to get macroscopic scale properties in terms of microstructural data. Hence, it is necessary to incorporate small length scale microstructural mechanisms including dislocation activities. The lower scales in nickel superalloys can be divided into sub-grain and grain scales, where in the sub-grain scale, the study includes the investigation of the dislocation mechanisms by explicit consideration of the γ-γ' morphology. In the grain scale, there will be a homogenized grain without explicit representation of the γ-γ' morphology. Therefore, the aim of this work is to develop a morphology-dependent flow stress from the sub-grain scale and bridge the homogenized scale through a multi-scale constitutive model that includes different dislocation activities at different temperatures. The methods yield a significant efficiency advantage, particularly for simulating polycrystals [9–11], since the microstructural RVE problem need not be solved anymore. In general, this multi-scale method can be applied to three scales; however, in this work, we will focus on just the sub-grain and homogenized single-crystal scales.

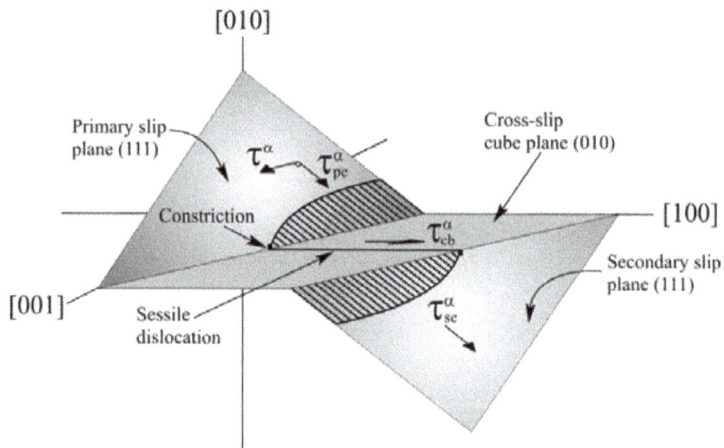

Figure 2. 3D configuration of the Kear–Wilsdorf lock.

Crystal plasticity finite element models [12,13] are applied to gather the information hierarchically at each scale to build constitutive models that can be implemented for microstructure-property relations, as well as microstructure design. Meso-scale analyses of superalloys, incorporating precipitate distributions, as well as the grain structure, have been conducted using phenomenological viscoplastic constitutive laws in [14,15]. Hardening parameters in many of the constitutive models have been expressed as assumed functions of the average precipitate size. Analytical models have been proposed using simplifying assumptions for dislocation distributions under uniaxial and monotonic loads in [16]. Crystal plasticity models with implicit dependencies on the grain size and precipitate size and volume fraction have been postulated for a random distribution of precipitates in [17]. Dislocation-density-based hierarchical crystal plasticity models of creep and fatigue have been proposed in [18,19], where the dependence of mechanical properties on microstructural characteristics like average γ' precipitate size and volume fraction are accommodated by parameters obtained by fitting with experimental data.

This paper is aimed at developing a functional form for the flow stress in nickel-based superalloys using a temperature- and orientation-dependent homogenized grain-scale crystal plasticity model with parametric representations of the sub-grain morphology in its evolution laws. The multi-scale approach, which is applied to develop the functional form of the yield stress, is fully presented in [8], and we have adopted that model to provide additional effects of the γ' morphology in this manuscript. This multi-scale approach incorporated in the current study, ranging from the sub-grain scale to the meso-scale polycrystalline ensemble, is shown in Figure 3. The cycle starts at the first step with the development of a crystal plasticity finite element or CPFE model of a sub-grain scale representative volume element or RVE, delineating the explicit γ-γ' morphology. The CPFE model incorporates a non-Schmid size-dependent dislocation density-based crystal plasticity model, in which signed dislocation densities are explicit variables [20,21]. The next step focuses on the homogenized single-crystal scale where an activation-energy-based crystal plasticity model from the homogenization of the dislocation density-based sub-grain model is developed. The homogenized model incorporates temperature variation from room temperature to 1200 K, orientation dependencies to capture asymmetry in tension-compression and activation of cubic slip systems along the effect of the discrete sub-grain morphology through critical morphological parameters. The resulting hierarchical model has the potential of significantly expediting crystal plasticity FE simulations, while retaining accuracy.

Section 2 of this paper introduces the sub-grain scale dislocation density crystal plasticity constitutive laws with anti-phase boundary (APB) shearing of γ' precipitates. Section 3 provides a framework for an activation energy-based model at the scale of single crystals. The homogenization procedure that yields morphology-dependent constitutive parameters and their calibration and validation with sub-grain RVE models, as well as experimental results are discussed in Section 4. The morphology dependencies of the mechanical properties in nickel-based superalloysis discussed in Section 5. A summary in Section 6 will conclude the paper.

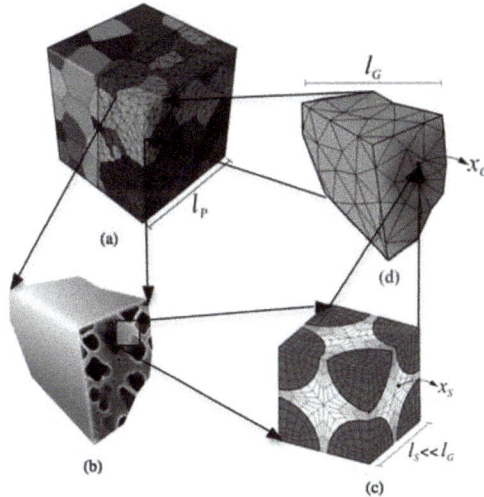

Figure 3. Schematic representation of three scales for Ni-based superalloys in the crystal plasticity finite element framework: (**a**) polycrystalline microstructure showing the grains and CPFEMmesh; (**b**) single grain description with the sub-grain γ-γ' microstructure; (**c**) discretized sub-grain γ-γ' microstructural representative volume element (RVE); and (**d**) homogenized crystal plasticity FE model for a single grain.

While this work provides a stand-alone solution to the multi-scale CPFEM problem it addresses, the model and techniques developed here are also intended to be incorporated into the Object-Oriented Finite Element code, or OOF [22], a general-purpose modeling code intended to assist materials scientists and materials engineers in undertaking computational investigations of structure-property relations in a large variety of systems, including mechanical systems whose behavior is dominated by crystal plasticity.

2. Sub-Grain Scale for the γ-γ' Microstructural of Nickel Superalloys

Binary nickel superalloy is a two-phase material consisting of γ (Ni) and γ' (Ni$_3$Al) phases. Plastic deformation is accumulated through crystallographic slip systems, which is different in the two phases, which results in plastic anisotropy in the sub-grain scale. A dislocation-density-based crystal plasticity model, proposed in [1,20,21], is used and implemented to model the rate-dependent plastic behavior. It incorporates the evolution of statistically-stored dislocations (SSDs) in both the γ -channels and the precipitates due to various dislocation generation and annihilation mechanisms, while cross-slip dislocations (CSDs) are considered just for γ' phase [8]. In order to take into account the gradient of plastic strain at the geometrically-incompatible locations such as the matrix-precipitate interface and grain boundaries, geometrically-necessary dislocations (GND) are also incorporated.

Plastic deformation of nickel-based superalloys can occur by activation of octahedral slip systems and cubic slip systems. However, the dislocation mechanisms are different in the γ-channel and in the precipitates. The length of a full dislocation or super-dislocation in precipitates dominant within Ni_3Al + XX compositions is $\langle 110 \rangle$, almost twice as large as a full dislocation in regular FCC crystals, where a full dislocation is $\frac{1}{2}\langle 110 \rangle$. The dominant deformation mechanism in precipitates for almost all ranges of temperatures is the dissociation of a screw super-dislocation into two super-partials, which have a Burgers vector of $\frac{1}{2}\langle 1\bar{1}0 \rangle$ and the corresponding creation of a planar fault anti-phase boundary or APB. Afterward, these superpartials basically split into two shockley partials, to bind a complex stacking fault (CSF) having Burgers vectors of $\frac{1}{6}\langle 11\bar{2} \rangle$ as shown in Figure 2. The non-Schmid resolved shear stresses (τ_{pe}^α and τ_{se}^α, τ_{cb}^α) associated with the shockley partials on primary, secondary octahedral and cube planes, as shown in Figure 2, are included in the constitutive models at both scales. The cross-slip mechanisms in Ni_3Al + XX compositions do not follow Schmid's law, commonly employed in crystal plasticity models [23], according to experimental observations. Materials that follow Schmid's law usually have symmetric evolution for hardness, while the evolution of hardness of the cross-slip mechanism is not symmetric and is different in tension and compression, and it also depends on the crystal orientations. The shear stress on the primary octahedral slip plane τ_{pe}^α constricts the Shockley partials and is partially responsible for the tension-compression asymmetry. For one of the tensile or compressive load direction, τ_{pe}^α constricts the Shockley partials to increase cross-slip rates resulting in the higher flow stress, while in the opposite direction, it hinders cross-slip with a decrease in flow stress.

2.1. Crystal Plasticity Model for the Sub-Grain Model

The constitutive model admits a multiplicative decomposition of $\mathbf{F} = \mathbf{F}^e\mathbf{F}^p$ where the total deformation gradient $\mathbf{F}$ contains an inelastic, incompressible part $\mathbf{F}^p$ associated with just slip without rotation and an elastic part $\mathbf{F}^e$ that accounts for rigid-body rotations and elastic stretching. For the plastic velocity gradient $\mathbf{L}^p$, the plastic shear strain rate $\dot{\gamma}^\alpha$ on the slip system α (including the slip direction $\mathbf{m}_0^\alpha$ and slip plane normal $\mathbf{n}_0^\alpha$ in the reference configuration) and the Schmid tensor $\mathbf{s}_0^\alpha = \mathbf{m}_0^\alpha \otimes \mathbf{n}_0^\alpha$ can be employed to calculate the evolution of plastic deformation as:

$$\mathbf{L}^p = \dot{\mathbf{F}}^p\mathbf{F}^{-p} = \sum_{\alpha=1}^{N}\dot{\gamma}^\alpha\mathbf{s}_0^\alpha = \sum_{\alpha=1}^{N}\dot{\gamma}^\alpha\mathbf{m}_0^\alpha \otimes \mathbf{n}_0^\alpha \tag{1}$$

The stress-strain relation invokes the second Piola–Kirchoff stress $\mathbf{S}$ and its work conjugate Green–Lagrange strain tensor $\mathbf{E}^e$ in the intermediate configuration:

$$\mathbf{S} = det(\mathbf{F}^e)\mathbf{F}^{e-1}\sigma\mathbf{F}^{e-T} = \mathbf{C} : \mathbf{E}^e \quad and \quad \mathbf{E}^e = \frac{1}{2}(\mathbf{F}^{e^T}\mathbf{F}^e - \mathbf{I}) \tag{2}$$

where $\mathbf{C}$ is a fourth order anisotropic elasticity tensor, σ is the Cauchy stress tensor and $\mathbf{I}$ is the identity tensor.

The plastic shear strain rate on a slip system is given by the Orowan equation as $\dot{\gamma}^\alpha = \rho_M^\alpha b^\alpha v^\alpha$ with the mobile dislocation density as ρ_M^α, Burgers vector as b^α and the dislocation velocity as v^α for a given slip system. The crystal plasticity framework incorporating the signed dislocation density used for superalloys in [1] is modified in the current study for rate-dependent plastic behavior. The modifications include adding cross-slip dislocation densities, the temperature dependency of cross-slip shear resistance and considering cubic slip systems in addition to octahedral slip systems. In general, the velocity of dislocations can be written as:

$$v^\alpha = v_0\exp\left(-\frac{Q}{K_B\theta}\right)\sinh\left(\frac{|\tau^\alpha| - \tau_{pass}^\alpha}{\tau_{cut}^\alpha}\right)^p sign(\tau^\alpha) \tag{3}$$

The initial dislocation velocity is considered as $v_0 = \lambda_\alpha f_0$ where f_0 is the attack frequency and λ_α is a temperature-dependent jump width. The jump width λ_α can be calculated in terms of parallel and forest dislocation densities as $\lambda_\alpha = \frac{c_0}{\sqrt{\rho_P^\alpha \rho_F^\alpha}}(\frac{\theta}{\theta_{ref}})^{c_1}$. The temperature-dependent velocity in this equation includes the absolute temperature θ and the activation energy Q. To provide a control to the velocity for a given hardening evolution, an exponent p is introduced in the form of the hyperbolic term in the current study. The dislocation velocity in this equation is a function of resolved shear stress τ^α, a component of applied load and slip system resistances parallel to the slip system or passing stress τ_{pass}^α and perpendicular to the slip system or cutting stress τ_{cut}^α. The passing stress is the result of the interaction of mobile dislocations with other dislocations and their networks in the slip plane, while the cutting stress is the result of the mobile dislocations cutting the forest dislocations with density ρ_F^α, which are perpendicular to the slip plane. The passing and cutting stresses are [24]:

$$\tau_{pass}^\alpha = c_2 Gb\sqrt{\rho_P^\alpha + \rho_F^\alpha} \quad , \quad \tau_{cut}^\alpha = \frac{c_3 K_B \theta}{b^2}\sqrt{\rho_F^\alpha} \tag{4}$$

where G is the shear modulus and c_2 and c_3 are material constants. Parallel and forest dislocation densities are due to statistically-stored dislocations or SSDs, which account for lock formation, dipole formation, athermal annihilation, thermal annihilation and geometrically-necessary dislocations or GNDs to account for the gradient of the plastic deformation between two phases [24]. Finally, the mobile dislocation density ρ_m^α can be written as a function of forest and parallel dislocation densities along with the temperature as:

$$\rho_M^\alpha = \frac{c_9 K_B \theta \sqrt{\rho_F^\alpha \rho_P^\alpha}}{Gb^3} \tag{5}$$

where c_9 can be evaluated from c_2 and c_3 as $c_9 = \frac{2c_3}{c_2}$.

In general, dislocation activities or plastic deformation in the two-phase nickel superalloys begin in the γ channel when the resolved shear stress is larger than the slip system resistance or passing stress. Then, the SSDs evolve, and due to gradients in plastic deformation in the γ channel and γ' precipitates, the GNDs also evolve; at some point, the dislocation in the channel has enough stress to cut through the precipitates, generating dislocation nucleation in the γ' phase. The dislocation nucleation criterion in the γ' phase can be divided into two categories, namely: (1) for octahedral slip systems with the non-Schmid effects and (2) for the cubic slip systems without the non-Schmid effects. To accommodate the criterion in the crystal plasticity framework, the APB shearing criterion in [1] is extended as follows:

$$\tau_{eff}^\alpha = |\tau^\alpha| - \tau_{pass}^\alpha > \tau_c \tag{6}$$

where:

$$\tau_{eff}^\alpha = \begin{cases} |\tau^\alpha| - \tau_{pass}^\alpha & for \ |\tau^\alpha| > \tau_{pass}^\alpha \\ 0 & for \ |\tau^\alpha| \le \tau_{pass}^\alpha \end{cases} \tag{7}$$

This criterion is valid for both octahedral and cubic slip systems; however, the critical shear stress for octahedral slip systems stated in Equation (7) is a function of three non-Schmid components of the shear stresses on the primary and secondary octahedral slip planes, as well as the cube plane, as well as the anti-phase boundary energy on both octahedral and planes. On the other hand, τ_c for the cubic slip system is just a function of temperature. Overall, the critical shear stress for both octahedral and cubic slip systems can be written as [8]:

$$\tau_c^\alpha = \begin{cases} \tau_{co}^\alpha = \tau_{co}^\alpha(\tau_{pe}^\alpha, \tau_{se}^\alpha, \tau_{cb}^\alpha, \theta, \Gamma_{111}, \Gamma_{010}) & \text{on octahedral slip systems} \\ \tau_{cc} = \tau_{cc}(\theta) & \text{on cube slip systems} \end{cases} \tag{8}$$

There are two factors, important in increasing dislocation densities of the cross-slip mechanism, creating thermally-activated constrictions and increasing temperature. The critical shear stress corresponding to Equation (8) evolves by increasing the dislocation densities of the cross-slip mechanism. However, the strength of obstacles decreases with an increase in temperature. Consequently, there is a competition between increasing strength due to the formation of KW locks and obstacle strength reduction with increasing temperature.

2.2. Material Constants in the Constitutive Law

There are two types of material constants in Equations (3)–(8). Constants in the first type can be found in the literature [25]. They include $h, \Gamma^{010}, \Gamma^{111}, b, \mu$ and ρ_0, which have values of 0.3, 0.083, 0.3, 2.49×10^{-10} m, 142.2 GPa and 5.0×10^{15} m/m^3, respectively. The statistically-stored dislocation density needs a proper initial value, which we derive based on the experiments and can be stated as a function of temperature:

$$\rho_{SSD0} = \begin{cases} 4.04 \times 10^{11} - 3.34 \times 10^8 \theta & \theta \leq 659K \\ 2.42 \times 10^{11} - 0.87 \times 10^8 \theta & 659K < \theta \leq 930K \\ 13.28 \times 10^{11} - 12.58 \times 10^8 \theta & 930K < \theta \leq 1000K \\ 1.53 \times 10^{11} - 0.8 \times 10^8 \theta & \theta > 1000K \end{cases}$$

As discussed, the critical shear stress in Equation (7) for cubic slip systems was just a function of temperature because the cross-slip mechanism only occurs for the octahedral slip systems. From the data given in [8], the cubic slip resistance can be calibrated as:

$$\tau_{cc} = \begin{cases} 460 \ MPa & \theta \leq 915K \\ 1558 - 1.2\theta \ MPa & otherwise \end{cases}$$

The elastic stiffness tensor $\mathbf{C}_{\alpha\beta} = \mathbf{C}_{\beta\alpha} (\alpha = 1, ..., 6, \beta = 1, ..., 6)$ is considered to have the cubic symmetry for both phases. The elastic stiffness tensor components are functions of temperature. For the γ phase, the non-zero components of the stiffness tensor can be derived [26]:

$$\mathbf{C}_{11} = \mathbf{C}_{22} = \mathbf{C}_{33} = (298 - 0.096\theta) \ GPa$$

$$\mathbf{C}_{44} = \mathbf{C}_{55} = \mathbf{C}_{66} = (139 - 0.035\theta) \ GPa$$

$$\mathbf{C}_{12} = \mathbf{C}_{13} = \mathbf{C}_{23} = (191 - 0.057\theta) \ GPa$$

For the γ' phase, the non-zero components of the stiffness tensor are:

$$\mathbf{C}_{11} = \mathbf{C}_{22} = \mathbf{C}_{33} = (325 - 0.096\theta) \ GPa$$

$$\mathbf{C}_{44} = \mathbf{C}_{55} = \mathbf{C}_{66} = (144 - 0.035\theta) \ GPa$$

$$\mathbf{C}_{12} = \mathbf{C}_{13} = \mathbf{C}_{23} = (209 - 0.057\theta) \ GPa$$

The rest of the material constants in the constitutive model, corresponding to the second type, are calibrated from experiments on single-crystal CMSX-4 in [8]. The alloy contains a 70% volume fraction of predominantly cuboidal γ' precipitates of an average size of 0.45 μm with the average size of the RVE of 0.5 μm.

The second type of material constants are the ones calibrated according to the constitutive model to capture the experimental data. In general, these constants can be divided into three categories: (1) yield state constitutive constants, (2) temperature state material constants; and (3) hardening state material constants. The yield state includes stresses corresponding to the onset of plastic deformation up to

0.2% offset strain. The temperature material constants are responsible for the anomalous behavior of Ni3Al alloys. The hardening material constants reflect the interaction of different dislocation mechanics, which result in hardening after the yield sate. The parameters corresponding to the yield state, temperature state and hardening state are listed in Tables 1–3.

Table 1. Calibrated material constants for the yield state of the constitutive model according to the experimental data.

Parameter	p_{oct}	p_{cub}	Q (J)	k_1	k_2
Value	1.1	1.2	1.1×10^{-20}	0.5	0.6

Table 2. Calibrated material constants for the temperature state of the constitutive model based on the experimental data.

Parameter	ζ_0	A	θ_c
Value	1.8	325	1400

Table 3. Calibrated material constants for the hardening state of the constitutive model based on experimental data.

Parameter	$f_0(1/s)$	$c_0(1/m)$	c_1	c_2	c_3	c_4	c_5	c_6	c_7	c_8
Value	10^8	0.078	−3.77	4	0.3	100	0.001	0.0001	10	10

2.3. Implementation of the Crystal Plasticity Constitutive Model into to the Code

The crystal plasticity constitutive model explained in Section 2 for the two-phase γ-γ' is implemented in a crystal plasticity FE (CPFE) code. The rate-dependent constitutive model requires the use of a time-integration scheme; therefore, an implicit time-integration scheme is implemented. In the implicit schemes developed in [25,27], backward Euler time integration methods are used to solve a set of nonlinear equations in the time interval $t \leq \tau \leq t + \Delta t$ using iterative Newton–Raphson methods. The algorithm proposed in [27] needs the solution of six equations corresponding to the number of second Piola–Kirchoff stress components, while that in [25] solves equations equal to the number of slip systems (>6 for the FCC systems). The integration algorithm in [27] is adopted in this work, which requires known deformation variables, e.g., $\mathbf{F}(t)$ and $\mathbf{F}^p(t)$, $\rho_{SSD}(t)$, $\rho_{CSD}(t)$ and $\rho_{GND}(t)$ and slip system deformation resistances $\tau_{pass}^\alpha(t)$, $\tau_{cut}^\alpha(t)$, $\tau_{co}^\alpha(t)$ and $\tau_{oc}^\alpha(t)$ at time t, as well as $\mathbf{F}(t + \Delta t)$, as inputs to a material update routine CPFEM-MAT. By, integrating Equation (1), the plastic part of the deformation gradient at time $t + \Delta t$ is expressed as:

$$\mathbf{F}^p(t + \Delta t) = \left(\mathbf{I} + \sum_{\alpha=1}^{N} \Delta\gamma^\alpha \mathbf{m}_0^\alpha \otimes \mathbf{n}_0^\alpha\right)\mathbf{F}^p(t) = \left(\mathbf{I} + \sum_{\alpha=1}^{N} \Delta\gamma^\alpha \mathbf{s}_0^\alpha\right)\mathbf{F}^p(t) \tag{9}$$

By substituting the expressions for $\mathbf{F}^p(t + \Delta t)$ and $\mathbf{F}(t + \Delta t)$ into Equations (1) and (9), the incremented second Piola–Kirchoff stress is calculated as:

$$\mathbf{S}(t + \Delta t) = \frac{1}{2}\mathbf{C} : \left(\mathbf{F}^{p^{-T}}(t + \Delta t)\mathbf{F}^T(t + \Delta t)\mathbf{F}(t + \Delta t)\mathbf{F}^{p^{-1}}(t + \Delta t) - \mathbf{I}\right)$$

$$= \frac{1}{2}\mathbf{C} : \left(\mathbf{A}(t + \Delta t) - \mathbf{I}\right) - \sum_{\alpha=1}^{N} \Delta\gamma^\alpha \left(\mathbf{S}(t + \Delta t), \tau_{pass}^\alpha(t), \tau_{cut}^\alpha(t), \tau_{co}^\alpha(t), \tau_{oc}^\alpha(t)\right)\mathbf{C}^\alpha \tag{10}$$

$$= \mathbf{S}^{tr} - \sum_{\alpha=1}^{N} \Delta\gamma^\alpha \left(\mathbf{S}(t + \Delta t), \tau_{pass}^\alpha(t), \tau_{cut}^\alpha(t), \tau_{co}^\alpha(t), \tau_{oc}^\alpha(t)\right)\overline{\mathbf{C}}^\alpha$$

where:

$$\mathbf{A}(t+\Delta t) = \mathbf{F}^{p-T}(t+\Delta t)\mathbf{F}^T(t+\Delta t)\mathbf{F}(t+\Delta t)\mathbf{F}^{p-1}(t+\Delta t) \text{ and } \overline{\mathbf{C}}^\alpha = \frac{1}{2}\mathbf{C}:\left(\mathbf{A}s_0^\alpha + s_0^{\alpha T}\mathbf{A}\right) \qquad (11)$$

A nonlinear Newton–Raphson iterative method is used to find the second Piola–Kirchoff stress stated in Equation (10):

$$\mathbf{S}^{i+1}(t+\Delta t) = \mathbf{S}^i(t+\Delta t) - \left(\mathbf{I} + \sum_{\alpha=1}^N \mathbf{C}^\alpha \otimes \frac{\partial \Delta\gamma^\alpha}{\partial \mathbf{S}^i}\right)^{-1}\mathbf{G}^i \text{ where } \mathbf{G} = \mathbf{S} - \mathbf{S}^{tr} + \sum_{\alpha=1}^N \Delta\gamma^\alpha \mathbf{C}^\alpha \qquad (12)$$

The plastic deformation gradient can be calculated by solving this equation. Subsequently, Cauchy stress and the tangent stiffness matrix $\mathbf{W}_{ijkl} = \frac{\partial \sigma_{ij}}{\partial \epsilon_{kl}}$ can be computed by having Cauchy stress and strain in CPFEM-MAT and passed on to the FE program for the equilibrium equation. The time integration scheme at the Gauss point level is detailed in Table 4.

Table 4. Time integration scheme in CPFEM-MAT. APB, anti-phase boundary.

A. For time increment from t to $t + \Delta t$ with known $\mathbf{F}(t + \Delta t)$ all known variables at time t

   i. Calculate $\mathbf{S}^{tr} = \frac{1}{2}\mathbf{C}:\left(\mathbf{A}(t+\Delta t) - \mathbf{I}\right)$ using Equations (10) and (11).

   ii. Evaluate the resolved shear stress due to trial stress $\tau^\alpha = \mathbf{S}^{tr}:s_0^\alpha$, and update deformation variables in Step **B**.

   iii. From Equation (10), calculate the first iterate $\mathbf{S}^1(t+\Delta t) = \mathbf{S}^{tr} - \sum_{\alpha=1}^N \Delta\gamma^\alpha(t+\Delta t)C^\alpha$

   iv. For the i-th iteration in the Newton–Raphson method

      (a) Evaluate $\tau^{i\alpha} = \mathbf{S}^i(t+\Delta t):s_0^\alpha$, and update deformation variables in Step **B**.

      (b) Using Equations (11) and (12), evaluate $\mathbf{S}^{i+1}(t+\Delta t) = \mathbf{S}^i(t+\Delta t) - (\mathbf{d}^i)^{-1}\mathbf{G}^i$

      where $(\mathbf{d}^i)^{-1} = \mathbf{I} + \sum_{\alpha=1}^N \mathbf{C}^\alpha \otimes \frac{\partial\Delta\gamma^{i\alpha}(t+\Delta t)}{\partial\mathbf{S}^i(t+\Delta t)}$ and $\mathbf{G}^i = \mathbf{S}^i(t+\Delta t) - \mathbf{S}^{tr} + \sum_{\alpha=1}^N \Delta\gamma^\alpha(t+\Delta t)C^\alpha$

      (c) Verify convergence: If no, go to Step (a); if yes, go to Step v.

   v. Evaluate $\tau^{(i+1)\alpha} = \mathbf{S}^i(t+\Delta t):s_0^\alpha$, and update deformation variables in Step **B**.

   vi. From Equation (9), evaluate $\mathbf{F}^p(t+\Delta t) = \left(\mathbf{I} + \sum_{\alpha=1}^N \Delta\gamma^\alpha(t+\Delta t)s_0^\alpha\right)\mathbf{F}^p(t)$

   vii. Calculate $\mathbf{F}^e(t+\Delta t) = \mathbf{F}(t+\Delta t)\mathbf{F}^{-p}(t+\Delta t)$,

      $\sigma(t+\Delta t) = \frac{1}{det(\mathbf{F}^{eT}(t+\Delta t))}\mathbf{F}^{eT}(t+\Delta t)\mathbf{S}(t+\Delta t)\mathbf{F}^e(t+\Delta t)$ and $\mathbf{W} = \frac{\partial\sigma}{\partial\epsilon}$

B. Update deformation variables at any stage

   I. Calculate dislocation density increments $\dot\rho_{SSD}^\alpha, \dot\rho_{CSD}^\alpha, \dot\rho_{GNDs}^\alpha, \dot\rho_{GNDet}^\alpha$ and $\dot\rho_{GNDen}^\alpha$

   II. Evaluate forest and parallel dislocation densities ρ_P, ρ_F and mobile dislocation density ρ_m from Equation (5)

   III. Check for the APB criterion given in Equation (6), then calculate cross-slip, passing and cutting shear resistances and the evolution of plastic shear strain from the Orowan equation by using Equation (3)

2.4. Validation of the Sub-Grain CPFEM Model

The results of the crystal plasticity constitutive model developed for the dislocation nucleation in both γ-γ' phases and stated in Section 2 in the CPFEM framework are compared with experimental data, which were performed by different experts [28–30]. These experiments are carried out on CMSX-4 nickel superalloys or on a very similar compound; therefore, the RVE is constructed for a regular array of cubic precipitates with a 70% precipitate volume fraction. The dimensions of the RVE are 0.5 μm × 0.5 μm × 0.5 μm . The size of cubic γ particles allocated symmetrically at the eight corners is 0.45 μm. The CPFE model of the microstructural RVE is discretized into 2200 elements using eight-noded trilinear brick elements. To emulate the experimental conditions, constant strain-rate and creep loads are applied to the top surface, while rigid body modes are suppressed by applying boundary conditions on the opposite bottom surface.

The model is able to predict the mechanical behaviors of nickel-based superalloys for a wide range of temperatures, different orientations and different strain rates in the quasi-static range and exhibits asymmetry in tension and compression. Tensile constant strain tests are performed for four temperatures including room temperature, 800 °C, 850 °C and 950 °C. These two sets of simulations are executed with respect to two [001] and [111] orientations at different temperatures. For orientations close to [001], three constant strain rate simulations, which correspond to the experiments, are performed at 800 °C [28], 850 °C [30] and 950 °C [30]. The tensile constant strain rate is 0.001 s^{-1}. The volume-averaged stress-strain responses are subsequently compared with the experimental data in Figure 4a. The simulations show a very good agreement with experimental data. It can be observed that the yield stress and hardening drop as temperature increases. The second set of comparisons is done for an orientation close to [111] where three constant strain rate simulations are performed at 25 °C [29], 850 °C [30] and 950 °C [30]. The constant tensile strain rate is 0.0001 s^{-1}. The volume-averaged stress-strain responses are subsequently compared with experimental data in Figure 4b. The simulations show a good agreement with experimental data. It can be observed that the yield stress at high temperature is much less than at room temperature due to the activation of cubic slip systems.

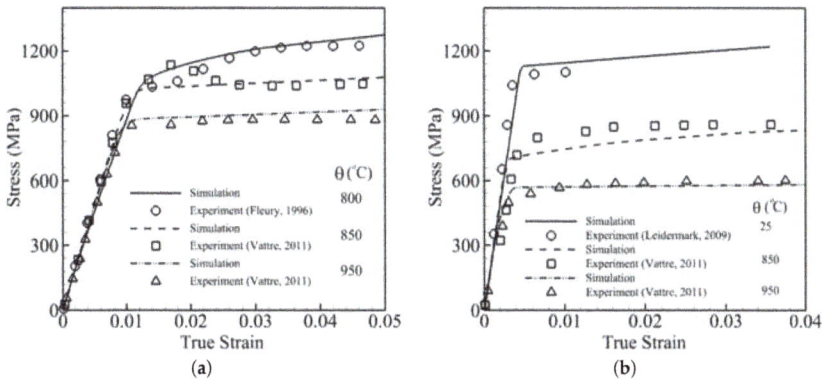

Figure 4. Volume-averaged true stress-logarithmic strain response by CPFEM and experiments [28–30] for different temperatures of a single crystal of nickel-based superalloy (CMSX-4): (**a**) [001] orientation under a tensile constant strain rate of 0.001 s^{-1} (**b**) [111] orientation under a tensile constant strain rate of 0.0001 s^{-1}.

3. Grain-Scale Crystal Plasticity Framework

The homogenized single-crystal grain-scale for nickel superalloys proposed in [1,31–33] is employed. The model is almost similar to the sub-grain model where hardening parameters are a function of plastic deformation instead of dislocation densities. The constitutive model incorporates an evolving thermal shear resistance, as well as an athermal shear resistance due to the plastic deformation. For a slip system α, the plastic shear strain rate can be calculated from the Orowan equation as:

$$\dot{\gamma}^\alpha = \begin{cases} 0 & \text{for } \tau_{\text{eff}}^\alpha \leq 0 \\ \dot{\gamma}_*^\alpha \exp\left\{-\dfrac{Q}{K_B \theta}\left[1 - \left(\dfrac{|\tau_{\text{eff}}^\alpha|}{\tau_{\text{cut}}^\alpha}\right)^p\right]^q\right\} \text{sign}(\tau^\alpha) & \text{for } 0 < \tau_{\text{eff}}^\alpha \leq \tau_{\text{cut}}^\alpha \end{cases} \tag{13}$$

For the slip system α, $\dot{\gamma}_*^\alpha$ is a reference strain-rate as a function of plastic strain and morphological parameters [8]. The temperature-dependent slip system resistance s_α is assumed to be a result of a thermally-activated obstacle to slip τ_{cut}^α or s_*^α and partly due to the athermal obstacles τ_{pass}^α or s_a^α

as defined in [8]. The driving force for dislocation motion on the slip system α is comprised of the difference between the athermal shear resistance and the resolved shear stress.

The athermal shear resistance reflecting the effect of parallel dislocations in the slip direction $\mathbf{m}^\alpha$ is defined as $\dot{s}_a^\alpha = \sum_{\beta=1}^N h_a^{\alpha\beta} |\dot{\gamma}^\beta sin(\mathbf{n}^\alpha, \mathbf{t}^\beta)|$ where $\mathbf{n}^\alpha$ is the slip-plane normal, $\mathbf{t}^\alpha = \mathbf{m}^\alpha \times \mathbf{n}^\alpha$. The thermal shear resistance or cutting stress incorporates two dislocation mechanisms. The first mechanism is employed in order to capture the effect of forest dislocations as $\dot{s}_*^\alpha = \sum_{\beta=1}^N h_*^{\alpha\beta} |\dot{\gamma}^\beta cos(\mathbf{n}^\alpha, \mathbf{t}^\beta)|$. The evolution of total shear slip resistance is $\dot{s}^\alpha = \sqrt{(\dot{s}_a^\alpha)^2 + (\dot{s}_*^\alpha)^2}$. The interactions between slip systems are taken to be isotropic; in other words, the coefficients are the same, i.e., $h_a^{\alpha\beta} = h_*^{\alpha\beta} = h^{\alpha\beta}$. Each component of $h^{\alpha\beta}$ is the deformation resistance on slip system α due to shearing on slip system β. It describes both self and latent hardening as:

$$h^{\alpha\beta} = q^{\alpha\beta} h^\beta, \quad \text{where} \quad h^\beta = \left[h_0 \left(1 - \frac{s^\beta}{s_{sat}^\beta} \right)^r \right] sign \left(1 - \frac{s^\beta}{s_{sat}^\beta} \right) \tag{14}$$

The parameter h^β is the resistance parameter for the dependent self-hardening rate; s_{sat}^β is the saturation value of reference shear stress; and exponent r is a material constant. The parameter $q^{\alpha\beta} = q + (1 - q)\delta^{\alpha\beta}$ or the interaction coefficient matrix includes q as a latent-hardening parameter and is chosen to be 1.4.

The activation enthalpy for cross-slip is extended in the same approach employed in the sub-grain scale presented in Equation (8) where the rate of cross-slip resistance is a function of the anti-phase boundary energies on the octahedral and cube planes, as well as on the non-Schmid components of the resolved shear stress. The non-Schmid components τ_{pe}^α, τ_{se}^α and τ_{cb}^α are considered to have the same duties in the dislocation dissociation and slip on the octahedral slip systems and contribute to their slip resistances. According to [34], the cross-slip shear resistance can be stated as:

$$\tau_{crossco}^\alpha = \xi_0 exp \left(\frac{A}{\theta - \theta_c} \right) \mu \sqrt{\rho_0 exp \left(-\frac{H^\alpha}{K_B \theta} \right)} \tag{15}$$

where $H^\alpha = c_H \left\{ h + k_1 (t_{pe}^\alpha - k_1 t_{se}^\alpha) + \sqrt{\left(\frac{1}{\sqrt{3}} - \frac{\Gamma^{010}}{\Gamma^{111}} + |t_{cb}^\alpha| \right) \frac{b}{B}} \right\}$

The total thermal shear resistance or cutting stress can be calculated as:

$$\tau_{cut}^\alpha = s_*^\alpha + s_{cross}^\alpha \tag{16}$$

Material parameters in the above homogenized constitutive model are calibrated for the superalloy CMSX-4 single crystals in [1] and listed in Table 5.

Table 5. Calibrated material constants for the single-crystal grain-scale activation-energy-based model.

Parameter	k_1	k_2	ξ_0	A	θ_c	Q (J)	p	q	$\dot{\gamma}$ (s^{-1})	h_0 (MPa)	r
Value	0.4	0.6	8	325	1600	6.5×10^{-19}	0.78	1.15	5×10^7	100	1.115

4. Homogenized Single-Crystal Model from the Sub-Grain RVE Model

The morphology-dependent constitutive parameters in Equations (13) and (14) for the activation-energy-based crystal plasticity model are considered to be governed by the Hill–Mandel principle of macro-micro energy equivalence [35], where the micromechanical analysis is conducted with the sub-grain RVE model. The constitutive model includes functional parameters, which are formulated in terms of critical morphological variables and are fitted by computational homogenization of the sub-grain RVE model response.

4.1. Morphological Parameters in the Sub-Grain Microstructural RVE

The sub-grain microstructural RVE consists of γ' precipitates homogeneously distributed in a matrix γ phase as shown in Figure 3c. The two-phase γ-γ' microstructure is characterized as three morphology parameters including: (i) the volume fraction of the γ' precipitates, (ii) the shape factor n of the γ' precipitates and (iii) the minimum channel width l_c between the γ' precipitates. The volume fraction is defined as the ratio of the γ' precipitate volume to the total RVE volume, i.e., $v_f = \frac{V_{\gamma'}}{V_{RVE}}$. The shape factor of the precipitates is described in terms of the exponent of a superellipsoid: $\left(\frac{x}{a}\right)^n + \left(\frac{y}{b}\right)^n + \left(\frac{z}{c}\right)^n = 1$, where a, b and c are the dimensions of the three principal axes and n is the shape exponent. Here, $a = b = c$ for equiaxed precipitates. A value $n = 2$ corresponds to spherical precipitates, while $n \to \infty$ corresponds to cubic ones. In the homogenization procedure, a transformed shape factor $n_1 = tan^{-1}(n)$ is used to avoid singularity.

4.2. Morphology-Dependent Constitutive Parameters in the CP Model

Plastic shear deformation and hardening constitutive parameters in the single crystal grain-scale of AE-CP model are functions of the statistically-stored dislocations and cross-slip dislocation densities. At this scale, the homogenized single-crystal scale, the distribution of the dislocations is uniform. At the sub-grain scale, the distribution is not uniform because in a two-phase material, there will be a gradient in the dislocation densities, which generates geometrically-necessary dislocations. GNDs can change significantly when the morphology of the RVE changes. In other words, when the shape, size and distance between precipitates change, which normally occurs during the heat treatment process, the mechanical response of the RVE will vary. Hence, morphological parameters should also be incorporated into the homogenization process through these functions to consider the gradient of plastic shear strain corresponding to GNDs. Sensitivity analyses in [1] show that the initial thermal shear resistance, the reference slip-rate, the saturation shear stress and the cross-slip shear resistance are functions of the morphology. Thus, in Equations (13) and (14), the parameters $s^\alpha_{*0}(n_1, v_p, l_c)$, $\dot{\gamma}_*(n_1, v_p, l_c)$, $s^\alpha_{sat}(n_1, v_p, l_c)$ and $s^\alpha_{cross}(n_1, v_p, l_c)$ can be derived in terms of morphology, as well as $(\gamma^\alpha, \nabla\gamma^\alpha)$.

4.3. Functional Forms of the Single-Crystal Homogenized Constitutive Parameters

Four constitutive parameters in the single-crystal grain scale, $s^\alpha_{*0}(n_1, v_p, l_c)$, $k_*(n_1, v_p, l_c)$, $k(n_1, v_p, l_p)$ $s^\alpha_{sat}(n_1, v_p, l_c)$ and $s^\alpha_{cross}(n_1, v_p, l_c)$ are represented as a functional forms in terms of the microstructural morphology. The functional forms are derived through the homogenization procedure. Therefore, a large number of sub-grain RVE model simulations with varying volume fractions, channel widths and shapes is generated and simulated in the dislocation density sub-grain scale where an explicit representation of the γ-γ' morphology is assigned in the RVE. For each morphology in the sub-grain scale, simulations in the homogenized single-crystal level are performed in order to satisfy macro-homogeneity [35,36]:

$$\langle S \rangle : \langle \dot{E} \rangle = \frac{1}{\Omega_{RVE}} \int_{\Omega_{RVE}} S \, dV : \frac{1}{\Omega_{RVE}} \int_{\Omega_{RVE}} \dot{E} \, dV = \frac{1}{\Omega_{RVE}} \int_{\Omega_{RVE}} S : \dot{E} \, dV = \langle S : \dot{E} \rangle \quad (17)$$

Here, S and $\dot{E}$ correspond to the second Piola–Kirchhoff stress and the Lagrangian strain rate, respectively, and the symbol $\langle X \rangle$ corresponds to volume averaging over the RVE domain.

This extensive set of simulations, as explained in detail in [1], results in the following functional forms of the single-crystal constitutive parameters by using the least square minimization method in order to find the coefficients. The final form of these four constitutive parameters is as follows:

$$s_{*0}^{\alpha}(n_1, v_p, l_c) = a_1(n_1, v_p) + \frac{b_1(n_1, v_p)}{\sqrt{l_c}} = -50v_p n_1 + 222v_p - 34n_1 + 384 +$$
$$\frac{-33.3v_p n_1 + 32.92v_p + 19.61n_1 - 0.037}{\sqrt{l_c}}$$

$$s_{sat}^{\alpha}(n_1, v_p, l_c) = a_2(n_1, v_p) + \frac{b_2(n_1, v_p)}{l_c} = 6680v_p n_1 - 8905v_p - 1648n_1 + 3185 +$$
$$\frac{-3359v_p n_1 + 5008v_p + 3631n_1 - 0.21}{\sqrt{l_c}}$$

$$k_*(n_1, v_p, l_c) = 19847v_p n_1 l_c + 12768v_p n_1 - 23120v_p l_c$$
$$+4080n_1 l_c - 7500v_p + 33n_1 - 2700l_c + 65$$

$$k(n_1, v_p, l_c) = a_3(n_1, v_p) + \frac{b_3(n_1, v_p)}{\sqrt{l_c}} = 221.4v_p n_1 - 327.6v_p + 31.5n_1 + 5.5 +$$
$$\frac{-176.5v_p n_1 + 281.2v_p - 2.44n_1 + 0.14}{\sqrt{l_c}}$$

$$s_{cross}^{\alpha}(n_1, v_p, l_c) = s_{cross}^{\alpha} * s_0(n_1, v_p, l_c) ,$$

$$vspace1pts_0(n_1, v_p, l_c) = a_1(n_1, v_p) + \frac{b_1(n_1, v_p)}{\sqrt{l_c}}$$

$$= -50.32v_p n_1 + 0.538v_p - 0.09528n_1 + 1 +$$
$$\frac{-0.08662v_p n_1 + 0.08566v_p + 0.051n_1 - 0.000096}{\sqrt{l_c}}$$

(18)

The size-dependent variable, the channel width l_c, can be seen in the parameters to reflect explicitly the size effect due to the presence of geometrically-necessary dislocations in the sub-grain scale of the model. In this equation, the unit of l_c is μm, while the units of initial thermal resistance and saturation shear resistance are MPa, and s_0 is a dimensionless function.

4.4. Validation of the Homogenized Single-Crystal Grain-Scale Model

In order to validate the parametric constitutive model at the single-crystal grain scale, tensile constant strain tests are simulated for two orientations [001] and [111] for three temperatures 25 °C [29], 800 °C [28] and 950 °C [30] shown in Figure 5. The simulation at room temperature is performed for the [111] orientation and shows high yield stress because cubic slip systems are not activated at this temperature. Furthermore, the transition from the elastic to plastic part is very sharp, which shows that dislocation activities in the channel and matrix start almost simultaneously. Two simulations under elevated temperature are done for [001] orientation. In both simulations, it can be observed that initially plastic deformation starts in the channel where the slope of the elastic part changes for the stress around 600 MPa. However, this change is not significant due to the very small volume fraction of the channel. The yield stress and hardening decrease dramatically from 800–950 °C. The simulations show a very good agreement with the experimental data.

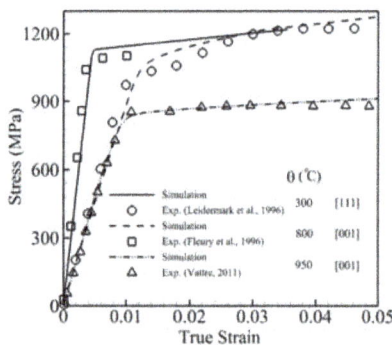

Figure 5. Volume-averaged true stress-logarithmic strain response by CPFEM and experiments [28–30] under a constant strain rate 0.001 s^{-1}.

5. Morphology Effect in Nickel-Based Superalloys

So far, all simulations are performed for CMSX-4, which is a single crystal of a nickel-based superalloy containing 70% of cuboidal precipitates with an average distance of 0.45 µm. The main idea of this work is to bridge the morphology effect from an explicit representation (sub-grain scale) to an implicit one (grain scale) in order to get the same response with significant savings in time and computation. Therefore, the multi-scale scheme could greatly benefit the design and optimization [37–39] for the next generation of these alloys. In this section, the effect of morphology on the mechanical behavior of these alloys is investigated as a function of three independent morphology parameters: precipitate size, volume fraction and shape. In the first set of simulations, the size of the precipitates is changed from a very small particle size (0.15 µm) to a large one of (1.35 µm), while the volume fraction and the shape of precipitates are kept constant at 70% and $n \to \infty$, respectively. In other words, for a unit cube of this material for one particle of γ' of a dimension of 1.35 µm, there will be 27 particles of 0.45 µm and 729 particles of a dimension 0.15 µm. The stress-strain curve under a tensile constant strain rate of 0.001 s^{-1} and 800 °C for these three sizes is shown in Figure 6. There is almost a 200-MPa difference between the 0.15 µm and 1.35 µm sizes. Smaller precipitate sizes for the same volume fraction and shape result in more particles, and more particles increase the dislocation densities and shear resistance.

In the second set of simulations, the volume fraction of the precipitates is altered from a very low volume fraction of 30% to a high volume fraction of 70%, while the size and the shape of precipitates are kept constant at 0.45 µm and $n \to \infty$, respectively. In other words, for a unit cube of this material, the channel width between precipitates in the case of 50% is 1.82-times and 30% is 2.95-times its width at a 70% volume fraction. The stress-strain curve under a tensile constant strain rate of 0.001 s^{-1} and 800 °C for these three volume fractions is shown in Figure 7. There is almost a 250-MPa difference between the 70% and 30% volume fractions. A larger volume fraction of the precipitate or a smaller channel width for the same size and shape increases the dislocation densities and shear resistance.

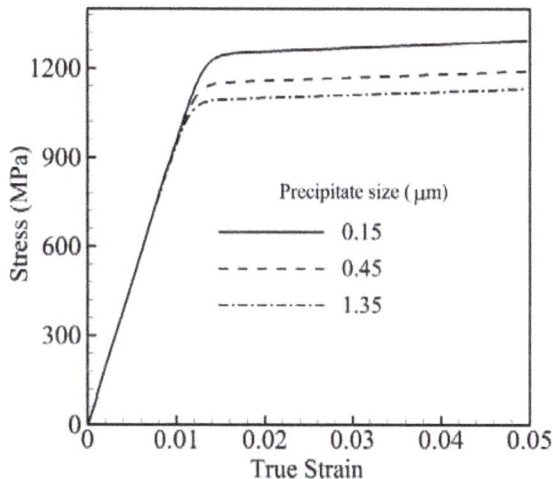

Figure 6. Volume-averaged true stress-logarithmic strain response by CPFEM under the strain rate of a single crystal of a nickel-based superalloy to investigate the effect of precipitate size at 800 °C under a tensile strain rate of 0.001 s^{-1}.

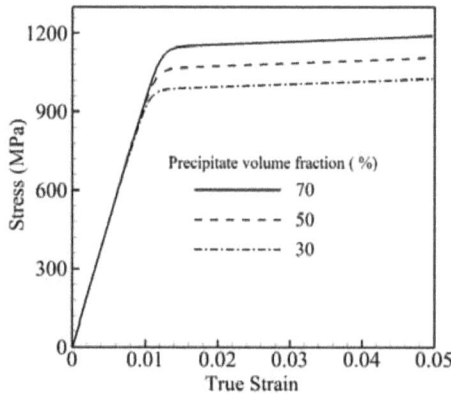

Figure 7. Volume-averaged true stress-logarithmic strain response by CPFEM under the strain rate of a single crystal of a nickel-based superalloy to investigate the effect of the precipitate volume fraction at 800 °C under a tensile strain rate of 0.001 s^{-1}.

In the last set of simulations, the shape of the precipitates is changed from a small shape factor $n = 1.1$ to a large shape factor $n = 1000$, while the size and volume fraction of precipitates are kept constant at 0.45 μm and 70%, respectively. In other words, the surface boundary between precipitates is very smooth in the case of a smaller shape factor, and it is very sharp in the case of $n = 1000$, which looks like a cube. The stress-strain curve under a tensile constant strain rate of 0.001 s^{-1} at 800 °C for these three shape factors is shown in Figure 8. The difference between the lowest and highest yield stresses is not as pronounced as for the size and volume effects; however, the transition between the elastic and plastic part is sharper for the bigger shape factors. From the figure, the dislocation activities or plastic deformation begin at the stress around 700 MPa where three curves start to have a slight divergence. The slope of the curve from this point is higher for higher shape factors, which indicates less plastic deformation in the channel due to the sharp precipitate edge. Therefore, round-shaped precipitates increase the dislocation density due to more dislocation bowing around precipitates and result in bigger yield stress.

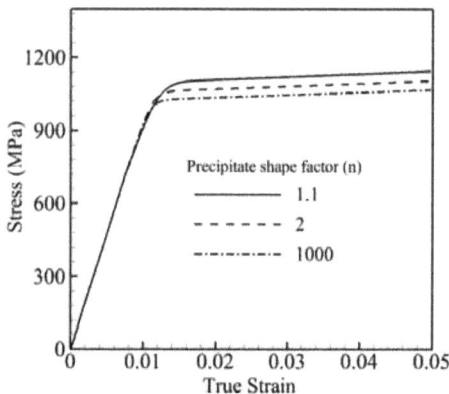

Figure 8. Volume-averaged true stress-logarithmic strain response by CPFEM under the strain rate of a single crystal of a nickel-based superalloy to investigate the effect of precipitate shape at 800 °C under a tensile strain rate of 0.001 s^{-1}.

5.1. Flow Stress Variations with Morphology in Nickel-Based Superalloys

The yield strength corresponding to 0.2% offset strain is considered to investigate the dependencies on the morphology of nickel-based superalloys. Accordingly, crystal plasticity simulations are performed for different morphologies including different shapes, volume fractions and sizes of precipitates.

5.1.1. Flow Stress Variations with the Shape of Precipitates

Crystal plasticity finite element simulations are performed at 1000 K under a strain rate of 0.001 s^{-1} for the crystal orientation [001] with the different shapes of precipitates while their volume fraction and size are constant for each set. In the first set of simulations, the volume fraction is constant and equal to 50%, while three sets of simulations are done for different sizes of precipitates, respectively 0.15 µm, 0.45 µm and 1.35 µm. In the second set, the size is fixed and equal to 0.45 µm, while the volume fraction changes as 30%, 50% and 70%, respectively. The results of the CPFEM simulations are shown in Figure 9 with symbols. The solid lines are plotted as the best trending functions to present these data. For all six plots, the functional form of $a + \frac{b}{n}$ is the best trending function to represent the data where n is the shape factor of precipitates defined in Section 4.1. In the the first set shown in Figure 9a, a and b are functions of the size of precipitates as $a(r)$ and $b(r)$, where in the second set shown in Figure 9b, they are functions of the volume fraction of the precipitates as $a(v_f)$ and $b(v_f)$. Hence, the flow stress in nickel-based superalloys will change proportionally with the shape factor of the morphology as $1/n$.

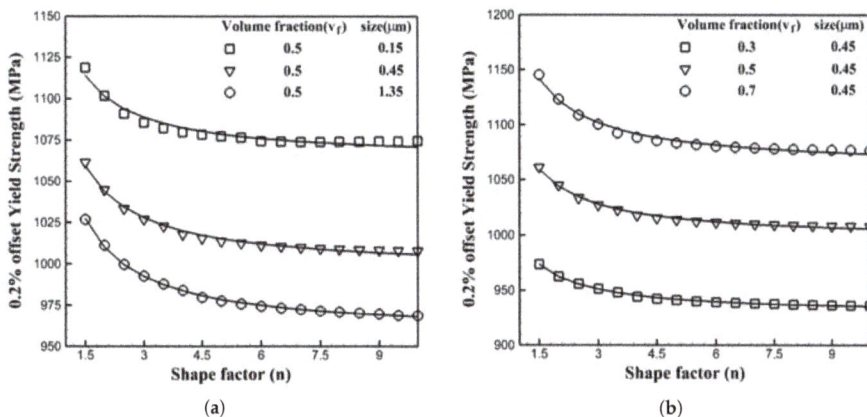

Figure 9. Variations of the 0.2% yield strength of a single crystal of a nickel-based superalloy with respect to the shape of precipitates at 1000 K under a tensile strain rate of 0.001 s^{-1}: (**a**) constant volume fraction for different sizes of precipitates; (**b**) different volume fractions for a constant size of precipitate.

5.1.2. Flow Stress Variations with the Size of Precipitates

Crystal plasticity finite element simulations are performed at 1000 K under a strain rate of 0.001 s^{-1} for the crystal orientation [001] with the different sizes of precipitates, while their shapes and volume fractions are constant for each set. In the first set of simulations, the volume fraction is constant and equal to 50%, while three sets of simulations are done for different shapes of precipitates: 1.5, 2 and 10, respectively. It should be mentioned that the shape factor of two corresponds to the spherical shape for precipitates, while 10 represents almost a cubic shape of the precipitates. In the second set, the shape factor is fixed and equal to four, while the volume fraction changes as 30%, 50% and 70%, respectively. The results of the CPFEM simulations are shown in Figure 10 with symbols. The plotted solid lines are the best trending functions to present CPFEM data. For all six plots, the functional form of $a + \dfrac{b}{\sqrt{r}}$ is the best trending function to represent the data where r is the size of precipitates. In the the first set shown in Figure 10a, a and b are functions of the shape factor of precipitates as $a(n)$ and $b(n)$, where in the second set shown in Figure 10b, they are functions of the volume fraction of the precipitates as $a(v_f)$ and $b(v_f)$. Hence, the flow stress in nickel-based superalloys will change proportionally with the size of the morphology as $1/\sqrt{r}$.

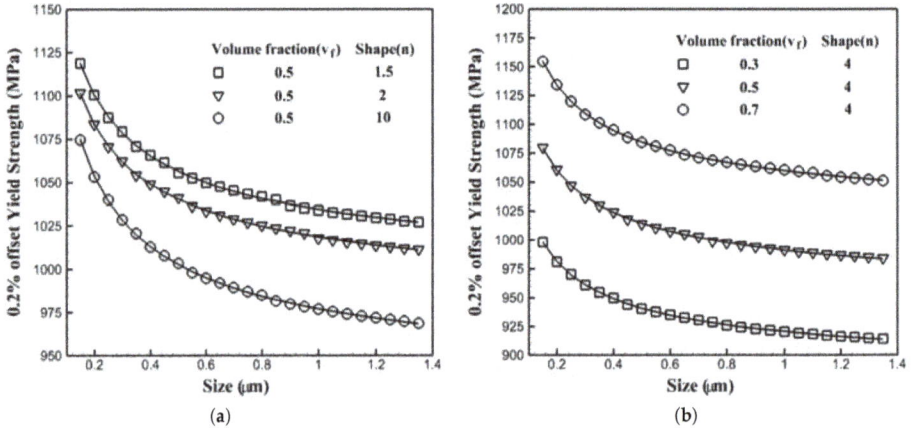

(a)   (b)

Figure 10. Variations of the 0.2% yield strength of a single crystal of a nickel-based superalloy with respect to the size of precipitates at 1000 K under a tensile strain rate of 0.001 s^{-1}: (**a**) constant volume fraction for different shape of precipitates (**b**) different volume fractions for a constant shape of precipitate.

5.1.3. Flow Stress Variations with the Volume Fraction of Precipitates

Crystal plasticity finite element simulations are performed at 1000 K under a strain rate of 0.001 s^{-1} for the crystal orientation [001] with the different volume fraction of precipitates while their shape and size are constant for each set. In the first set of simulations, the size is constant and equal to 50%, while three sets of simulations are done for different shapes of precipitates: 1.5, 2 and 10, respectively. In the second set, the shape factor is fixed and equal to four, while the size of precipitates changes as 0.15 µm, 0.45 µm and 1.35 µm, respectively. The results of the CPFEM simulations are shown in Figure 11 with symbols. The solid plotted lines are the best trending functions to present these data. For all six plots, the functional form of $a + bv_f$ is the best trending function to represent the data where v_f is the volume fraction of precipitates defined in Section 4.1. In the the first set shown in Figure 11a, a and b are functions of the shape of precipitates as $a(n)$ and $b(n)$, where in the second set shown in Figure 11b, they are functions of the size of the precipitates as $a(r)$ and $b(r)$. Thus, the flow stress in nickel-based superalloys will change linearly with the volume fraction of the morphology as v_f.

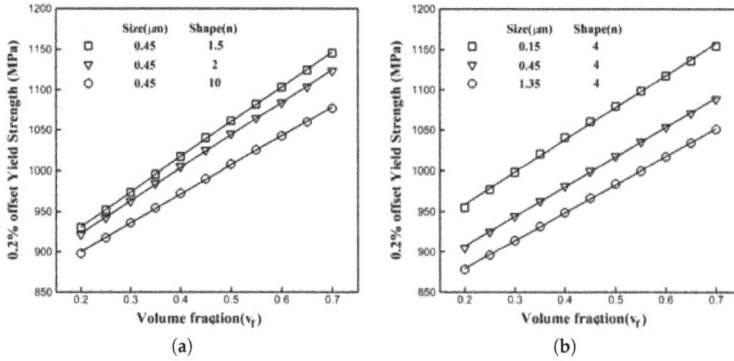

Figure 11. Variations of the 0.2% yield strength of a single crystal of a nickel-based superalloy with respect to the volume fraction of precipitates at 1000 K under a tensile strain rate of 0.001 s^{-1}: (**a**) constant size for different shapes of precipitates; (**b**) different sizes for a constant shape of precipitate.

5.1.4. Functional Form of the Flow Stress with Respect to the Morphology of Precipitates

As presented in the last section, the flow stress changes by changing the morphology of the microstructure. The change is proportional with $1/n$ with the shape of the precipitates while it changes as $1/\sqrt{r}$ with the size of the precipitates and linearly with the volume fraction of the precipitates. Therefore, the flow stress in the single crystals of nickel-based superalloys can be stated as:

$$\sigma_y = \left(c_1 + \frac{c_2}{n}\right)\left(c_3 + \frac{c_4}{\sqrt{r}}\right)(c_5 + c_6 v_f) \tag{19}$$

According to this equation, the size of precipitates has the major effect on the yield stress, where the behavior shows a sort of Hall–Petch effect. The Hall–Petch effect represents the variation of the yield stress with respect to the size of the grain in a polycrystalline microstructure. The three variables in this equation, the shape, size and volume fraction of the precipitates, affect the channel width between precipitates. While the channel width between precipitates becomes narrower, the dislocation will have a difficult time passing through the channel. The size of precipitates in the nickel-based superalloys is related to the distance between precipitates for a specific shape and volume fraction. For example, if we have a cubic two-phase γ-γ' microstructure with the unit dimension, the average size of precipitates would be 0.8 of the unit for the volume fraction of 51.2% with cubic precipitates. In order to change the size of precipitates to 0.4 units while the volume fraction is 51.2% and the shape of the precipitates is cubic, we have to change the size of the microstructure; therefore, the size of the cubic microstructure becomes half of a unit. When the size of the microstructure decreases, the precipitates become closer, and the channel width becomes narrower. Hence, for a specific volume fraction and shape, finer precipitates mean less size in the microstructure, which results in less distance between precipitates or less channel width. Dislocations have a difficult time passing through the channel with less distance between precipitates, so they start to bow around the precipitates and result in the generation of another source of hardening, which is geometrically-necessary dislocations (GNDs). The generation of these dislocations accordingly results in increasing the yield stress. The decrease in the channel width, when the shape of precipitates changes, is not as sharp as changing the size of precipitates; therefore, the increment in the yield stress for lesser shape factor or more curvy precipitates is not the same as the increment seen in the size; however, the explanation stays the same. Dislocation will have more space to bow around precipitates and creates additional sources of dislocations to harden materials with a lesser shape factor. There will be the same explanation for increasing the yield stress by increasing the volume fraction. Having a greater volume fraction for the same size and shape of precipitates means less channel width between precipitates, which directly results in increasing GNDs.

Different optimization methods can be used in order to find the unknown coefficient. However, the least square method is employed in this work in order to calculate the constants c_1–c_6. The input date are obtained from 324 crystal plasticity finite element simulations, which are performed at 1000 K under a tensile strain rate of 0.001 s^{-1} as shown in Figures 9–11 and used for the least square method as the input data. The functional form of the flow stress as a function of shape, size and volume fraction of precipitates is calculated as:

$$\sigma_y = 1069.88 - 345.39 v_f + \frac{34.67}{n} + \frac{181.14 v_f}{n} - \frac{146.97}{\sqrt{r}} + \frac{425.51 v_f}{\sqrt{r}} - \frac{19.15}{n\sqrt{r}} - \frac{0.411 v_f}{n\sqrt{r}} \qquad (20)$$

In this equation, the volume fraction of precipitates can change from 0.2–0.7; the shape factor of precipitates can change from 1.5–10; and the size of precipitates can vary from 0.15 µm–1.35 µm. The result of the above equation will be the flow stress in MPa.

5.1.5. Validation of the Functional Form of Flow Stress with Respect to the Morphology of Precipitates

In order to validate the functional form of the flow stress given in Equation (20), 10 random morphologies are created and first simulated with the crystal plasticity finite element model, then the results are compared with the ones obtained from Equation (20). All simulations are performed at 1000 K under a tensile strain rate of 0.001 s^{-1}. The results and comparison are shown in Table 6. As can be seen, the functional form of the flow stress obtained from the multi-scale framework gives almost identical flow stress compared with the crystal plasticity finite element models.

Table 6. The CPFEM and Equation (20) results and comparison for the flow stress for different morphologies.

Shape Factor (n)	Volume Fraction (v$_f$)	Size (r) (µm)	CPFEM Flow Stress (MPa)	Equation (20) (MPa)
1.7	0.31	0.45	973.6	976.9
2.2	0.32	0.37	975.1	969.3
2.5	0.36	0.43	980.6	983.2
2.3	0.44	0.57	1006.1	1009.8
2.9	0.48	0.55	1012.7	1014.3
3.5	0.52	0.31	1046.3	1050.6
4.2	0.57	0.38	1049.6	1053.4
6.7	0.61	0.45	1048.1	1044.4
7.3	0.41	0.37	986.9	984.0
8.1	0.44	0.31	1006.1	1000.1

6. Conclusions

This paper proposes a functional form for the flow stress of single crystals of nickel-based superalloys as a result of a multi-scale crystal plasticity finite element framework. The multi-scale scheme bridges two scales in a hierarchical framework from the two-phase γ-γ' sub-grain scale to a homogenized single-crystal grain-scale constitutive model that can be augmented to model polycrystalline microstructures of nickel superalloys. The non-Schmid constitutive models for two scales include the main dislocation mechanism active in these materials. For the single-crystal grain scale, an activation-energy-based crystal plasticity finite element model is developed that incorporates non-Schmid effects along with the characteristic parameters of the sub-grain scale γ-γ' morphology. For the next scale, a crystal plasticity homogenized model is used, which includes the effect of the morphology implicitly through the homogenized constitutive parameters. A notable advantage of this multi-scale model is that its high efficiency enables it to be effectively incorporated in the polycrystalline grain scale of crystal plasticity finite element simulations, while retaining the accuracy of detailed RVE models. The homogenized model incorporates the effect of important characteristics of the sub-grain γ-γ' morphology, viz. the size, shape and volume fraction of the precipitates. The uniformly-distributed precipitates in the sub-grain RVEs in the shape of generalized ellipsoidal particles provide a platform for a modeling framework connecting the two scales, one with explicit representation and the other with their respective parametric forms. There is a size dependency in the two-scale model, where it naturally occurs in the sub-grain scale due to the presence of geometrically-necessary dislocations or GNDs, and it is reflected in the homogenized single-crystal grain-scale model through the explicit dependence on the channel width. The homogenized activation-energy-based crystal plasticity model is found to accurately reproduce the stress-strain response of the detailed RVE for a range of microstructural variations. It is also found to agree quite well with the results of experimental studies on single-crystal superalloys in the literature. The morphology parameters include the shape factor, volume fraction and size of the precipitates, which have different impacts on the flow stress. The functional form of the flow stress obtained from the multi-scale framework gives almost identical flow stress in comparison with the crystal plasticity finite element models.

Acknowledgments: Shahriyar Keshavarz acknowledges support from the U.S. Department of Commerce, National Institute of Standards and Technology under Financial Assistance Award 70NANB15H240.

Author Contributions: Shahriyar Keshavarz and Zara Molaeinia developed the model for the micro-structural design of Nickel-based superallys and analyzed the data and parameters for the test cases however all of the authors contributed equally to adapting the multi-scale framework used in the developing design parameter (flow stress) for these superalloys.

Conflicts of Interest: The authors declare no conflicts of interest.

References

1. Keshavarz, S.; Ghosh, S. Multi-scale crystal plasticity finite element model approach to modeling nickel-based superalloys. *Acta Mater.* **2013**, *61*, 6549–6561.
2. Unocic, R.R.; Zhou, N.; Kovarik, L.; Shen, C.; Wang, Y.; Mills, M.J. Dislocation decorrelation and relationship to deformation microtwins during creep of a $\gamma\prime$ precipitate strengthened Ni-based superalloy. *Acta Mater.* **2011**, *59*, 7325–7339.
3. Pollock, T.M.; Tin, S. nickel-based superalloys for advanced turbine engines: Chemistry, microstructure and properties. *J. Propuls. Power* **2006**, *22*, 361–374.
4. Unocic, R.; Kovarik, L.; Shen, C.; Sarosi, P.; Wang, Y.; Li, J.; Ghosh, S.; Mills, M. Deformation mechanisms in Ni-base disk superalloys at higher temperatures. *Superalloys* **2008**, *8*, 377.
5. Cormier, J.; Milhet, X.; Mendez, J. Non-isothermal creep at very high temperature of the nickel-based single crystal superalloy MC2. *Acta Mater.* **2007**, *55*, 6250–6259.
6. Hong, H.; Kim, I.; Choi, B.; Kim, M.; Jo, C. The effect of grain boundary serration on creep resistance in a wrought nickel-based superalloy. *Mater. Sci. Eng. A* **2009**, *517*, 125–131.
7. Lours, P.; Coujou, A.; Coulomb, P. On the deformation of the <100> oriented $\gamma\prime$ strengthening phase of the CMSX2 superalloy. *Acta Metall. Mater.* **1991**, *39*, 1787–1797.
8. Keshavarz, S.; Ghosh, S.; Reid, A.C.; Langer, S.A. A non-Schmid crystal plasticity finite element approach to multi-scale modeling of nickel-based superalloys. *Acta Mater.* **2016**, *114*, 106–115.
9. Keshavarz, S.; Ghosh, S. Hierarchical crystal plasticity fe model for nickel-based superalloys: Sub-grain microstructures to polycrystalline aggregates. *Int. J. Solids Struct.* **2015**, *55*, 17–31.
10. Ghosh, S.; Keshavarz, S.; Weber, G. Computational multiscale modeling of nickel-based superalloys containing gamma-gamma'precipitates. In *Inelastic Behavior of Materials and Structures Under Monotonic and Cyclic Loading*; Springer: Cham, Switzerland, 2015; pp. 67–96.
11. Ghosh, S.; Weber, G.; Keshavarz, S. Multiscale modeling of polycrystalline nickel-based superalloys accounting for subgrain microstructures. *Mech. Res. Commun.* **2016**, *78*, 34–46.
12. Beyerlein, I.; Tomé, C. A dislocation-based constitutive law for pure Zr including temperature effects. *Int. J. Plast.* **2008**, *24*, 867–895.
13. Knezevic, M.; Beyerlein, I.J.; Lovato, M.L.; Tomé, C.N.; Richards, A.W.; McCabe, R.J. A strain-rate and temperature dependent constitutive model for BCC metals incorporating non-Schmid effects: Application to tantalum–tungsten alloys. *Int. J. Plast.* **2014**, *62*, 93–104.

14. Pollock, T.; Argon, A. Creep resistance of CMSX-3 nickel base superalloy single crystals. *Acta Metall. Mater.* **1992**, *40*, 1–30.

15. Busso, E.; Meissonnier, F.; O'dowd, N. Gradient-dependent deformation of two-phase single crystals. *J. Mech. Phys. Solids* **2000**, *48*, 2333–2361.

16. Probst-Hein, M.; Dlouhy, A.; Eggeler, G. Interface dislocations in superalloy single crystals. *Acta Mater.* **1999**, *47*, 2497–2510.

17. Fedelich, B. A microstructural model for the monotonic and the cyclic mechanical behavior of single crystals of superalloys at high temperatures. *Int. J. Plast.* **2002**, *18*, 1–49.

18. Shenoy, M.; Kumar, R.; McDowell, D. Modeling effects of nonmetallic inclusions on LCF in DS nickel-base superalloys. *Int. J. Fatigue* **2005**, *27*, 113–127.

19. Fromm, B.S.; Chang, K.; McDowell, D.L.; Chen, L.Q.; Garmestani, H. Linking phase-field and finite-element modeling for process–structure–property relations of a Ni-base superalloy. *Acta Mater.* **2012**, *60*, 5984–5999.

20. Ma, A.; Roters, F. A constitutive model for fcc single crystals based on dislocation densities and its application to uniaxial compression of aluminium single crystals. *Acta Mater.* **2004**, *52*, 3603–3612.

21. Ma, A.; Roters, F.; Raabe, D. A dislocation density based constitutive model for crystal plasticity FEM including geometrically necessary dislocations. *Acta Mater.* **2006**, *54*, 2169–2179.

22. Reid, A.C.; Langer, S.A.; Lua, R.C.; Coffman, V.R.; Haan, S.I.; García, E. Image-based Finite Element Mesh Construction for Material Microstructures. *Comput. Mater. Sci.* **2008**, *43*, 989–999.

23. Schmid, E.; Boas, W. *Plasticity of Crystals*; F. A. Hughes & Co., Limited: London, UK, 1950.

24. Roters, F.; Eisenlohr, P.; Hantcherli, L.; Tjahjanto, D.; Bieler, T.; Raabe, D. Overview of constitutive laws, kinematics, homogenization and multiscale methods in crystal plasticity finite-element modeling: Theory, experiments, applications. *Acta Mater.* **2010**, *58*, 1152–1211.

25. Cuitino, A.M.; Ortiz, M. Computational modelling of single crystals. *Model. Simul. Mater. Sci. Eng.* **1993**, *1*, 225.

26. Tinga, T.; Brekelmans, W.; Geers, M. Time-incremental creep–fatigue damage rule for single crystal Ni-base superalloys. *Mater. Sci. Eng. A* **2009**, *508*, 200–208.

27. Kalidindi, S.R.; Bronkhorst, C.A.; Anand, L. Crystallographic texture evolution in bulk deformation processing of FCC metals. *J. Mech. Phys. Solids* **1992**, *40*, 537–569.

28. Fleury, G.; Schubert, F.; nickel, H. Modelling of the thermo-mechanical behavior of the single crystal superalloy CMSX-4. *Comput. Mater. Sci.* **1996**, *7*, 187–193.

29. Leidermark, D.; Moverare, J.J.; Simonsson, K.; Sjöström, S.; Johansson, S. Room temperature yield behavior of a single-crystal nickel-base superalloy with tension/compression asymmetry. *Comput. Mater. Sci.* **2009**, *47*, 366–372.

30. Vattré, A.; Fedelich, B. On the relationship between anisotropic yield strength and internal stresses in single crystal superalloys. *Mech. Mater.* **2011**, *43*, 930–951.

31. Kocks, U.; Argon, A.; Ashby, M. Progress in materials science. *Thermodyn. Kinet. Slip* **1975**, *19*, 110–170.

32. Frost, H.J.; Ashby, M.F. *Deformation Mechanism Maps: The Plasticity and Creep of Metals and Ceramics*; Pergamon Press: Oxford, UK, 1982.

33. Xie, C.; Ghosh, S.; Groeber, M. Modeling cyclic deformation of HSLA steels using crystal plasticity. *J. Eng. Mater. Technol.* **2004**, *126*, 339–352.

34. Keshavarz, S.; Ghosh, S. A crystal plasticity finite element model for flow stress anomalies in ni3al single crystals. *Philos. Mag.* **2015**, *95*, 2639–2660.

35. Hill, R. On macroscopic effects of heterogeneity in elastoplastic media at finite strain. *Math. Proc. Camb. Philos. Soc.* **1984**, *95*, 481–494.

36. Nemat-Nasser, S. Averaging theorems in finite deformation plasticity. *Mech. Mater.* **1999**, *31*, 493–523.

37. Keshavarz, S.; Khoei, A.R.; Molaeinia, Z. Genetic algorithm-based numerical optimization of powder compaction process with temperature-dependent cap plasticity model. *Int. J. Adv. Manuf. Technol.* **2013**, *64*, 1057–1072.

38. Khoei, A.R.; Keshavarz, S.; Biabanaki, S.O.R. Optimal design of powder compaction processes via genetic algorithm technique. *Finite Elem. Anal. Des.* **2010**, *46*, 843–861.

39. Khoei, A.R.; Molaeinia, Z.; Keshavarz, S. Modeling of hot isostatic pressing of metal powder with temperature–dependent cap plasticity model. *Int. J. Mater. Form.* **2013**, *6*, 363–376.

crystals

MDPI

Article

Phase Transformation and Hydrogen Storage Properties of an La$_{7.0}$Mg$_{75.5}$Ni$_{17.5}$ Hydrogen Storage Alloy

Lin Hu *, Rui-hua Nan *, Jian-ping Li, Ling Gao and Yu-jing Wang

School of Materials and Chemical Engineering, Xi'an Technological University, Xi'an 710021, China; jpli_0416@163.com (J.L.); Gaoling@xatu.edu.cn (L.G.); Wangyujing@xatu.edu.cn (Y.W.)
* Correspondence: hulin@xatu.edu.cn (L.H.); nanrh@xatu.edu.cn (R.N.) Tel.: +86-186-0292-5246 (L.H.)

Academic Editor: Peter Lagerlof
Received: 25 September 2017; Accepted: 16 October 2017; Published: 18 October 2017

Abstract: X-ray diffraction showed that an La$_{7.0}$Mg$_{75.5}$Ni$_{17.5}$ alloy prepared via inductive melting was composed of an La$_2$Mg$_{17}$ phase, an LaMg$_2$Ni phase, and an Mg$_2$Ni phase. After the first hydrogen absorption/desorption process, the phases of the alloy turned into an La–H phase, an Mg phase, and an Mg$_2$Ni phase. The enthalpy and entropy derived from the van't Hoff equation for hydriding were −42.30 kJ·mol^{-1} and −69.76 J·K^{-1}·mol^{-1}, respectively. The hydride formed in the absorption step was less stable than MgH$_2$ (−74.50 kJ·mol^{-1} and −132.3 J·K^{-1}·mol^{-1}) and Mg$_2$NiH$_4$ (−64.50 kJ·mol^{-1} and −123.1 J·K^{-1}·mol^{-1}). Differential thermal analysis showed that the initial hydrogen desorption temperature of its hydride was 531 K. Compared to Mg and Mg$_2$Ni, La$_{7.0}$Mg$_{75.5}$Ni$_{17.5}$ is a promising hydrogen storage material that demonstrates fast adsorption/desorption kinetics as a result of the formation of an La–H compound and the synergetic effect of multiphase.

Keywords: hydrogen storage alloy; phase transition; La hydride compound; hydriding kinetic

1. Introduction

Hydrogen is considered a promising energy carrier to replace the traditional fossil fuels because it is abundant, lightweight, and environmentally friendly, and has high energy content (142 MJ·kg^{-1}) [1]. As an ideal energy carrier, hydrogen can be easily converted into a needed form of energy without releasing harmful emissions [2]. However, up to now, highly efficient hydrogen storage technology remains challenging, as the practical application of hydrogen energy is constrained [3]. More attention has been paid to magnesium hydride (MgH$_2$), which is a highly promising material for hydrogen storage given its high hydrogen storage capacity (7.6 wt %), low density, good reversibility, and low cost [4–8]. Unfortunately, its practical application in hydrogen storage is limited by slow absorption/desorption kinetics and high thermodynamic stability (ΔH = −74.50 kJ·mol^{-1} H$_2$) [9,10]. In the past several decades, Mg has been alloyed with various metals to form Mg-based alloys to improve its hydrogenation properties. In numerous Mg-based alloys, rare-earth (RE) Mg alloys and Mg-transition metal alloys have been intensively investigated. An La–Mg alloy was first reported by Hagenmull et al. in 1980 with admirable hydrogen storage properties [11], and Zou et al. [12] found that RE (RE = Nd, Gd, Er) in solid state solutions in Mg contribute to the improved hydrogenation thermodynamics and kinetics of Mg ultrafine particles. Considering the relatively fine hydrogen storage properties, Mg$_2$Ni alloys are a classic material in Mg-transition metal alloys [13].

In recent studies, ternary alloys defined as La–Mg–Ni alloys was successfully developed by introducing Ni into La$_2$Mg$_{17}$. This material with the optimized composition of LaMg$_2$Ni exhibited a relatively low hydrogen desorption temperature and fine kinetic properties [14]. However, the hydrogen storage capacity of this alloy was dramatically reduced by excessive un-hydrogenation elements such as La and Ni. A feasible method of improving the hydrogenation capacity of an La–Mg–Ni alloy is to increase the account of Mg in the alloy. Balcerzak investigated the hydrogen storage capacity of $La_{2-x}Mg_xNi_7$ alloys (x = 0, 0.25, 0.5, 0.75, 1). It was observed that the gaseous hydrogen storage capacity of La–Mg–Ni alloys increases with Mg content to a maximum in $La_{1.5}Mg_{0.5}Ni_7$ alloys [15]. In this paper, an $La_{7.0}Mg_{75.5}Ni_{17.5}$ alloy was prepared by inductive melting to achieve a uniform phase distribution. Their hydrogenation properties and phase transformation in the hydrogen storage cycles were further investigated to evaluate their potential for application in hydrogen storage.

2. Experimental Section

An $La_{7.0}Mg_{75.5}Ni_{17.5}$ alloy ingot was prepared by inductive melting of high-purity La, Mg, and Ni (purity more than 99.9%) in a magnesia crucible under an argon atmosphere. The ingots were mechanically crushed and ground in air into fine powders. Powders with particle sizes of 38–74 µm were used in a P–C–T test, and those with particle sizes less than 38 µm were used in X-ray diffraction analysis. The phase structure of the as-cast alloy and the hydrogenated alloy were measured with a D/max-2500/PC X-ray diffractometer (Rigaku, Tokyo, Japan) with Cu Kα radiation and analyzed using Jade-5.0 software. Scanning electron microscopy (SEM) images of the $La_{7.0}Mg_{75.5}Ni_{17.5}$ alloy were obtained with a HITACHI S-4800 scanning electron microscope (Hitachi, Tokyo, Japan) with an energy dispersive X-ray spectrometer (EDS) (Shimadzu, Kyoto, Japan). Hydrogen storage property measurements were carried out using P–C–T characteristic measurement equipment (Suzuki Shokan, Tokyo, Japan). The measurement conditions were as follows: delay time: 300 s; maximum pressure: 3 MPa. The hydriding kinetic of the as-cast alloy was also tested via P–C–T characteristic measurement equipment under a hydrogen pressure of 3 MPa.

3. Results and Discussion

No standard powder diffraction pattern for $LaMg_2Ni$ is available in any of the ICDD (International Center for Diffraction Data) files. For a clear, standard XRD pattern of $LaMg_2Ni$, the cell parameters and atom positions for single crystal alloys are listed in Table 1. Its crystal structure was drawn with Diamond 3.0 software in Figure 1. It can be seen that the $La_{7.0}Mg_{75.5}Ni_{17.5}$ alloy had a multiphase structure containing an $LaMg_2Ni$ phase with a $CuMgAl_2$-type structure [16–18], an La_2Mg_{17} phase with an $Ni_{17}Th_2$-type structure, and an Mg_2Ni phase. No La–Ni phase was observed, which is ascribed to the abundant Mg atoms in the alloy; thus, the La and Ni atoms are a solid solute and readily alloy with the Mg.

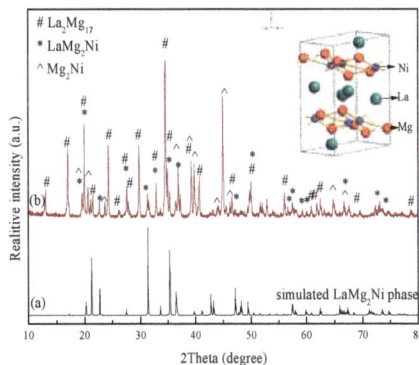

Figure 1. XRD patterns of the $La_{7.0}Mg_{75.5}Ni_{17.5}$ alloy: (**a**) simulated $LaMg_2Ni$ phase; (**b**) as-cast and crystal structure of $LaMg_2Ni$.

Table 1. X-ray structure refinement results on a single crystal of LaMg$_2$Ni.

Phase	Cell Parameters (Å)			Atom Position					References
	a	b	c	Atoms	x	y	z	Site	
LaMg$_2$Ni	4.2266	10.3031	8.3601	La	0	0.4402	1/4	4c	[10]
				Mg	0	0.1543	0.0552	8f	
				Ni	0	0.7266	1/4	4c	

Figure 2 shows the SEM micrographs of the La$_{7.0}$Mg$_{75.5}$Ni$_{17.5}$ alloy, and its EDS analysis is listed in Table 2. The La$_{7.0}$Mg$_{75.5}$Ni$_{17.5}$ alloy contains three distinct crystallography phases: the first is the La$_2$Mg$_{17}$ phase in the black region (identified as A), the second is the LaMg$_2$Ni phase in the bright region (identified as B), and the third is the Mg$_2$Ni phase in the dark region (identified as C). According to the EDS analysis results of the phases in the alloy, which are listed in Table 2, the determinations of the phases correspond to the XRD analyses.

The phase constituent of the alloys after dehydrogenation is presented in Figure 3. The peaks of LaH$_{2.46}$, Mg, Mg$_2$Ni, and MgH$_2$ were observed after the hydrogen desorption process. When hydrogenation is carried out, the LaMg$_2$Ni phase decompounds to an La phase and an Mg$_2$Ni phase, the La phase then forms a stable La hydride, and the Mg$_2$Ni turns into Mg$_2$NiH$_4$ in the hydrogen atmosphere [19,20]. The reaction of the LaMg$_2$Ni phase can be summarized as follows:

$$LaMg_2Ni + x/2H_2 \xrightarrow[\text{absorption}]{\text{hydrogen}} LaH_x + Mg_2NiH_4 \xrightarrow[\text{desorption}]{\text{hydrogen}} LaH_x + Mg_2Ni + H_2 \qquad (1)$$

The La$_2$Mg$_{17}$ phase decompounded to an La phase and an Mg phase. Then, the La phase turned into La–H, and Mg changed into MgH$_2$. The phase transformation of La$_2$Mg$_{17}$ can be described as

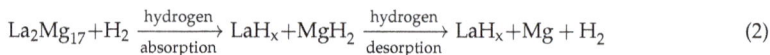

$$La_2Mg_{17} + H_2 \xrightarrow[\text{absorption}]{\text{hydrogen}} LaH_x + MgH_2 \xrightarrow[\text{desorption}]{\text{hydrogen}} LaH_x + Mg + H_2 \qquad (2)$$

The Mg$_2$Ni phase turns into Mg$_2$NiH$_4$ when it absorbs hydrogen, and turns back into Mg$_2$Ni in the hydrogen desorption process. The MgH$_2$ phase can be seen in Figure 3, which shows that the hydrogen desorption process of the alloy is incomplete.

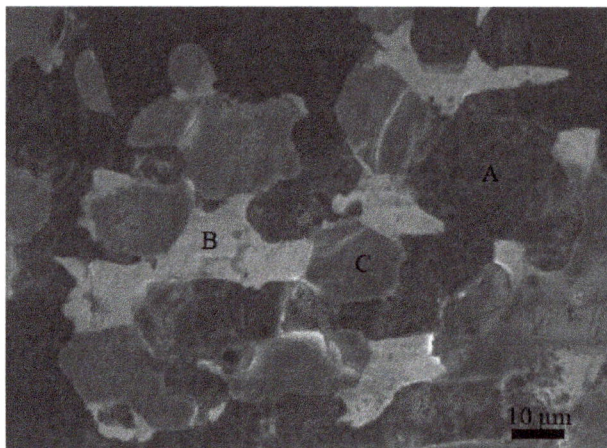

Figure 2. SEM micrographs and elements distribution images on the surface of the La$_{7.0}$Mg$_{75.5}$Cu$_{17.5}$ alloy.

Table 2. EDS analysis of the La$_{7.0}$Mg$_{75.5}$Ni$_{17.5}$ alloy.

Elements (at.%)	La	Mg	Ni
A	11.01	88.04	0.95
B	26.08	49.04	24.88
C	0.89	65.05	34.06

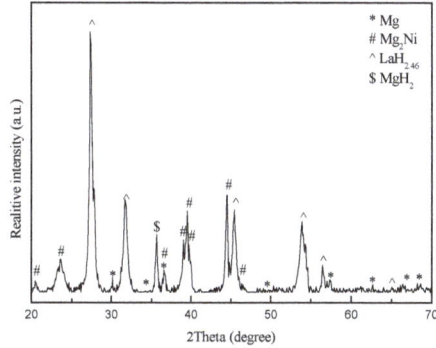

Figure 3. XRD pattern of the dehydrogenated La$_{7.0}$Mg$_{75.5}$Ni$_{17.5}$ alloy.

The P–C–T curves of the La$_{7.0}$Mg$_{75.5}$Ni$_{17.5}$ alloy at different temperatures are shown in Figure 4. Because of the decomposition of the LaMg$_2$Ni phase and the La$_2$Mg$_{17}$ phase, the actual absorb/desorb hydrogen phase are the Mg phase and the Mg$_2$Ni phase, which is consistent with the XRD pattern of the dehydrogenated alloy in Figure 3. A slant plateau can be observed in each hydrogen absorb/desorb process, indicating that the amount of H solute in the alloy increases with the rise of H pressure. This is due to the multiphase structure of the alloy, which is abundant in phase interfaces. H atoms readily enter the interfaces and then compose with Mg or Mg$_2$Ni to form the MgH$_2$ and Mg$_2$NiH$_4$. With the reduction of hydrogenation temperature, the maximum capacity of the La$_{7.0}$Mg$_{75.5}$Ni$_{17.5}$ alloy decreased from 3.18 wt % (588 K) to 2.45 wt % (523 K), suggesting that the activity of the interface is restricted by a relatively low temperature.

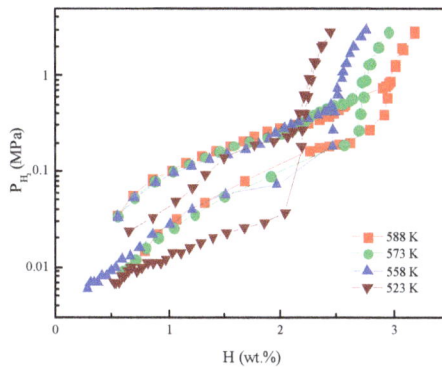

Figure 4. The P–C–T curves of the La$_{7.0}$Mg$_{75.5}$Ni$_{17.5}$ alloy at different temperatures.

In order to obtain the thermodynamic parameters of the hydriding reaction of the La$_{7.0}$Mg$_{75.5}$Ni$_{17.5}$ alloy, the plateau pressure (P, absolute atmosphere) and temperature (T, in K) are plotted according to the van't Hoff equation (Equation (3)).

$$\ln K = \frac{\Delta H}{RT} - \frac{\Delta S}{R} \tag{3}$$

where K is the equilibrium constant (K = $1/P_{H_2}$ in the hydriding process) and the van't Hoff plots are demonstrated in Figure 5. According to the van't Hoff equation, the enthalpy and entropy for the hydriding reaction are calculated to be -42.30 kJ·mol^{-1} and -69.76 J·K^{-1}·mol^{-1}, respectively. The results show that the alloy has a low enthalpy, its absolute value is lower than those of MgH$_2$ and Mg$_2$NiH$_4$. The improvement on the hydrogen storage properties of the La$_{7.0}$Mg$_{75.5}$Ni$_{17.5}$ alloy is because of the existence of LaH, which can reduce the enthalpy change of the hydriding process of the Mg and Mg$_2$Ni phase in this alloy. The main reason for these experimental results is that LaH can increase the reactive surface area and dramatically reduce the diffusion length of hydrogen.

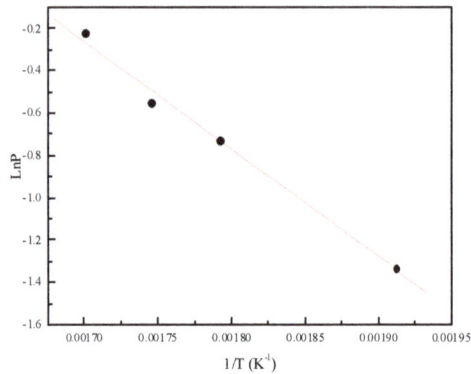

Figure 5. The van't Hoff plot of the alloy in hydrogenation processes.

The hydriding/dehydriding processes kinetic curves of the La$_{7.0}$Mg$_{75.5}$Ni$_{17.5}$ alloy at different temperatures are shown in Figure 6. Figure 6a shows that the uptake time for the hydrogen content to reach 90% of the maximum storage capacity was less than 60 s at various temperatures. The amount of hydrogen desorption increased as the temperature increased. Figure 7 compares the dehydriding kinetic curves of La$_{7.0}$Mg$_{75.5}$Ni$_{17.5}$ with that of pure MgH$_2$ powder at 573 K. The hydride of the alloy presented a significant improvement on both hydrogen desorption kinetics and hydrogen absorption capacities. During 1800 s, 0.99 wt % hydrogen desorbed from the hydride of the La$_{7.0}$Mg$_{755}$Ni$_{17.5}$ alloy, while pure MgH$_2$ powder only desorbed 0.39 wt % hydrogen.

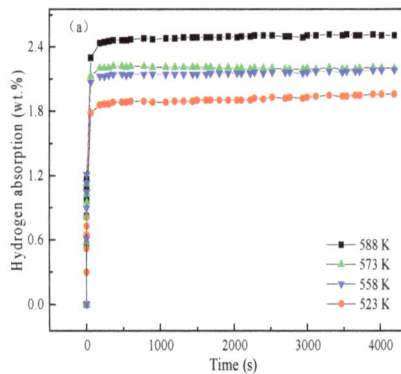

Figure 6. *Cont.*

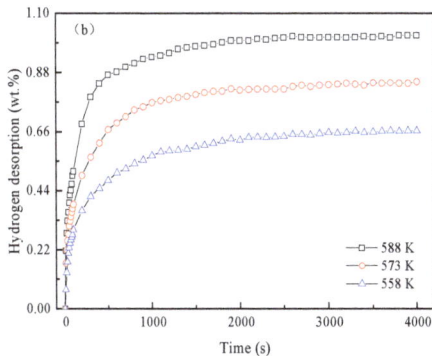

Figure 6. Kinetics of the La$_{7.0}$Mg$_{75.5}$Ni$_{17.5}$ alloy at different temperatures: (**a**) hydriding processes, (**b**) dehydriding processes.

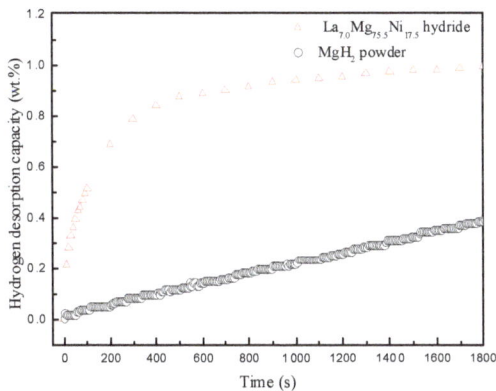

Figure 7. Dehydriding kinetics of the La$_{7.0}$Mg$_{75.5}$Ni$_{17.5}$ alloy and MgH$_2$ at 573 K.

The differential thermal analysis (DTA) of MgH$_2$, Mg$_2$NiH$_4$, and hydride of the La$_{7.0}$Mg$_{755}$Ni$_{17.5}$ alloy are presented in Figure 8. It can be seen that the initial hydrogen desorption temperature of the alloy hydride of 531 K was lower than that of MgH$_2$ (714 K) and Mg$_2$NiH$_4$ (549 K). The ameliorations on the thermo and kinetic properties of La$_{7.0}$Mg$_{75.5}$Ni$_{17.5}$ are attributed to the facts that hydrogen atoms in its hydride increase the crystalline parameters and further enhance the interaction between the MgH$_2$ phase and the Mg$_2$NiH$_4$ phase. This interaction evolves into a synergetic effect in the multiphase structure and facilitates its hydrogen desorption [21,22]. During dehydrogenation, Mg$_2$NiH$_4$ first desorbed hydrogen and exhibited a significant volume contraction, causing a significant contraction strain of MgH$_2$. Therefore, it increased the energy of the MgH$_2$ phase and advantaged the dehydrogenation of the MgH$_2$ phase [23–25].

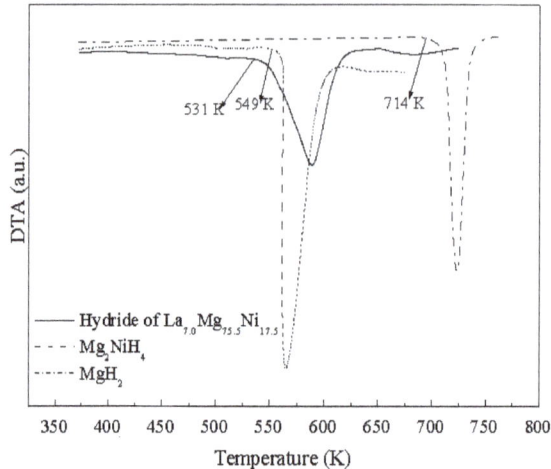

Figure 8. DTA curves of MgH_2, Mg_2NiH_4, and hydride of the $La7.0Mg_{755}Ni_{17.5}$ alloy.

4. Conclusions

The phase transformation and hydrogen storage properties of a multiphase $La_{7.0}Mg_{17.5}Ni_{17.5}$ alloy were investigated in this work. The hydriding process leads to the formation of the La–H, Mg_2NiH_4, and MgH_2 phases. The hydrogen absorption capacity is 2.45 wt % at 523 K. A low enthalpy (-42.30 kJ·mol^{-1}) that is lower than that of MgH_2 (-74.50 kJ·mol^{-1}) and Mg_2NiH_4 (-74.50 kJ·mol^{-1}) was obtained. The uptake time for hydrogen content to reach 90% of the maximum storage capacity for the $La_{7.0}Mg_{75.5}Ni_{17.5}$ alloy was less than 60 s at all testing temperatures, and the initial hydrogen desorption temperature of the alloy hydride (531 K) was lower than both MgH_2 (714 K) and Mg_2NiH_4 (549 K). Generally, the dissociation of magnesium hydride does not occur below 573 K. Improvement in the hydrogen storage properties of the $La_{7.0}Mg_{75.5}Ni_{17.5}$ alloy is due to the presence of La–H and the synergetic effect between the phases during the hydriding/dehydriding process.

Acknowledgments: This research was supported by the National Natural Science Foundation of China (51502235 and 51502234), the President's Fund of Xi'an Technological University (XAGDXJJ5011), and the Natural Science Basic Research Plan in Shaanxi Province of China (2016JQ5011).

Author Contributions: Hu Lin designed the research and wrote the manuscript. Nan Ruihua, Gao Ling, and Wang Yujing performed experiments, collected data, and generated the figures. Both authors contributed to editing and reviewing the manuscript.

Conflicts of Interest: The authors declare no conflict of interest.

References

1. Liu, T.; Shen, H.L.; Liu, Y.; Xie, L. Scaled-up synthesis of nanostructured Mg-based compounds and their hydrogen storage properties. *J. Power Sources* **2013**, 227, 86–93. [CrossRef]
2. Liu, Y.F.; Yang, Y.X.; Gao, M.X.; Pan, H.G. Tailoring Thermodynamics and Kinetics for Hydrogen Storage in Complex Hydrides towards Applications. *Chem. Rec.* **2016**, 16, 189–204. [CrossRef]
3. Ma, X.J.; Xie, X.B.; Liu, P.; Xu, L.; Liu, T. Synergic catalytic effect of Ti hydride and Nb nanoparticles for improving hydrogenation and dehydrogenation kinetics of Mg-based nanocomposite. *Prog. Nat. Sci-Mater.* **2017**, 27, 99–104. [CrossRef]
4. Zhu, X.L.; Pei, L.C.; Zhao, Z.Y. The catalysis mechanism of La hydrides on hydrogen storage properties of MgH_2 in MgH_2 + x wt.% LaH_3 (x = 0, 10, 20, and 30) composites. *J. Alloys Compd.* **2013**, 557, 64–69. [CrossRef]
5. Hu, L.; Han, M.; Li, J.H. Phase structure and hydrogen absorption property of $LaMg_2Cu$. *Mater. Sci. Eng. B.* **2010**, 166, 209–212. [CrossRef]

6. Hu, L.; Han, S.M.; Yang, C. Phase Structure and Hydrogen Storage Property of LaMg$_2$Cu$_{1-x}$Ni$_x$ (x = 0~0.90) Alloys. *Chin. J. Inorg. Chem.* **2010**, *26*, 1044–1048. [CrossRef]

7. Abdellatief, M.; Campostrini, R.; Leoni, M. Effects of SnO$_2$ on hydrogen desorption of MgH$_2$. *Int. J. Hydrog. Energy* **2013**, *38*, 4664–4669. [CrossRef]

8. Pei, P.; Song, X.P.; Liu, J. Study on the hydrogen desorption mechanism of a Mg-V composite prepared by SPS. *Int. J. Hydrog. Energy* **2012**, *37*, 984–989. [CrossRef]

9. Ares, J.R.; Leardini, F.; Díaz-Chao, P. Hydrogen desorption in nanocrystalline MgH$_2$ thin films at room temperature. *J. Alloys Compd.* **2010**, *495*, 650–654. [CrossRef]

10. Wang, H.; Zhang, J.; Liu, J.W. Improving hydrogen storage properties of MgH$_2$ by addition of alkali hydroxides. *Int. J. Hydrog. Energy* **2013**, *38*, 10932–10938. [CrossRef]

11. Darriet, B.; Pezat, M.; Hbika, A. Application of magnesium rich rare-earth alloys to hydrogen storage. *Int. J. Hydrog. Energy* **1980**, *5*, 173–178. [CrossRef]

12. Zou, J.X.; Zeng, X.Q.; Ying, Y.J. Study on the hydrogen storage properties of core-shell structured Mg-RE (RE = Nd, Gd, Er) nano-composites synthesized through arc plasma method. *Int. J. Hydrog. Energy* **2013**, *19*, 2337–2346. [CrossRef]

13. Atias-Adrian, I.C.; Deorsola, F.A.; Ortigoza-Villalba, G.A. Development of nanostructured Mg$_2$Ni alloys for hydrogen storage applications. *Int. J. Hydrog. Energy* **2011**, *36*, 7897–7910. [CrossRef]

14. Li, X.; Yang, T.; Zhang, Y.H. Kinetic properties of La$_2$Mg$_{17-x}$ wt.% Ni (x = 0–200) hydrogen storage alloys prepared by ball milling. *Int. J. Hydrog. Energy* **2014**, *39*, 13557–13563. [CrossRef]

15. Balcerzak, M.; Nowak, M.; Jurczyk, M. Hydrogenation and electrochemical studies of La–Mg–Ni alloys. *Int. J. Hydrog. Energy* **2017**, *42*, 1436–1443. [CrossRef]

16. Lin, H.J.; Ouyang, L.Z.; Wang, H. Phase transition and hydrogen storage properties of melt-spun Mg$_3$LaNi$_{0.1}$ alloy. *Int. J. Hydrog. Energy* **2012**, *37*, 1145–1150. [CrossRef]

17. Renaudin, G.; Guenee, L.; Yvon, K. LaMg$_2$NiH$_7$, a novel quaternary metal hydride containing tetrahedral [NiH$_4$]$^{4-}$ complexes and hydride anions. *J. Alloys Compd.* **2003**, *350*, 145–150. [CrossRef]

18. Rodewald, U.C.; Chevalier, B.; Pöttgen, R. Rare earth-transition metal-magnesium compounds—An overview. *J Solid State Chem.* **2007**, *180*, 1720–1736. [CrossRef]

19. Negri, S.D.; Giovannini, M.; Saccone, A. Constitutional properties of the La–Cu–Mg system at 400 °C. *J. Alloys Compd.* **2006**, *427*, 134–141. [CrossRef]

20. Chio, M.D.; Ziggiotti, A.; Baricco, M. Effect of microstructure on hydrogen absorption in LaMg$_2$Ni. *Intermetallics* **2008**, *16*, 102–106. [CrossRef]

21. Teresiak, A.; Uhlemann, M.; Thomas, J. Influence of Co and Pd on the formation of nanostructured LaMg$_2$Ni and its hydrogen reactivity. *J. Alloys Compd.* **2014**, *582*, 647–658. [CrossRef]

22. Pei, L.C.; Han, S.M.; Zhu, X.L. Effect of La hydride Compound on Hydriding Process of Mg$_2$Ni Phase in LaMg$_2$Ni Alloy. *Chin. J. Inorg. Chem.* **2012**, *28*, 1489–1494.

23. Zaluska, A.; Zaluski, L.; Ström-Olsen, J.O. Synergy of hydrogen sorption in ball-milled hydrides of Mg and Mg2Ni. *J. Alloys Compd.* **1999**, *289*, 197–206. [CrossRef]

24. Li, Z.N.; Jiang, L.J.; Liu, X.P. Dehydriding properities of Mg-20%(RE-Ni)(RE = La, Y, Mm) composites. *J. Chin. Rare Earth Soc.* **2008**, *26*, 624–628.

25. Zhang, H.G.; Lv, P.; Wang, Z.M. Effect of Mg content on structure, hydrogen storage properties and thermal stability of melt-spun Mg$_x$(LaNi$_3$)$_{100-x}$ alloys. *Int. J. Hydrog. Energy* **2014**, *39*, 9267–9275. [CrossRef]

Review

Dislocation-Free SiGe/Si Heterostructures

Francesco Montalenti [1],*, Fabrizio Rovaris [1] (iD), Roberto Bergamaschini [1] (iD), Leo Miglio [1], Marco Salvalaglio [2] (iD), Giovanni Isella [3], Fabio Isa [4,5,†] (iD) and Hans von Känel [4,5]

[1] L-NESS and Dipartimento di Scienza dei Materiali, Università di Milano-Bicocca, Via R. Cozzi 55, I-20125 Milano, Italy; f.rovaris1@campus.unimib.it (F.R.); roberto.bergamaschini@unimib.it (R.B.); leo.miglio@unimib.it (L.M.)

[2] Institute of Scientific Computing, Technische Universitat Dresden, 01062 Dresden, Germany; marco.salvalaglio@tu-dresden.de

[3] L-NESS and Dipartimento di Fisica, Politecnico di Milano, Via Anzani 42, I-22100 Como, Italy; giovanni.isella@polimi.it

[4] Laboratory for Solid State Physics, ETH Zürich, Otto-Stern-Weg 1, CH-8093 Zürich, Switzerland; fabio.isa@csiro.au (F.I.); vkaenel@solid.phys.ethz.ch (H.v.K.)

[5] Electron Microscopy Center Empa, Swiss Federal Laboratories for Materials Science and Technology, Überlandstrasse 129, CH-8600 Dübendorf, Switzerland

*   Correspondence: francesco.montalenti@unimib.it; Tel.: +39-02-6448-5226

†   Present address: CSIRO Manufacturing, 36 Bradfield Road, Lindfield, NSW 2070, Australia.

Received: 5 June 2018; Accepted: 16 June 2018; Published: 19 June 2018

Abstract: Ge vertical heterostructures grown on deeply-patterned Si(001) were first obtained in 2012 (C.V. Falub et al., *Science* **2012**, *335*, 1330–1334), immediately capturing attention due to the appealing possibility of growing micron-sized Ge crystals largely free of thermal stress and hosting dislocations only in a small fraction of their volume. Since then, considerable progress has been made in terms of extending the technique to several other systems, and of developing further strategies to lower the dislocation density. In this review, we shall mainly focus on the latter aspect, discussing in detail 100% dislocation-free, micron-sized vertical heterostructures obtained by exploiting compositional grading in the epitaxial crystals. Furthermore, we shall also analyze the role played by the shape of the pre-patterned substrate in directly influencing the dislocation distribution.

Keywords: heteroepitaxy; defects; semiconductors; elasticity; plasma-enhanced chemical vapour deposition

---

1. Introduction

Integration of materials with superior optical and/or electronic properties on Si [1] is extremely appealing as it leads to a wealth of new possible devices and applications while maintaining mainstream silicon technology. As both the lattice parameter and the elastic constants of the deposited material generally differ from the Si ones, some fundamental issues are encountered in terms of misfit- and thermal-stress fields unavoidably originated during growth and/or annealing. Misfit strain is typically relaxed via the introduction of a suitable network of dislocations which can thread through the whole film, deteriorating the performances of devices built on the heterostructure. Thermal stress, instead, can lead to cracking of thick films.

Reduction of the threading dislocation density (TDD) has been the subject of countless studies. A direct way for reducing the TDD is to grow thick films (see [2] and references therein), as this increases the probability that two opposite threading dislocations fall within their interaction cutoff, eventually annihilating each other [3]. The same effect can be obtained by extended annealing cycles, where thermal stress is exploited to promote dislocation motions. For Ge/Si systems, such strategies were demonstrated to lower the original TDD by several (depending on the film thickness) orders of magnitudes, saturating around the limit TDD $\approx 10^7/\text{cm}^2$. In order to lower the defect density by a further order of magnitude different approaches must be used. A key one was introduced by Fitzgerald and coworkers. They demonstrated [4,5] the possibility to lower typical TDDs in Ge/Si films down to only $\approx 10^6/cm^2$ by growing "graded layers", i.e., by actually depositing $Si_{1-x}Ge_x$ alloys with x gradually increasing during deposition. Grading allows for two main advantages with respect to constant-composition films: (a) threading arms are always subject to a nonzero gliding driving force [6] (b) the character of a dislocation can change during growth, threading segments bending and allowing for further strain relaxation without the need for nucleating new dislocations. While nowadays graded layers are still the main route to produce substrates with low dislocation densities, it is worth to emphasize that recent attempts have shown that $10^6/cm^2$ density values can be reached also by direct deposition at the desired final composition, provided that a suitable annealing and "etch back" procedure is exploited [7].

All the aforementioned techniques focused on deposition on unpatterned/unmasked wafers. In the last decade, considerable research has been instead devoted to defect control by suitable substrate design. These include "epitaxial necking" [8] and the similar "aspect ratio trapping" [9] techniques, in which the epitaxial material is selectively deposited in oxide mask windows and TDs are geometrically confined close to the heterointerface, "pendeoepitaxy" (leading to the formation of a suspended film starting from a suitable seed layer) [10], and 3D heteroepitaxy [11]. Here we shall focus on the latter, summarizing the main results obtained so far and focusing on last achievements and perspectives. In particular, in Section 2 we briefly review 3D heteroepitaxy obtained by depositing materials with constant composition, in Section 3 we discuss the key role played by grading, in Section 4 we analyze the influence of the shape of the pillar on the dislocation distribution. Finally, Section 5 is devoted to Conclusions and Perspectives.

2. Vertical Heterostructures with Constant Composition

2.1. Vertical Growth of Ge/Si by LEPECVD

3D heteroepitaxy [11] of constant-composition Ge/Si heterostructures was already reviewed in [12], so that here we shall only recall a few key results.

Two examples of Ge/Si heterostructures are reported in Figure 1a,c. In both cases, prior to deposition, a Si(001) substrate was deeply patterned, exploiting a Bosch process [13] resulting in ordered arrays of square-based $2 \times 2 \ \mu m^2$ Si pillars, separated by 2 and 4 μm trenches. Ge was subsequently deposited by Low-Energy Plasma-Enhanced Chemical Vapour Deposition (LEPECVD). LEPECVD [14] allows for the grow of crystalline Ge under strong out-of-equilibrium conditions, determined by the high deposition rate (≈ 4 nm/s for both structures in Figure 1) and by the low deposition temperature (440 °C in Figure 1a; 490 °C in Figure 1c). Under such conditions, typical diffusion lengths are much smaller than the micrometric pillar sizes, so that the Ge crystal has a tendency to grow vertically [11] from the very beginning. Some lateral enlargement also takes place, leading to a progressive shrinking of the lateral distance between crystals growing on adjacent pillars. This causes a strong self-shielding effect ultimately leading to (almost) perfect vertical growth. Once vertical growth is established, the crystals can be grown for several dozens of microns without ever touching, separated by a very small gap (Figure 1b). The vertical morphology offers two key advantages with respect to common 2D layers. On one hand, the free surface surrounding the crystals allows for very efficient relaxation of the thermal-stress field [15–18], therefore avoiding cracking. On the other hand, 60° dislocations forming at the Ge/Si interface and laying on (111) planes, are confined to the bottom of the crystal only (no 60° defect can reach the region located at a height h >1.4 b, where b is the Si pillar base width). While 60° dislocations are the dominant linear defects in Ge/Si systems grown under typical conditions, the LEPECVD out-of-equilibrium conditions lead to the formation of perfectly vertical defects [19,20] which can thread through the whole pillar, reaching the topmost surface as shown in Figure 2a,b. However, also these dislocations can be expelled laterally, provided that the top facet of the pillar is not parallel to the (001) substrate during growth. This can be easily achieved by raising the growth temperature [12], as shown in Figure 2c. Fortunately, indeed, linear defects tend to follow the growth front, so that the problem of vertical dislocations can be easily solved by properly tuning the growth conditions eliminating the top (001) facet.

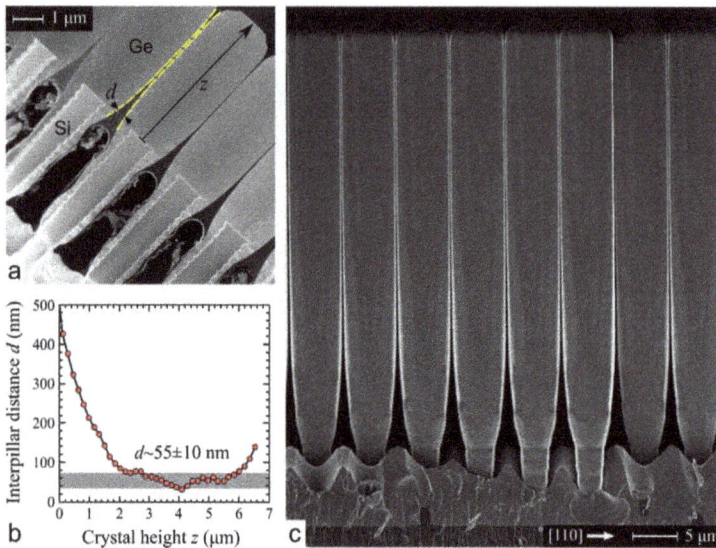

Figure 1. Self-aligned vertical growth of Ge micro-crystals on 2×2 μm^2 Si pillars, 8 μm deep. (a) Dark-field STEM view of 7 μm tall Ge crystals grown at 440 °C on pillars spaced by 2 μm. (b) Plot of the distance between the adjacent crystals in panel a. (c). SEM lateral view of 50 μm tall Ge crystals grown at 490 °C on pillars spaced by 4 μm. Reproduced with permission from [12].

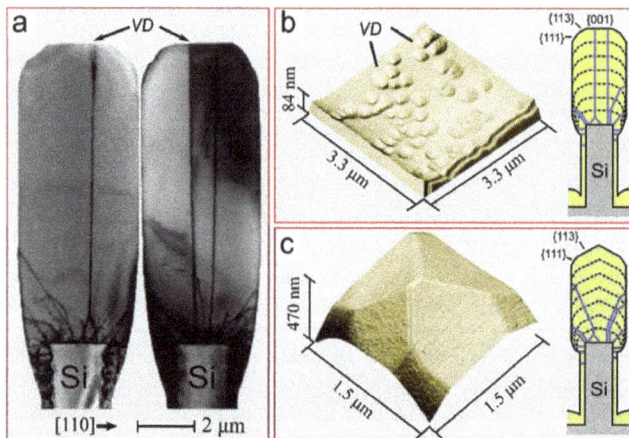

Figure 2. Dislocations in Ge micro-crystals on Si pillars. (a) Bright-field TEM cross-section in the Ge[220] Bragg condition showing both 60° and vertical dislocations. AFM view of the Ge crystal top after defect etching in iodine solution for (b) a (001) flat-top morphology obtained by growing at 440 °C and (c) a {113} pyramidal shape obtained at 560 °C. The schematics illustrate the different propagation of dislocation lines with respect to the faceting of the growth front. Reproduced with permission from [12].

3D LEPECVD growth of vertical heterostructures is not limited to pure Ge: upon properly tuning the growth conditions, analogous crystals have been obtained for Ge_xSi_{1-x} alloys of various Ge-content x. Actually, growth at low x is even more convenient in terms of lateral expulsion of defects, as the aforementioned vertical dislocations are not formed [20]. This is explicitly shown in Figure 3, where the etch-pit distribution clearly shows the lateral expulsion of defects.

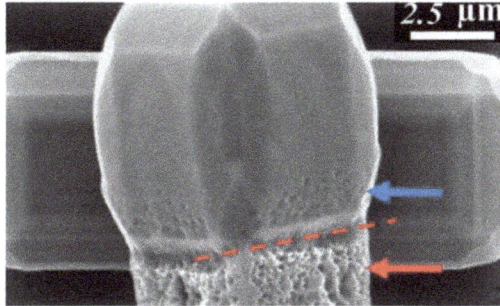

Figure 3. Lateral expulsion of dislocations. SEM image showing confinement of etch-pits in the bottom pillar region for a $Ge_{0.2}Si_{0.8}$/Si(001) crystal. Reproduced with permission from [20]. American Institute of Physics.

2.2. Other Deposition Techniques and Other Materials

The aim of this subsection is to show that (a) 3D heteroepitaxy is not achievable only by LEPECVD and that (b) a wide class of materials different from Ge/SiGe alloys can be grown in similar ways.

Let us first consider point (a). In Figure 4a a few representative snapshots of the morphology of a Ge crystal grown by Reduced-Pressure Chemical Vapour Deposition (RPVCD) on a 2×2 μm^2 Si pillar are reported [21]. As the typical growth conditions in RPCVD are very different from LEPECVD in both terms of flux (lower) and temperature (higher), longer diffusion lengths make vertical growth more difficult. To limit horizontal growth and material diffusing from the top regions to the pillar bottom, growth was therefore performed on pillars whose lateral walls were oxidized.

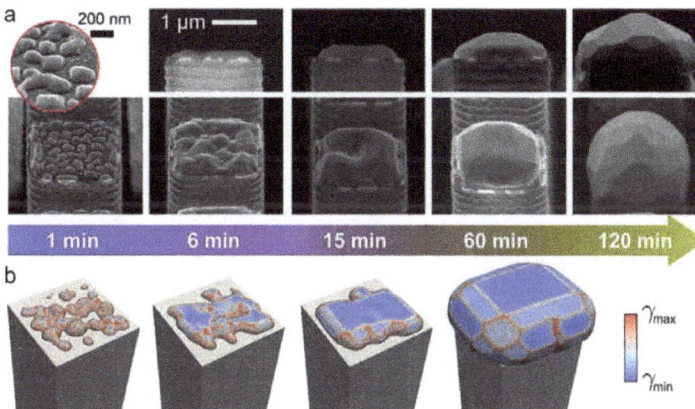

Figure 4. Growth sequence of a Ge micro-crystal on a 2×2 μm^2 Si pillar by RPCVD. (**a**) Lateral and perspective SEM views of samples after deposition of different duration. Multiple crystal seeds are observed at the early stages, coalescing into a single faceted structure later on. (**b**) Profiles obtained by a phase-field simulation of crystal growth, matching the experimental behavior. Reproduced with permission from [21]. American Chemical Society.

In Figure 4b, results of a phase-field [22] growth simulation are displayed. Material deposition is modeled in terms of condensation from the gas phase, according to the local chemical potential μ, including anisotropies in the surface energy γ to induce spontaneous faceting [23]. The nice agreement between the model and the experiments of Figure 4a allowed for a detailed analysis, reported in Ref. [21], of both thermodynamic and kinetic factors influencing the crystal morphology, confirming that growth by RPCVD is closer to equilibrium with respect to LEPECVD.

Let us now analyze the possibility to grow other materials. In Figure 5, the final morphology of SiC [24] (panel a), GaN [25] (panel b), GaAs [26,27] (panel c), and GaAs/Ge [28,29] (panel d) crystals grown on Si pillars is displayed. The various deposition techniques are listed in the caption. All such materials are extremely interesting for applications (in fields such as power electronics and optics), and the urge of lowering the typical defect density while attempting integration on Si is perhaps even more important than the already discussed Ge case.

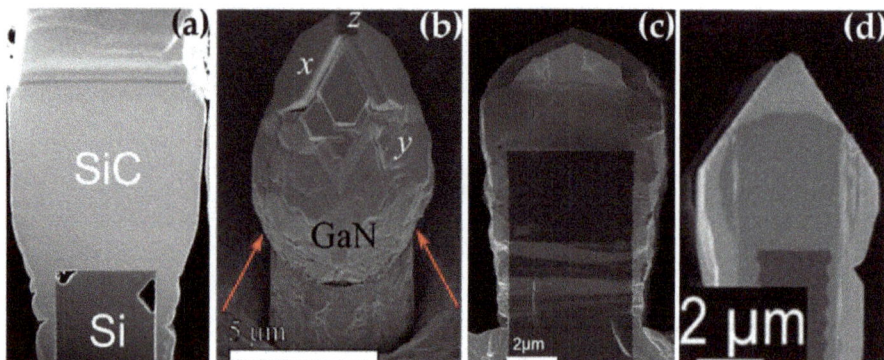

Figure 5. Deposition of several different materials on Si pillars. (**a**) SiC/Si, deposited by hot-wall Chemical Vapour Deposition. Reproduced with permission from [24]. Electrochemical Society (**b**) GaN/Si, deposited by plasma-assisted molecular beam epitaxy. Reproduced with permission from [25]. American Chemical Society (**c**) GaAs/Si, deposited by molecular beam epitaxy. Reproduced with permission from [26]. Copyright 2013, the American Institute of Physics (**d**) GaAs/Ge/Si, where Ge was deposited by LEPECVD and GaAs by metal-organic vapour-phase epitaxy. Reproduced with permission from [28]. American Institute of Physics.

Despite the successful growth of 3D crystals reported in Figure 5, it is important to emphasize that 3D heteroepitaxy of binary materials such as SiC or GaAs still demands for a significant effort in order to control the additional defects, such as stacking faults and/or anti-phase domains, which are typical of these systems while playing a lesser role in SiGe.

2.3. Suspended Films

We have pointed out that micron-size vertical crystals can be grown stress-free and without dislocations in the majority of their volume. These are surely appealing features from the point of view of applications. However, mainstream technology is mainly developed to handle planar 2D layers. With this in mind, researchers have also started to look at the possibility to form suspended films, formed by merging of adjacent vertical crystals. Note that if crystals are first vertically grown above the height needed to laterally expel dislocations (see Figure 3) and merging occurs only at the topmost regions, the suspended film would not inherit dislocations formed at the pillar/crystal interface.

The suspended film displayed in Figure 6 was obtained [30] by prolonged high-temperature annealing of vertical Ge crystals on Si pillars. Temporal snapshots of the evolution are also displayed, together with a corresponding continuum simulation [31] (panel b). The latter was performed exploiting a diffusion-equation approach, implemented in a phase-field framework [22], where the only contribution to the chemical potential μ (determining material flow) was the Mullins [32] term μ = −kγ, where k is the local curvature and γ the surface-energy density. Therefore, the almost perfect agreement between experiments and theory signals that the whole material redistribution leading to the merging process is determined solely by surface-curvature. The evolution displayed in Figure 6 was subsequently simulated [33] by also taking into account surface-energy anisotropy [23], yielding results even closer to the experimental ones.

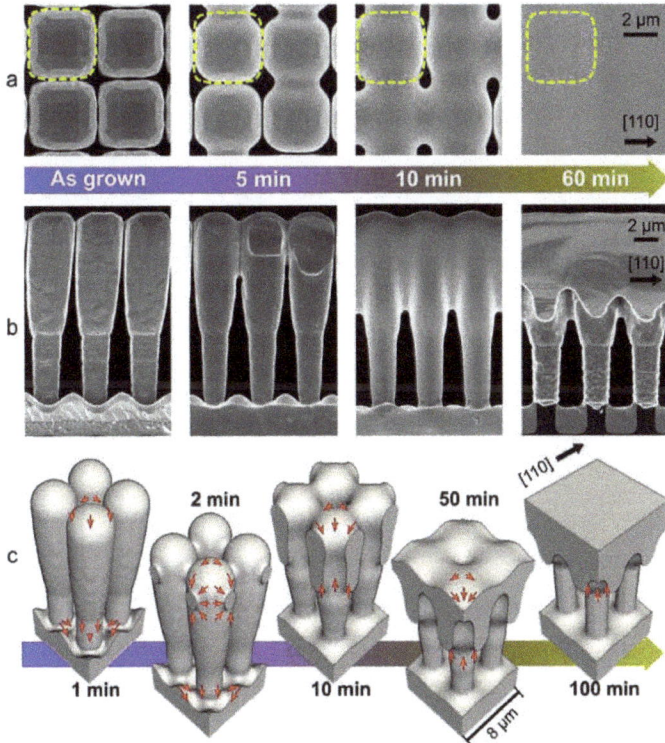

Figure 6. Coalescence of vertically-aligned Ge micro-crystals into a suspended film. (**a**) Top and (**b**) lateral SEM images of the as-grown crystals and of identical samples after annealing of different duration. (**c**) Evolution sequence obtained by a phase-field simulation of surface diffusion. First, local rounding of the facets occurs; then, connection bridges form between neighboring crystals leaving holes, that are finally filled by material flow. Reproduced with permission from [30]. American Chemical Society.

It has been further demonstrated that merging can also be obtained directly during growth (by raising the growth temperature) [21,31]. The suspended film is found [18] to still profit of the ability of the underlying pillars to release thermal strain by tilting. Despite displaying a good crystal quality, suspended films do display some defects, likely to be created during actual merging [21]. Present efforts are dedicated to minimizing such defects by adding further control on growth conditions and by exploiting fully dislocation-free micro-crystals using the technique described in the next Section.

3. Graded, Vertical Heterostructures

3.1. Elastic vs. Plastic Relaxation: Theory

Vertical heterostructures (VHEs) are characterized by the presence of free surfaces allowing for the partial elastic relaxation of in-plane strains/stresses. Indeed, at variance with heteroepitaxial films, these structures can expand/contract laterally to partially accommodate lattice misfits. This also leads to a significant compliance mechanism, i.e., a redistribution of the misfit strain between layers having a different lattice parameter. As a result, the tendency towards plastic relaxation is expected to be inhibited.

These properties of VHEs allow for dislocation-free structures, provided that the lateral size is smaller than critical values depending on the materials and on the misfit strain as discussed in [34]. In Ref. [35] these critical sizes were computed for the specific case of $Si_{1-x}Ge_x$/Si VHEs by means of a quasi-3D approach allowing for detailed estimation of thermodynamic plasticity onsets. This method exploits isotropic linear elasticity theory and Finite Element Method (FEM) calculations (see also [36]) while VHEs are modeled by simplified geometries as in Figure 7a. However, notice that it can be straightforwardly used to investigate realistic shapes as shown in Refs. [37–39].

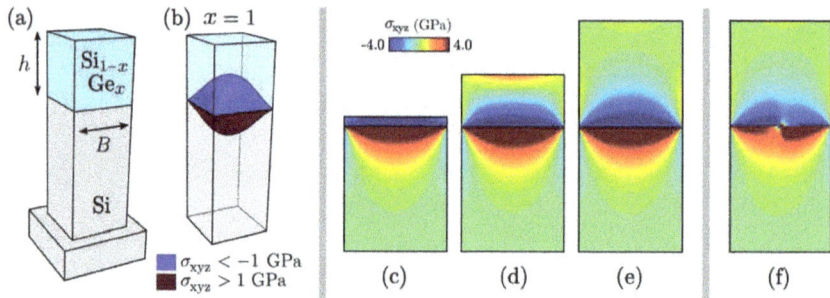

Figure 7. Theoretical modeling of VHEs and stress fields. (**a**) Simplified geometry of VHEs used to perform the theoretical analysis. (**b**) Illustrative map of the hydrostatic stress σ_{xyz} in a VHE as in panel (**a**) with x = 1. (**c**–**e**) σ_{xyz} in the central slice of a coherent 3D VHE as in panel (**a**,**b**) with an aspect-ratio R = h/B equal to 0.1, 0.5 and 1.0 respectively. (**f**) σ_{xyz} as in panel (**e**) superposed to the hydrostatic stress field induced by a dislocation lying at the interface (B = 25 nm). Reproduced with permission from [35]. American Institute of Physics.

Following the work in [35], the typical elastic field of VHEs as in Figure 7a can be computed by FEM calculations and it is illustrated in Figure 7b for a pure Ge epilayer by means of the hydrostatic stress σ_{xyz}. It consists of a compressive and a tensile region above and below the $Si_{1-x}Ge_x$/Si interface, respectively. The resulting elastic field is self-similar, i.e., its qualitative features depend on the height-to-base aspect-ratio R = h/B and it can be adapted to any specific size upon proper rescaling. Figure 7c–e shows the different distributions of σ_{xyz} in the central slice of a VHE for R = 0.1, R = 0.5, and R = 1.0, respectively.

The tendency towards plastic relaxation can be quantified by the formation energy, $\Delta E = E_{dislo} - E_{coh}$, computed in the 2D central slice of the full 3D geometry representing the VHE. E_{coh} is the elastic energy of the coherent system, i.e., without dislocations. E_{dislo} is the elastic energy accounting for the presence of a 60° dislocation at the interface, known to be always favored in $Si_{1-x}Ge_x$/Si systems with respect to other dislocation types [40]. In particular, the stress field of the coherent system is evaluated in the 3D structure, as in Figure 7b, and it is extracted in the central slice to compute E_{coh}. Then, the same elastic field is superimposed to the one of a straight (perpendicular to the 2D slice) 60° dislocation lying at the center of the $Si_{1-x}Ge_x$/Si interface, as in Figure 7f, to compute E_{dislo}. When $\Delta E < 0$, plastic relaxation is energetically favored, and dislocations are expected. The central position for the dislocation is the minimum energy configuration for relatively small and large values of R. For intermediate aspect-ratios, the minimum energy configuration may be shifted towards the sidewalls. However, the central position gives a very good approximation of the global energy minimum. Notice that computing ΔE in the central slice corresponds to evaluate an energy per unit length of the dislocation misfit segment. Therefore, this approach describes the tendency of a dislocation to elongate, resembling the classical method used to evaluate the critical thickness for the insertion of dislocation, h_c, in planar structures [5,41]. Further details about the method, its assessment, and the simulation setup can be found in Refs. [35,36].

With the approach described above, h_c can be calculated as a function of the Ge content in the $Si_{1-x}Ge_x$ epilayer, x, and the lateral size of the VHE, B. Moreover, it is possible to determine the critical lateral size as a function of x, namely $B_c(x)$. These quantities are shown in Figure 7: $h_c(B,x)$ is illustrated by means of dashed black isolines, while $B_c(x)$ corresponds to the solid red line. Above the red curve, $\Delta E > 0$ for thicknesses smaller than h_c, at which plasticity is expected to set in. Below the red curve, $\Delta E > 0$ for any thickness of the epilayer, i.e., the corresponding VHE is predicted to be always coherent.

For a fixed Ge content value x, we then expect a coherent structure when $B < B_c(x)$. VHEs with a larger B are predicted to have dislocations with a critical thickness that is larger than the corresponding one in a planar configuration, achieved in the $B \to \infty$ limit. Similar arguments apply for a targeted lateral size B. A VHE with a Ge content x smaller than the critical one as a function of the base, $x_c(B)$, is predicted to be coherent, while dislocations are expected for $x > x_c(B)$. From the results reported in Figure 8, such critical curves can be fitted as follows:

$$B_c(x) = \frac{\alpha}{x} + \frac{\beta}{x^2} + (B_c^{x=1} - \alpha - \beta) \tag{1}$$

with $\alpha = 55.2$, $\beta = 2.52$, $B_c^{x=1} = 35.6$ nm (the latter corresponding to the predicted critical base for a pure-Ge pillar epilayer) and

$$x_c(B) = \frac{B_c^{x=1}}{B}\alpha' + \frac{B_c^{x=1}}{B^2}\beta' + (1 - \alpha' - \beta') \tag{2}$$

with $\alpha' = 1.31$, $\beta' = -0.38$. Being these values based on a purely energetic balance, and due to the assumption of the model, they are expected to be quantitative lower bounds indicating the worst possible scenario to growth coherent VHEs.

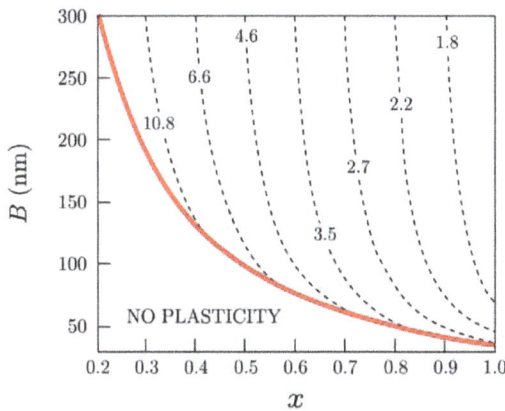

Figure 8. Model results. Critical pillar base B_c as a function of the Ge content x (solid red curve) and critical thickness of $Si_{1-x}Ge_x/Si$ VHEs as function of B and x (dashed black isolines, numbers correspond to h_c expressed in nm). Reproduced with permission from [35]. American Institute of Physics.

From a technological point of view, $B_c(x)$ and $x_c(B)$ set the limits for the realization of fully-coherent VHE. Let us focus on the target of a $Si_{1-x}Ge_x$ epilayer on Si with Ge content x_{epi}. With the configuration of VHEs as discussed so far, this can be achieved only up to a lateral size of $\sim B_c(x_{epi})$, that is less than 40 nm for a pure-Ge epilayer. However, looking at the elastic field of the VHEs in Figure 7, one can easily notice that for a thickness h $\sim$ B, i.e., for an aspect ratio R $\sim$ 1, the top of the epilayer is fully relaxed. Moreover, this is a purely geometric effect that is not dependent on B or x [34]. So that, even for basis larger than $B_c(x_{epi})$, one can think about growing a first coherent layer with $x_1 \leq x_c(B)$ up to h $\sim$ B. Then, as far as the top of the structure is relaxed, a second layer can be considered, and the vertical structure is predicted to be coherent provided that the increase in the Ge content with respect to the first layer is lower than $x_c(B)$, i.e., the second layer can have a Ge content x_2 such as $x_2 - x_1 \leq x_c(B)$. So that a coherent structure with a Ge content up to $2x_c(B)$ can be grown. If this Ge content is smaller then x_{epi}, further layers can be considered exploiting the same idea up to the desired target. The discrete number of layers $n(B)$ required to reach x_{epi} for a given B is then given by the simple relation $n(B) = x_{epi}/x_c(B)$.

This concept has been explicitly verified in Ref. [35] by checking the formation energy at different interfaces and also investigating in the detail the elastic relaxation of layers with different thicknesses. In experiments aiming at the growth of SiGe structures, a continuous grading of the Ge content is often performed. From the aforementioned results, and under the assumption of h $\sim$ B ideal layer, the grading rate $r(B)$ to achieve a fully coherent structure up to a Ge content x_{epi} can be easily calculated as

$$r(B) = \frac{x_{epi}}{n(B)B} = \frac{x_c(B)}{B} \tag{3}$$

where the second equal sign holds true when removing the ceiling function in $n(B)$ as the constraint of a discrete number of buffer layers is not required when speaking about the continuous grading rate. Notice that, the resulting expression of the grading rate is a function of the size but it is independent of x_{epi}. The calculations reported in Figure 8 are limited to structures with B $\leq$ 300 nm. However, by assuming that the equation for $B_c(x)$ and $x_c(B)$ are still valid for larger basis, recipes to grow fully-coherent VHEs can be provided for any size. In Table 1 the prediction about $x_c(B)$ and $r(B)$, as well as $n(B)$ and the total thickness computed as $t(B) = n(B)B$ to obtain pure Ge ($x_{epi} = 1$), are reported for some representative values of B (larger than B $\sim$ 100 nm for which $n(B) \sim 1$). It is worth mentioning that once a dislocation-free structure is obtained following the recipes in Table 1, any thickness of additional pure Ge can be deposited, still resulting in a dislocation-free structure. Indeed, the top of this structure is fully relaxed and made of pure Ge, so that further Ge deposition would result, *de facto*, in a homoepitaxial growth. However, notice that when approaching values of B in the order of ten microns, $t(B)$ become very large and it is practically unfeasible for applications.

Table 1. Critical parameters and design of fully-coherent VHEs as function of B. The values of critical Ge content $x_c(B)$, grading rate $r(B)$, number of layers $n(B)$ to achieve $x_{epi} = 1$ and corresponding total thickness $t(B)$ under the assumption of h $\sim$ B are reported.

B	$x_c(B)$	$r(B)$	$n(B)$	$t(B)$
150 nm	0.360	0.23%/nm	2	300 nm
200 nm	0.291	0.45%/nm	3	600 nm
300 nm	0.210	73.37%/μm	4	1.2 μm
500 nm	0.161	32.27%/μm	6	3.0 μm
1.0 μm	0.116	11.62%/μm	8	8.0 μm
1.5 μm	0.101	6.73%/μm	9	13.5 μm
2.0 μm	0.093	4.66%/μm	10	20.0 μm
3.0 μm	0.085	2.85%/μm	11	33.0 μm
5.0 μm	0.079	1.59%/μm	12	60.0 μm
7.5 μm	0.076	1.01%/μm	13	97.5 μm
10.0 μm	0.075	0.75%/μm	13	130.0 μm

3.2. Dislocation-Free Graded Heterostructures

The theoretical recipe for growing dislocation-free crystals (see Table 1) was experimentally demonstrated in Ref. [42] and further analyzed in Ref. [43]. The target Ge content was set to 40%, and a Ge grading rate GR = 1.5%/μm was employed. Two sets of samples were considered. In case 1 (2) the initial Si pillar was 2 μm (5 μm in case 2) wide, while the deposited SiGe crystal grew laterally reaching a final width of 5.5 μm (8.0 μm). According to the theoretical estimate reported in Table 1, case 1 should lead to dislocation-free crystals, at variance with case 2. This was exactly the case, as demonstrated in Figure 9.

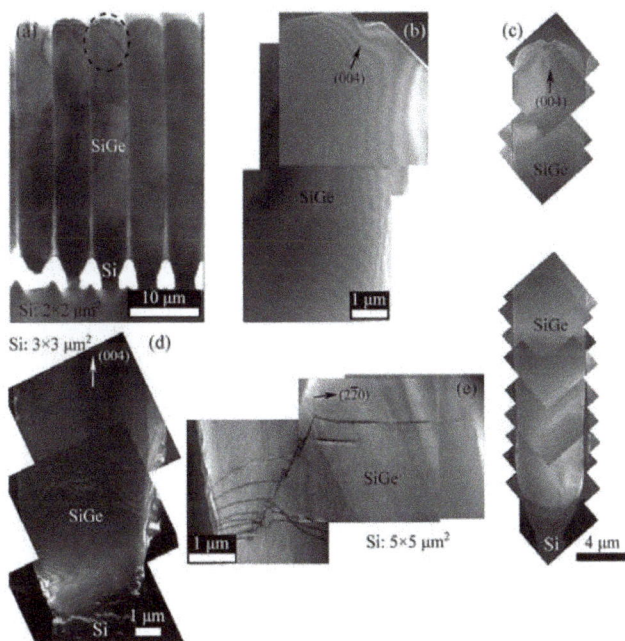

Figure 9. Graded vertical heterostructures. From Ref. [42]: TEM images of Dislocation-free (panels (**a**–**c**)) vs. plastically relaxed (**d,e**) SiGe crystals. Panel b reports the encircled region of panel (**a**) with a higher magnification. Reproduced with permission from [42]. Copyright 2016, Wiley.

The result displayed in Figure 9 is of particular importance as it directly demonstrates the possibility to grow micron-sized heterostructures completely free of dislocations, in spite of the large lattice mismatch. The misfit strain is only accommodated by lateral elastic relaxation. More recently, the same result was achieved also for higher lattice mismatches (up to 80% Ge content) [44]. In principle, there is no limitation in the maximum lattice mismatch that can be relaxed or in the width of the SiGe crystal. In practice, however, one is limited by the need to grow very tall crystals (Table 1). To overcome this problem, attempts were made to lower the tendency towards inserting dislocations by changing the shape of the Si pillars. This led to interesting and unexpected results, reported in the next Section.

4. Deposition on Under-Etched Pillars

4.1. Graded Heterostructures on Under-Etched Pillars: Experimental Results

As discussed in the previous section, vertical heterostructures can exploit lateral surfaces to relax the misfit strain also in the in-plane direction at variance with planar films. With the aim of increasing the compliance of the Si pillars, vertical Si structures (Figure 10a) were suitably under-etched (exploiting a two-step dry etching process [45]) prior to SiGe deposition, leading to the necked structures reported in Figure 10b. Graded vertical heterostructures where then grown on both "standard" (Figure 10d) and under-etched (Figure 10c) pillars of different width, using a grading rate GR = 1.5%Ge/μm and reaching a final 40% Ge content. The dislocation density was inferred by etch-pits counts, leading to the results displayed in Figure 11.

Figure 10. Growth on vertical and under-etched pillars. From Ref. [45]: SEM images of vertical (**a**) and under-etched (**b**) Si pillars. Cross-sectional SEM images of SiGe crystals grown on under-etched (**c**) and vertical (**d**) pillars. Reproduced with permission from [46]. American Physical Society.

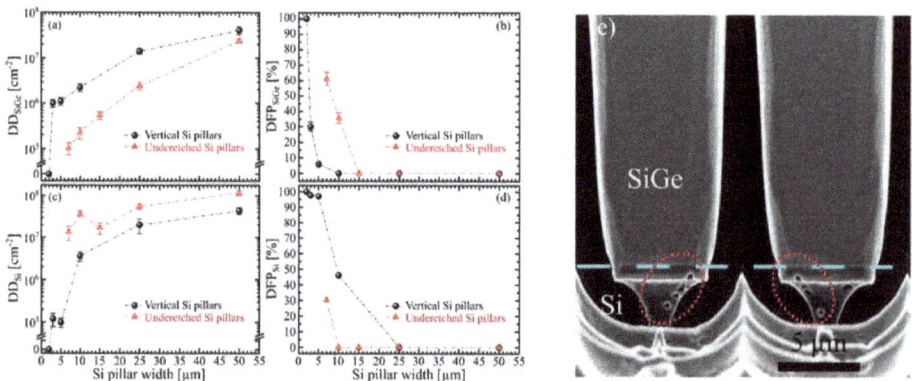

Figure 11. Experimental characterization of dislocations in SiGe crystals. (**a**) Average dislocation density in SiGe crystals (DDSiGe) deposited on vertical (black spheres) and under-etched (red triangles) Si pillars with different widths. (**b**) Probability of having dislocation-free SiGe crystals (DFPSiGe) as a function of the vertical (black spheres) and under-etched (red triangles) Si pillars width. (**c**,**d**) Analogous to (**a**,**b**), respectively, but for dislocations located in the Si pillars. The strong tendency towards dislocations piling-up in the Si region is well evident in the SEM image reported in panel (**e**). Reproduced with permission from [46]. American Physical Society.

Measurements of the dislocation density (DD) and the dislocation-free probability (DFP) are reported in Figure 11, distinguishing between estimate relative to the actual SiGe crystal (panel a and b) and to the Si pillar (panel c and d). Results clearly show the key role played by morphology. In the presence of the "neck" (under-etched morphology), indeed, a large fraction of defects is transferred from the SiGe crystal to the underlying Si pillar, the effect being present also for large pillar widths. The effect is nicely evidences by the SEM image in panel (e), where etch pits are clearly distinguished. Notice that also for a base as large of 10 μm, a considerable fraction of the crystals grown on under-etched pillars is dislocation free, as evidenced by the value of the DFP reported in Figure 11b.

Summarizing the results discussed so far, by using a GR of 1.5%Ge/μm one can grow fully dislocation-free heterostructures, where "fully" means that defects are not present in the deposited crystal as well as in the pillar, up to to ~5 μm. If one is willing to grow larger vertical heterostructures, then suitable under-etching allows for enlarging the critical width to 10 μm. In this case, however, only the crystal is dislocation-free, while defects are introduced in the Si pillar. Still, this could be still very appealing for applications exploiting only the SiGe crystal as active region. In the following the theoretical justification of the effect produced by introducing substrate necking is discussed.

4.2. Graded Heterostructures on Under-Etched Pillars: Theoretical Interpretation

The role played by necking was theoretically investigated in Ref. [46]. Here we briefly summarize the main findings. In Figure 12 we report the hydrostatic stress maps computed by solving isotropic elasticity theory by finite element methods, simulating the same grading rate used in the experiments (GR = 1.5%Ge/μm).

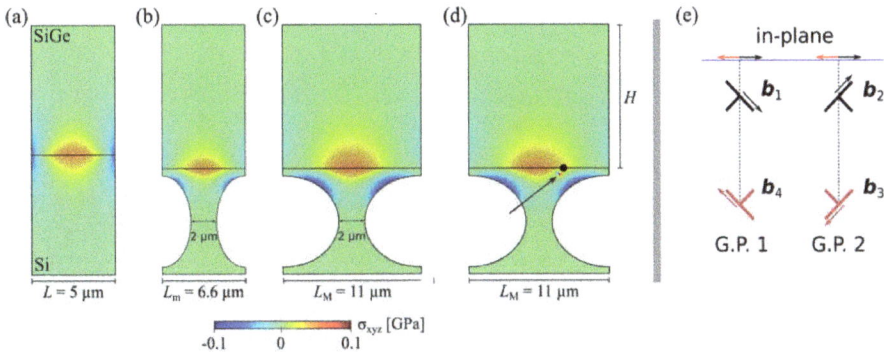

Figure 12. Model results. Hydrostatic stress maps (σxyz) and considered geometry for the vertical (**a**) and under-etched (**b**), (**c**) Si pillars. The SiGe/Si interface is marked with a black line. Two considered pillar bases L_m (**b**) and L_M (**c**) are taken to mimic the extreme values of bases measured on the tapered geometry of the grown pillar (Figure 9c). In (**d**) is reported the stress map for an under-etched pillar once the first dislocation is introduced. In panel (**e**) a schematic representation of the four possible Burgers vectors is reported along with their in-plane projection. Reproduced with permission from [46]. American Physical Society.

In Figure 12a, a standard "vertical" pillar morphology is considered, while in all other panels the effect of under-etching was taken into account. As in Section 3.1, the pillar morphology is simplified: instead of considering the enlargement of the width (Figure 10c) caused by deposition, we fix it and repeat calculations for the lowest (L_m) and largest (L_M) dimensions. Figure 12 shows that necking changes the local misfit relaxation both in Si and in SiGe, introducing a strongly compressed region (blue color in the maps) within Si. This has a profound influence on the nucleation and distribution of dislocations.

In SiGe/Si systems, misfit is typically relaxed by 60° dislocations. In the 2D model used to generate all results discussed in this Section, the dislocation line is assumed to run in the third direction (perpendicular to the Figure, in Figure 12), and there exist two possible glide planes (G.P. 1 and G.P. 2 in Figure 12e) and four Burgers vectors, called $\mathbf{b}_1$, $\mathbf{b}_2$, $\mathbf{b}_3$, and $\mathbf{b}_4$ in Figure 12e. Two of them ($\mathbf{b}_1$ and $\mathbf{b}_2$) provide expansion of the region above the core, the others have the opposite effect. These are the ones more often encountered in SiGe/Si planar films or in vertical pillars, as the tensile strain introduced by the dislocations reduces the lattice compression due to lattice-parameter misfit. The presence of the strong compressive stress in under-etched Si pillars (Figure 12b–d), instead, reverses the sign of the lowest-energy defects. This is shown in Figure 13: insertion of the same dislocations relaxing SiGe/Si films or SiGe on vertical pillars ("normal" case in Figure 13) raises the energy of the system for both explored sizes. On the contrary, the introduction of dislocations with opposite sign of the in-plane component of the Burgers vector ($\mathbf{b}_3$, or $\mathbf{b}_4$, helping to relax compression in the Si region) becomes energetically favored beyond a critical height. The difference in energy between the system with and without a dislocation located at its minimum-energy position (ΔE_{min}), indeed, becomes negative. In Ref. [46] it was shown by dislocation dynamics simulations that accounting for the change in Burgers vector orientation is fundamental in order to explain the typical dislocation distributions experimentally observed in SiGe crystals grown on underetched Si pillars. The typical pile-up in the Si region, well evident in Figure 11e, is indeed compatible only with dislocations removing half atomic plane in the Si region (Burgers vector $\mathbf{b}_3$).

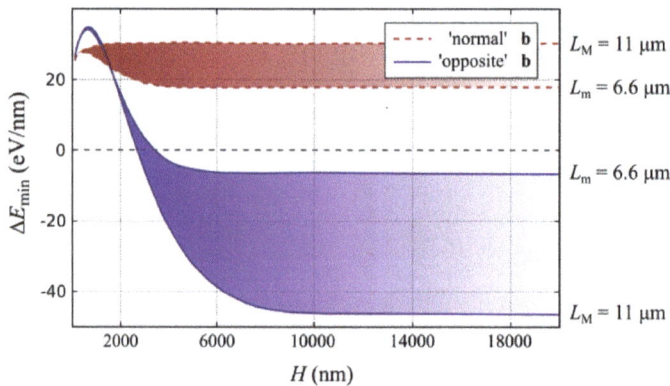

Figure 13. Model results. Energy gain for the introduction of the first 'normal' (red-dashed curve) or 'opposite' (purple solid curve) dislocations in under-etched pillars with base L with respect to the pillar height H. The formation energy is negative only for dislocations with the 'opposite' Burgers vector **b**. Reproduced with permission from [46]. American Physical Society.

5. Conclusions and Perspectives

In this paper we reviewed a five-years-long investigation on the 3D growth of Ge, GeSi and Ge-graded crystals on deeply patterned substrates, where proper deposition conditions and patterning geometry allows for a dense array of vertical structures. This high aspect ratio strategy, in turn, exploits the termination of misfit dislocations on lateral sidewalls, the accommodation of the thermal strain, and even the elimination of plastic relaxation by fully elastic misfit-strain relaxation, in case of graded deposition. The important result is that, whenever a device application requires a dense array of Ge, or GeSi, individual crystals, such as the case of X-ray detectors, or infrared single-photon-avalanche-detectors (SPAD), this proof-of-concept is ready to be transferred to the stage of the industrial prototype. In fact, for both the applications mentioned above are currently the subject of projects devoted to increase the technology readiness level.

However, more widespread applications may arise if the concept is extended to other heteroepitaxial binary materials, such as the cubic SiC for power electronics, GaN and GaAs for optoelectronics. In Section 2.2 we have reviewed a set of promising, preliminary attempts to grow such materials. However, a convincing proof-of-concept is still missing, due to the fact that other extended defects, such as stacking faults and twins, or anti-phase domains may occur, and that the extended lateral sidewalls may act as a source of them. To this end, a merging of the individual crystals into a continuous suspended layer should be eventually obtained, taking the risk that in the merging process additional dislocations, or other extended defects, may originate, as it has been shown for the case of Ge, probably because of some crystal tilting. The efforts and the timescale required for such an improvement are probably much larger than the ones devoted in obtaining defect-free Ge crystals.

Still, a lot of new scientific knowledge is likely to be produced when exploring the features of 3D heteroepitaxy of binary materials on deeply patterned substrates, and vertical growth by selected area depositions. In our opinion, the morphologies arising from tuning the competitive growth of neighboring facets by suitable deposition conditions may result in exciting and, possibly, promising issue to extend this Ge-based technique to other materials.

6. Patents

H. von Känel and L. Miglio, Dislocation and stress management by mask-less processes using substrate patterning and methods for device fabrication, PCT/IB2011/000895".

Funding: This research was funded by Regione Lombardia, under the TEINVEIN project, Call Accordi per la Ricerca e l'Innovazione, co-funded by POR FESR 2014-2020 (ID: 242092), by the European Union (Horizon-2020 FET "μ-Spire" project, ID: 766955 and CHALLENGE project, ID: 720827, NMBP-02-2016), by the Swiss National Science Foundation (Sinergia Project NOVIPIX CRSII2 147639), and by the Alexander von Humboldt Foundation (Postdoctoral Research Fellowship awarded to M.S.).

Conflicts of Interest: The authors declare no conflict of interest.

References

1. Kazior, T.E. Beyond CMOS: Heterogeneous integration of III–V devices, RF MEMS and other dissimilar materials/devices with Si CMOS to create intelligent microsystems. *Philos. Trans. R. Soc. A* **2014**, *372*. [CrossRef] [PubMed]
2. Wang, G.; Loo, R.; Simoen, G.; Souriau, L.; Caymax, M.; Heyns, M.M.; Blanpain, P. A model of threading dislocation density in strain-relaxed Ge and GaAs epitaxial films on Si(100). *Appl. Phys. Lett.* **2009**, *94*, 102115. [CrossRef]
3. Speck, J.S.; Brewer, M.A.; Beltz, G. Scaling laws for the reduction of threading dislocation densities in homogeneous buffer layers. *J. Appl. Phys.* **1996**, *80*, 3808. [CrossRef]
4. Fitzgerald, E.A.; Xie, Y.H.; Green, M.L.; Brasen, D.; Kortan, A.R.; Michel, J.; Mii, Y.J.; Weir, B.E. Totally relaxed Ge$_x$Si$_{1-x}$ layers with low threading dislocation densities grown on Si substrates. *Appl. Phys. Lett.* **1991**, *59*, 811. [CrossRef]
5. Fitzgerald, E.A. Dislocations in strained-layer epitaxy: Theory, experiment, and applications. *Mater. Sci. Rep.* **1991**, *7*, 87–142. [CrossRef]
6. Tersoff, J. Dislocations and strain relief in compositionally graded layers. *Appl. Phys. Lett.* **1993**, *62*, 693. [CrossRef]
7. Yamamoto, Y.; Kozlowski, G.; Zaumseil, P.; Tillack, B. Low threading dislocation Ge on Si by combining deposition and etching. *Thin Solid Films* **2012**, *520*, 3216–3221. [CrossRef]
8. Langdo, T.A.; Leitz, C.W.; Currie, M.T.; Fitzgerald, E.A. High quality Ge on Si by epitaxial necking. *Appl. Phys. Lett.* **2000**, *76*, 3700. [CrossRef]
9. Park, J.-S.; Curtin, M.; Adekore, B.; Carroll, M.; Lochtefeld, A. Defect reduction of selective Ge epitaxy in trenches on Si(001) substrates using aspect ratio trapping. *Appl. Phys. Lett.* **2007**, *90*, 052133. [CrossRef]
10. Linthicum, K.; Gehrke, T.; Thomson, D.; Carlson, E.; Rajagopal, P.; Smith, T.; Batchelor, D.; Davis, R. Pendeoepitaxy of gallium nitride thin films. *Appl. Phys. Lett.* **1999**, *75*, 196. [CrossRef]

11. Falub, C.V.; von Känel, H.; Isa, F.; Bergamaschini, R.; Marzegalli, A.; Chrastina, D.; Isella, G.; Müller, E.; Niedermann, P.; Miglio, L. Scaling hetero-epitaxy from layers to three-dimensional crystals. *Science* **2012**, *335*, 1330–1334. [CrossRef] [PubMed]

12. Bergamaschini, R.; Isa, F.; Falub, C.V.; Niedermann, P.; Müller, E.; Isella, G.; von Känel, H.; Miglio, L. Self-aligned Ge and SiGe three-dimensional epitaxy on dense Si pillar arrays. *Surf. Sci. Rep.* **2013**, *68*, 390–417. [CrossRef]

13. Laermer, F.; Schilp, A. Method of Anisotropically Etching Silicon. U.S. Patent No. 5501893, 26 March 1996.

14. Rosenblad, C.; von Känel, H.; Kummer, M.; Dommann, A.; Müller, E. A plasma process for ultrafast deposition of SiGe graded buffer layers. *Appl. Phys. Lett.* **2000**, *76*, 427. [CrossRef]

15. Falub, C.V.; Meduňa, M.; Chrastina, D.; Isa, F.; Marzegalli, A.; Kreiliger, T.; Taboada, A.G.; Isella, G.; Miglio, L.; Dommann, A.; et al. Perfect crystals grown from imperfect interfaces. *Sci. Rep.* **2013**, *3*, 2276. [CrossRef] [PubMed]

16. Meduna, M.; Falub, C.V.; Isa, F.; Marzegalli, A.; Chrastina, D.; Isella, G.; Miglio, L.; Dommann, A.; von Känel, H. Lattice bending in three-dimensional Ge microcrystals studied by X-ray nanodiffraction and modelling. *J. Appl. Cryst.* **2016**, *49*, 976–986. [CrossRef]

17. Meduna, M.; Isa, F.; Jung, A.; Marzegalli, A.; Albani, M.; Isella, G.; Zwelacker, K.; Miglio, L.; von Känel, H. Lattice tilt and strain mapped by X-ray scanning nanodiffraction in compositionally graded SiGe/Si microcrystals. *J. Appl. Cryst.* **2018**, *51*, 368–385. [CrossRef]

18. Marzegalli, A.; Cortinovis, A.; Basset, F.B.; Bonera, E.; Pezzoli, F.; Scaccabarozzi, A.; Isa, F.; Isella, G.; Zaumseil, P.; Capellini, G.; et al. Exceptional thermal strain reduction by a tilting pillar architecture: Suspended Ge layers on Si(001). *Mater. Des.* **2017**, *116*, 144–151. [CrossRef]

19. Marzegalli, A.; Isa, F.; Groiss, H.; Müller, E.; Falub, C.V.; Taboada, A.G.; Niedermann, P.; Isella, G.; Schaeffler, F.; Montalenti, F.; et al. Unexpected Dominance of Vertical Dislocations in High-Misfit Ge/Si(001) Films and Their Elimination by Deep Substrate Patterning. *Adv. Mater.* **2013**, *25*, 4408–4412. [CrossRef] [PubMed]

20. Isa, F.; Marzegalli, A.; Taboada, A.G.; Falub, C.V.; Isella, G.; Montalenti, F.; von Känel, H.; Miglio, L. Onset of vertical threading dislocations in $Si_{1−x}Ge_x$/Si(001) at a critical Ge concentration. *APL Mater.* **2013**, *1*, 052109. [CrossRef]

21. Skibitzki, O.; Capellini, G.; Yamamoto, Y.; Zaumseil, P.; Schubert, M.A.; Schroeder, T.; Ballabio, A.; Bergamaschini, R.; Salvalaglio, M.; Miglio, L.; et al. Reduced-Pressure Chemical Vapor Deposition Growth of Isolated Ge Crystals and Suspended Layers on Micrometric Si Pillars. *ACS Appl. Mater. Interfaces* **2016**, *8*, 26374–26380. [CrossRef] [PubMed]

22. Li, B.; Lowengrub, J.; Ratz, A.; Voigt, A. Geometric evolution laws for thin crystalline films: Modeling and numeric. *Commun. Comput. Phys.* **2009**, *6*, 433.

23. Salvalaglio, M.; Backofen, R.; Bergamaschini, R.; Montalenti, F.; Voigt, A. Faceting of Equilibrium and Metastable Nanostructures: A Phase-Field Model of Surface Diffusion Tackling Realistic Shapes. *Cryst. Growth Des.* **2015**, *15*, 2787–2794. [CrossRef]

24. Von Känel, H.; Isa, F.; Falub, C.V.; Barthazy, E.J.; Müller, E.; Chrastina, D.; Isella, G.; Kreiliger, T.; Taboada, A.G.; Meduna, M.; et al. Three-dimensional Epitaxial $Si_{1−x}Ge_x$, Ge and SiC Crystals on Deeply Patterned Si Substrates. *ECS Trans.* **2014**, *64*, 631–648. [CrossRef]

25. Isa, F.; Cheze, C.; Siekacz, M.; Hauswald, C.; Laehnemann, J.; Fernandez-Garrido, S.; Kreliger, T.; Ramsteiner, M.; Arroyo Rojas Da Silva, A.; Brandt, O.; Isella, G.; et al. Integration of GaN Crystals on Micropatterned Si(0 0 1) Substrates by Plasma-Assisted Molecular Beam Epitaxy. *Cryst. Growth Des.* **2015**, *15*, 4886–4892. [CrossRef]

26. Bietti, S.; Scaccabarozzi, A.; Frigeri, C.; Bollani, M.; Bonera, E.; Falub, C.V.; von Känel, H.; Miglio, L.; Sanguinetti, S. Monolithic integration of optical grade GaAs on Si (001) substrates deeply patterned at a micron scale. *Appl. Phys. Lett.* **2013**, *103*, 262106. [CrossRef]

27. Bergamaschini, R.; Bietti, S.; Castellano, A.; Frigeri, C.; Falub, C.V.; Scaccabarozzi, A.; Bollani, M.; von Känel, H.; Miglio, L.; Sanguinetti, S. Kinetic growth mode of epitaxial GaAs on Si(001) micro-pillars. *J. Appl. Phys.* **2016**, *120*, 245702. [CrossRef]

28. Taboada, A.G.; Kreiliger, T.; Falub, C.V.; Isa, F.; Salvalaglio, M.; Wevior, L.; Fuster, D.; Richter, M.; Uccelli, E.; Niedermann, P.; et al. Strain relaxation of GaAs/Ge crystals on patterned Si substrates. *Appl. Phys. Lett.* **2014**, *104*, 022112. [CrossRef]

29. Taboada, A.G.; Meduna, M.; Salvalaglio, M.; Isa, F.; Kreiliger, T.; Falub, C.V.; Barthazy Meier, E.; Müller, E.; Miglio, L.; Isella, G.; et al. GaAs/Ge crystals grown on Si substrates patterned down to the micron scale. *J. Appl. Phys.* **2016**, *119*, 055301. [CrossRef]

30. Salvalaglio, M.; Bergamaschini, R.; Isa, F.; Scaccabarozzi, A.; Isella, G.; Backofen, R.; Voigt, A.; Montalenti, F.; Capellini, G.; Schroeder, T.; et al. Engineered Coalescence by Annealing 3D Ge Microstructures into High-Quality Suspended Layers on Si. *ACS Appl. Mater. Interfaces* **2015**, *7*, 19219–19225. [CrossRef] [PubMed]

31. Bergamaschini, R.; Salvalaglio, M.; Backofen, R.; Voigt, A.; Montalenti, F. Continuum modelling of semiconductor heteroepitaxy: An applied perspective. *Adv. Phys. X* **2016**, *1*, 331–367. [CrossRef]

32. Mullins, W.W. Theory of Thermal Grooving. *J. Appl. Phys.* **1957**, *28*, 333. [CrossRef]

33. Salvalaglio, M.; Bergamaschini, R.; Backofen, R.; Voigt, A.; Montalenti, F.; Miglio, L. Phase-field simulations of faceted Ge/Si-crystal arrays, merging into a suspended film. *Appl. Surf. Sci.* **2017**, *391*, 33–38. [CrossRef]

34. Glas, F. Critical dimensions for the plastic relaxation of strained axial heterostructures in free-standing nanowires. *Phys. Rev. B* **2006**, *74*, 121302. [CrossRef]

35. Salvalaglio, M.; Montalenti, F. Fine control of plastic and elastic relaxation in Ge/Si vertical heterostructures. *J. Appl. Phys.* **2014**, *116*, 104306. [CrossRef]

36. Gatti, R.; Marzegalli, A.; Zinovyev, V.A.; Montalenti, F.; Miglio, L. Modeling the plastic relaxation onset in realistic SiGe islands on Si(001). *Phys. Rev. B* **2008**, *78*, 184104. [CrossRef]

37. Montalenti, F.; Salvalaglio, M.; Marzegalli, A.; Zaumseil, P.; Capellini, G.; Schülli, T.U.; Schubert, M.A.; Yamamoto, Y.; Tillack, B.; Schroeder, T. Fully coherent growth of Ge on free-standing Si(001) nanomesas. *Phys. Rev. B* **2014**, *89*, 014101. [CrossRef]

38. Scarpellini, D.; Somaschini, C.; Fedorov, A.; Bietti, S.; Frigeri, C.; Grillo, V.; Esposito, L.; Salvalaglio, M.; Marzegalli, A.; Montalenti, F.; et al. InAs/GaAs Sharply Defined Axial Heterostructures in Self-Assisted Nanowires. *Nano Lett.* **2015**, *15*, 3677–3683. [CrossRef] [PubMed]

39. Niu, G.; Capellini, G.; Lupina, G.; Niermann, T.; Salvalaglio, M.; Marzegalli, A.; Schubert, M.A.; Zaumseil, P.; Krause, H.-M.; Skibitzki, O.; et al. Photodetection in Hybrid Single-Layer Graphene/Fully Coherent Germanium Island Nanostructures Selectively Grown on Silicon Nanotip Patterns. *ACS Appl. Mater. Interfaces* **2016**, *8*, 2017–2026. [CrossRef] [PubMed]

40. Hirth, J.P.; Lothe, J. *Theory of Dislocations*; Krieger Publishing Company: Malabar, FL, USA, 1992; ISBN 0894646176.

41. Matthews, J.W.; Blakeslee, A.E. Defects in epitaxial multilayers. *J. Cryst. Growth* **1974**, *27*, 118–125. [CrossRef]

42. Isa, F.; Salvalaglio, M.; Arroyo Rojas Dasilva, Y.; Meduna, M.; Barget, M.; Jung, A.; Kreiliger, T.; Isella, G.; Erni, R.; Pezzoli, F.; et al. Highly Mismatched, Dislocation-Free SiGe/Si Heterostructures. *Adv. Mater.* **2016**, *28*, 884–888. [CrossRef] [PubMed]

43. Isa, F.; Salvalaglio, M.; Arroyo Rojas Dasilva, Y.; Jung, A.; Isella, G.; Erni, R.; Niedermann, P.; Groening, P.; Montalenti, F.; von Känel, H. From plastic to elastic stress relaxation in highly mismatched SiGe/Si heterostructures. *Acta Mater.* **2016**, *114*, 97–105. [CrossRef]

44. Isa, F.; Jung, A.; Salvalaglio, M.; Arroyo Rojas Dasilva, Y.; Marozau, I.; Meduna, M.; Barget, M.; Marzegalli, M.; Isella, G.; Erni, R.; et al. Strain Engineering in Highly Mismatched SiGe/Si Heterostructures. *Mater. Sci. Semicond. Process.* **2017**, *70*, 117–122. [CrossRef]

45. Isa, F.; Salvalaglio, M.; Arroyo Rojas Dasilva, Y.; Jung, A.; Isella, G.; Erni, R.; Timotijevic, B.; Niedermann, P.; Groening, P.; Montalenti, F.; et al. Enhancing elastic stress relaxation in SiGe/Si heterostructures by Si pillar necking. *Appl. Phys. Lett.* **2016**, *109*, 182112. [CrossRef]

46. Rovaris, F.; Isa, F.; Gatti, R.; Jung, A.; Isella, G.; Montalenti, F.; von Känel, H. Three-dimensional SiGe/Si heterostructures: Switching the dislocation sign by substrate under-etching. *Phys. Rev. Mater.* **2017**, *1*, 073602. [CrossRef]

crystals

MDPI

Review

Influence of Dislocations in Transition Metal Oxides on Selected Physical and Chemical Properties

Kristof Szot [1,2,3,*], **Christian Rodenbücher** [1,2] , **Gustav Bihlmayer** [2,4] , **Wolfgang Speier** [2] , **Ryo Ishikawa** [5] , **Naoya Shibata** [5] and **Yuichi Ikuhara** [5]

1   Peter Grünberg Institute (PGI-7) Forschungszentrum Jülich, 52425 Jülich, Germany; c.rodenbuecher@fz-juelich.de
2   JARA-FIT, Forschungszentrum Jülich, 52425 Jülich, Germany; g.bihlmayer@fz-juelich.de (G.B.); w.speier@fz-juelich.de (W.S.)
3   Institute of Physics, University of Silesia, 40-007 Katowice, Poland
4   Peter Grünberg Institute (PGI-1) Forschungszentrum Jülich, 52425 Jülich, Germany
5   Institute of Engineering Innovation, School of Engineering, The University of Tokyo, Tokyo 113-8656, Japan; ishikawa@sigma.t.u-tokyo.ac.jp (R.I.); shibata@sigma.t.u-tokyo.ac.jp (N.S.); ikuhara@sigma.t.u-tokyo.ac.jp (Y.I.)
*   Correspondence: k.szot@fz-juelich.de

Received: 6 March 2018; Accepted: 27 May 2018; Published: 4 June 2018

Abstract: Studies on dislocations in prototypic binary and ternary oxides (here TiO_2 and $SrTiO_3$) using modern TEM and scanning probe microscopy (SPM) techniques, combined with classical etch pits methods, are reviewed. Our review focuses on the important role of dislocations in the insulator-to-metal transition and for redox processes, which can be preferentially induced along dislocations using chemical and electrical gradients. It is surprising that, independently of the growth techniques, the density of dislocations in the surface layers of both prototypical oxides is high ($10^9/cm^2$ for epipolished surfaces and up to $10^{12}/cm^2$ for the rough surface). The TEM and locally-conducting atomic force microscopy (LCAFM) measurements show that the dislocations create a network with the character of a hierarchical tree. The distribution of the dislocations in the plane of the surface is, in principle, inhomogeneous, namely a strong tendency for the bundling and creation of arrays or bands in the crystallographic <100> and <110> directions can be observed. The analysis of the core of dislocations using scanning transmission electron microscopy (STEM) techniques (such as EDX with atomic resolution, electron-energy loss spectroscopy (EELS)) shows unequivocally that the core of dislocations possesses a different crystallographic structure, electronic structure and chemical composition relative to the matrix. Because the Burgers vector of dislocations is *per se* invariant, the network of dislocations (with additional d^1 electrons) causes an electrical short-circuit of the matrix. This behavior is confirmed by LCAFM measurements for the stoichiometric crystals, moreover a similar dominant role of dislocations in channeling of the current after thermal reduction of the crystals or during resistive switching can be observed. In our opinion, the easy transformation of the chemical composition of the surface layers of both model oxides should be associated with the high concentration of extended defects in this region. Another important insight for the analysis of the physical properties in real oxide crystals (matrix + dislocations) comes from the studies of the nucleation of dislocations via in situ STEM indentation, namely that the dislocations can be simply nucleated under mechanical stimulus and can be easily moved at room temperature.

Keywords: dislocations; TiO_2; $SrTiO_3$; STEM; EELS; ChemiSTEM; SPM; etch pits; electrical properties; mechanical properties; resistive switching

1. Introduction

The challenge of an extensive treatise about the influence of dislocations in oxides on the physical and chemical properties in form of a short review is, without doubt, demanding. For instance, the analysis of the modification of the mechanical properties (especially plasticity) in multinary oxides as an effect of the existence of a high density of dislocations in crystalline or polycrystalline oxides can be found in many articles (see, e.g., [1–46] and references therein). Similarly, the literature about the role of filaments and their possible relation to dislocations in the resistive switching phenomena in oxides is continuously growing (e.g., [30–33,47–59] and their references). Therefore, it is necessary to select a group of oxides and a list of properties which could be presented in such a review, by taking into account the actual state-of-the-art in this subject. In our paper, we will focus on two prototypical oxide materials as representative model systems for the binary and ternary oxides with early transition metals, here TiO_2 and $SrTiO_3$. For our study of the physical and chemical properties we will furthermore concentrate the attention on those properties which can only be understood on the basis of a comprehensive knowledge of the electronic, chemical, and crystallographic structure of the core of dislocations as derived from modern microscopic techniques such as high-resolution transmission electron microscopy (HRTEM, with spherical aberration correction), electron-energy loss spectroscopy (EELS), scanning transmission electron microscopy with chemical resolution (ChemiSTEM) studies and scanning probe microscopy (SPM) analysis, such as atomic force microscopy (AFM), Kelvin-probe force microscopy (KPFM), piezo force microscopy (PFM), or locally-conducting AFM (LCAFM). However, that this does not mean that we will neglect the other more classical techniques as, for example, etch pits technique or X-ray topography in our presentation. The microscopic analysis, which has become available during recent years, especially by HRTEM (e.g., [4–13,29,33,46,60–65]) and SPM analysis (e.g., [53,55,66–68]), allows for us to underpin our hypothesis that the dislocations in band-insulating transition metal oxides take the role of semi-conducting or metallic nano-wires (e.g., [3,13,32,40,50,51,55,68]). We place special emphasis on this problem, namely that the properties of the local stoichiometry and specifically the valence of the transition metal cations close to the core of dislocations is responsible for the kind of specific electric properties of the dislocations encountered in these materials and that allows for considering the dislocation lines as a kind of nano-wire extending throughout the crystal. In fact, it is the specific electronic structure of the core of dislocations in connection with the invariance of the Burges vector of dislocations which generates a kind of macroscopically relevant "entity" in the real oxide crystals, here the perfect matrix with a network of short-circuit dislocations. We will also show that the dislocations can be electrically addressed and individually modified by external means as well as macroscopically aligned in these materials. A further important topic of our paper is connected with existence of an enhanced concentration of dislocations in the surface layer of TiO_2 and $SrTiO_3$ oxides which gives rise to the ability of chemical and physical transformations of the surface layer at high temperature for different oxygen partial pressures ([68] and references therein). Finally, we also briefly discuss the plastic properties of the oxides in presence of dislocations. Of course, it is self-evident that the variety of properties of both model oxides, which can be potentially modified by dislocations, is clearly broader than the mentioned points of our list.

2. Edge Dislocations in TiO_2 and $SrTiO_3$

From a topological point of view, edge dislocations, screw dislocations and dislocation loops belong to the class of "1"-dimensional extended defects. In the classical presentation of edge dislocations (see Figure 1) they should be imagined as the end of an extra half plane which is shifted into the perfect matrix [1,21,25–27,32,34,37]. The end of the half plane defines, in fact, the position of the dislocation line. This representation can be used without modification for a schematic visualization of the distribution of atoms along the dislocations in TiO_2. In contrast to the classical presentation of dislocation lines in TiO_2, dislocation lines in case of a ternary oxide, such as $SrTiO_3$, should be represented by the end of two half planes (namely TiO_2 and SrO), which have been shifted into the matrix of SrTiO$_3$ (Figure 2) [39,69]. In the other situation with a shift of only one half plane (i.e., TiO$_2$ or SrO), the configuration corresponds to the generation of a stacking fault, and the dislocation in its character will be partial [65,68].

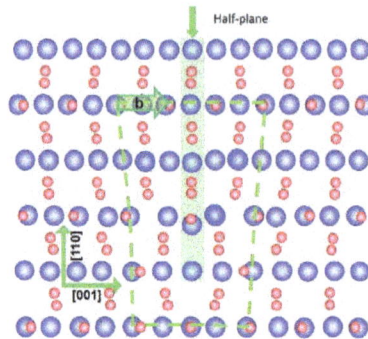

Figure 1. Relaxed atomic model of an edge dislocation in TiO$_2$ with Burgers vector b = ¼ [001] viewed from the [1$\bar{1}$0] projection. The model is constructed based on experimental images. The larger balls are Ti atoms and the smaller ones are O atoms (Adapted from [32]).

For rutile crystals, which at room temperature possess a tetragonal structure, the orientation of the edge dislocation can be realized in different directions. Because each of the edge dislocations with defined Burgers vector can lie in different planes (so-called slip system) we can observe different combinations between the mentioned crystallographic directions of the dislocations and the plane, e.g., {101}<101>, {110}<001> [70,71], (101){010}, (111){101}, and (100){101} [37].

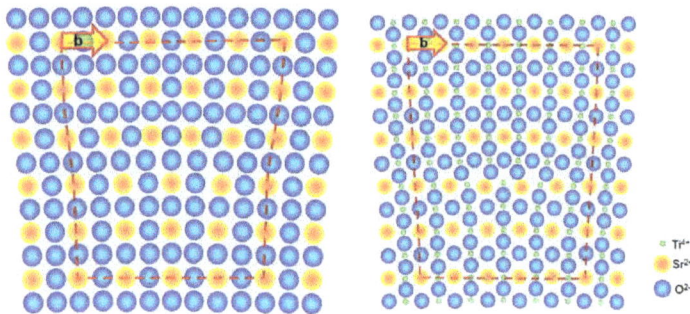

Figure 2. Atomistic models for edge dislocations in the SrTiO$_3$ perovskite for the <100>{010} (**left**) and <100>{011} (**right**) slip systems. The indicated Burgers vector is b = <100>. Adapted from [72].

For the SrTiO$_3$ crystal, which has a cubic structure (Pm3m) above 110 K, it has been observed that the edge dislocations can, in principle, be arranged with Burgers vector in direction <100>, <110>, and <111>. Those directions, in combination with the appropriate plane, give the following possibilities of slip systems for SrTiO$_3$: <100>{110}, <001>{110}, <011>{100}, and <010>{100} [73].

The mentioned set of the dislocation directions does not describe all of the possibilities concerning the crystallographic orientation of dislocation lines and assignment of the slip system. It is possible that in SrTiO$_3$ crystals, dislocations can be generated with higher index, for example [012] [74]. In this case, the dislocation core is more complicated than for dislocations with a low index, that is, the structure should be analyzed as an overlap of many partial dislocations (with lower indexes) and other non-invariant extended defects such as stacking faults (see Figure 3).

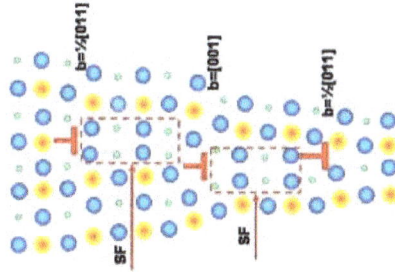

Figure 3. Schematic illustration of the dissociation structure of the [012] dislocation in SrTiO$_3$: configuration of the partial dislocations at the boundary. The dashed lines indicate stacking faults (SF). The color coding follows Figure 2. Adapted from [74].

3. Arrangement of Dislocations

Dislocations do not only exist as individual objects which extend throughout the crystalline material, but the dislocations in real systems can accumulate in bundles, line arrays, bands as well as in three-dimensional networks [75]. One aspect of the arrangement of dislocations is the dislocation-dislocation interaction and, in the oxide materials considered here, the issue of the electrostatics of dislocations and their surroundings comes into play. In the sense of Kröger–Vink defect chemistry, charged defects should interact with other charged defects because the crystal as a whole has to fulfill the electroneutrality condition demanding that the sum of all charged defects such as electrons, holes, and vacancies has to be zero in steady state [70,76–89]. Because some of the d^0 Ti states along the dislocations in TiO$_2$ and SrTiO$_3$ are transformed into d^1 configuration, such dislocations with a surplus of d electrons cannot be electrically neutral (e.g., [31]). Hence, a screening of the electrical charge which has accumulated along the core of the dislocations is needed. Therefore, such screening leads to the creation of a space charge zone, the extension of which depends (in the first approximation) on the concentration of the charge (the mentioned concentration of d^1 electrons) in the core of the dislocations, the concentration of the charged defects in the matrix and its dielectric properties. [31,32,90–95]. This screening, in fact, induces a kind of electrostatic repulsion forces between similarly charged dislocations. Such Coulomb repulsion can hinder a potential accumulation, for instance, a bundling of dislocations. Of course, it is possible that in case of oppositely charged cores of dislocations the interaction can be attractive and an annihilation of dislocations takes place [96,97]. Notice: for the analyses of the relaxed configuration of the dislocations in the matrix (that means with a fixed position) of ionic or ionic covalent oxides one should not only take into account the minimum of the electrostatic energy but also the minimization of the total free energy of the crystal, which in this case means the minimization of the mechanical energy (here the stress). Therefore, when the density of generated and annihilated dislocations does not change (at constant temperature and constant internal or external stress), $\dot{\rho}_+ = \dot{\rho}_-$ (ρ is the density of dislocations with opposite sign), the total density of dislocations is given by:

$$\rho(\varepsilon) = \rho_0 + \int_0^\varepsilon (\dot{\rho}_+ - \dot{\rho}_-)\dot{\varepsilon}^{-1}d\varepsilon$$

and the deformation rate $\dot{\varepsilon}$ should be constant. This classical description for the generation of the edge dislocations in metal or single oxides (binary oxides) cannot be one-to-one transferred on the situation in SrTiO$_3$, where the dislocations can possess a different chemical core (see e.g., [39]) which *per se* induces different local stress fields, and the interaction of edge dislocations with the same Burgers vector, cannot be simply interpreted [98].

Despite this electrostatic "inadequacy", one can clearly see, based on empirical findings by different techniques, that the edge dislocations in case of SrTiO$_3$ and TiO$_2$ are often found to be accumulated in bundles, line arrays, bands and in three dimensional networks of dislocations (see, e.g., [10,42,99–104]). All processes which can lead to new arrangements of dislocations in the matrix are connected with two basic possibilities to move the dislocations in the matrix. The first one is a conservative moving (gliding) of dislocations, which occurs when, by a change of the position of the dislocations in the crystalline structure, its Burgers vector and the dislocations line belong to the same plane. The other possibility for the moving of dislocations (so-called non-conservative motion or climbing) occurs when the mentioned condition for gliding is not fulfilled. This standard description of the conservative and non-conservative motion of dislocations is, in fact, valid for metals and nonmetals, but for ionic oxides there is an additional "hidden dimension" of the motion, especially for climbing, namely a complicated non-synchronous evolution (on the time scale) of the positions of ions or charged vacancies (see Figure 4). Because the diffusion of oxygen vacancies is easier than that of cation vacancies, this process cannot be realized in only one step. The detailed scenario of all steps of the climbing of the dislocations in this electrical complicated surrounding has been described by Hirel et al. [105].

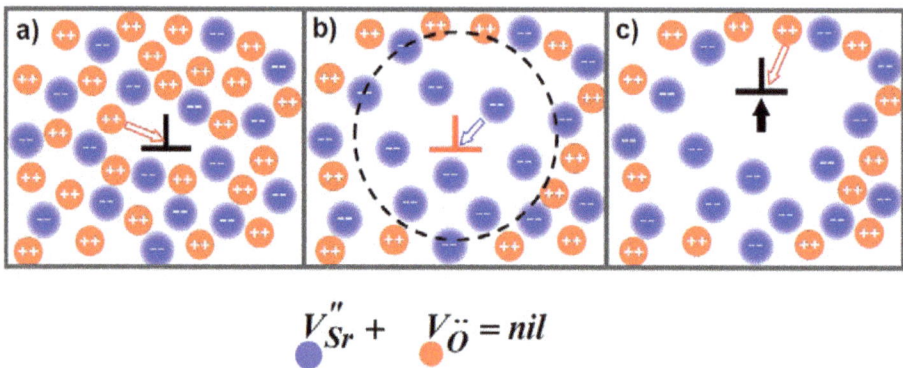

$$V''_{Sr} + V\ddot{}_O = nil$$

Figure 4. Illustration of dislocation climb in SrTiO$_3$ crystal according to a scenario from Hirel et al. [106]. In this scenario, (**a**) the core of dislocation is originally in a neutral state, i.e., both points of gravity of the positively charged and negatively charged cloud of vacancies overlap. Because positively-charged oxygen vacancies migrate faster than cation vacancies (here V$_{Sr}$) towards the dislocation, the dislocation becomes charged; (**b**) After absorbing neighboring oxygen vacancies, this leads to a positive charging of the dislocation (red) and development of a oppositely charged space-charge region (dashed line), where only cation vacancies remain and form the so-called Debye–Hückel cloud. In this situation, cation vacancies do migrate slowly towards the dislocation core; (**c**) After absorbing cation vacancies, the dislocation then climbs and becomes again attractive to oxygen vacancies. Adapted from [106].

The motion of dislocations in oxide crystals cannot only be analyzed under electrical (electrostatic) aspects, but also gliding or climbing mechanisms should be discussed in terms of strain and temperature. This set of thermodynamic parameters can, for example, radically change the character of the gliding of dislocations in SrTiO$_3$ crystals. At room temperature (RT) easy gliding of dislocations in SrTiO$_3$ allows for dramatic plastic deformation (in fact, the crystal can be bent by bare hands [107]). For this temperature, it should be accepted that the coordinated gliding of dislocations (as an origin of the significant plastic deformation) occurs without a change of the core structure (see Figure 5A). In contrast, at higher temperature the ductility of the crystal is lower and the crystal will become brittle [16,33,108,109]. This is connected with the transformation of segments of dislocations into climb-dissociated configuration, which is responsible for pinning and eventually for slowing down the motion of these linear defects (see Figure 5B). For high temperatures, which allow an interaction between atoms (ions) from the core of dislocations with point defects in the matrix, the moving (here the gliding) cannot *per se* be conservative (without change of the atomic structure of dislocations), and in this temperature range a non-conservative climbing will dominate (see Figure 5C). Notice that, for SrTiO$_3$ and TiO$_2$ it was shown that the oxygen occupancy in the core of dislocations at high temperatures under reducing conditions can be decreased (see, e.g., [110,111]).

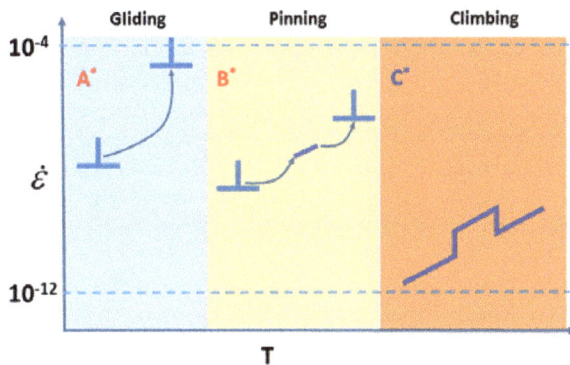

Figure 5. Schematic illustration of the expected behavior of <110>$_{pc}$ dislocations in perovskite materials as a function of temperature, T, and strain rate $\dot{\varepsilon}$. Thin blue lines represent the dislocation in its glide-dissociated configuration. Adapted from [105].

4. Annihilations or Multiplications of Dislocations

The simplest configuration for the annihilation of two dislocations is given by the dipole configuration with the opposite sign. In the case of annihilation of the dipoles from two neighboring half planes, which are aligned close together in the matrix, a perfect plane in the matrix will be created. Mechanical or thermal stress can induce such annihilation processes [27,112–114]. In fact, for rutile crystals (as grown) a reduction of the density of dislocations from 4×10^5 per cm^2 to about 1×10^5 per cm^2 was observed after thermal treatment at 1375 °C for 45 min [112]. Similar influence on the density of dislocations, namely the small lowering of their density [44,110], is known for a thermal treatment for SrTiO$_3$ crystals. In principle, the reduction of the density of the edge dislocations in the matrix of the crystal can be realized by gliding or climbing of dislocations to the free surface. This process causes a shifting of the half plane to the surface and thus generates an additional step on the surface. Therefore, for the verification of the often-postulated extremal annihilation of the dislocations (that means a high reduction of the density of dislocations) in TiO$_2$ or SrTiO$_3$ (via thermal treatment at high temperature around 1000 °C [105,112,113,115]), the difference in the morphology of the surface before and after the thermal treatment should be checked. Only in the case of the unequivocal increase of the step concentration in the plane of the surface one would be able to postulate that the reduction of dislocations density via thermal treatment has actually occurred. However, such an increase of the step density is generally not observed, if the chemical composition of the surface layer stays unchanged. At this point it should be noted that, under reducing and oxidizing conditions at high temperatures, changes of the crystal geometry of the surface layer can be induced, due to modifications of the stoichiometry. This has been found by X-ray and TEM investigations showing the evolution of Magnéli phases or low Ti oxides of reduced TiO$_2$ [53,116,117] and reduced SrTiO$_3$ or BaTiO$_3$ [118–121] or the creation of Ruddlesden Popper phases under oxidizing/reducing conditions, or after electrical polarization at room temperature [118,119,122]. Those effects can dramatically modify the height of the steps on the surface (see. e.g., [120,123]).

Thermal treatment does not significantly change the density of dislocations, if the stoichiometric of the surface layers of TiO$_2$ or SrTiO$_3$ is maintained. At first sight, this information seems to be in conflict with the statement in the literature and experimental confirmation (using TEM technique) that the density of dislocations can be significantly reduced (by about two orders of magnitude) via thermal treatment. In the part of our review which is dedicated to the identification of the exit of dislocations with the etch pits technique, we provide an answer to the origin of this discrepancy, namely that the distribution of the dislocations in the crystal after thermal treatment relative to the statistical distribution of the dislocations after polishing can be changed due to the reduction of the free energy of the crystal. The total elastic strain energy of the same number of dislocations in the random configuration in the matrix give a higher contribution to the total energy of the crystal than the same amount of dislocations agglomerated in bundles, lines or bands (of course, in this case one should bear in mind that the other opposite reaction can occur, e.g., electrostatic repulsion for the potential approach of dislocations with the same sign). Therefore, the density of dislocations for the thermally treated crystal cannot be checked using only a simple TEM inspection of one lamella, but for a representative (statistically correct) study of the potential reduction of the density of dislocations via TEM measurements foils for many dozen positions in the plane of the investigated surface layer need to be prepared.

On the other hand, the density of dislocations in the TiO_2 and $SrTiO_3$ crystals can be increased (multiplied). Although the native concentration of dislocations in the bulk after growth can vary from $10^5/cm^2$ (for crystals which have been grown using Czochralski method) to $10^7/cm^2$ (for crystals which have been produced with Verneuil technique, see, e.g., [68]) the real concentration of dislocations using different treatment can be extremely multiplied, nearly to the rendering amorphous limit, that means to about $10^{13}/cm^2$. This increase of the density of dislocations can be generated via mechanical polishing, cleaving, scraping, axial or hydrostatic pressure, electrostatic pressure (e.g., which accompanies electrodegradation or breakdown processes), thermal gradient, or redox processes at high temperature (see, e.g., [68,101] and references therein). This last formulation about potential multiplication of dislocations at high temperature during reduction concerns the creation of so-called hairpin dislocations, which eventually allows for the transformation of TiO_{2-x} into Magnéli phases [124].

5. Experimental Techniques for the Investigation of Dislocations in Real Oxides Crystals

5.1. Etch Pits Technique (Optical, SEM and AFM Investigation)

Inspection of the exit of the edge dislocations in the plane of the free surface of the crystal can be obtained using an appropriate etchant. Although the connotation "appropriate" does not imply a negative meaning, it should be noted that the search for appropriate etchant has been classified as "a black art" of science. In fact, it is very difficult to define the chemical composition of the etchant for the selective decoration of the dislocations solely on the basis of theoretical considerations. Since 1953, when F. J. Vogel et al. [125] presented for the first time an unequivocal correlation between the position of dislocations and the position of the etch pits, the etch techniques has been established as a simple and popular method for the investigation of the density of dislocations, for their character (e.g., edge or spiral dislocations), crystallographic orientation and their arrangement (see, e.g., [126]). The most complicated problem which is connected with creation of etch pits concerns the nature of the selective dissolution of the surface region close to the core of dislocations. The best introduction in this problem of etch pits formation can be found in the book by Sangwal [127] who analyzed different models such as: kinematic theories, thermodynamic theories, diffusion theories, and topochemical absorption theories. Our short review cannot concentrate on a detailed analysis of each step of the etchant interaction with the core of dislocations and their surroundings. Therefore, we will only mention here that there is no plausible theory for TiO_2 or $SrTiO_3$ available which tries to correlate the creation of the nuclei (by dissolutions in etchant) caused by the preferential etching of dislocations with the available data from HRTEM techniques about of the local electronic and crystallographic structure. The important statement for this part of our review is however that etching of both model oxides does actually work!

Etching of TiO_2 crystals can be realized using different etchants: HF, H_3PO_4, KOH, $KHSO_4$, NaOH in a temperature regime between RT and 400 °C [28,128–132]. The typical wet etching with HF acid leads to the following chemical reaction, which in its character allows for preferentially decorating the exits of the dislocations on the surface of TiO_2 [(100) or (110)] [132]:

Reaction:

$$TiO_{2(s)} + 6HF_{(sol)} \rightarrow TiF^{2-}{}_{6(sol)} + 2H^+{}_{(sol)} + 2H_2O_{(sol)}$$

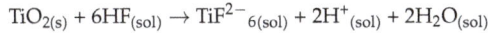

The dissolution of TiO_2 in molten KOH can be described in terms of a Lux–Flood transformation (TiO_2 in contact with OH^- ions can be fragmented into simple titanate ions) and soluble acid titanate, here $K_{2m}Ti_nO_{2n+m}$ will be formed after intercalation with K^+ cations, being known as stable crystal phases in the $K_2O\text{-}TiO_2$ system [133,134].

Reaction:

$$TiO_2 + 2\,KOH \rightarrow K_2TiO_3 + H_2O$$

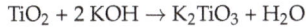

However, a simple optical inspection underestimates the density of dislocations in the surface layer of TiO_2 crystal (Figure 6). In contrast, the topographic measurement using AFM shows that the real concentration of etch pits is two orders of magnitude higher than for the macroscopic (here, optical) measurement. A macroscopic inspection of the etched surface reveals different distribution of dislocations such as statistical distribution (Figure 7a) or linear distributed bundles along scratches (Figure 7b). The nano-investigation (using AFM) of the etched surface indicates that the shape of the etch pits can be round and ellipsoidal [68,128]. Notice that the AFM data show a clear tendency towards bundling of dislocations.

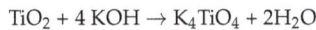

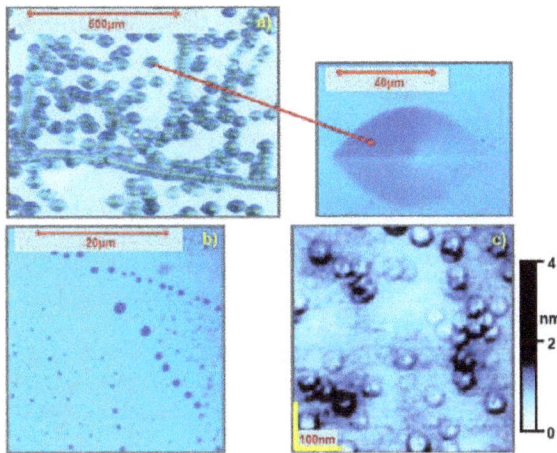

$$TiO_2 + 4\,KOH \rightarrow K_4TiO_4 + 2H_2O$$

Figure 6. Two examples of etch pits characterized with optical inspection: (**a**) of the surface of a TiO_2 (110) etched in molten KOH (400 °C, 10 min), magnification shows a typical shape of individual etch pits; and (**b**) of a TiO_2 (100) surface etched using buffered HF. In the latter case, a much higher density of dislocations is visible due to the fact that the HF-buffered etching is less effective than using KOH etchant, which gives an indication of different dislocation densities with respect to etching depth; (**c**) atomic force microscopy (AFM)-topographic picture of etch pits (with similar shapes) on TiO_2 (110) surface (for the decoration of the exit of dislocation only a short etching using HF was applied, adapted from [68]). This gives to support to a tendency of agglomeration or so-called bundling of dislocations. The determined density of dislocation is here 1.5×10^{10} per cm^2.

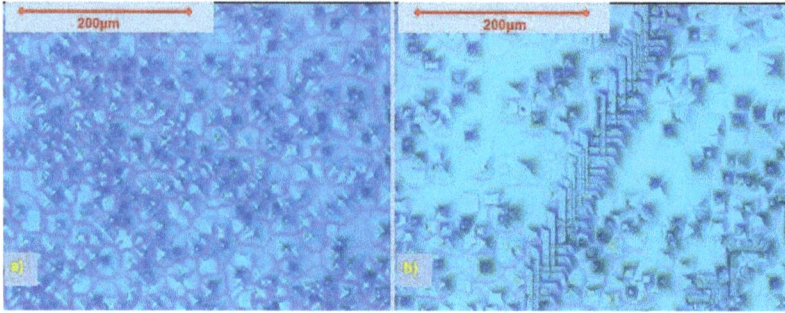

Figure 7. SrTiO$_3$ crystal, example of the distribution of the etch pits in the plane of (100) for the central region of the crystal (**a**). Photography (**b**) shows an agglomeration of dislocations along both orthogonal directions <100>.

The best results of a selective decoration of the end of dislocation lines for SrTiO$_3$ crystals can be reached using HCl, HF or HNO$_3$ etchants for short etching at 80–90 °C [44,45,68,135–137]:

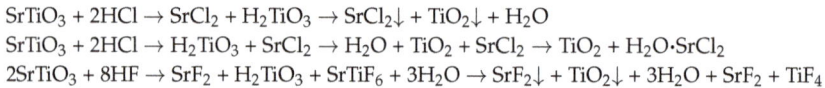

$$SrTiO_3 + 2HCl \rightarrow SrCl_2 + H_2TiO_3 \rightarrow SrCl_2\downarrow + TiO_2\downarrow + H_2O$$
$$SrTiO_3 + 2HCl \rightarrow H_2TiO_3 + SrCl_2 \rightarrow H_2O + TiO_2 + SrCl_2 \rightarrow TiO_2 + H_2O{\cdot}SrCl_2$$
$$2SrTiO_3 + 8HF \rightarrow SrF_2 + H_2TiO_3 + SrTiF_6 + 3H_2O \rightarrow SrF_2\downarrow + TiO_2\downarrow + 3H_2O + SrF_2 + TiF_4$$

In the last step of this chemical treatment the crystal should be flushed with deionized water and for stopping the leaching the water should be bonded with, e.g., methanol. It should be realized that for the etching of SrTiO$_3$ using HCl the chemical attack is in principle limited to the dislocation lines which contain a surplus of SrO. Therefore, the shape of the etch pits is round for short etching time and in fact will create a kind of pipe (see Figure 8); this shape can be easily identified by topographic investigation of the etched surface using AFM. The etching using HF produces etch pits with typical shapes of inverted pyramids (see Figure 9). The density of dislocations (for the first stage of etching), which can be calculated on the basis AFM measurements, is about 6×10^9/cm^2. An important conclusion which can be derived from an AFM study concerns the distribution of dislocations in-plane and out-of-plane, namely the fact that the dislocations can bundle and accumulate in <100> or <010> directions. The dramatic change of the length of segments of dislocations (analysis of the depth of the etch pits, see Figure 8) suggests that the dislocations out-of-plane are connected in a network which has a similar character than a hierarchical tree or hierarchical structure of dislocations [10,45,68]. This hierarchical structure is characterized by a high concentration of dislocations in the upper part of the surface layer and a decrease of the density towards the bulk. Further evidence of such an arrangement in a network with hierarchical structure (or alternatively hierarchical tree) is given by TEM studies out-of-plane (see Section 5.3.3), the analysis of the density of etch pits and their shapes as a function of etching time (Figure 10) and the observation of the electrical conductivity (out-of-plane) of the reduced TiO$_2$ and SrTiO$_3$ crystal (see below, Section 7). In the upper part of the surface layer the density of dislocations determined by TEM measurement (see e.g., [10]) is similar to the concentration of etch pits after short etching (see e.g., [45,68]). Because for both model oxides (TiO$_2$ and SrTiO$_3$) we were able to correlate the position of the etch pits and the position of the conducting filaments (see, e.g., [51,68]) we are convinced that the character of the change of the density of dislocations in surface region (as a function of the distance from the actual surface) corresponds with the hierarchical structure in the network. Of course, such networks have to follow the condition that in all nodes the sum of the Burgers vector should be zero. Finally, also notice that the distribution of the etch pits on the nanoscale shows a very high inhomogeneity, especially along dislocation bands (Figure 9). In such a region, the density of dislocations can even be one or two orders of magnitude higher than for the rest of the surface.

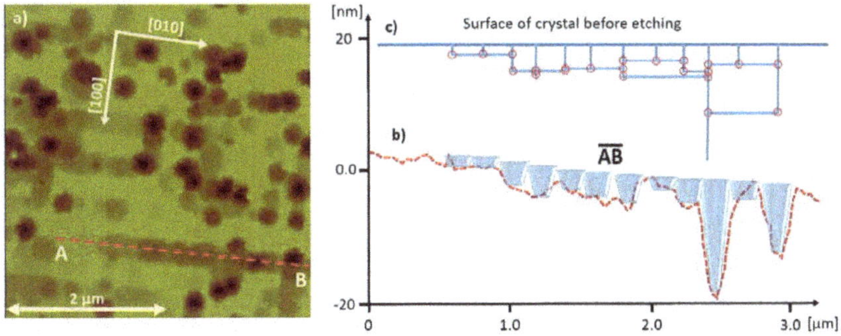

Figure 8. (a) AFM-topographic picture of etch pits on $SrTiO_3$ (100) surface (for the decoration of the exit of dislocation only a short etching was applied). The determined density of dislocation is 1.5×10^{10} per cm^2. (b) Cross-section (A–B) shows a variation of the depths of the etch pits. Inset (c) shows a possible 2D projection of a three-dimensional arrangement of the dislocations in a network of dislocations in surface layer of $SrTiO_3$; red circles mark the nodes of dislocations network, which fulfill the criterion that in such nodes the sum of Burgers vectors has to vanish. This scheme illustrates a dramatic reduction of the density of dislocations from the outer surface to the deeper parts of the surface layer (region), which has been proved by TEM study [10,42] (for details see, e.g., [68] and sub-Section 5.3.3). Such a kind of network is typical for a so-called hierarchical tree of dislocation arrangement.

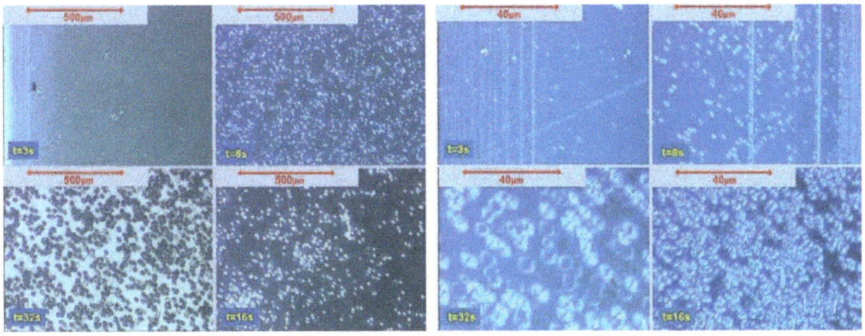

Figure 9. Collection of the optical photographs which shows the evolution of the etch pits distribution on $SrTiO_3$ (100) crystal for different etching time of 3s (**a,e**), 8s (**b,f**) 16s (**c,g**) and 32s (**d,h**) in low magnifications (**a–d**) and high magnifications (**e–h**). Note, that for short etching time only the region close to the edge of the crystal with dislocations (accumulated in bands) can be identified via optical (here microscopic) inspection. Note further, that time-dependent etching of the $SrTiO_3$ crystals was obtained on different small pieces prepared from the same large crystal. This was necessary since the optical characterization requires clean surfaces (in a bath of hot deionized water and washed in ethanol) and could influence the kinetics of the etch pits creation in subsequent etching periods due to preferential leaching of SrO in hot water.

In a similar way as noted for the TiO_2 investigation, an optical inspection of the etch pits density of $SrTiO_3$ underestimates the concentration of etch pits, which is not only an effect of the resolution of an optical microscope but is connected again with the bundling of dislocations since this gives only one etch pit for a group of dislocations (Figure 10) or is caused by the hierarchical construction of the network [48]. In this case, after a short etching time, the etchant reaches a node of two or three short dislocation segments, and a prolonged etching leads to the creation of only a single etch pit. Despite the disadvantage of a microscopic investigation of etch pits in the plane of the etched surface of the SrTiO$_3$ crystals, this technique is clearly extremely useful for a fast analysis of the distribution (homogenous, see Figure 7, or inhomogeneous as illustrated, e.g., in Figure 11) of etch pit for large areas of an etched surface.

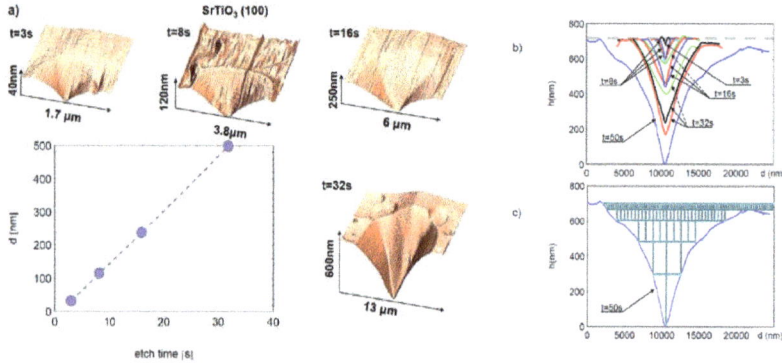

Figure 10. AFM "3D" topography of the representative etch pits on the surface of SrTiO$_3$ (100) for different interaction time between the crystal and etchant (here HF) (**a**). The diagram on the upper right (**a**) shows the linear increase of the average depths of the etch pits as a function of etching time. Cross sections of selected etch pits (for different etching time (**b**)) do not correspond with classical linear inclination (slope) of the facets, rather the cross-sections have a typical shape for a funnel (see insert on right). Only for two etch pits the mentioned linear inclination has been found (cross sections marked with green lines). The deviation from the planarity of the shape of etch pits facets is a consequence of the change of the etching yield for the hierarchical tree of dislocations (**c**), where the concentration of dislocations is dramatic reduced between surface layer and interior of the crystal (for details see [10,42,68]).

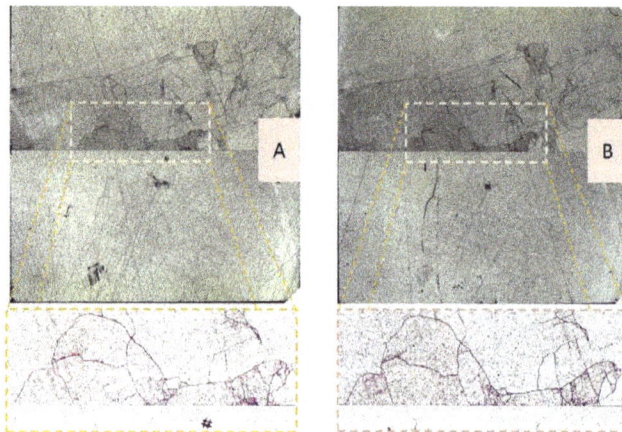

Figure 11. Distribution of etch pits on both sides (**A** and **B**) of a SrTiO$_3$ bicrystal (boundary 36.8°, dimension of the crystal 1 × 1 × 0.05 cm^3). The magnifications from the same regions of the crystals (upper and lower sides) show a similar position of the etch pits close to the boundary of the mosaic structure, which has been created during sythesis of the bicrystal. Notice: the photography B was mirrored.

The most impressive application of the etch technique for SrTiO$_3$ crystals can be visualized by the distribution of the etch pits for bicrystals (see Figures 12–16). Bicrystals of SrTiO$_3$ are typically prepared by a hot-joining technique. Before joining the contact planes of the both crystals are epi-polished. The kind of temperatures used are 1400 °C and joining may last as long as 10 h, for details see [138]. For the ideal case of the distribution of dislocations along the boundary, one should only observe (after etching of the bicrystal) a kind of overlap of etch pits along a line with a distance of a few nm. *Nota bene*, such a configuration of the regular distribution on the nanoscale can be directly derived from the HRTEM measurement (see, e.g., [65]). Therefore, for such a position of etch pits one can expect the creation of a continuous canyon with triangular shape (considering the resolution given by an optical microscope) (see Figures 14 and 15). A collection of photographs of etch pits for a bicrystal shows instead the real distributions of etch pits along the boundary and in its surroundings. For a representative presentation of etched surface, we have decided to use bicrystals of SrTiO$_3$ with the same boundary (36.4°) originating from three different suppliers (here: Shinkosha, MaTeck and CrysTec). In our opinion, this collection demonstrates how imperfect real bicrystals can be on macroscopic dimensions. From Figure 12 one can immediately see that the densities of dislocations of both parts of the crystals are not the same (which means, that during the synthesis of both tilted parts of the crystals the dislocations in one part were more easily generated than in the other). For a simple check that the inhomogeneity has its origin in the synthesis and not by polishing one can compare the "averse" and "reverse" of the etched bicrystal (see Figure 11). Since the position of the etch pits on the top and the bottom of the crystal is the same or at least very similar, the array of dislocations seems to go through the whole thickness of the crystal. In fact, the distribution of etch pits represents a position of low and high angle boundary of the mosaic structure which has been created during synthesis (see Figures 12 and 14). Such an analysis of the etch pattern is obviously helpful to extract information on the quality of real bicrystalline boundaries, revealing in the present cases the unpleasant fact that new dislocations can be created during bicrystal production (besides the boundary), dislocations also show a tendency to bundle (see Figures 13–15), and, moreover, dislocations along the bicrystalline boundary and close to the boundary of the mosaic structure can agglomerate in bands. We presented these data to remind the readers that great care and experience is required to generate high-quality crystals with bicrystalline boundaries.

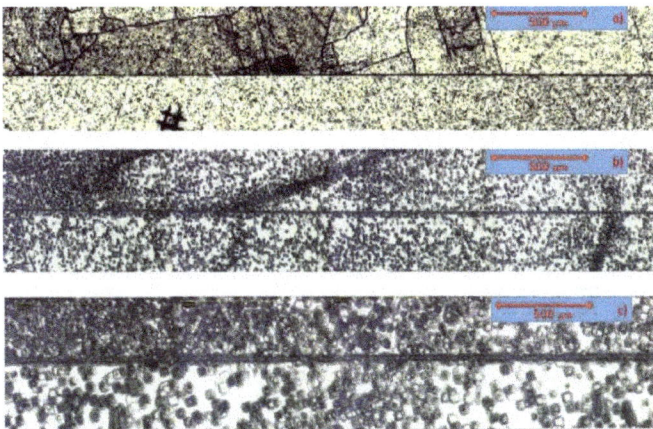

Figure 12. Collections of the etch pits distributions, close to the bicrystalline boundary (36.8°), of SrTiO$_3$ bicrystals from different suppliers: (**a**) Shinkosha; (**b**) CrysTec; (**c**) MaTecK.

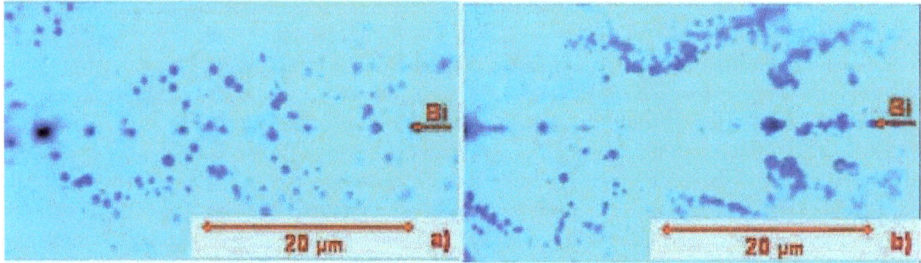

Figure 13. Example of the distribution of the bundles of dislocations along the bicrystalline boundary (36.8°) in SrTiO$_3$ show a "dramaturgy of the imperfection on the microscale" of the real boundary (**a**). Photography of etch pits (after short etching in HF) was taken using maximal magnification (**b**) with Nikon Optiphot metallurgical microscope with ND 32 filter.

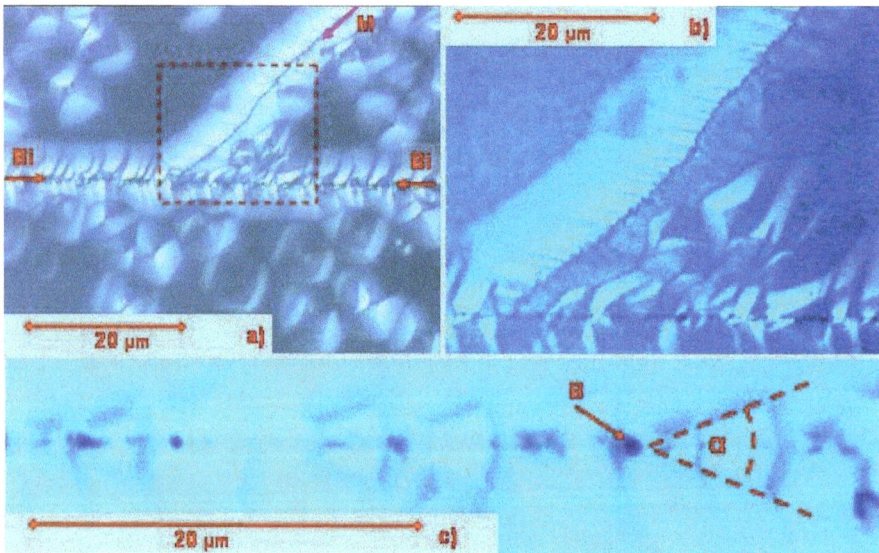

Figure 14. Optical photographs of etched SrTiO$_3$ bicrystal (boundary 36.8°). (**a,b**) (zoom) show the arrangement of etch pits along of the bicrystalline boundary (Bi) and a similar grouping of etch pits along of the boundary between mosaic structures (M). It should be noted that the distribution of etch pits along of the "native" boundary (M) is much more regular than for the "artificial" (Bi). Optical photography (**c**)) with maximal optical magnifications of etch pits along of the bicrystalline boundary present the variations of the shape of the pits and their irregular distribution. The local angle (α) which can be determined from the inclination of the etch pits along of the boundary is not the same and can vary for different etch pits from 32° to 38°. Using the adjustment of the focus of the microscope it can be clearly verified that the dark spots along of the boundary (B) lie deeper than the bottom of the etch pits. There are exits of the bundle of dislocations.

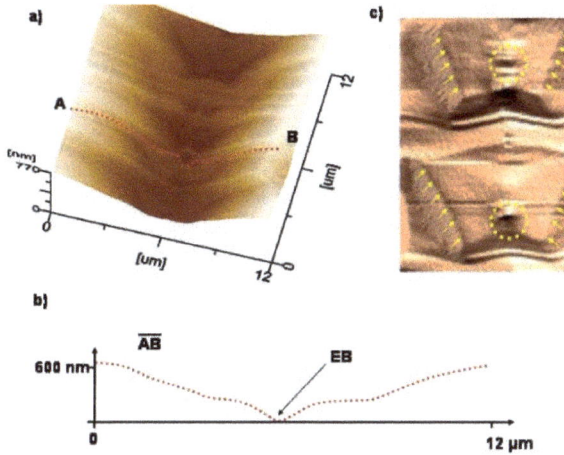

Figure 15. AFM 3D topography of etch pits along of the boundary (36.8°) (a) shows complicated shapes of the individual etch pits, which are dependent on bundle of dislocations and irregular distribution of dislocations on both sides of the boundary (see Figure 11). EB: etch pits of the bundle (b), EG: etch pits of the band of dislocations (c).

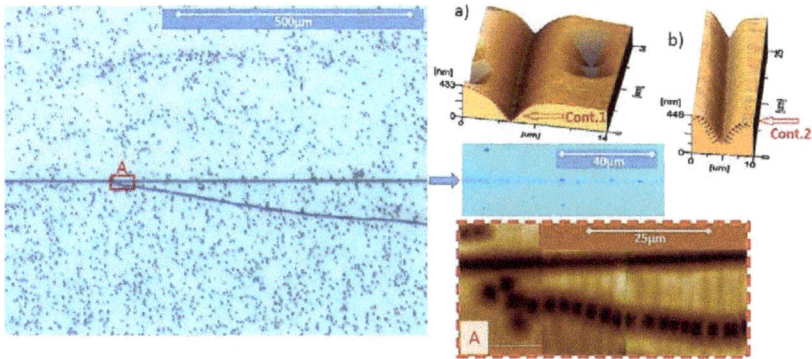

Figure 16. Distribution of etch pits for the 24.6° bicrystalline boundary in SrTiO$_3$ crystal shows similar irregularity than for the boundary 36.8°, namely a broad band of etch pits along of the boundary and their bundling (see magnification of region A). The shape of the "canyon" (see contours 1 and 2) of etch pits accumulated along of the boundary (**a,b**) suggests the existence of a hierarchical network (see discussion about etch pits for SrTiO$_3$ crystal given above in Figure 10). Probably the stress which accompanies the synthesis of the bicrystals is responsible for the increase of the mosaicity in both parts of the crystal and has generated a new boundary.

The classical etch technique can be successfully used for TiO$_2$ and SrTiO$_3$ crystals for an analysis of the change of the distribution of dislocations in-plane caused by thermal treatment (Figure 17) or mechanical stress (see below, Figure 17, right).

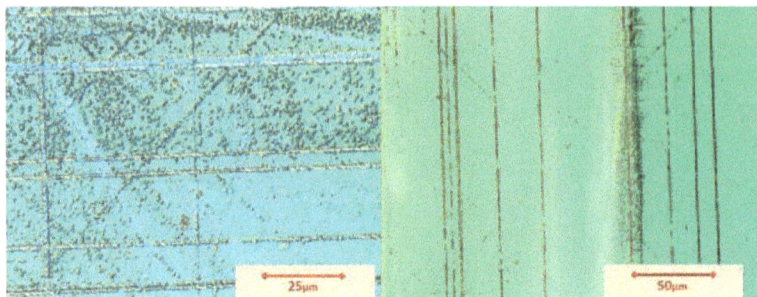

Figure 17. Thermal annealing of SrTiO$_3$ (100) crystals (oxidation at 1000 °C, 200 mbar O$_2$, t = 45 h) (**left**); forced arrangement of dislocations along <100> and <110> directions (**right**).

5.2. X-ray Topography

X-ray topography using either labor X-ray sources or synchrotron irradiation is a powerful X-ray diffraction technique for the non-destructive study of the nature and the distribution of extended defects (especially dislocations) in the plane of the surface layer and in the bulk of real crystals [139,140]. In this method, a collimated X-ray (with a shape of a ribbon) irradiates single crystals for different (selected) Bragg angles. The diffracted beam (in fact, a diffraction spot is called an X-ray topography) is analyzed via a classical X-ray sensitive photo-film or using X-ray sensitive position detectors. The influence on the local diffraction power coming from the imperfect (distorted) regions generates a variation of the intensity relatively to the perfect matrix. The manipulation of the Bragg conditions can for example be useful for the analysis of the Burgers vector of the dislocations etc. The nature of the contrast on the X-ray topography is described by the kinematical and dynamical theories of X-ray diffraction [141].

This very sensitive method for the analysis of the extended defects and strained regions has often been used for SrTiO$_3$ crystals, especially for studies of the mosaicity and the density of dislocations which are generated during growth of the crystals with different methods [99–104,142,143]. It should be noted that the X-ray topography studies for SrTiO$_3$ crystals have revealed that these crystals cannot be (in the regular phase) classified as a monocrystals, because the low- and high-angle boundary are typical for the mosaic-like structure, which in turn means that such crystals can be defined as single crystals.

In contrast to the detailed study using X-ray topography of SrTiO$_3$ crystals there is no literature about this topic for TiO$_2$ crystals.

5.3. Imaging and Spectroscopy by Atomic-Resolution STEM

In order to investigate the structure and chemistry of dislocation cores, it is necessary to employ a new tool to directly observe the local atomic configurations, chemical types and their electronic structures. One of the most versatile experimental systems for this purpose is an electron microscope and we briefly describe in this section the concepts and the ability of modern electron microscopy. Following the successful implementation of 3rd or 5th order aberration correctors in electron microscopy [144,145], the spatial resolution has dramatically improved and the current attainable lateral resolution is 40.5 pm—less than half an Ångström [146]. In addition to the progress on electron optics, the mechanics and electronics and detectors have been significantly improved which allows us to perform electron energy-loss spectroscopy (EELS) [147–149] and energy dispersive X-ray spectroscopy (EDS) [150,151] at atomic dimensions. There is a wide variety of microscopy, and two major systems in atomic-resolution are well known: transmission electron microscopy (TEM) and scanning transmission electron microscopy (STEM). In the nature of great flexibility of detectors, we here introduce the currently widespread imaging and spectroscopy of modern STEM.

In history, STEM was invented and designed by M. von Ardenne in 1937, which is only several years behind the development of TEM by M. Knoll and E. Ruska in 1932 (detailed can be found in this book [152]). However, there were many technical difficulties (mostly limited by noise) of STEM and then most of efforts were dedicated on the development of TEM. In the late 1960s, A. Crewe newly designed and constructed STEM with a field emission gun, annular detector and spectrometer for EELS, where the transmitted electrons through the center hole of annular detector were collected by the EELS spectrometer [153]. The basic design of modern STEM is referred to Crewe's microscope. In 1970s, Crewe firstly achieved the visualization of single Uranium atoms on thin amorphous carbon films [154]. In late 1980s, STEM technique was applied to the crystal viewed along the low index crystal orientation and firstly realized the chemical sensitive imaging in the crystal, where the annular dark field (ADF) detector was set at high-angle to efficiently remove elastic scattering contrast [155]. This imaging mode is known as high-angle annular dark-field (HAADF), where the contrast is sensitive to the constituent atomic number (Z) and therefore the mode is also known as Z-contrast imaging.

Hardware of the modern STEM. Figure 18 shows the schematic view of STEM configuration and detector geometry. The imaging electrons are generated by a field emission gun, and the electrons are very finely focused by the aberration correctors and probe forming objective lens. Then, the sub-ångström electron probe is scanning over the specimen and the atomically resolved images are formed by collecting the transmitted electrons at each probe position with detectors beneath the sample.

Typical description for each component is given as follows. Incident electrons are typically accelerated at 60–300 keV and relatively higher accelerate voltages such as 200–300 keV are typically selected for oxide materials because of higher spatial resolution. Lower accelerate voltages of 60–80 keV are also commercially available and used for the observation of two-dimensional materials, i.e., graphene or transition metal dichalcogenides [156]. The energy spread of the incident electrons are typically ΔE ~0.4 eV with a cold field emission gun, which is enough to perform core-loss EELS measurements. Recent progress on the monochromator of an electron source, the energy resolution is greatly improved up to 10–30 meV (10 times better than that of a cold field emission gun), which makes it possible to measure the local bandgap and phonon excitation (vibrational spectroscopy) [157,158]. The most limiting factor for atomic-resolution imaging was 3rd order spherical aberration but at the end of last century the combination of multi-pole lens system (quadra-octa poles or hexapole lens system) solved the problem, the invention of aberration correctors. Using an aberration corrector, the illumination angle in STEM is typically increased 20–30 mrad (semi-angle), providing sub-ångström probe. For detection of transmitted electrons, scintillators (Ce-doped YAP) combined with photomultiplier tubes are mostly adapted for the imaging detectors because short life time (~30 ns) at room temperature and moreover it is sensitive to single electron levels [159]. Recently, the shape of detectors is reconsidered, and segmented-type [160,161] and pixelated-type detectors [162] are now used for electromagnetic field imaging and ptychographic imaging. In this section, we will briefly describe the imaging and spectroscopy of HAADF, ABF (annular bright-field), EDS and EELS, respectively.

High-Angle Annular Dark-Field: HAADF. Using an aberration corrector and objective lens, the image of a gun tip or probe is formed on the specimen surface or preferable location. The wave function of the probe at the position r_p in real space (r) is given by the illumination angle ($\alpha_{max} = \lambda k_{max}$) and the aberration wave function $\exp[-i\chi(k)]$ in reciprocal space (k) [163]:

$$\Psi(r, r_p) = A \int_0^{k_{max}} \exp\left[-i\chi(k) - 2\pi i k \cdot (r - r_p)\right] dk.$$

The aberration correctors are operated to maximize the aberration wave function in a wide range of k-frequency and the probe size becomes smaller than 0.1 nm. The probe is then scanned across the sample and the transmitted electrons are used to form atomic-resolution images. The image intensity and accessible information depend on the location of the detector in reciprocal space:

$$I(r_p) = \int |\Psi_t(k, r_p)|^2 D(k) dk,$$

where $\Psi_t(k, r_p)$. is a Fourier transformation of the transmitted wave function, and $D(k)$ is detector function (1: on the detector and 0: otherwise). The detector shape is usually radially symmetric such as circular or annular detectors. The inner angle of the HAADF detector is typically twice higher than the illumination angle, e.g., the integration range is 60–200 mrad for the illumination angle of 30 mrad. In such high-angle condition, the amount of elastically scattered electrons is progressively reduced and incoherently scattered electrons are the dominant contribution of the image: the thermal diffuse scattering related to Debye-Waller factor rather than coherent Bragg scattering [164,165]. The electrons at high-angles were scattered nearly from the nuclei and the contrast in HAADF shows the Z^2-dependence of unscreened Rutherford scattering and the HAADF STEM is called Z-contrast imaging. In this sense, the intensity of HAADF image may be simply described by the straightforward convolution of point spread function $P(r_p)$ and objective function $O(r_p)$:

$$I_{HAADF}(r_p) = P(r_p) \otimes O(r_p).$$

Figure 18b shows the HAADF STEM image of $SrTiO_3$ viewed along the [001] direction, and one can clearly recognize the locations of Sr and Ti-O atomic columns, as respective bright contrasts [166]. Owing to the Z-contrast nature, the heavier Sr atomic column is brighter than Ti-O atomic column, and therefore Z-contrast imaging can be considered as a chemical sensitive imaging. The other advantage of HAADF image is non-reversal contrast, i.e., the atom position appears as a bright dot contrast even with relatively thick specimen of 50–100 nm, providing a simple interpretation of the image. Although it requires full dynamical image simulation, it is now possible to quantify the intensity of HAADF image or even count the number of atoms in projection [167–170].

Annular Bright-Field: ABF. HAADF STEM is quite powerful and solves a number of materials' problems [152]. It is however hard to recognize light atom positions in HAADF STEM because of the low scattering power of the light elements into high-angles. In oxide materials community, the identification of oxygen atom position has long been desired to understand the origin of physical properties such as ionic conductivity, superconducting, ferroelectricity and ferromagnetism. After the implementation of aberration corrector, the bright field region (the smaller scattering angle than the illumination angle) has been reconsidered and a new method of annular bright-field imaging was proposed with the aid of comprehensive theoretical investigations [171,172]. As shown in Figure 18a, ABF detector is placed at the bright-field region but the half of central regime is not used for imaging, i.e., detection angle is 15–30 mrad with the illumination angle of 30 mrad. Figure 18c shows the ABF STEM image of $SrTiO_3$ viewed along the [100] orientation, simultaneously recorded with HAADF STEM. In ABF image, the atomic column can be seen as dark contrast that is opposite contrast of HAADF image, but the atom positions in ABF image are exactly the same as HAADF image. In addition to Sr and Ti–O atomic column, new dark contrasts appear in ABF image, corresponding to the oxygen atom positions. In bright-field region, there appears electron scattering interference between transmitted and scattered waves, which may produce the artefact contrast eventually at the oxygen atom positions. To establish the robustness of ABF STEM imaging, deliberate theoretical investigations have been performed, including various crystal systems, defocus-thickness dependence, specimen tilt, detector misalignment and non-uniformity of the detector response [172,173]. Although the image formation mechanism of ABF STEM is not as simple as HAADF STEM, it realized that the dark contrasts can universally be observed at the light element atomic positions. It is noteworthy that, under the in-focus condition, the atom positions of both heavy and light elements can be seen as non-reversal dark contrasts, and therefore ABF STEM imaging is directly interpretable and as robust as HAADF STEM imaging.

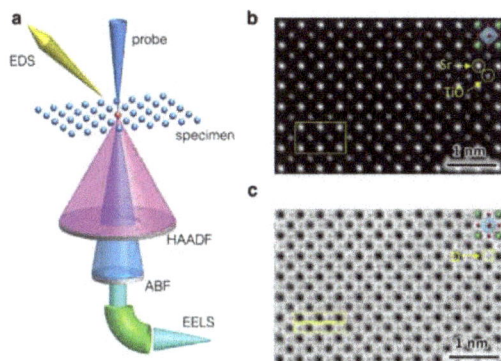

Figure 18. (**a**) Schematic view of ray-path of scanning transmission electron microscopy (STEM) mode and detector geometry of high-angle annular dark-field (HAADF), annular bright-field (ABF), electron-energy loss spectroscopy (EELS) and EDS. Simultaneously recorded (**b**) HAADF and (**c**) ABF STEM images of $SrTiO_3$ viewed along the [001] orientation.

Here we describe the image formation mechanism of ABF STEM in a simplified manner [174]. When the probe is traveling through the heavy atomic column, the electrons are channeling along the column and most of electrons are scattered into dark-field regime, visualizing the heavy atomic column as a strong dark contrast. While at the light atomic column, the electrons are weakly channeling along the column and a part of electrons are scattered forward and go through the central hole, and therefore the light atomic column becomes faint dark contrast.

Although we explain the robustness of ABF STEM imaging but the remaining intriguing question is whether the lighter atoms than oxygen is possible to detect in ABF imaging. As an example, Figure 19 shows the simultaneously recorded HAADF and ABF STEM images of $LiCoO_2$ viewed along the [100] crystallographic orientation [174]. In HAADF image, one can see only the Co atoms as a bright contrast and the other atoms of oxygen and Li are difficult to recognize. However, in ABF image, all the chemical types of atoms are visible as dark contrast including Li atoms. ABF imaging has been used to lithium ion battery and hydrogen storage materials and successfully achieved the visualization of lithium [175] and even the lightest element of hydrogen atomic column [176,177]. Therefore, it is possible to visualize all the chemical type of elements at atomic dimension.

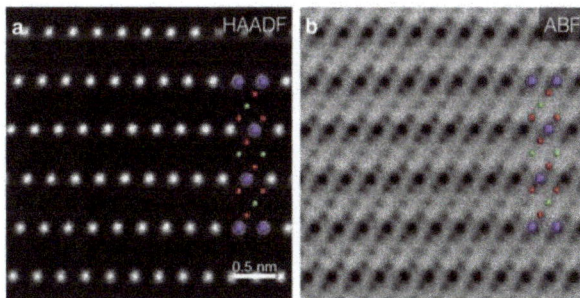

Figure 19. Simultaneously recorded (**a**) HAADF and (**b**) ABF STEM images of LiCoO2 viewed along the [100] direction. The structure models are overlaid and the colors are: Li (green), O (red), Co (purple).

Electron Energy-Loss Spectroscopy: EELS. In metal oxides, transition metals possess a wide variety of valence states that directly dictate the physical and chemical properties. However, the intensity of HAADF and ABF images is insensitive to the valence state of transition metals and another approach is required to identify the local electronic structures. When high-energy incident electrons pass through the specimen, the incident electron may interact with atomic electrons surrounding a nucleus, which process is known as inelastic scattering. The core electrons of the constituent atom will then be excited to the outer shell and the incident electrons conversely lose the corresponding energy. EELS can typically measure the loss-energy of the incident electron in the range of less than 3000 eV, which allows us to determine the valence state (>200 eV), plasma excitation (10–30 eV), band gap (<10 eV), and phonon vibration (10~100 meV). The cross-section of inelastic scattering may be written as a dynamic form factor $S(q, E)$ [178]:

$$\frac{\partial^2 \sigma}{\partial E \partial \Omega} \propto S(q, E) = \sum_{i,f} \left| \phi_f \left| e^{iqr} \right| \phi_i \right| \delta \left(\varepsilon_f - \varepsilon_i - E \right),$$

where $|\phi_i>$ is the initial core state of energy and momentum (ε_i, k_0), $|\phi_f>$ is the final conduction state of energy and momentum (ε_f, k) for the electron of the constituent atom, the transferred momentum is $q = k_0 - k$. It is noteworthy that, as illustrated in Figure 18a, EELS measurement in STEM is compatible with HAADF STEM imaging because the most of inelastic (forward) scattering electrons can go through the central hole of HAADF detector.

Figure 20 shows an example of the line scan of Ti-$L_{2,3}$ edges (STEM-EELS) obtained from the photocatalytic materials of LaTiO$_2$N from the surface to the bulk (1 nm step) [179]. In the bulk, two peaks split into four peaks (t_{2g} and e_g states) and the valence state of Ti is identified to be 4+. While in the near surface (<2 nm), two peaks are not split and therefore the valence state of Ti is 3+ only at the surface and this region is active in photocatalytic reaction.

In STEM mode, the EELS data acquisition is not limited to line scan profile but is possible to perform in 2D scan. Although the acquisition time is at least three orders of magnitudes slower than ADF imaging, the higher mechanical stability of modern STEM makes it possible to acquire spatially resolved 3D data cube at atomic-scale, where EEL spectrum is obtained at 2D probe position. On the basis of post data analysis, atomically resolved chemical mapping and moreover valence mapping are possible, although the channeling condition is carefully considered for the interpretation of data set [147–149]. In this section, we have discussed the valence state of transition metal, but the EELS has more capability: measurement of the specimen thickness (log-ratio method), estimation of local concentration of vacancy, valence EELS, vibrational EELS [178].

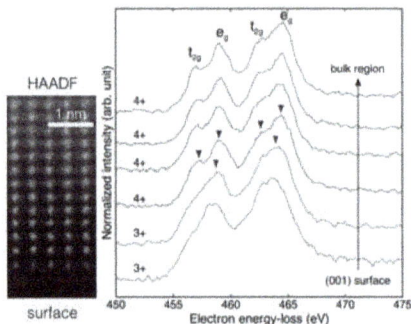

Figure 20. HAADF STEM image and core-loss EELS of Ti-L edges from the surface to the bulk with 1 nm step, where the specimen is LaTiO$_2$N and the orientation is [100] direction.

Energy Dispersive X-ray Spectroscopy: EDS. When the constituent atom of a specimen is excited by the high energy electron beam, an electron hole is created in the inner shell. The excited electron then refills the hole and the specific energy (difference between inner and outer shells) is then released in the form of an X-ray. Generated specific-energy X-ray is directly related to the chemical type and therefore it is useful to determine the chemical composition of a specimen. The cross-section in EDS is typically three orders of magnitudes smaller than that of EELS and moreover a collection solid angle of EDS detector was ~0.1 sr (a sphere is 4π sr), which has long been the technical difficulty to perform atomic-resolution EDS mapping. Recently, a considerably large size of silicon drift detector (>100 mm^2) has been introduced and the solid angle is significantly increased to 1.0 sr. Furthermore, using multiple silicon drift detectors, the total solid angle is now larger than 2.0 sr. To understand the measurement system, we here compare the methodology of EDS to EELS. The measurable energy range in EELS is less than 2 or 3 keV, whereas EDS can perform in much wider range from 0.1 eV to 30 keV, and therefore EDS can detect all the chemical types, except for the lighter elements like lithium. While in the energy resolution, EELS has much high energy resolution than that of EDS, and EELS is suitable to investigate fine electronic structure of materials. In EDS, silicon drift detector is connected to a pulse processer (incoming voltages convert into counts) and processing is very fast. The EDS detector can wait for eventual X-rays, and therefore multiple-frame scanning can easily perform as used in STEM imaging to enhance signal-to-noise ratio. While in EELS, the read-out time is considerably slow in charged-couple devise and moreover a reading-out noise is introduced for each time and therefore one-time measurement is suitable. Considering the beam damage issue, EDS with fast multiple scanning would be suitable for the compositional mapping at atomic-scale.

Figure 21 shows atomic-resolution HAADF and EDS images of a single dislocation core at the low-angle tilt grain boundary of Mn-doped SrTiO$_3$, where the crystallographic orientation is (100)/[001] with a 5° mistilt angle [180]. After the acquisition of EDS spectra at each probe position, Mn-L, Sr-L, Ti-K edges were integrated and the then EDS maps were obtained. It can be clearly seen the Sr and Ti columns are spatially resolved in EDS maps. At the dislocation core, Sr becomes slightly depleted and it is then realized that Mn dopants substituted Sr combining with EELS analysis.

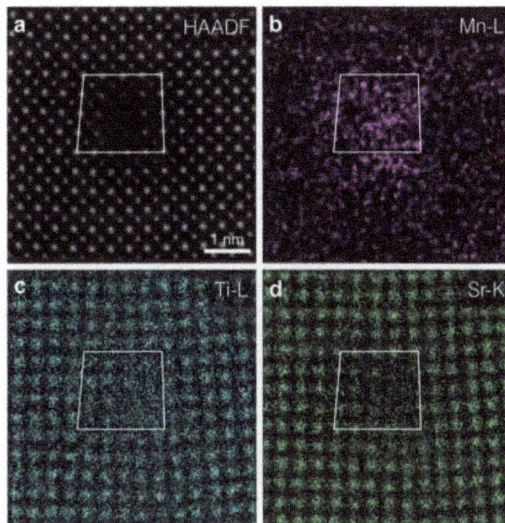

Figure 21. (a) HAADF STEM image and EDS elemental maps of (b) Mn; (c) Ti; and (d) Sr obtained from the dislocation core formed at the low-angle tilt grain boundary of SrTiO$_3$.

Crystals **2018**, *8*, 241

5.3.1. Dislocation Structure and Chemistry in the Grain Boundary of SrTiO₃

A dislocation is a one-dimensional crystal defect and has so far been investigated to understand the mechanism of deformation or mechanical properties in metals. In semiconductor and insulator materials, however, we may have an opportunity to embed a new functionality in dislocations, especially electric [3], magnetic [180,181] and optoelectronic properties. For further development of functional dislocations, it is a prerequisite to identify the chemistry and atomic/electronic structures of dislocation cores. To perform comprehensive investigations on a dislocation, it is desirable to fabricate well-defined dislocations. One appreciable approach is the bicrystal method, where two single crystals are thermally bonded in a specific crystal orientation. In a low-angle tilt grain boundary, dislocations are periodically introduced along the grain boundary to compensate the mis-tilt. In this section, we will discuss a specific example of SrTiO₃ dislocation core structure, especially on the dislocation core structure in the grain boundary rather than the dislocations in the bulk, by using modern atomic-resolution STEM [95].

The SrTiO₃ bicrystal with a [001]/(100) 10° tilt grain boundary was prepared by a thermal diffusion bonding of two single crystals, where the bicrystal was finally annealed at 1000 °C for 80 h after the pre-annealing at 700 °C for 20 h. The electron transparent TEM sample was prepared by conventional mechanical polishing and subsequent Ar ion milling with lower voltage of 0.2 kV. Figure 22a shows HAADF STEM image obtained from the bicrystal viewed along the [001] crystal orientation. There are two types of dislocation cores are alternatively arrayed along the grain boundary, and we will denote these dislocation core structures as core-A (SrO-termination) and core-B (TiO₂ termination), respectively. The distance between these cores are about 6 unit cells which is consistent with Frank's formula calculated from the mistilt of 10°. The Burgers vector of these cores are the same as *a* [100], but the observed Z-contrast has much difference: the contrast of the core-B is much darker than that of the core-A. Figure 22b–e shows simultaneously recorded enlarged HAADF and ABF STEM images of core-A and core-B, respectively, where oxygen atomic columns in ABF images are also visible as a dark faint contrast even at the dislocation cores. The corners at the dashed white lines are TiO and Sr atomic columns for core-A and core-B, respectively. Some oxygen atomic columns have low occupations and moreover at the center of core-A, the oxygen atomic column is completely missing. At the cation site in these cores, some atomic columns exhibit the contrast elongation or split, suggesting local structure inhomogeneity. To precisely identify the dislocation cores, it is necessary to determine the elemental distribution at the dislocation cores.

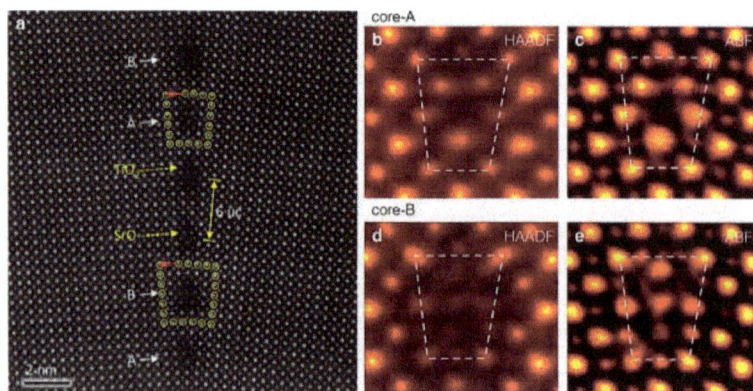

Figure 22. (a) Atomic-resolution HAADF STEM image obtained from SrTiO₃ bicrystal with a [001]/(100) 10° tilt grain boundary; (b–e) enlarged HAADF and ABF STEM images of core-A and core-B, respectively.

Figure 23a shows atomic-resolution EDS maps at the dislocation cores (red: Sr, green: Ti), where 10 pair of dislocations core maps were averaged to minimize the spatial variety and improve signal-to-noise ratio. The enlarged EDS maps of core-A, -B are shown in Figure 23b,c, respectively, where the dashed lines correspond to the location in Figure 22. As expected from the HAADF and ABF images, the atomic structure of core-A seems to be simple: no mixture of Sr and Ti and the three Sr columns at the core-A have slightly small occupation. While in the core-B, Sr occupation seems to be poor and Ti signals are widespread at the dislocation core region (not sharp contrast). Combining Ti EDS map with ABF image, it is realized that the reconstruction is related to a transition of TiO_6 octahedrons from corner sharing in perovskite to edge sharing: the local formation of rocksalt-like structure. Since the probe channeling effects along the atomic column always disturb the quantification of EDS signals, we here use a linear approximation of EDS signals [182], which provides a rough estimation of the local composition. The elemental occupations of the dislocation cores are estimated to be (Sr, Ti) $\simeq$ (0.82, 085) for the core-A and (Sr, Ti) $\simeq$ (0.63, 0.90) for the core-B, respectively. On the basis of these analysis, we identify the local composition of at each atomic column for both dislocation cores, and the results are given in Figure 23b,c, as overlaid structure models, where the numbers are indices for atomic columns and the color corresponds to the occupation of elements (red: Sr, green: Ti, yellow: O, while: unoccupied). At the core-A, the oxygen columns of #11 and #12 have very low occupations. At the core-B, most of Sr columns of #13, #14, #18, #19, #21 are partially occupied by Ti atoms and some oxygen atomic columns, expected from the bulk extrapolation, are also occupied by Ti, which gives an impression of very green-colored EDS map at the core-B.

In addition to the composition, the electrical activity of dislocation core is also strongly affected by the local electronic structure of Ti, which can be measured in EELS. Figure 24a shows EEL spectra obtained from the core-A, core-B, gap between A and B (continuously connected two single crystals at the grain boundary), and the bulk region, respectively. In $SrTiO_3$, the two peaks of the Ti-L edge for Ti^{4+} are well splitting into four peaks, corresponding to the e_g and t_{2g} states respectively. At both the dislocation core regions, Ti-L edges show slight energy shift towards lower energy and less pronounced peak splitting, suggesting the mixture of Ti^{3+} and Ti^{4+}. Using multiple linear least squares fitting, the spatial distribution of Ti valence states are determined and given in Figure 24b,c for Ti^{3+} and Ti^{4+}, respectively. The slightly reduced Ti^{3+} regions are concentrated at the dislocation core regions, while the bulk and gap regions maintain the valence state of Ti^{4+}. Combining atomic-resolution HAADF, ABF imaging and EDS, EELS spectroscopy, the dislocation core-A and core-B is estimated to be $Sr_{0.82}Ti_{0.85}O_{3-x}$ ($Ti^{3.67+}$; $0.48 < x < 0.91$), and $Sr_{0.63}Ti_{0.90}O_{3-y}$ ($Ti^{3.6+}$, $0.57 < y < 1$), respectively. To possess the charge neutrality at the dislocation cores, the amount of the oxygen vacancies is estimated to be $x = 0.63$, and $y = 0.75$.

In general, the oxygen vacancy formation energy is significantly reduced at the dislocation core, which is true in this case of core-A and core-B. For the dislocation core-A, the occupation of the oxygen columns of #11, #12 is considerably small because of strong Coulomb repulsive interaction between these two neighboring oxygen columns. For the core-B, oxygen columns are replaced by Ti, which is originated to the TiO_2 plane termination at this core. The observed oxygen deficiency results in the reduction of Ti valence, which is also confirmed by EELS. Compared with bulk $SrTiO_3$, the reduced Ti ion may contribute to the local electrical conductivity along the dislocation cores. Oxygen vacancies also introduce some spaces and therefore oxygen ion conductivity may be enhanced along the dislocation cores. The composition of dislocation core can be expressed as $\alpha SrO \cdot \beta TiO_2 \cdot \gamma Ti_2O_3$ (Ti^{3+} for Ti_2O_3, Ti^{4+} for TiO_2), where α and $\beta + \gamma$ can be measured from the EDS mapping of Sr and Ti and the ratio of β to γ can be estimated from the EELS. From the experimental results, the respective components for the core-A and core-B are determined to be $(\alpha, \beta, \gamma)_A = (0.82, 0.57, 0.14)$, $(\alpha, \beta, \gamma)_B = (0.63, 0.54, 0.18)$. Since the dislocation cores are different from the bulk in the sense of crystallography and chemistry, the dislocation cores should be probably considered as foreign objects when we discuss the local properties of electric or ionic conductivity. Therefore, a simple space charge model without consideration of the presence of these foreign objects is no longer sufficient to precisely interpret the local properties at the dislocation cores.

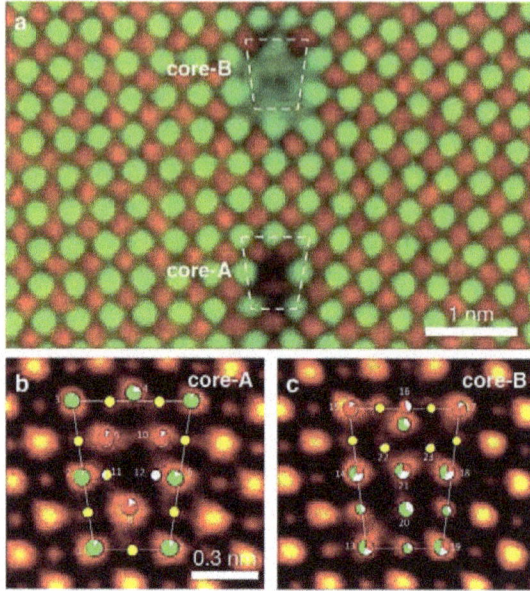

Figure 23. (**a**) Atomic-resolution EDS mapping obtained from the SrTiO$_3$ grain boundary. (**b,c**) Structure models of core-A and core-B are overlaid on the respective ABF STEM images.

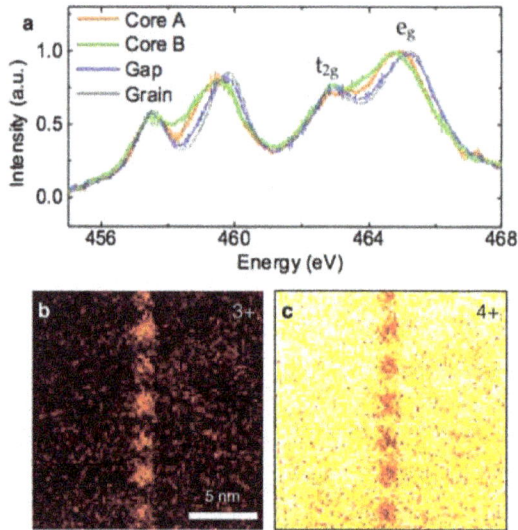

Figure 24. (**a**) EELS spectra obtained from the core-A, core-B, gap between cores along the boundary and the bulk. (**b,c**) show the fitted maps of Ti^{3+} and Ti^{4+} along the 10° grain boundary in a SrTiO$_3$ bi-crystal, respectively. The valence mapping of Ti obtained by multiple linear least squares fitting.

5.3.2. Dislocation Structure and Chemistry of Titanium Dioxide (TiO$_2$)

TiO_2 has been intensively investigated for long time because it shows potential applications in a wide range of industrial fields, including catalysis, photocatalysis, sensors, solar cells [183–186]. It also shows the properties as a wide-band-gap (3 eV) semiconductor, but many of its applications can be enhanced by additional electrons that originate from the defect states in the forbidden band gap [187,188]. Different explanations of these states are possible, e.g., such in-gap states have been observed in DFT calculations originating from edge dislocations or extended defects [32,48]. An understanding of their origin is therefore critical to improve the functional properties of TiO$_2$ because dislocations are implicated in many applications [189,190]. It has been reported that dislocations are existent on the {101}, {110} and {100} planes in rutile TiO$_2$ [22,27,191,192]. The plastic deformation in rutile TiO$_2$ was also found to take place by activating {101}<10$\bar{1}$> and {100}<0$\bar{1}$0> slip systems [192]. The dislocations have been confirmed to affect the electrical properties of TiO$_2$ single crystals [19] and the ionic conductivity of polycrystalline TiO$_2$ [20]. However, the relationship between the dislocation core structures and properties is still unclear because the general knowledge on the atomic-structures of dislocations has not been well characterized. In this chapter, atom-resolved imaging is demonstrated for the core structure of <001> dislocation in TiO$_2$ bicrystals [32] by Cs-corrected STEM.

It has been known that the <001> and <101> dislocations can be introduced into TiO$_2$ crystals [37]. However, the naturally or plastically introduced dislocations are usually wavy and it is not so straight to directly resolve their core structures at the edge-on condition by TEM. In this study, by using bicrystal technique, the low angle tilt grain boundaries, which contain periodic array of straight edge dislocations, were fabricated to directly observe the core structure of TiO$_2$ [32,193,194]. In this case, all atomic columns can be identified from the [1$\bar{1}$0] direction. For these reasons, the <001> edge dislocation was selected as a model system. The TiO$_2$ bicrystals with a small tilt angle of about 5 degree were fabricated by precisely joining two pristine single-crystal blocks of high purity rutile TiO$_2$ from the orientation relationships (110)[001]//(110)[001] [32]. To prepare the bicrystals with an array of periodic dislocations, two single crystals were cut by 2.5 degree from the (110) plane of the TiO$_2$ single crystal, followed by the mechanical grinding and polishing of the contact planes to a mirror finish. The two surfaces were bonded with the [001] direction of one crystal almost parallel to [001] direction of the other by a small tilt angle of 5 degree (2.5 degree + 2.5 degree) under hot pressing at 1773 K for 10 h in air. TEM and STEM specimens were prepared by cutting, grinding, dimpling and argon ion-beam thinning. HAADF and ABF images were taken using the Cs-corrected ARM-200FC (JEOL, Ltd., Tokyo, Japan) STEM operated at 200 kV, which provides a small probe size less than 0.1 nm. The ABF STEM images were observed with a detector of 12–24 mrad, and EELS spectra were recorded using the Gatan Enfina system equipped on the STEM with an energy resolution, at full-width of half-maximum, of about 0.5 eV.

To confirm the orientation relationship of the bicrystal and clarify whether dislocations are introduced into the boundary area, the selected-area diffraction pattern (SADP) was taken (Figure 25a) together with STEM image (Figure 25b) of the boundary area in the 5 degrees [001] bicrystal, in which the incident electron beam is parallel the [110] direction. An overall symmetry is clearly seen in the pattern, and a slight visible splitting is revealed for some diffraction spots (Figure 25a). In light of the splitting spots in the SADP, the tilt angle (θ) of the bicrystal is estimated to be 5.1 degree, indicating a successful joining of the bicrystal with the prescribed orientation relationships. As seen in Figure 25b, the periodic image contrast exactly on the mirror plane of the (110)[001] boundary, suggesting the formation of an array of dislocations at the boundary. The dislocation core structures are expected to be observed at edge-on condition. It can be thus said that the two crystals are well bonded with no secondary-phase layers, amorphous layers, contaminants, or transitional areas at the boundary region.

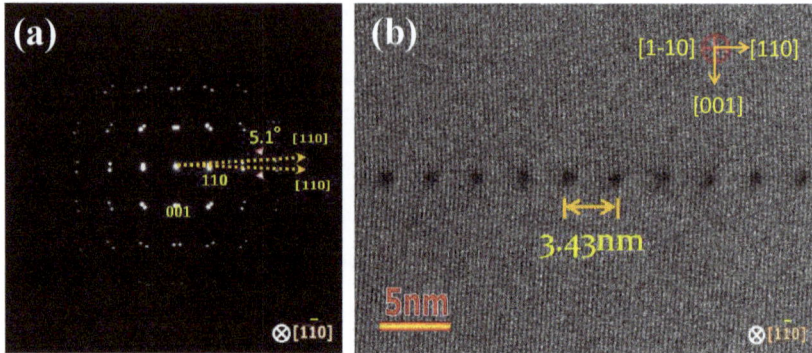

Figure 25. (**a**) Selected-area diffraction pattern (SADP) taken from the boundary area. An overall symmetry is seen in the pattern, indicating a perfect joining of the two single crystals; (**b**) Low-magnification ADF STEM image of the boundary in the bicrystal, suggesting the formation of a tperiodic array of dislocations at the boundary with 3.4 nm separation [32].

Figure 26a shows HAADF STEM image of a typical dislocation projected along the [110] direction. Considering the contrast and the crystal structure, the brightest spots represent Ti–O mixed columns, which unravel an almost symmetric core geometry for the edge dislocation of TiO_2. It was found that all of the dislocations showed the identical core configuration. The core region is slightly darkened owing to the strain associated with the dislocation. In addition, the considerably lighter O atoms are not scattered strongly enough to be visualized. As mentioned in the previous section, ABF STEM imaging technique has an advantage to directly observe light atoms. Figure 26b shows ABF STEM image of the same area of Figure 26a, where only the Ti–O mixed column and the pure Ti column are observed from this projective direction.

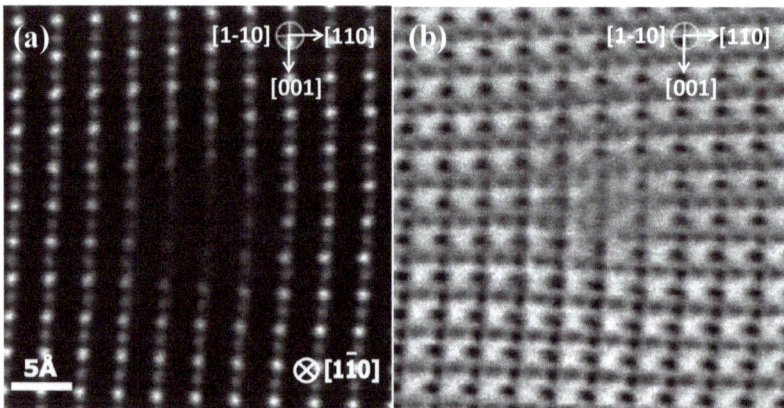

Figure 26. (**a**) Atomic-resolution HAADF STEM image of the dislocation viewed from the [110] direction, and (**b**) the corresponding ABF STEM image obtained simultaneously with the HAADF STEM image [32].

In order to evaluate the electronic structures of a single dislocation, atomic resolution EELS analyses were conducted for the Ti-$L_{2,3}$ edge for the dislocation core region together with that of the bulk as a reference (figure 27). It is clearly shown that Ti-$L_{2,3}$ EELS spectrum in the region away from the dislocation core (close to the bulk) mainly consists of four peaks (i.e., two doublets), where one doublet at lower energy loss comprises the Ti-L$_3$ edge, whereas the other at the higher energy loss Ti-L$_2$ edge shows the characteristic of having a valence state of +4 for the Ti atoms away from the core [195]. In contrast, the Ti-L$_{2,3}$ EELS spectrum taken at the core area exhibits a broader profile (marked by arrows in Figure 27), suggesting that they are of a mixed valence state of +3/+4 because the main difference in the Ti-L$_{2,3}$ EELS spectrum between Ti^{4+} and Ti^{3+} rests with the broadening of the t$_{2g}$ peaks in each doublet. We further conducted EELS measurements over a broad energy span and found that only the signatures of Ti-L$_{2,3}$ and O-K edges are detected both at the core and in bulk, providing evidence that a substantial segregation of impurity to the core can be ruled out, i.e., the dislocation core is impurity free. Such a shift in electronic states indicates the formation of a conducting channel at an individual dislocation and points to the existence of novel impacts associated with the impurity-free dislocations of TiO$_2$ [32].

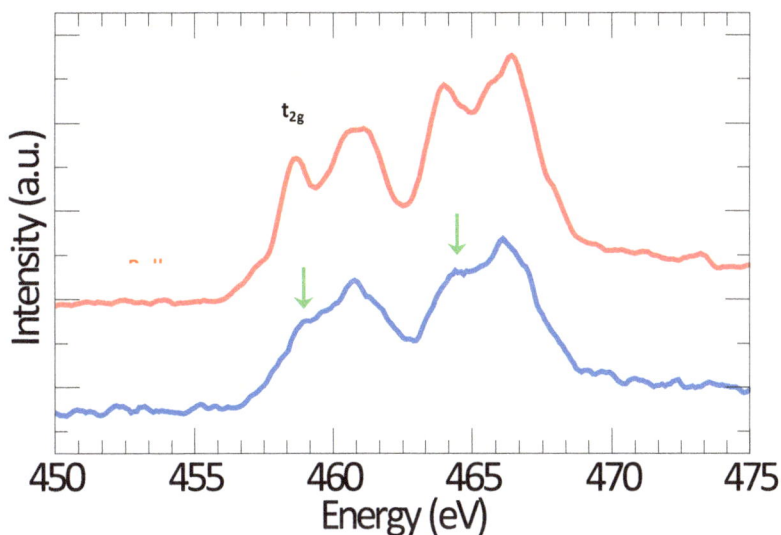

Figure 27. The Ti-L$_{2,3}$ edge EELS profiles obtained from the dislocation core and in the bulk region (i.e., away from the core). The region used to represent the TiO$_2$ bulk is 10 nm far from the exact dislocation core [32].

5.3.3. Hierarchical Distribution of Dislocations in the Surface Layer of SrTiO$_3$

The TEM investigations on the distribution of the dislocations in the surface region of SrTiO$_3$ crystal (as received) show a progressive lowering of the dislocations density (Figure 28) going from the surface towards the interior [10]. Depending on the polishing procedure the density can vary from $4 \times 10^9 - 6 \times 10^9$ (for the epi-ready polished surface [10,68]) to 10^{12} for the rough surface [42]. Such tendency in the distribution of the dislocations in the surface region (thickness of about few dozen μm) allows to analyze this system in terms of a hierarchical network, which fulfills the criterion that the sum of the Burgers vector of the dislocations in each of the nodes is zero. On the basis of the depth dependence of the density of dislocations a schematic view of the mentioned network has been created [68]. Strong reduction of the density of dislocations in the upper part of the network forces that the dislocations near the surface should be very short (only a few nm) while the length of the dislocations line between deeper nodes becomes successively longer. Jin et al. [42] showed that the dislocations in surface region exist not only as individual dislocations, but that these dislocations have a tendency to create a pair or agglomerate in form of bundles.

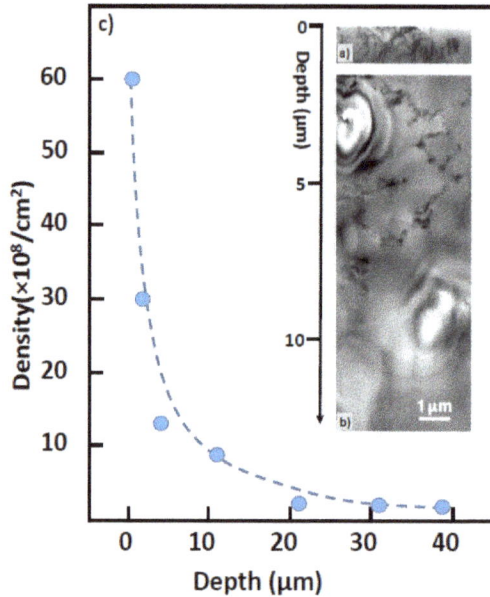

Figure 28. Hierarchical distribution of dislocations in the surface region (thickness ~30 μm) of SrTiO$_3$ crystal (**a,b**) (TEM photography). The density of dislocations has been reduced from 6×10^9/cm^2 (**c**) to ~10^8/cm^2 in the deeper part of the surface region (**c**). Adapted from [10].

5.4. Electron Channeling Contrast Imaging (ECCI)

Electron channeling contrast imaging (ECCI) combines the classical scanning electron microscopy (SEM) with the simultaneous analysis of backscattered electrons. This technique can be used for the investigation of extended defects such as dislocations or stacking faults, in crystalline samples or in polycrystalline materials (only for single grains). In principle, the penetration depth of this method is limited by the escape depth of backscattered electrons. The interaction of the electron beam can be described in terms of the standing electron-density waves which are formed in the matrix of the electron-irradiated crystal. The intensity of the backscattered electrons is influenced by the position of the nuclei in the lattice and the orientation of the lattice relatively to the primary beam. The minimum of the backscattering can be observed for the orientation between beam and lattice plane when the Bragg conditions are fulfilled. That means that similar than for TEM techniques for the identification of the dislocations the classical invisibility criterion namely ($\mathbf{g \cdot b} = 0$) can be used [196–199]. In other cases, the electron can be channeled in deeper parts of the crystal without strong interaction with the matrix. For extended defects the channeling of electrons into the interior of the crystal is strongly reduced; therefore, the backscattered electrons are carriers of information about the position of the defects and their type (Figure 29).

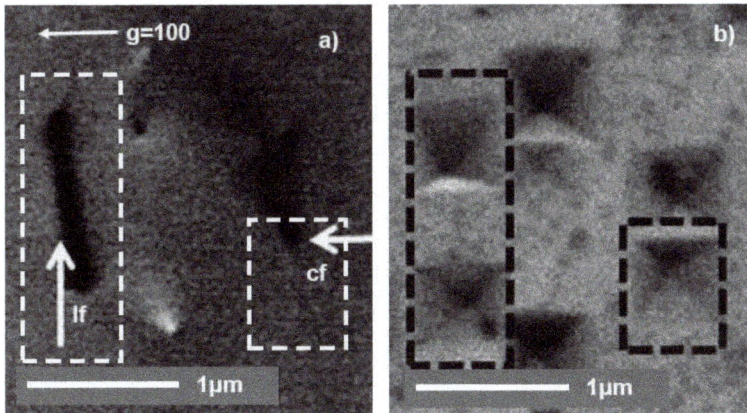

Figure 29. Image of the dislocations of pre-etched SrTiO$_3$ crystal (SEM photography (**b**)) using channeling contrast imaging (ECCI photography (**a**)) shows a correlation (marked by dashed line) between the position of the etch pits and the electronic contrast. Adapted from [200], for the details please see mentioned reference.

5.5. Electron Holography

In electron holography, the complex wave function (that means the amplitude and the phase) of fast electrons which are transmitted through a thin foil during TEM investigation is analyzed [92]. Because it is impossible to detect the phase of the electron wave function directly, since it can be modified by local electrostatic and magnetic fields, two indirect techniques for the reconstruction of the phase modulation are used namely the off-axis and the inline method. In the off-axis method, the electron beam is split by an electron bi-prism into one part traveling through the specimen and one part passing through vacuum which are then brought to interference. In inline holography, the specimen is placed in the divergent electron beam such that only one part of the beam interacts with the object and an interference with the undistorted part of the same beam occurs. In both cases, the phase then can be calculated from the recorded interference pattern [201].

Because this technique has a very high lateral resolution, it can be used for the study of the local potential distribution especially in the interface [90–93]. The prime example for the application of this technique is the analysis of the potential change close to the small boundary in SrTiO$_3$ crystals with and without doping. It is surprising that the thickness of the barrier is only 1–2 nm (without influence of the doping) (Figure 30). The potential drop is about 0.45 V for the barrier in the Nb doped bicrystal and −0.6 V for the barrier in the undoped crystal [92]. This barrier can be annihilated using appropriate polarization (perpendicular to the barrier) probably via local breakdown. Notice, such a change of the potential (+0.45 V or −0.6 V) on the small distances of just 2 nm actually creates a local electrical field of the order of 10^6 V/cm, high enough for a cold emission of electronic carriers according to literature.

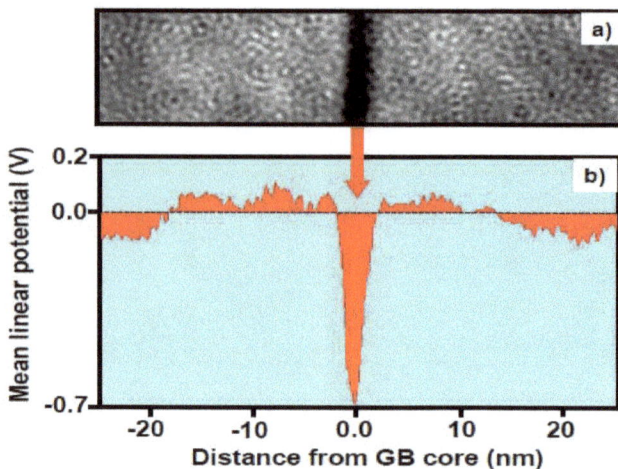

Figure 30. Local map of the variations in the mean inner potential across the bicrystalline boundary Σ3(112) of SrTiO$_3$ crystal (a). The colour scale: −0.7 V (black) and +0.4 V (white). The cross section (b) shows a spatial variation of the local potential profile extracted from the map (a). Adapted from [91].

5.6. AFM STM Study on Dislocations in TiO$_2$ and SrTiO$_3$ Crystals with Atomic Resolution

SPM techniques allow studying the surface with atomic resolution [66,67]. These very sensitive techniques can be used for a precise analysis of the ordering of atoms close to the exit of dislocations in the plane of the surface. The lion's share of AFM or STM studies with atomic resolution was obtained on surfaces which have been prepared via sputtering and subsequent thermal treatment. The STM or AFM pictures from such surfaces show a perfect reconstruction of the surface on the nanoscale and only a few numbers of point defects. Here it is worth emphasizing that there only very few AFM or STM measurements of dislocations exist in the literature with the highest resolution available [68,120].

Figure 31 presents the distributions of the atoms close to the cores of two partial dislocations.

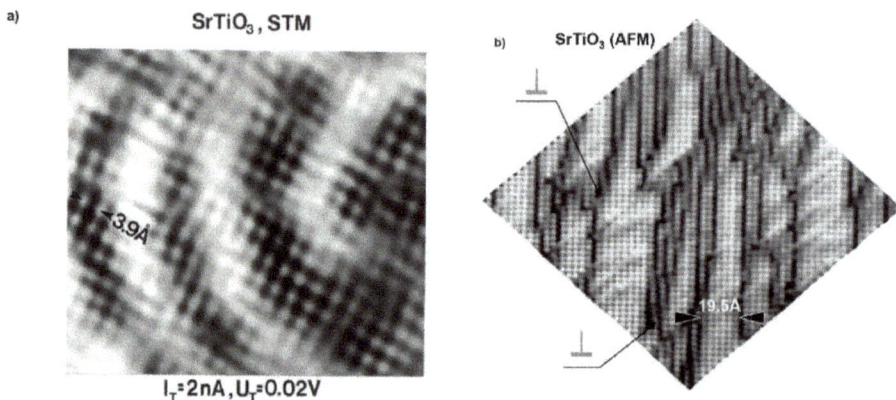

Figure 31. Scanning tunneling microscopy (STM) image (low pass filtering of the STM data from [120]) of a growth spiral observed on the surface of a SrTiO$_3$ crystal after reduction (a). Topography of the surface of an oxidized SrTiO$_3$ crystal, obtained using AFM, shows the two exits of the dislocations (b), adapted from [68].

6. Calculations of the Electronic Structure and Molecular Atomic Modelling of the Cores of Dislocations in TiO$_2$ or SrTiO$_3$ Crystals

6.1. DFT Calculation and Atomic Simulations

We have discussed above that in edge dislocations planes or half-planes terminate inside the perfect crystal lattice. In metallic systems, where the valence electron density is free-electron like, such extended defect will not drastically modify the electronic properties of the system while, of course, the mechanical properties will be affected. In the insulating oxide, where covalent or ionic bonding dominates, a terminated plane will necessarily exhibit dangling bonds or, if possible, a local reconstruction of the lattice with modified electronic properties.

To understand the basic mechanisms that appear when a defect is introduced in a crystal like SrTiO$_3$ or TiO$_2$, it is useful to look at point defects first. Density functional theory (DFT) [202] is a widely-used tool to study the electronic rearrangements in these materials, but care has to be taken when dealing with defects in oxides. Early calculations by Shanthi and Sarma [203] indicated that oxygen deficient SrTiO$_3$ is in a metallic state as the defect levels overlap with the bottom of the conduction band of the oxide. This is caused by the underestimation of the band gap of most materials in DFT and later calculations using hybrid functionals found that these defect states are narrow and about 0.8 eV below the conduction band, as expected [204]. The situation is similar in TiO$_2$ where oxygen vacancies (V$_O$) create defect levels at an energy of 0.5 to 0.9 eV below the conduction band edge [205].

Using DFT it is also possible to calculate formation energies of single defects and these studies have been performed on different levels of the theory [206,207]. One has to keep in mind that the creation of a V$_O$ by removing ½ O$_2$ leaves behind two electrons that localize on neighboring Ti atoms. In SrTiO$_3$ these electrons occupy d_{z^2}-type orbitals oriented in the Ti-V$_O$-Ti axis. It is well known that in 3d atoms in d^1 configuration strong Coulomb interaction plays an important role and the occupancy of the levels (and their conducting properties) are determined by correlation effects [208]. Therefore, often the so-called DFT + U method is used, that incorporates additional correlation effects in an approximate scheme [209]. The dominance of localized, directed orbitals also affects the vacancy-vacancy interaction in these oxides. It is, therefore, not surprising that in some of these calculations a tendency of clustering of defects has been observed, both for SrTiO$_3$ [210] and for TiO$_2$ [211]. If a Ti atom is surrounded by two oxygen vacancies, e.g., in a Ti-V$_O$-Ti-V$_O$-Ti configuration, the middle atom is in d^2 configuration and the second electron can occupy the conduction band of the oxide [210]. This induces a complex interplay between defect arrangement, charge state, and electronic properties.

In an extended defect, e.g., when a full plane of SrO is removed from SrTiO$_3$, the resulting structure resembles an interface between two TiO$_2$-terminated surfaces. At this interface, it is energetically favorable to displace the two sides by an ½ [010] translation. Such arrangements have been found in HAADF-STEM images of in low-angle tilt grain boundaries [74] discussed above.

In the same manner it is possible to remove a full TiO$_2$ plane from the perovskite lattice and displace by an ½ [110] translation to arrange the SrO layers in a rocksalt like structure. Such SrO-SrO arrangements form under Sr rich conditions in a periodic arrangement giving rise to Ruddlesden-Popper (RP) phases. DFT calculations predict that these extra SrO layers are energetically easy to incorporate and do not affect the electronic properties of SrTiO$_3$ significantly [212].

The situation is more complex if a half-plane (or, as in some calculations a one-dimensional ribbon) of SrO is removed from SrTiO$_3$. In general, the electronic structure will depend on the boundary conditions of the removed part: if a ribbon is removed that is on one side oxygen- on the other strontium-terminated (Figure 32) the charge neutrality is retained; no defect states are created and only a reduction of the band-gap due to "internal surface-states" at the defect site is obtained. Due to the non-vanishing dipole moment of the removed part of the crystal, strong local distortions of the lattice accompany the defect. If the removed ribbon is Sr-terminated at both edges, defect states appear in the gap, again d electrons localized on the Ti atoms at the edges, that can form conduction channels [55]. In a more realistic model of the dislocation core [39] there will be more effects to consider, e.g., an off-stoichiometry around the defect and long-range relaxations that are difficult to model in general.

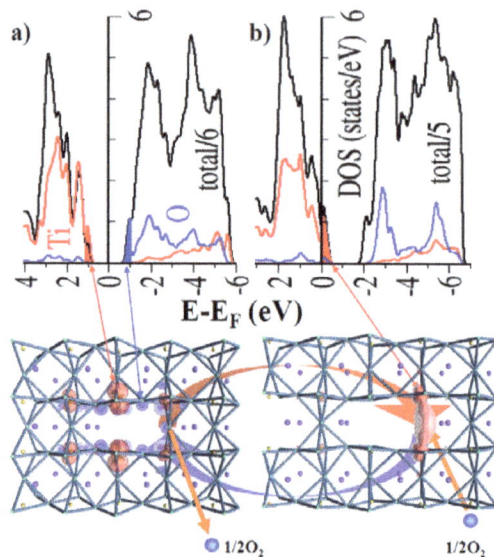

Figure 32. Ab initio calculations of the electronic structure of an extended SrO defect in SrTiO$_3$. (a) The density of states (DOS) shows a reduction of the band gap due to states near the defect. Occupied O states are shown in blue, unoccupied Ti states in red isosurfaces. Yellow, green and purple spheres represent Ti, O, and Sr atoms, respectively; (b) An increase of the oxygen nonstoichiometry leads to the generation of metallic state on the bottom of the conducting band. Reverse oxidation can restore the electronic structure to the semiconducting situation (blue arrow).

It was already mentioned that in TiO$_2$ oxygen vacancies have a tendency to form linear arrangements [211]. In this arrangement, the Fermi level cuts the conduction band formed by Ti *d* states. Calculations show that the associated charge density decays exponentially with distance from the linear defect and is reduced by four orders of magnitude within a nanometer [48]. This behavior is in line with LC-AFM measurements of electrically active channels on a TiO$_2$ surface. Also, other larger one-dimensional defect arrangements that show oxygen deficiencies have conductive states confined within a few nanometers.

Although several DFT studies were concerned with the structure and defect energetics of grain boundaries and dislocation cores, not much is reported on the electronic structure in these defects. Recent work on bicrystalline boundaries in SrTiO$_3$ and TiO$_2$ confirmed good agreement between the structures obtained in DFT calculations and observed in HAADF-STEM measurements. Calculations of the dislocation core as shown in Figure 1 show that the local inherent non-stoichiometry gives rise to spin-polarized states at the Fermi level that lead to a mixed +3/+4 valence state of the Ti atoms in the core [32]. Again, these states are rather confined (see Figure 33) and can serve as localized conduction channel.

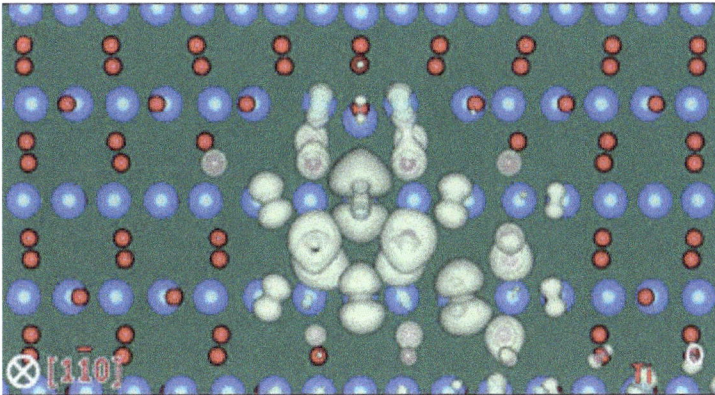

Figure 33. Charge density distribution at the edge dislocation in TiO$_2$ with b = ¼ [001] viewed from the [1$\bar{1}$0] projection. From the (grey) isosurfaces it can be seen that the charge induced by the local non-stoichiometry remains rather localized around the core. Adapted from [32].

Although the exact structure and electronic configuration at these extended defects is hard to capture (and probably in reality quite diverse) there are certain features that can be predicted by DFT calculations: Whenever electrons are localized in defect levels at the edges of dislocations or they will form states that are confined within a few nm around the defect. This behavior is very similar to the one observed in charged domain walls in perovskites, e.g., Pb(Ti,Zr)O$_3$, where the structural distortions around the longitudinal domain wall are limited to a few unit cells [213]. Also in this case DFT + U calculations have shown that correlation effects have the tendency to localize Ti d electrons ensuring a rather sharp domain wall [214]. At least when $3d$ transition metal states are involved in the formation of the conduction band, this localization seems to be a general trend in oxides. Also at the (001) interface of LaAlO$_3$ with SrTiO$_3$ it was reported that the two-dimensional electron gas formed there is confined within a few nanometers [215].

6.2. Molecular Atomic Simulation

In recent years, atomistic simulations have become a versatile tool for the investigation of dislocations in SrTiO$_3$ thanks to the enhanced computational power and the availability of supercomputers to the research community and a variety of simulations of the influence of dislocations on the structural behavior as well as on the ionic transport has been performed. In molecular dynamics simulations, the forces between the atoms within the simulation cell are calculated and the resulting positions of the atoms are determined in an iterative process. Using indentation in combination with etch-pit analysis the structure and evolution of dislocations under external loads were investigated and corresponding molecular dynamics simulations were conducted showing that the evolution of dislocations as consequence of a mechanical stress is strongly influenced by preexisting dislocations [113]. When a (001) oriented SrTiO$_3$ crystal is plastically deformed by indentation, at first dislocation pile-ups are aligned in <100> directions while at higher loads an orientation in <110> directions is preferred revealing that the evolution and arrangement of the network of dislocations is related to a glide plane mechanism determined by the crystal structure. The simulations indicate that within the glide plane the dislocations tend to dissociate in two partial dislocations with distances of 4-5 nm separated by a strongly distorted antiphase boundary stacking fault [216]. The gliding of dislocations is also strongly temperature-dependent since it was calculated that at high temperatures the dislocations lose their mobility leading to brittle behavior of the crystal [105]. By employing a concurrent atomistic continuum simulation Yang et al. [95] also confirmed the importance of easy glide planes in SrTiO$_3$ resulting in a preference of the <110> directions for dislocation nucleation and revealed that the presence of an agglomeration of dislocations in form of bicrystalline boundaries in SrTiO₃ has a significant effect on the mechanical properties and the crack evolution [97]. Bicrystals have also been widely investigated in order to determine whether dislocations have an impact on the ionic transport of oxygen [45] or not [17]. By molecular dynamics simulations of dislocations it was found that the formation enthalpy of oxygen vacancies is significantly lower compared to the bulk [31]. Hence, in a non-equilibrium situation as is typically present during vacuum annealing, the dislocations will be reduced at first, accompanied by a local insulator-metal-transition resulting in the evolution of a network of conducting nanowires within a less-conductive matrix. On the other hand, it was calculated that in thermodynamic equilibrium the self-diffusion (tracer diffusion) of vacancies along the dislocation is suppressed [30] and Marrocchelli et al. [31] concluded that dislocations are "easy to reduce but not so fast for oxygen transport".

7. Electrical Properties of the Dislocations in TiO₂ and SrTiO₃

The crystallographic and electronic structure as well as the chemical composition of the core of dislocations in SrTiO₃ and TiO₂, as presented in subchapters 5.3.1 and 5.3.2, disclose the fundamentally different physical properties of these objects in the matrix of the two band insulators of our review. In contrast to dislocations in a metal, which show only small modifications in the electronic structure and can be regarded as neutral relative to the rest of the matrix from the electrical point of view (in the matrix we have a free electron gas), the electrical neutrality of non-metallic ionic or ionic-covalent oxides is not given. Before detailed microscopic data with atomic resolution were available, models about the electrical properties of the crystals with dislocations and especially the screening along dislocations (e.g., polycrystalline boundary) were proposed, for instance on the basis of impedance spectroscopy measurements or the analysis of electrical transport phenomena [30,217–219]. Already back in 1978, it was reported that the dislocations in the paraelectric phase of BaTiO₃ single crystals are responsible for "an increased mobility of charge carriers" [220]. The high potential of ceramics on the basis of the transition metal oxides, such as BTO, PZT, BST, for applications in microelectronics has aroused great interest in the understanding of the influence of the polycrystalline boundary (especially the small angle boundary, which was associated in the literature with an array of dislocations with distribution in equidistance along the boundary line) on the global dielectric answer of the ceramic capacitor. Early models about the special role of the boundary in the ceramic materials or artificial bicrystalline boundaries, which have been obtained on the basis of the dielectric spectroscopic studies, have put in the foreground the screening phenomena close to the boundary (with dislocations). In their spirit, these models were a kind of direct adaptation of the space charge model known from semiconductor physics. It should be accepted that, on the basis of the data from impedance spectroscopy measurements of crystals with dislocations and ceramic grain boundaries, these models were a plausible and valuable approach for the description of the neutralization process of the surplus charge in the core of dislocations which can, e.g., agglomerate along a small angle and high angle crystalline boundary. Thanks to the knowledge from EELS spectroscopy about the local valence of the transition metal ions along the core of edge dislocations in TiO₂ and SrTiO₃ (see above), the existence of a high concentration of Ti³⁺ or Ti²⁺ ions in the core of edge dislocations became evident in both materials, and new data from the electron-holographic investigations, considering the charging of the core of dislocations (e.g.,: [11,12,30,40]), started a broad discussion about how the "impedance spectroscopic" description and the "atomic" models (which are directly extracted from TEM-based techniques) could be correlated. The best illustration of the discrepancies encountered for these two perspectives was presented in the paper by Alfthan et al. [92]. The authors proofed, on the one hand, the validity of the classical space charge model by impedance spectroscopy measurements but demonstrated, on the other hand, by use of electron holographic measurements that the core of dislocations has an opposite polarity (here a negative charging of the core of the dislocations) relatively to the mentioned impedance models. However, the discrepancy in this work was connected with an estimation of the spreading of the potential profile (from impedance studies) of about 30 nm (at the grain boundary; here the bicrystalline boundary) with respect to a direct measurement of the dimension of the potential profile of the negatively charged core of single dislocation (by electron holography) resulting in ~2 nm. Note: such holographic results could be considered as a direct projection of the potential distribution in the core of dislocations (~2 nm) [221]. The difference in the extension of the space charge zone presented by Alfthan et al. [91] is here just 2 nm to 30 nm, but measurements of the surface potential distribution, which have been obtained for a similar system (here bicrystalline SrTiO3) using non-contact KPM by Kalinin et al. [94], even indicated that the dimension of the space charge region may be as large as 1.5 μm.

Instead of discussing how the electrical properties of isolated dislocations and their contribution can be "extracted" from the global electric or "dielectric answer" in a real crystal, we should focus our attention at this point of our review on the possibility of investigating individual dislocations under different electrical or mechanical stress.

If we accept that the core of dislocations possesses the typical characteristics of low titanium oxides (including a high concentration of d^1 or d^2 delocalized electrons) and that the Burgers vector of dislocations is invariant (i.e., dislocations cannot simply end within the crystal but only on the outer or an inner surface or by crossing another dislocation) we should observe an inhomogeneity in the distribution of the electrical conductivity in plane of the surface using modern LCAFM studies. The advantage of the LCAFM technique is connected with the very small contact area between the apex of the conducting tip of the cantilever and the investigated surface, which in fact determines 90% of the potential drop and thereby contributes in similar scale to the spreading resistance. The real mechanical contact area of the tip and the galvanic contact area are not the same, which can be simply proofed using Fowler-Nordheim tunneling through very thin SiO2 films [222]. It is surprising that for the tip with typical radius 10 nm the electrical contact can be reduced to only ~1 nm^2. Therefore, it is possible to analyze using LCAFM the map of the electrical conductivity with atomic resolution. Notice: for the correctly prepared surface (after removal of the physisorbates and chemisorbates) of a homogenously conducting material the uniformity of the electrical conductivity is indeed measureable by LCAFM in the plane of the scanned surface (e.g., Au thin films [223]). Other test measurements, here of highly oriented pyrolytic graphite HOPG, reveal small but regular modulations of conductivity on an atomic scale related to its crystallographic structure. Turning to the analysis of the electrical conductivity of the oxides (here band insulators) on the nano-scale poses a further challenge, in that measurements have to be conducted at extremal low values of currents down to 10^{-12}–10^{-15} A. Facilitating such measurements at ultra-low conductivity does require a very sensitive LCAFM current-to-voltage converter (with reduced parasitic capacity in feedback of the I/V converter [224]), see Figure 34.

Spatial variations of the electrical conductivity of stoichiometric crystalline oxides have been observed by LCAFM on the surface of a number of binary and ternary band insulators. This includes our oxides of choice in this review: TiO2 [48,53], SrTiO3 crystals [50,51,55,68,225,226], and doped SrTiO3 such as Nb:SrTiO3 [227], La:SrTiO3 [228], Fe:SrTiO3 [58], as well as related oxides such as BaTiO3 [229,230], PbZrO3 [231], KTaO3 [232], NaNbO3 [233,234], and BiFeO3 [235]. Similar effects have also been observed for thin films, see, e.g., TiO2 [236], SrTiO3, Fe:SrTiO3 [225,237–239], BaTiO3 [230,240], HfO2 [241–244], and NiO [245]. Here, examples of the current (resistance) mapping of SrTiO3 and TiO2 are given in Figures 35 and 36, respectively. It should also be noticed that TEM images of SrTiO3:Nb give a rather homogeneous picture of the samples [246], while LCAFM measurements indicate spatial fluctuations of the conductivity at the surface [227].

LC-AFM: Principle of the measurement of the local conductivity

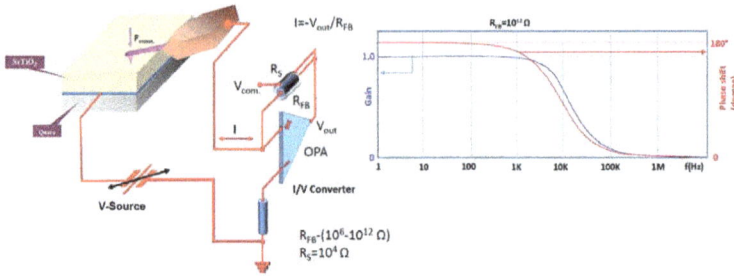

Figure 34. Schematics of the LCAFM method for determining the resistance distribution in the surface layer. An important element of the electronical circuit is the voltage-to-current converter, whose sensibility for current detection of high-ohmic materials should be extremal high (~1–10 fA). Notice: for this ultra-low current the high-ohmic resistor (R_{FB}) in the feedback of the operation amplifier (OPA) needs an additional compensation of the parasitic capacity (see [224]). Only for this compensated configuration the bandwidth of the system is broad enough (see inset) to scan in "finite time" a surface of crystal with resistance of many hundred teraohms.

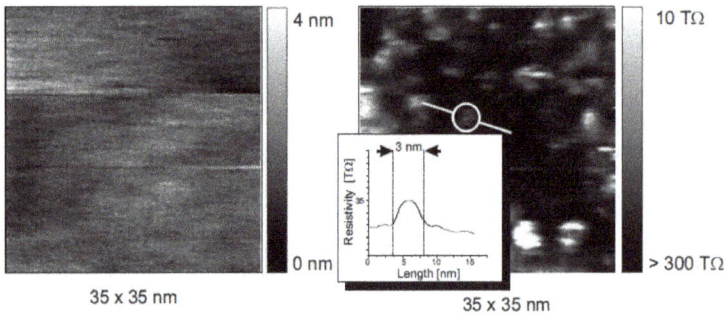

Figure 35. Topography (left) and resistance (right) measured on SrTiO$_3$ (100) crystal using LCAFM. Adapted from [230].

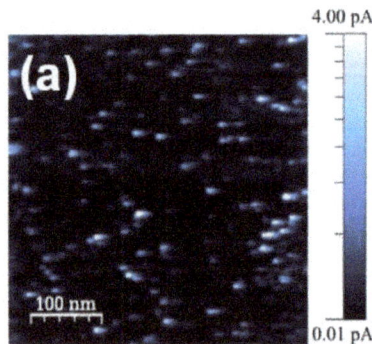

Figure 36. LCAFM image of 500 nm × 500 nm of a 'pristine' TiO$_2$(110) surface recorded at 20 V. Adapted from [53].

Of course, the observation of a local variation of the current by LCAFM studies in the surface plane of the stoichiometric crystals does not necessarily imply that the dislocations play an important role in the electrical transport. Although the measurements were obtained under vacuum condition and after thermal desorption of water (at 250 °C), one should take into account that the rest of the physisorbates or chemisorbates can still exert an influence on the local potential distribution and may automatically modify the local conductivity [230]. A clear evidence for the role of dislocations stems from LCAFM analysis of regions with high concentration of dislocations which have been artificially induced, for example, along steps in a plastically deformed crystal (Figure 37) or at a bicrystalline boundary (Figure 38). Here, the current (or resistance) maps show unequivocally a correlation between the position of conducting filaments and the locations of dislocations. From these data, one can extract that the electrical conductivity of the dislocations for stoichiometric SrTiO$_3$ and TiO$_2$ is at least of 2–3 order of magnitude higher than the conductivity of the matrix. However, it is not possible to precisely determine the actual electrical resistance of the matrix due to the finite current sensitivity of the I/V converter in use (see Figures 35, 37 and 38). Therefore, one can only state at this point that the local resistance is higher than 1 PΩ (Peta Ohm).

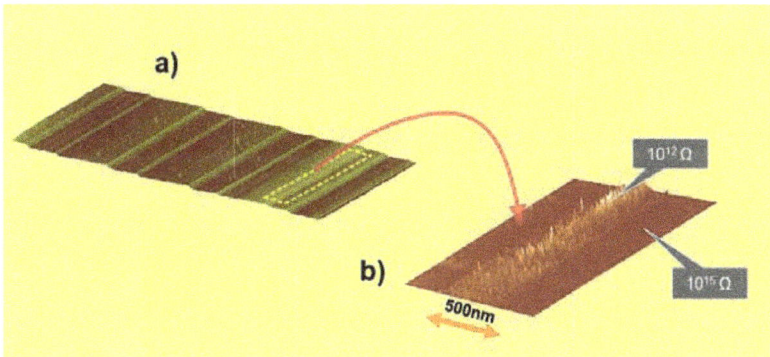

Figure 37. AFM and LCAFM measurements of plastically deformed single crystals of SrTiO$_3$. (**a**) AFM topography after plastic deformation; (**b**) LCAFM conductivity maps at a deformed step. Adapted from [68].

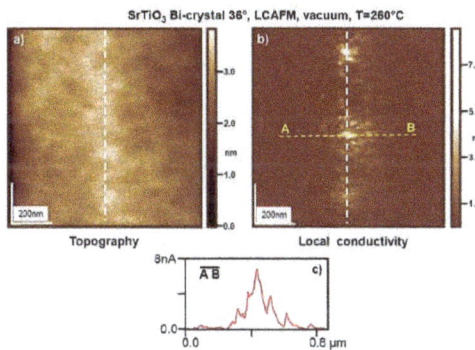

Figure 38. AFM topography (**a**) and LCAFM conducting map (**b**) of the region close to the bicrystalline boundary (36.8°) of stoichiometric SrTiO$_3$ (100). Cross section (**c**) shows a drastic variation of the local conductivity in a broad region perpendicular to the bicrystalline boundary. Notice: Measurement was obtained under HV conditions and elevated temperature. These conditions provided desorption of water molecules.

An additional thermal reduction of the crystal enhances the concentration of the oxygen vacancies [48,50,51,53,55,68,222,225,226] along the dislocations, which in turn leads to a further local increase of the electrical conductivity (with dominance of the electronic conductivity). In contrast to the electrical conductivity of dislocations in stoichiometric crystals, LCAFM studies of reduced crystal do indeed show an enhancement of the conductivity of the core of dislocations relative to the surrounding matrix, leading now to values higher than 3–6 orders of magnitude with respect to the background (here notice also from an experimental point of view that the dynamic of the LCAFM measurements can be increased by a parallel investigation of the regions between dislocations using a higher sensitivity of the I/V converter). This gives an added opportunity to correlate the position of the conducting filaments and the core of dislocations by LCAFM investigations for reduced SrTiO$_3$ [51], reduced bi-crystalline SrTiO$_3$ and TiO$_2$ crystals [48]. For this, pre-etched crystals were investigated after thermal reduction under vacuum conditions, showing that the position of filaments is localized at the center of etch pits which mark the position where the dislocation line crosses the surface (Figures 39–41).

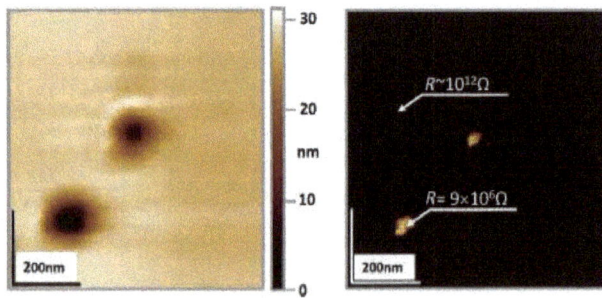

Figure 39. Topography of a slightly reduced TiO$_2$ crystal with beforehand etched surface (**left**). The characteristic etch pits and the map of the electrical conductivity (**right**) show correlation between the position of the etch pits and the exits of the conducting filaments. Notice the contour of the etch pits is not sharply defined due to the smoothing effect induced via thermal segregations of materials in the plane of the surface, adapted from [48].

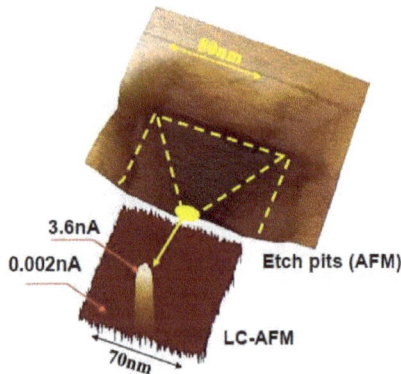

Figure 40. The position of etch pits determined by AFM topography and the coinciding position of filaments with enhanced conductivity (by LCAFM) shown in the surface layer of the (100) face of a reduced SrTiO$_3$ crystal. The SrTiO$_3$ has been extensively etched and subsequently annealed at 1023 K for 30 min at p$_{O2}$ = 10^{-9} mbar (adapted from [51]).

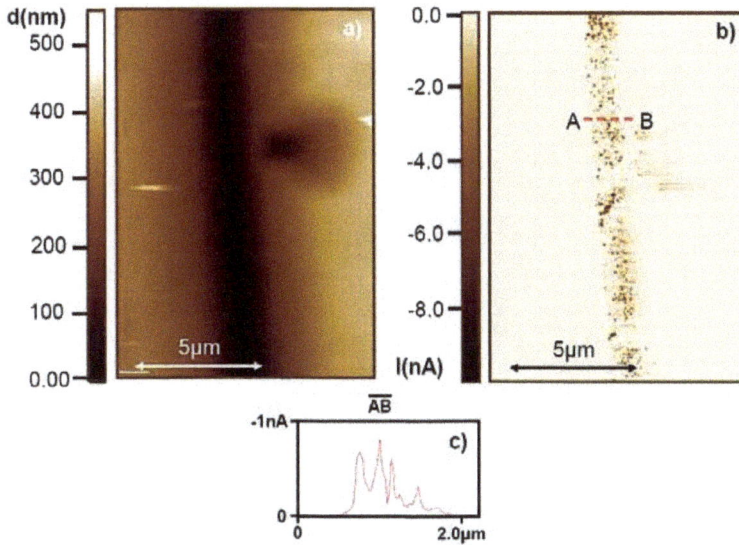

Figure 41. AFM (**a**) and LCAFM (**b**) investigations of the reduced and pre-etched SrTiO$_3$ bicrystal shows correlations between positions of the etch pits (along the boundary) and bands of the conducting filaments. This finding in fact confirms the hypothesis about easy reductions of the core of dislocations. Cross-section (**c**), obtained perpendicular to the boundary, presents that the broad conducting channels are accumulated along the boundary.

So far, LCAFM was only used to map the electrical conductivity of the surface. In order to analyze the electrical conductivity below the surface in an out-of-plane direction, one can additionally cleave the crystals to obtain cross-sectional LCAFM maps.

The fold-up maps of the conductivity in-plane and out-of-plane indicate that a 3D network of filaments exists whose structure is a hierarchical network with high density of the extended defects very close to the surface and a declining concentration of defects in deeper parts of the surface region. A prominent example of such a hierarchical structure with locally enhanced conductivity is presented for a TiO$_2$ single crystal (see Figure 42) which was subjected to a thermal reduction strong enough to establish a pronounced, well-structured network of filaments (here, probably due to a chemical restructuring of the dislocation cores) [48]. We have found a similar density distribution of filaments in the surface region of reduced SrTiO$_3$ crystals (see Figure 43) with a high concentration of well-conducting filaments near the surface and a decline of the density towards the interior of the reduced crystal. Notice: to avoid the interaction with rest gases cleaving of the crystals was undertaken in situ under very low oxygen partial pressure (here 10 mbar of H$_2$).

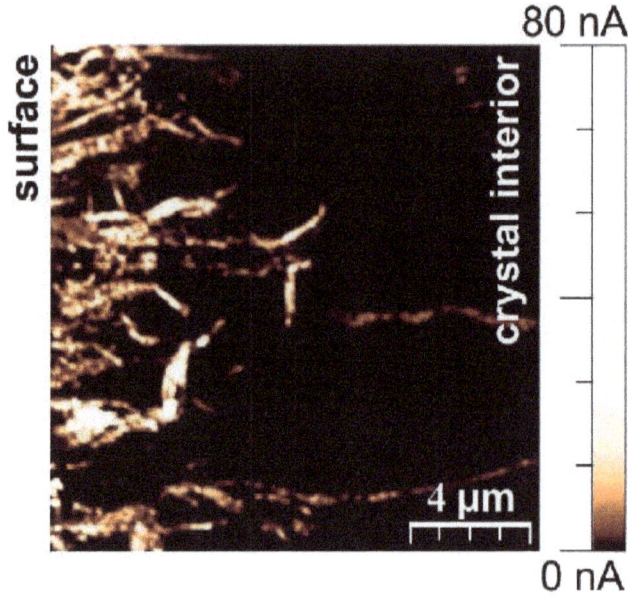

Figure 42. LCAFM on a cross-section of a reduced TiO$_2$ (110) crystal (adapted from [48]).

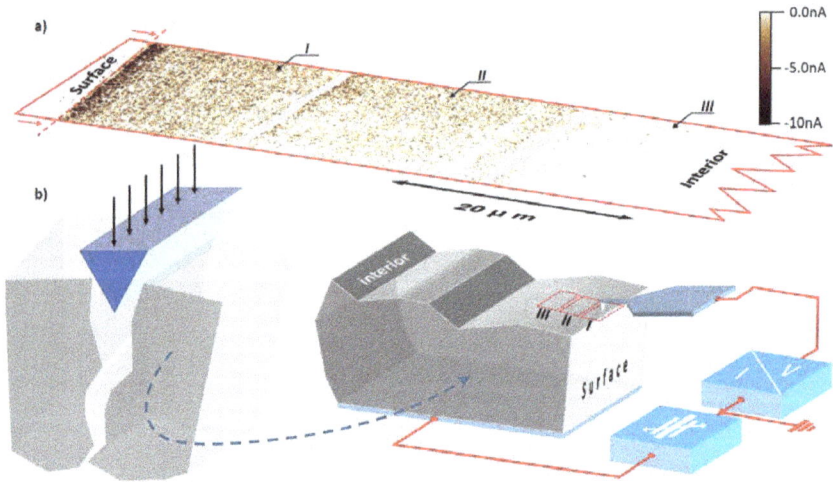

Figure 43. Distribution of the conducting filaments in the surface region in the reduced SrTiO$_3$ crystals revealed by LCAFM mapping (**a**) of the cleaved crystal. The distribution of the conducting filaments shows a similarity to the hierarchical distributions of dislocations, which have been analyzed using TEM technique (see Figure 18). The cleaving of the reduced crystal was obtained at RT in situ (in 4%H$_2$ in Ar) (**b**).

All our results indicate that the hierarchical structure of conducting filaments (Figures 42 and 43) shows similarities to the distribution of the density of dislocations as obtained from electron microscopy (see [10,42]). We are aware that we have generated new populations of dislocations during cleaving our samples for the LCAFM studies of the out-of-plane distribution of conducting filaments in reduced

SrTiO$_3$ or TiO$_2$. However, their conductivity is equivalent to the very low conductivity of dislocations in the crystals without additional reduction. This in turn implies that the "electrical traces" of such mechanically created, new dislocations are invisible for the typical dynamical range of LCAFM measurements for reduced crystals (with current resolution of 4 orders of magnitude). It is important to emphasize that the "electrical trace" of the hierarchical structure in the surface region does not show a noticeable influence of the cleaving conditions for both reduced materials (in situ under vacuum or in H$_2$, as well as ex situ). The most important effect is the preferential reduction of the core of dislocations [31], and "cylindrical distribution of oxygen vacancies around dislocation lines in the crystal" [247]. For a such cylindrical Cottrell atmosphere (with a typical radius between 1 to 10 nm) we can estimate (after determination of the total concentration of the removed oxygen during reduction; see Figure 44) the local concentration of d^1 electrons of strongly reduced SrTiO$_3$ and TiO$_2$ crystals be in the order of 10^{20}–10^{22} /cm^3 (see, e.g., [45]). Notice: the EELS measurement of the core of dislocations for stoichiometric crystal shows that the original concentration of d electrons is a few percent, which is in fact very high close to the metallic limit [39,40,94]. Thus, the enormous additional doping by reduction (or preferential self-doping induced by removing of oxygen from the core of dislocations) is in our opinion connected with a delocalization of the high concentration of d^1 electrons along such extended defects in a pre-existing network (caused by the invariance of the Burgers vectors) and could be made responsible for the insulator-to-metal transition in reduced SrTiO$_3$ [55] and TiO$_2$ crystals [53]. Notice: we have already mentioned above that Buban et al. [40] have clearly presented, using EELS measurement, that the dislocations in SrTiO$_3$ can be associated with TiO nanowires and R. Sun et al. [32] have made a similar statement about the role of the core of dislocations in TiO$_2$ as conducting nano-wires. The insulator-to-metal transition is a long-standing problem in the literature and is often associated with the famous Mott criterion [248] which connects the stability of the metallic state with the carrier concentration (n) and the effective Bohr radius (a_H) of an electron-hole pair, where $a_H = \frac{\varepsilon \hbar}{m^* e^2}$ (Equation (1)). The critical concentration of the carriers can be calculated from the formula $n^{1/3} \times a_H = 0.2 \ (0.254)$ (Equation (2)). The Mott criterion is not only valid for doped semiconductors but also for transition metal oxides. Cox in his book [203] has pointed out that the critical concentration of the electrons for the insulator-to-metal transition in SrTiO$_3$ crystal should be of order of 4.1×10^{18} cm^3 (the calculations were obtained using Equation (1), for $\varepsilon = 300$ (at RT), and $m^* = 10 \ m_0$ (see, e.g., [249]) as well as the collection of data in [250]). In the meantime, an upper bound of the critical concentration was even estimated to be as low as 10^{16}/cm^3 based on electrical measurements [251]. If we neglect for the time being the idea that the d^1 electrons are selectively agglomerated along dislocation lines and are responsible for the local insulator-to-metal transition and postulate instead that the oxygen vacancies are distributed statistically in the reduced crystal we can calculate an effective Bohr radius a_H for the electron carriers. Using in our case 10^{15}/cm^3 as obtained from the above-mentioned effusion measurements, we can determine the Bohr radius according to Equation (2) to be around 26 nm. Since the Bohr radius is connected with the dielectric constant and the effective mass via Equation (1), we can then estimate using these numbers that the dielectric constant has to be $\varepsilon = 4000$ when the effective mass, as specified in literature, is ~10 m_0 (a maximal value). However, this overestimates the dielectric constant of SrTiO$_3$ at RT by one order of magnitude. On the other hand, Spinelli et al. [251] estimated values for a_H of 200 nm at RT and as large as 10 µm for low temperatures, where ε is enhanced by two orders of magnitude. To cope with the very low value of the carrier concentration limit in the case of the classical interpretation of the insulator-to-metal transition for SrTiO$_3$ other authors have suggested resolving the contradiction by applying, e.g., the Mott–Ioffe–Regel (MIR) limit [252]. This model can be used when the mean-free-path of a carrier falls below its Fermi wavelength or the length is smaller than the lattice constant (interatomic distance). Thus, in the literature values of a_H ranging from 1 nm [253] to several micrometers can be found.

This brings us back to the dislocations and their role for an understanding of the nanoscopic nature of insulator-to-metal transition. Maybe a simple experiment provides convincing arguments for the important role of the dislocations by considering two pieces from the same substrate with different density of dislocations and exposing them simultaneously to the same reduction conditions. For this, we used one piece with a high concentration of dislocations, here around $10^{12}/cm^2$ in the surface region, which was generated by mechanical polishing, and another piece with the original epi-polished surface (here the concentration of dislocations is of order $10^{10}/cm^2$). These two samples were simultaneously reduced to the minimum of the resistance of the bathtub curve (see, e.g., [45,226]) and subsequently cooled down to room temperature. Under these conditions, the electrical measurements revealed that the temperature dependence of the resistance for the crystal with high concentration of dislocations is typical for a metallic behavior while it exhibits only semiconducting behavior in the case of the epi-polished crystal (see Figure 45). This does not necessarily imply that individual dislocations in the latter (semiconducting) case may not turn metallic in character but their contribution is simply not sufficient to produce an overall (macroscopic) metallic behavior. It should be noted that the probability of a bundling of dislocations is higher for crystals with high density than for the crystal with low density of dislocations (see, e.g., above in the case of bicrystalline boundaries in Figures 37, 38 and 41). As an effect, such bundles may result in a higher degree of oxygen nonstoichiometry and may, therefore, be more easily transformed to become metallic in character. The simple experiment is also a fine illustration that further systematic and quantitative studies are necessary to compare nanoscopic and macroscopic measurements and to take the inhomogeneity of the crystalline material and the surface region into account.

A further support for the notion of the inhomogeneity of the electrical conductivity in $SrTiO_3$ crystals and the important contribution from the core of dislocations stems from Electron Beam Induced Current (EBIC) studies on reduced $SrTiO_3$ crystals. Using appropriate Schottky contacts (in this case, a thin Pt electrode becomes deposited on the surface of reduced $SrTiO_3$ crystal [241,242]) the contributions of different locations in the surface region (i.e., below the contact) for recombination processes of the incident electron beam can be analyzed. Measurement of the current at the bottom electrode synchronously to the scanning position of the incident electron (SEM) beam gives then the planar distribution of electronically active defects in the form of a current contrast. The EBIC maps for a reduced $SrTiO_3$ crystal with a Pt/Schottky contact reveals an orthogonal arrangement of dislocations and two kinds of reactions to the electron irradiation (Figure 46). On the one hand, linear objects with reduced contrast (it corresponds probably to the electron interaction with dislocations with lower conductivity than the rest of the matrix) can be observed. On the other hand, dislocations with higher conductivity than the rest of the crystal can be identified by a higher EBIC contrast. This result is not surprising considering that two types of dislocations (Sr-rich and Ti-rich) exist in stoichiometric and reduced $SrTiO_3$. The local electronic and crystallographic structure of these types of dislocation does indeed show dramatic differences. Similar EBIC investigation of the contact Pt/Nb:$SrTiO_3$ (with Nb concentration of 0.01 weight percent) demonstrates that, even in a system with metallic conductivity, one can identify the bright contrast of a relatively high population of dislocations in the interface at an appropriate voltage, here V < 5 kV [13]. This implies that dislocations in such a system may even show an enhancement of the electrical activity though the surrounding crystalline structure is expected to be metallic in character.

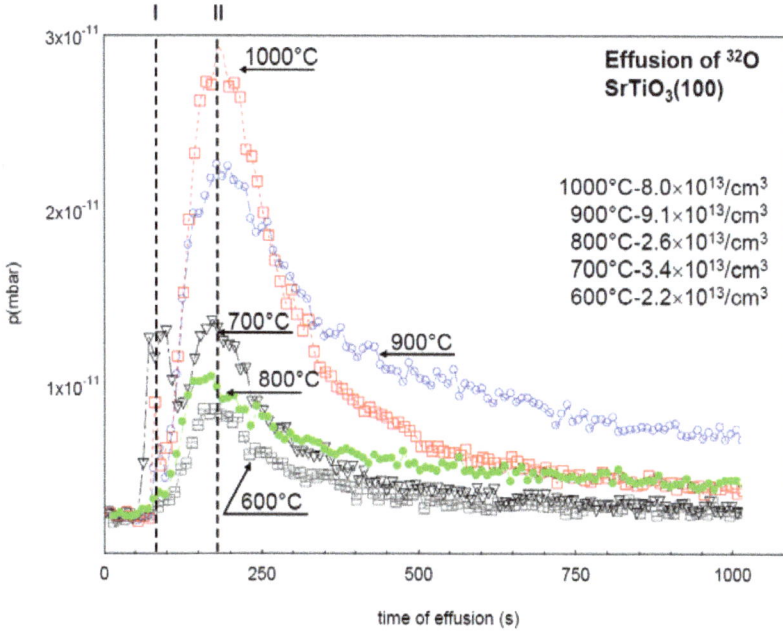

Figure 44. Measurement of the thermo-stimulated desorption of oxygen (obtained using mass spectrometry) during successive reduction of SrTiO$_3$ crystal under UHV conditions (T = 600–1000 °C) shows that the total concentration of the oxygen molecules, which have left the crystal is smaller than 10^{15}/cm^3. Notice: the apparatus was calibrated using controlled pump out under isobaric conditions (p$_{const.}$ = 10^{-6} mbar) of the vessel with exactly defined concentration of oxygen particles. The calibration procedure was the same as has been used for effusion measurement in our previous paper [45].

The open question for LCAFM and EBIC investigations, here especially for donator doped crystals, is the influence of the surface layer. On the one hand, we know that the bulk of the Nb:SrTiO$_3$ crystals possess a homogenous distribution of aliovalent Nb ions and, without doubt, the dislocations play only a minor role for the global metallic conductivity of the matrix induced by such chemical doping, as presented by Rodenbücher et al. [246]. On the other hand, in situ and *in operando* investigations using surface sensitive techniques of the surface layer of Nb:SrTiO$_3$ [227] give evidence that the stoichiometry of Ti and Sr can dramatically change and the typical so-called metallic pick at the bottom of the conducting band actually disappears, which in turn "switches" the electrical properties of the surface region to become semiconducting in character. Therefore, a higher electrical activity of the dislocations can be observed in the surface region.

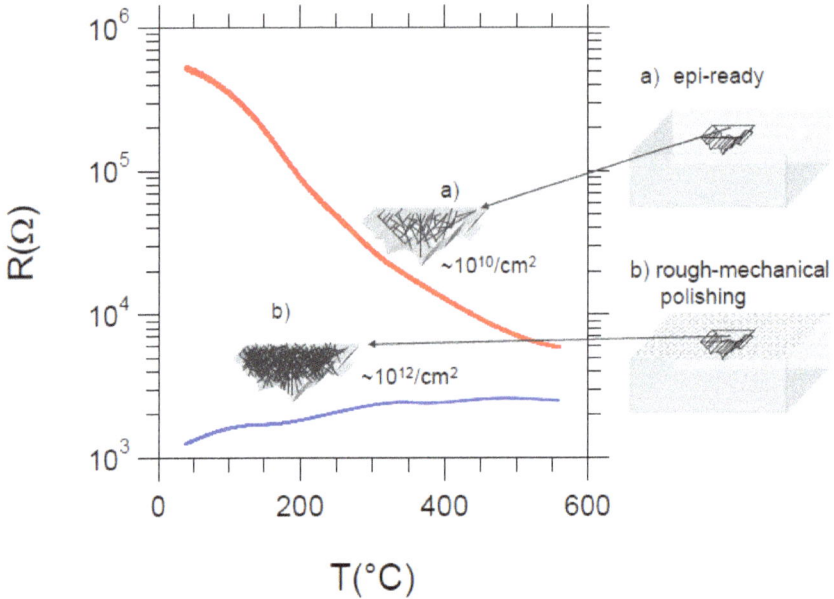

Figure 45. The simultaneous reduction of two pieces (one with original epi-polished surface and one with higher roughness, at T = 550 °C and vacuum -10^{-7} mbar) from the same $SrTiO_3$ (100) crystal under UHV conditions shows that the thermal dependence of electrical conductivity of the piece with original epi-polished surface gives semiconducting behavior while in contrast the electrical resistance of the piece of the crystal with the rough surfaces (generated via scratching) reveals the characteristics of metallic conductivity. Note 1: mentioned mechanical preparation of the original (epi-polished surface) leads to the increase of the concentration of dislocations from 4×10^9/cm^2 to higher than 10^{12}/cm^2. Note 2: the thermal reduction of both samples was obtained for an optimal reduction time, namely for the time which allowed reaching the minimum of the resistance curve (for more details please see reference [68]).

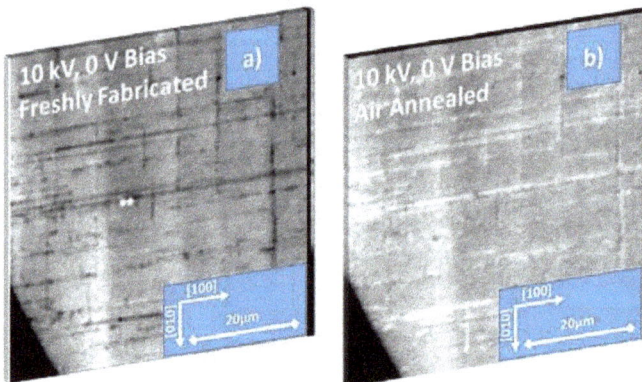

Figure 46. Maps of the electronic beam induced contrast in the surface region between reduced $SrTiO_{3-x}$ and Pt thin films (~100 nm). Electron Beam Induced Current (EBIC) investigation was obtained with energy 10 keV. (**a**) was imaged immediately after fabrication; (**b**) after annealing the contact in air at 120 °C for 10 min. Adapted from [242].

In essence, studies of the local conductivity of oxide surfaces, here summarized for the surface layer of SrTiO$_3$ and TiO$_2$, show that the presence of dislocations can lead to modifications of the electrical transport on the nano- and micro-scale. This effect could be connected with the frequently observed inhomogeneity of the electrical surface properties of oxide materials.

8. Role of Dislocations in Resistive Switching of TiO$_2$ and SrTiO$_3$ Crystals

Easy and preferential reduction of the dislocations cores in oxides cannot only be induced by thermal reduction at high temperatures and low oxygen partial pressures [45,51,55]. Another way to switch dislocations in both materials from being semiconducting (corresponding with their original state in the stoichiometric crystal) to become metallic filaments is possible using electrical polarization with appropriate voltage and current. Although it was already presented in 1962 that under external electrical stimuli the resistance of polycrystalline BaTiO$_3$ can be reduced [254], only as recently as 1992 it was observed for KNbO$_3$ that the resistance of single crystals can be easily increased by many orders of magnitude up to metallic behavior by the application of a constant voltage or constant current polarization [59]. As an effect, a dense set of filaments are being created between the position of the geometrical position of the cathode and a progressing virtual cathode. As a result of the electrical stimulus, the electrical conductivity of the whole crystal gets "switched" from insulating to metallic or even superconducting behavior [4] (see Figures 47 and 48). Such transformation, i.e., switching, of the macroscopic crystal into an object with metallic conductivity (see the typical dependence of the resistance as a temperature function, Figure 47 or Figure 49), has been observed for many crystals with perovskite structure (SrTiO$_3$ [45,51,55,68,225,226,255], see also Figure 49), Nb:SrTiO$_3$ [227], Fe:SrTiO$_3$ [58] (see also Figures 50 and 51), La:SrTiO$_3$ [228,256], BaTiO$_3$ [257], KTaO$_3$ [232], NaNbO$_3$ [233,234], PbZrO$_3$ [231], BiFeO$_3$ [235], and binary crystals such as TiO$_2$ [48,53] (see also Figures 52 and 53). The electrically induced insulator-to-metal transformation is in its character very similar to the above described thermal reduction process (see previous chapter). The dynamics of the electrically induced change of the resistance for these ternary and binary crystals, relatively to the resistance of the stoichiometric crystal at room temperature, is highly effective and can be as large as 6 to 11 orders of magnitude.

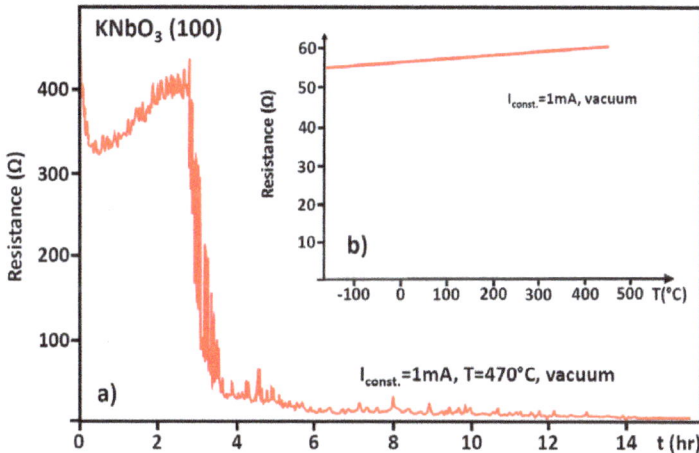

Figure 47. Electroreduction of KNbO3 single crystal (**a**), by electrical polarization using a constant current in the paraelectric phase (T = 470 °C, vacuum)). Inset (**b**) shows the typical metallic temperature dependence of the resistivity for the completely reduced sample. Adapted from [59].

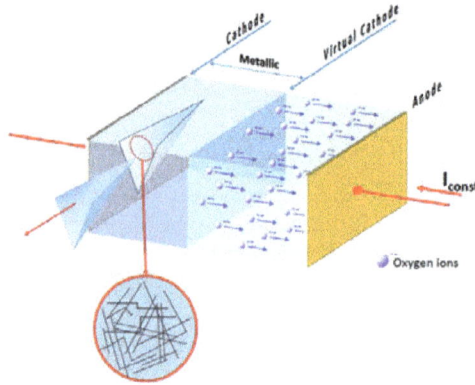

Figure 48. Schematic illustration of the electroreduction process. Adapted from [59].

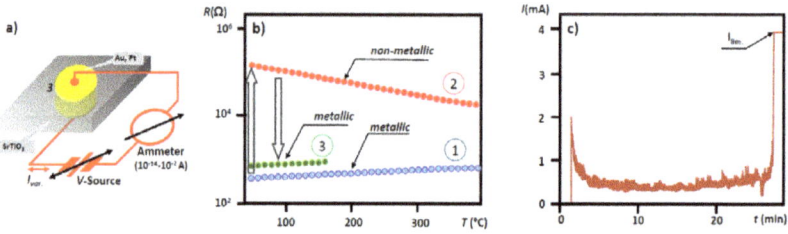

Figure 49. Insulator-to-metal transition and resistance switching in undoped SrTiO$_3$ single crystals. (**a**), Schematic of the reduction and electroreduction apparatus. (**b**), Measurements of the resistance versus temperature (1) crystal after thermal treatment at elevated temperatures under reducing conditions, indicating metallic behavior; (2) after re-oxidation under ambient conditions with an activation energy of 0.1 eV; and (3) after electroreduction at room temperatures under high-vacuum conditions in the constant-voltage mode (*V*-source set at ±100 V and the current flow, *I* $_{var}$, limited to *I* = ±4 mA), restoring metallic conductance. (**c**), Evolution of current for the final electroreduction process (from step 2 to step 3) in a sandwich metal–insulator–metal structure. The current initially shows classical dielectric relaxation with non-regular fluctuations superimposed on it and suddenly jumps to a current above the set current limit I_{lim}, corresponding to the LRS with 800 Ω (adapted from [55]).

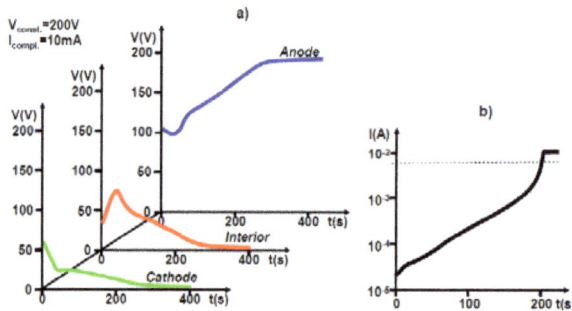

Figure 50. Potential difference change for region close to the anode, cathode and for the bulk during electroreduction of Fe:SrTiO$_3$ single crystal, by electrical polarization using constant voltage (**a**). Inset (**b**) shows the increase of the current flow through the crystal. Adapted from [58].

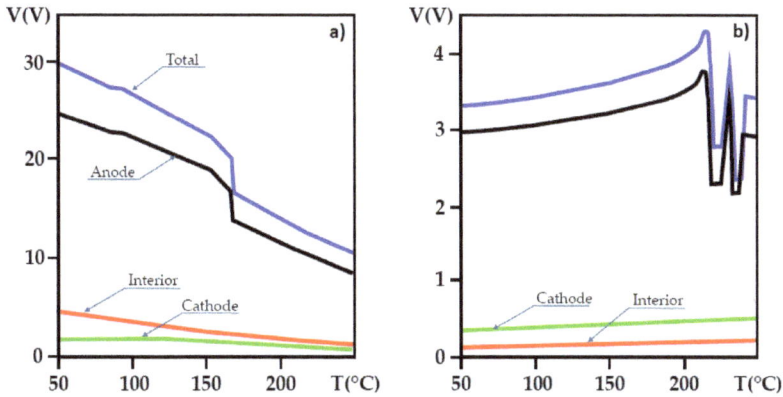

Figure 51. Temperature dependence of the potential drop along the electrodegraded Fe:SrTiO$_3$ crystal for constant current polarization. The measurements have been obtained for different electrodegradation (electro-coloration) times (**a**) t = 20 h; (**b**) t = 80 h. Adapted from [58].

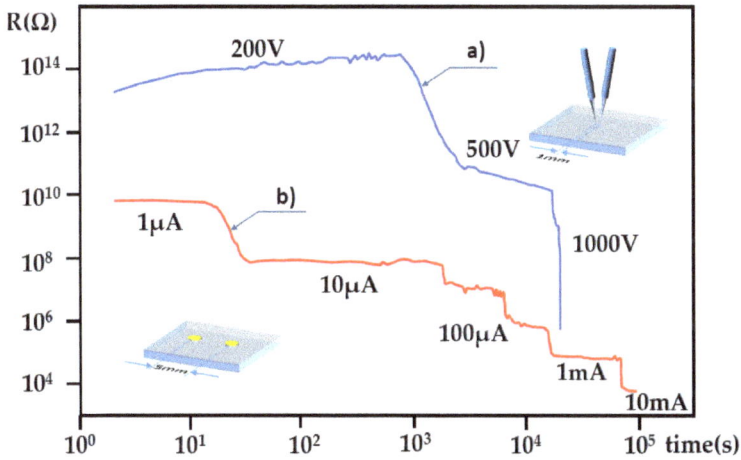

Figure 52. The resistivity of a TiO$_2$ single crystal during electrodegradation experiment. The (**a**) curve presents a voltage-controlled course with the setup presented in the right inset (two platinum needles were in mechanical contact with the crystal surface). The (**b**) curve presents the current-controlled course obtained with the setup from the left inset (the current was applied to the two platinum electrodes applied on the TiO$_2$ surface by sputtering). Adapted from [53].

a)

b)

c)

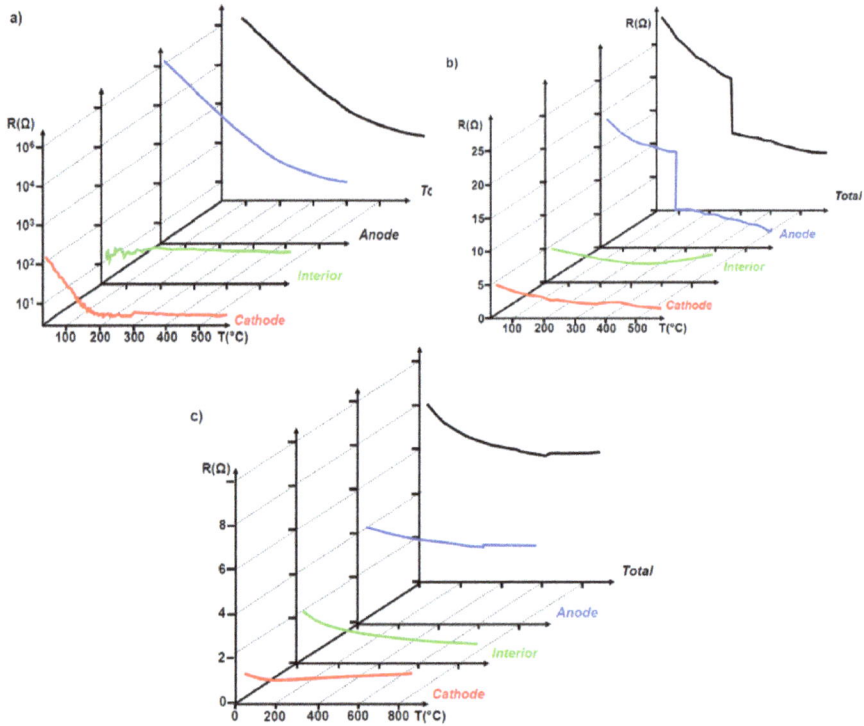

Figure 53. The thermal dependence of resistance for the cathode, interior and anode regions of the TiO$_2$ sample after (**a**) 'soft'; (**b**) 'heavy'; and (**c**) 'extra heavy' electroforming. Adapted from [53].

The electrodegradation process is accompanied by two visual effects: on the one hand, one observes a modification of the color due to a network of bands with better electrical conductivity compared to the rest of the crystal (often the bands possess an orthogonal orientation, see Figures 54 and 55). On the other hand, bubbles develop on the metal film of the anode in the course of the electrical stimulus. Even more, the bubbles do not disappear after switching "off" of the electrical polarization. The bubbles decorate the positions on the anode at which oxygen ions or molecules leave the crystal (see Figure 56). Because the electrodes are in principle obtained from a hermetic material (here, thin uniform deposited metallic films, e.g., Pt or Au) the effusion of the oxygen into the surroundings is hindered. After having reached a critical pressure, this causes a local plastic deformation of the electrode [48,51,53,55,58] (see Figures 54–56). A similar electrodegradation process undertaken for the system Pt/SrTiO$_3$/Pt with porous electrodes shows that the system is in fact open and effusion of oxygen becomes detectable. The effusion of oxygen from the crystals can be monitored by mass spectroscopy in the course of the electrodegradation processes. The thus measured, electrically induced increase of the oxygen nonstoichiometry in the SrTiO$_3$ crystal is of the order of 10^{15}/cm^3 (for 1 mA) or 10^{16}/cm^3 (for 10 mA) (see Figure 57) and is thus one or two orders of magnitude higher than for thermal reduction, respectively (see Figure 44). Therefore, it should not be surprising that the system for this electrically enforced oxygen nonstoichiometry has been switched into the metallic state (Figure 49). At this point it is interesting to raise the question whether the nature of the electrical insulator-to-metal transition is similar to the processes involved in thermal treatment; here in particular considering the aspect whether the electrically induced removal of oxygen is limited to the core of dislocations in the surface layer of SrTiO$_3$ or TiO$_2$. The first indicator that this assertion may be true is the fact that there exists a correlation between the arrangement of the bubbles and the crystallographic direction of the material, which can easily be checked using simple optical microscopic inspection (see, e.g., Figures 54–56). Another convincing argument for this hypothesis is the observation of bubbles evolving along the bicrystalline boundary (see Figure 56).

A further, though indirect, effect which suggests the important role of dislocations in the electrodegradation phenomena is related to the easy chemical transformations of the cathode region of, e.g., TiO_2 crystal into Magnéli phases, as confirmed by X-ray and electrical measurement [53]. The argument is, that, following Landuyt, and Amelickx [124,258], dislocations (so-called hairpin dislocations) are a necessary requisite for this kind of solid state reaction to occur.

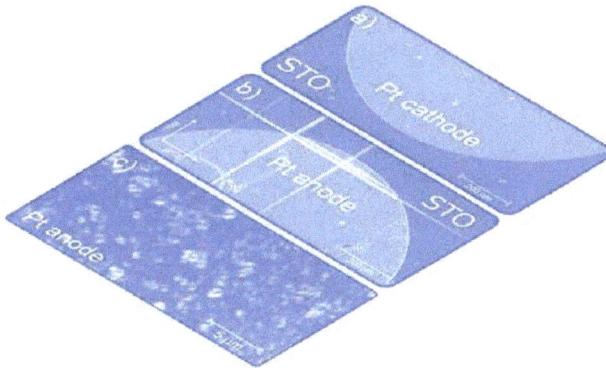

Figure 54. Images from the optical microscope (phase contrast) showing the sample after electrocoloration: (**a**) in the cathode region; (**b**) in the anode region; (**c**) magnification of the anode region with micro-bubbles. Adapted from [58].

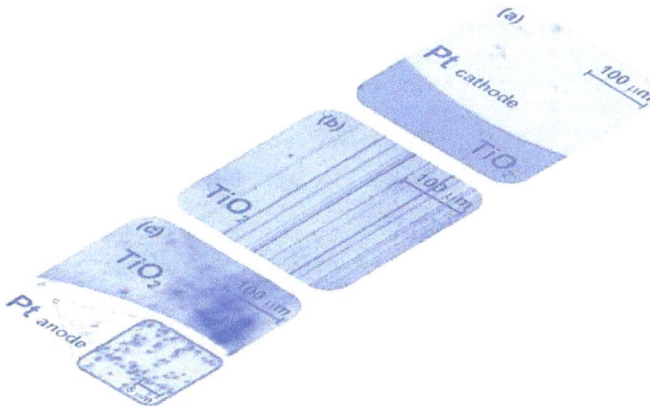

Figure 55. The optical microscope images of the TiO_2 single crystal sample after the electrodegradation process carried out with the setup presented in the left inset in Figure 17. The images (**a–c**) show the cathode region, the central region (between the electrodes) with visible parallel stripes (agglomeration of oxygen vacancies) and the anode region with bubble-like changes visible in the magnification in the inset in the (**c**) part. The average dimension of observed bubbles is close to 800 nm. Adapted from [53].

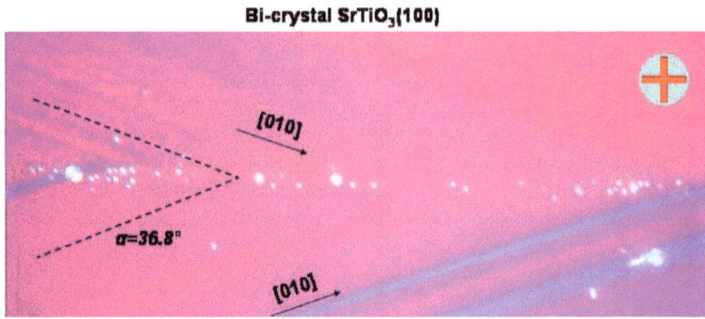

Figure 56. Creation of bubbles on the anode (30 nm-Pt thin film) of a SrTiO₃ (100) bicrystal during electrodegradation. It should be mentioned that along such a boundary a broad band of the dislocations and dislocation bundles has been found (see Figures 12–16). Notice: a simple identification of the tilting angles of both parts of the bicrystal on the non-transparent Pt electrode is an effect of the plastic deformation of the electrode which leads to the interference contrast.

A participation of an array of dislocations or their bundles (agglomerated along a bicrystalline boundary) can not only be identified for electro-migration processes, but the self-diffusion of oxygen in bicrystalline SrTiO₃ at high temperatures is similar in character. An illustration of a selective incorporation of ¹⁸O isotope (under equilibrium condition) is given in Figure 58, where the exchange of isotopes takes place preferentially along the boundary as well as in regions with a high concentration of dislocations originating from local mechanical damage [259].

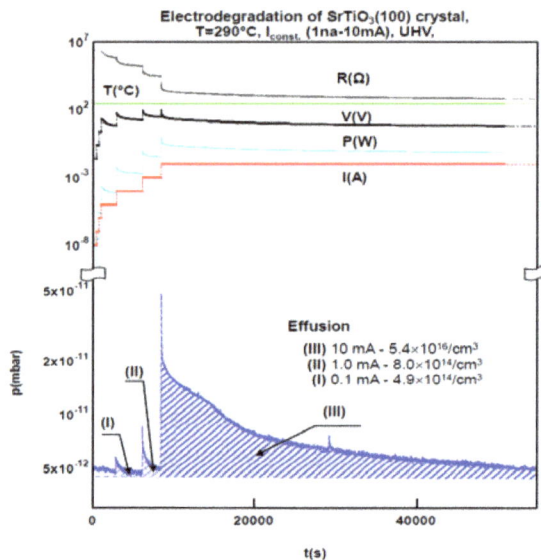

Figure 57. Study of electro-degradation (I-const.) and effusion processes shows that the decrease of the resistance of a SrTiO₃ crystal can even be observed for very low currents. A measurable effusion of oxygen from the crystal can be identified for currents as low as 0.1 mA. Yet notice, although electro-degradation with I ~0.1 mA already leads to a "switching" of the electrical properties of the crystal into metallic properties the total concentration of removed oxygen is still much lower than predicted from Mott criterion.

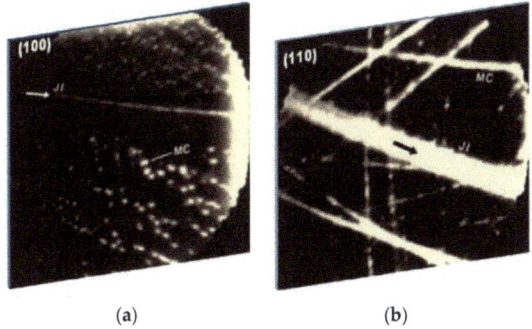

(a)   (b)

Figure 58. ^{18}O images at a depth of 1 μm in (**a**) (100) and (**b**) (110) bicrystals annealed at 1325 K for 1800 s. JI and MC indicate the joining interface and mechanical damage. The integration time of ^{18}O is 300 s. The size of ^{18}O images is 100 mm. Adapted from [259].

The insulator-to-metal transition induced by electrodegradation cannot only be studied on the macroscopic scale using two or four-point electrical measurement or optical inspection, but also the nanoscopic dimension of this transformation can again be investigated using LCAFM techniques. As illustrated in Figures 59–61, one can clearly discern a fine structure of conducting filaments along bands between the cathode and the anode on the surface of electro-reduced crystals (here $SrTiO_3$, Fe $SrTiO_3$, TiO_2) with a planar geometry of the electrodes. This linear arrangement indicates a correlation between the position of the filaments and the position of dislocations or dislocations bands. Moreover, etch pits investigation on such electrodegraded crystals support this notion in that the etch pits accumulate in just the same way along the bands with high concentration of filaments. It should be mentioned that the local enhancement of the conductivity of filaments with respect to the matrix amounts to three to four orders of magnitude.

Additional supporting argument about the important role of dislocations in the electric activity of the surface layer of $SrTiO_3$ can be obtained from the analysis of the already above mentioned EBIC measurements for stoichiometric [242] (see also Figure 46) and as well as for doped $SrTiO_3$ crystals [241]. One can see that the brightness of the EBIC contrast in the vicinity of the straight lines which marking the segments of dislocations can be changed by applying a bias voltage in the case of undoped $SrTiO_3$ crystals (here −10 V, Figure 62) and Nb-doped $SrTiO_3$ (0.01%) (−2 V, Figure 63), respectively.

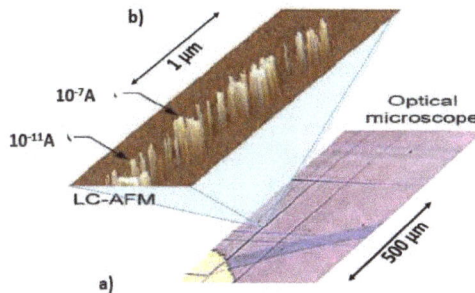

Figure 59. Electroformation of $SrTiO_3$ single crystals showing an orthogonal structure of filaments induced in the surface layer of $SrTiO_3$ (100). (**a**) Optical image of the filaments network close to the cathode; (**b**) Conductivity distribution measurement (LCAFM) along the filaments which cross the surface of the $SrTiO_3$ crystal. Adapted from [45].

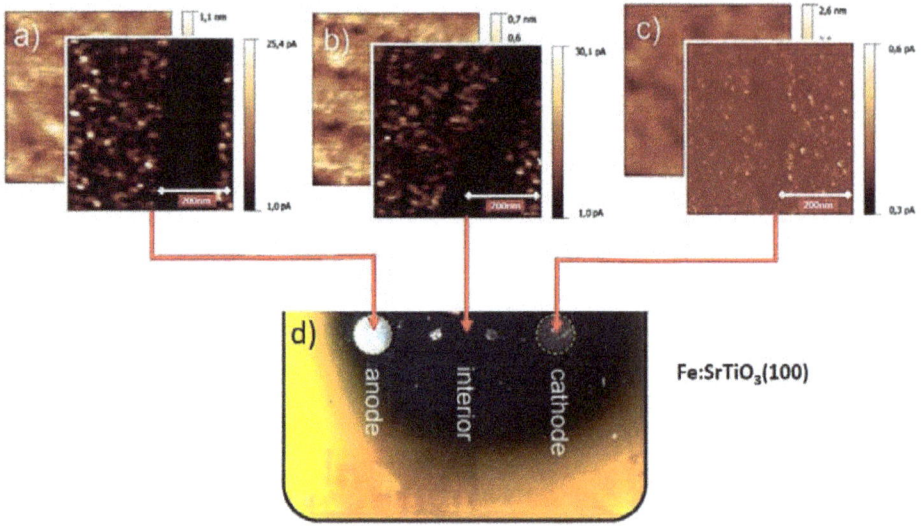

Figure 60. LCAFM measurements of a Fe:SrTiO₃ (100) crystal after electro-reduction. The local topography and corresponding local conductivity maps for: (**a**) cathode; (**b**) interior; and (**c**) anode are presented. Optical image of the sample after electroreduction and reoxidation in ambient atmosphere (**d**). Adapted from [58].

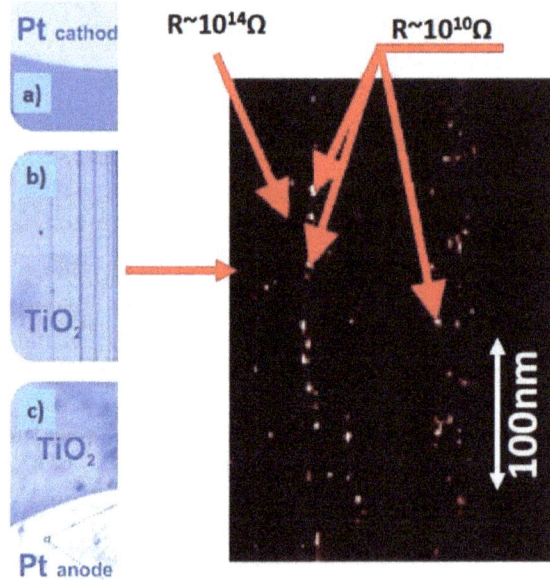

Figure 61. LCAFM measurements (in-plane) on electrodegraded TiO₂ (110) crystals. Polarization ~100 mV temperature below 300 °C. Adapted from [53,68].

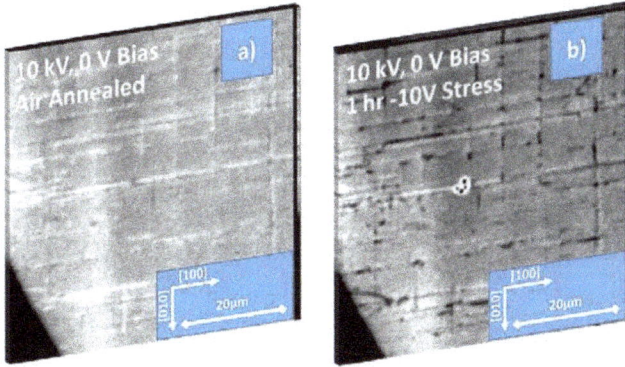

Figure 62. EBIC images of a Pt/SrTiO$_{3-x}$ contact. All images are taken with a 10 kV beam acceleration. (**a**) was taken at 0 V applied bias to the Pt contact; (**b**) after stressing the air annealed contact at −10 V for 1 h.

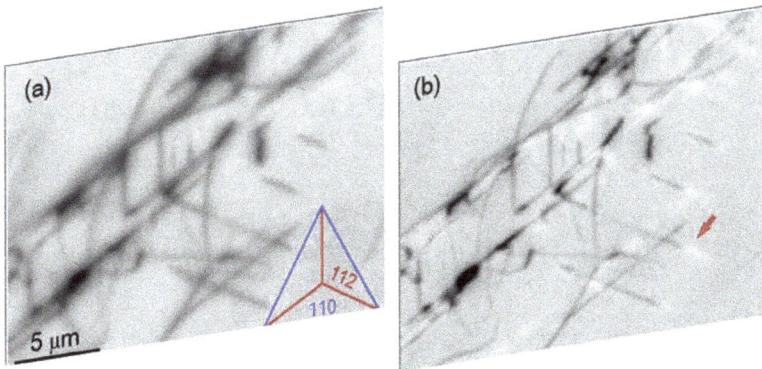

Figure 63. EBIC images of dislocations in SrTiO$_3$ (0.01%Nb) at zero bias (**a**) and at −2 V (**b**). All the dislocations exhibited dark EBIC contrast at zero bias. When applying bias, bright contrast appeared in the vicinity of some straight lines with the directions along <112>. Adapted from [241].

An important aspect of the foregoing analysis is the possibility to locate and electrically address individual dislocations crossing the surface, as demonstrated in the case of SrTiO$_3$ [51,55,68,225,260] and TiO$_2$ [48,53]. In fact, LCAFM has proved to be an invaluable tool for this, as it allows us to address single dislocations crossing the surface with adequate spatial resolution, and use, at the same time, the conducting cantilever as a nanoelectrode through galvanic point contact. Some preparatory procedures by brief thermal pre-treatment allowed to easily locate single dislocations with radius of about 2 nm and then influence and modulate the electrical state with currents smaller than 1 µA [68] (Figure 64). Due to the low current necessary, the process cannot be classified as only a thermally stimulated process (e.g., [261]). In our scenario of the resistive switching of individual dislocations in SrTiO$_3$ via the conducting tip of the LCAFM cantilever the following parameters play an important role:

- the polarization (negative or positive) of the tip,
- the voltage, i.e., local strength of the electrical field, and
- the low oxygen partial pressure available in the AFM microscope.

For the negatively polarized tip of the cantilever it is possible to reduce the concentration of oxygen in the last segment of the dislocations crossing the surface by a cathode attack of hot electrons which will be emitted via field emission from the tip [55,68]. This way, the dislocations can be reduced successively on the full length by the movement of the "virtual cathode". This electrical state can be defined as a low resistance state (LRS) (Figure 64a–c). DFT analysis demonstrates that an increase of the oxygen non-stoichiometry along a dislocation core induces a metallic state close to the bottom of the conductions band. For the opposite polarization of the LCAFM tip (here positive), oxygen ions can be pulled from the inner network of dislocations and moved to those parts of the dislocations which contain a high concentration of oxygen vacancies. This then resembles an oxidation process of the core of dislocations which leads to an "annihilation" of the metallic state in the band gap. (This oxidation process restores the original semiconducting properties of the core of dislocations, the so-called high resistance state (HRS)). Also, on the TiO_2 surface, highly localized regions such as conducting grains with diameters of several tens of nanometers can be switched between states of different conductivity [53] (see Figure 65).

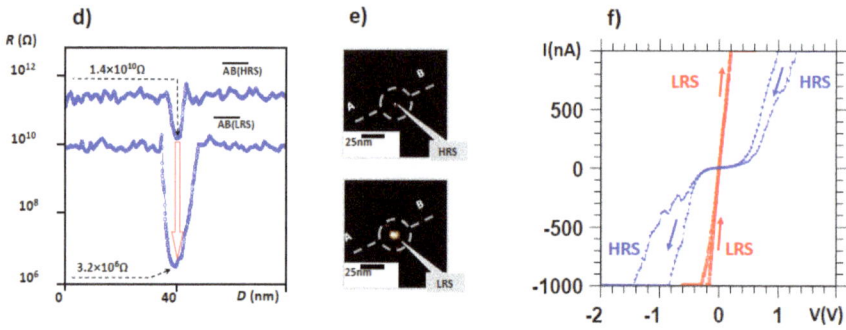

Figure 64. The electrical addressing of an individual dislocation in $SrTiO_3$, using the tip of LCAFM, clearly demonstrated the possibility for a static manipulation of its resistance (**a–c**). For the dynamical polarization, typical cyclic switching I/V curves with two branches can be observed. The branch with linear dependence between current and voltage represents the LRS state. For the second branch, the I/V characteristic has typical marks of a nonlinear behavior; it is a HRS state. The switching between the HRS and the LRS state occurs in the III. quadrant. Restauration of the high ohmic state takes place in the first quadrant. Adapted from [55].

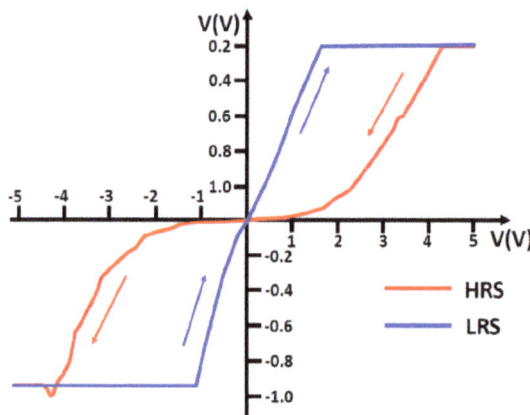

Figure 65. The I/V curve recorded during reversible switching inside of the electric-modified grain of the TiO_2 surface. Adapted from [53].

This effect of a "preferential addressing" of dislocations during electrical polarization can be highlighted by inspecting the case of a bicrystalline boundary which is macroscopically electrodegraded employing a planar geometry of the electrodes (Figure 66a).   After the electrodegradation, the local conductivity between the electrodes was measured by LCAFM under ultra-high vacuum (UHV) conditions. Although a broad area of the crystal was electrically polarized, a dramatic change of the electrical properties was only identifiable (on the micro- and nano-scale) along the bicrystalline boundary (Figure 66c). The nature of electroformation and switching along nano-filaments in single crystals is still under discussion [29,47–55,262,263]. One issue, which has not been dealt with in this chapter, is related to the possibility of a local transformation of the model binary and ternary oxides (here TiO_2 and $SrTiO_3$ crystal or thin films) into new chemical phases [e.g., Ti_4O_7, Ti_5O_9, Ti_2O_3, TiO, $SrO \cdot SrTiO_3$, or $Sr_2Ti_6O_{13}$ or $Sr_1Ti_{11}O_{20}$, [57,121,182,264–267]]) under electrical stimuli. Such solid-state transformations are most likely connected with extended defects. This holds particularly true for dislocation cores which can be considered as foreign objects in the matrix (see above and [94]) due to their different crystallographic structure and chemical properties with respect to the surrounding matrix. They could thus play an important role as a "seed" for the thermally or electrically induced transformation. In fact, thermodynamic arguments show [268] that if the resistive switching would only be induced by a change of the center of gravity of the population of point defects in the crystal the induced, new resistance state would not be stable on a long time scale. We therefore argue that the chemical and crystallographic transformation of the core of dislocations seems to be a plausible alternative to models which solely rely on a Schottky-type disorder [267,269,270].

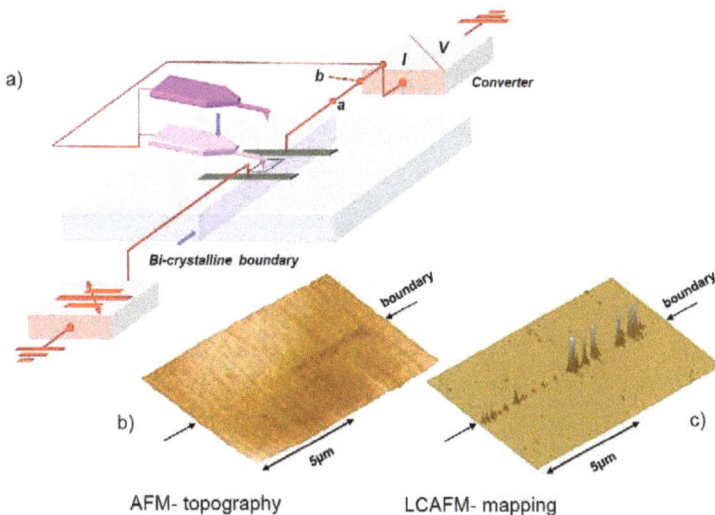

Figure 66. (a) Schematic view of an electrodegradation experiment with a bicrystalline boundary where planar electrodes have been mounted on both sides of the crystal and the contacting and electrodegradating the bicrystalline boundary (switcher in position "**a**"). For the switcher in position "**b**" (open), the mapping of the electrical conductivity on the nanoscale using LCAFM could be measured; (**b**) shows a slight change of the topography along the bicrystalline boundary, probably as an effect of an electrically induced migration. Conducting filaments have been preferentially created along the boundary (**c**)).

9. Mechanical Properties versus Dislocations

It may sound unrealistic if someone says that he can plastically deform a single SrTiO$_3$ crystal with bare hands, but, as already mentioned above, this statement is indeed true. In fact, the ductility of the SrTiO$_3$ crystal at room temperature is just about 10% [68,107]. Using the device which is illustrated in Figure 67 it is possible to deform (plastically) a SrTiO$_3$ crystal in a controlled manner [68,104,132]. Depending on the radius of the prism and external forces F$_1$ and F$_2$, the density of the steps and their height close to the contact lines (bands) can be manipulated (see Figure 67). This effect of an extraordinary shift of whole blocks of the crystal is connected with the polygonization of the edge dislocations in the plane {100} and {110}. An impressive confirmation that the ductility of the mechanically deformed SrTiO$_3$ crystal is connected with the linear arrangement of dislocations is given by the perfect ordering of the etch pits observed along the steps (Figure 68d–f). It should be mentioned that by mechanical deformation two different mechanism for the new ordering of dislocations become activated. The first one is connected with a shifting of the dislocations (via gliding) which exist in original crystal to new positions and their grouping in the appropriate gliding plane. The second mechanism has two steps. During the first step the new population of dislocations will be nucleated and subsequently, in the next step, the dislocations will be polygonized. The possibility to generate new dislocations is in fact an effect of the increase of the local pressure along the contact line (Figure 67) to values which allow their multiplication. Therefore, the density of the dislocations along the contact bands is much higher than the density of dislocations in the original crystal.

The plastic deformation of SrTiO$_3$ crystals seems to takes place in principle along the <110>directions [15,16,33]. In our investigation with the mentioned configuration of the acting forces as presented in Figure 67, we found that the dislocations occasionally assemble in an arrangement along <110> while numerous groups of dislocations were observed with orientation in the direction <100> (see Figure 69a). At the same time, a crossing of these different directional arrangements can be observed and seems to be "independent", that means without mutual interaction of both planes with accumulated dislocations (Figure 69b,c).

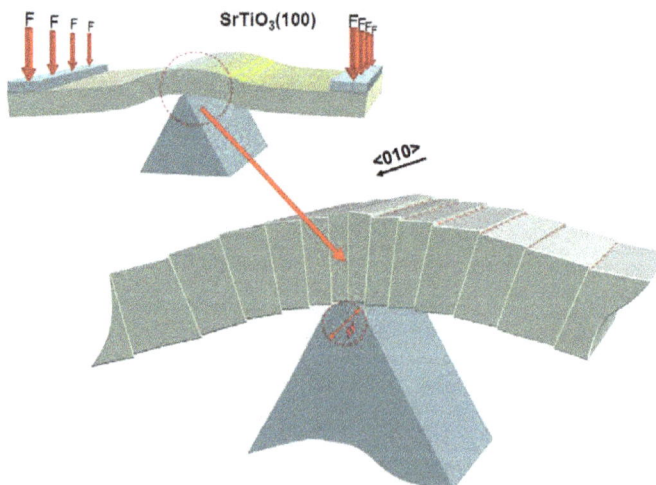

Figure 67. Visualization of the device which has been used for the plastic deformation of the SrTiO$_3$ crystals. The strength of the deformation can be controlled by external forces and curvature of the part of prism which is in a direct contact with the bottom of the crystal.

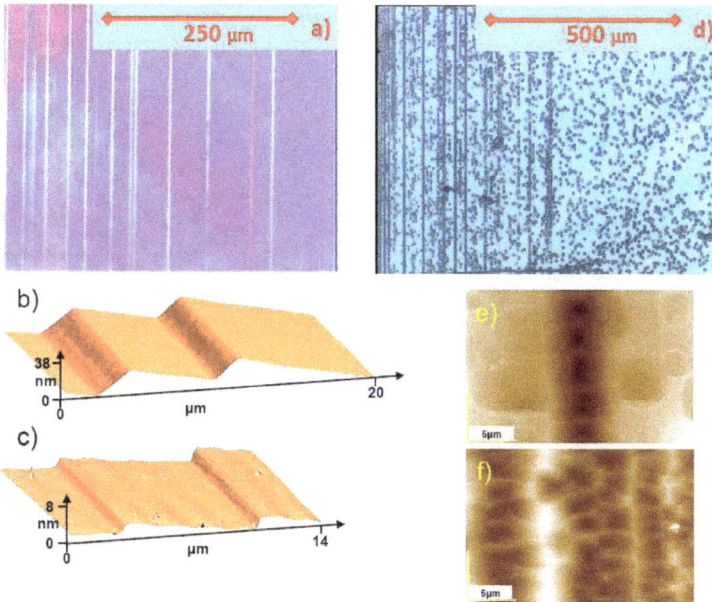

Figure 68. Optical inspection (**a**) and AFM topography (**b,c**) of mechanical deformed SrTiO$_3$ crystal (100) along [010] directions confirmed a discrete shift of the crystal blocks (on the whole width of the crystals) in the plastically deformed region. Relative shift of the blocks is about 8–40 nm (**b,c**) Using etch pits techniques it can be confirmed that a high concentration of edge dislocations and their bundle is present along the linear boundary between shifted blocks (**d**). Notice: because the density of etch pits outside of the region with plastic deformation is similar in comparison to the shifted blocks, it is possible that the dislocations array between the blocks are not only created via gliding of existing dislocations of the original crystal but a new population of dislocations could be nucleated under the influence of the external forces. AFM measurement of the topography of etched crystals shows linear arrangement of etch pits along the boundary between plastically deformed regions of the crystal (**e,f**).

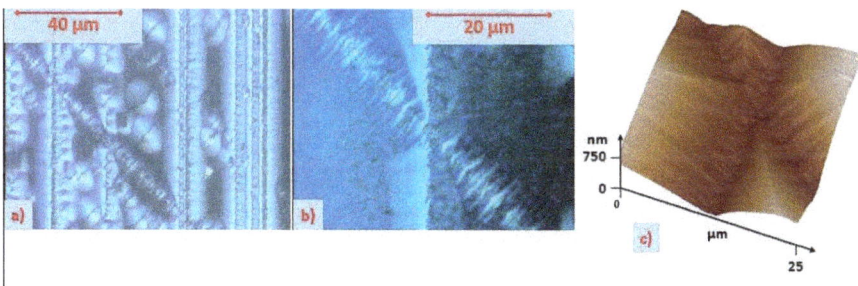

Figure 69. During plastical deformation of the SrTiO$_3$ crystal the bands of dislocations along <100> and <110> are independently generated (**a**). Optical photograph of the etch pits distribution in the plastically deformed region shows a perfect crossing of both bands of dislocations. Optical magnifications (**b**) and AFM topography (**c**) suggest that the dislocations along the <110> direction have a stronger tendency for bundling than for the <100> direction.

10. Dynamic Behavior of Dislocations in SrTiO₃ by In Situ Nanoindentation Experiments inside TEM

In recent years, in situ observation techniques for transmission electron microscopy (TEM) have been developed for directly observing dynamic dislocation motions in crystals under stresses [271–273]. Especially, in situ TEM nanoindentation techniques [108,273–276] now allow us to dynamically observe dislocation behavior at an arbitrary local area in a TEM specimen even at nanometer dimensions. Moreover, if we use well-controlled bicrystals for TEM specimens, it should be possible to directly observe the interaction processes of dislocations with a well-defined grain boundary during deformation. In the present study, we use the in situ TEM nanoindentation technique to directly and dynamically observe the nanoscale deformation processes of SrTiO₃ at room temperature. By using the suitable dark-field imaging conditions, we clearly highlight the dynamic behaviors of individual dislocations such as nucleation, propagation and interactions with other defects during the nanoindentation experiments [108,276].

First, we show the results of in situ TEM nanoindentation experiments of single crystalline SrTiO₃ [108]. Figure 70a gives an overview of the in situ TEM nanoindentation system. We used a TEM-NanoIndenter holder (Nanofactory Instrument AB.) with a double-tilt capability as schematically shown in Figure 70b. The specimen motions can be precisely controlled by a piezo actuator and the specimen is gradually moved toward the fixed indenter tip as the nanoindentation proceeds. Here, a wedge-shaped diamond indenter tip was selected. Figure 70c shows the dark-field TEM image (using $\mathbf{g}$ = 002) of the specimen edge and the indenter tip just before the nanoindentation experiments. Then, the indenter tip was manually inserted along the [001] direction. Figure 71 shows the sequential dark-field TEM images captured from the movie recorded during the nanoindentation experiment. The complete movie is available elsewhere [108]. In the figure, 0 s corresponds to the moment when the indenter tip has just contacted with the SrTiO₃ specimen edge. The indenter tip was gradually inserted into the specimen during 0 s to 65 s, and then extracted from the specimen during 65 s to 97 s. Here, we inserted the tip into a relatively thick specimen region in order to minimize the thin TEM specimen effects. It is noted that dislocations were generated from the specimen edge where the indenter tip contacted. The introduced dislocations may be classified into three different types, type-I, -II and -III dislocations. The type-I dislocations initially propagated in a semicircle shape, and then the dislocations split into two segments. This indicates that a part of the dislocation line reached to the specimen surface, leaving two dislocation segments. Then, each dislocation segment moved toward the ±[100] directions. The type-II and type-III dislocations were also nucleated near the indented area, but they moved toward the [101] and [$\bar{1}$01] directions, respectively. In order to characterize the slip systems of the introduced dislocations, the static TEM analyses after the TEM nanoindentation experiment were carried out. The Burgers vectors were determined by the conventional $\mathbf{g} \cdot \mathbf{b}$ = 0 invisibility criterion [277], and the slip planes were determined by the observations from the different crystallographic orientations. After the careful image contrast analyses, we found that there are two types of dislocation characters in the type-I dislocations: one is near-screw type dislocation with the slip system of [011]($0\bar{1}1$) type, while the other is also near-screw type dislocation but with the slip system of [$0\bar{1}1$](011) type. The type-II and type-III dislocations belong to the different slip systems of [101]($\bar{1}01$) and [$\bar{1}01$](101), respectively. These four slip systems belong to the same family of the <110>{1$\bar{1}$0} slip system. The present results are consistent with the previous reports of Vickers indentations or compression tests of bulk SrTiO₃ crystals at room temperature [33,61,278,279]. Thus, the microscopic plastic deformation processes of SrTiO₃ single crystals can be now directly observed by the in situ TEM nanoindentation technique.

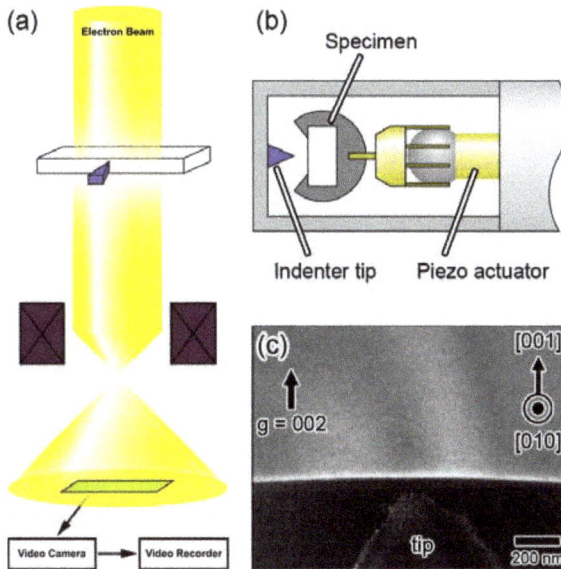

Figure 70. Schematic illustration of the in situ TEM nanoindentation system. (**a**) Overview of the in situ nanoindentation experiment inside TEM; (**b**) Schematic of the TEM nanoindentation holder used in this study; (**c**) Dark-field TEM image (**g** = 002) showing the specimen and the indenter tip before indentation. Adapted with permission from Ref. [108].

Second, we focus on the dynamic observations of the interaction processes of individual lattice dislocations with well-defined grain boundaries of SrTiO$_3$ by using the above-mentioned techniques. We start the in situ TEM nanoindentation experiments with the symmetric tilt [010](30$\bar{1}$) Σ5 grain boundary [280,281] as a model case for high-angle grain boundaries. Figure 72a shows the geometric set-up of the in situ TEM nanoindentation experiments. In this experiment, we also use the diamond indenter tip with wedge-type shape. Figure 72b shows a dark-field TEM image observed from the [010] direction just before nanoindentation. Here, the indentation direction is chosen to be parallel to the grain boundary plane. We inserted the indenter tip near the grain boundary and the emitted lattice dislocations were interacted with the grain boundary. The sequential dark-field TEM images captured from the movie during the nanoindentation are shown in Figure 73. Here, 0 s corresponds to the moment when the indenter tip has just contacted with the SrTiO$_3$ specimen edge, and each time shown in the respective images corresponds to the passing time from the start. The complete movie is available elsewhere [276]. It is clearly seen that the dislocations were emitted from the specimen edge and propagated along the [100] direction one by one. From the detailed **g·b** dislocation contrast analyses (**g·b** = 0 analysis for determining Burgers vectors of dislocations) after the nanoindentation experiment, it was confirmed that the introduced dislocations belong to the glide system of the [0$\bar{1}$1](011), consistent with the case of single crystals discussed above. With the continuous insertion of the indenter tip, the leading dislocation (termed the 1st dislocation in the image) propagated and impinged to the grain boundary. It was found that the 1st dislocation was completely attached to the grain boundary plane after 19 s. Then, the 2nd dislocation was also attached to the grain boundary plane as seen in the 26 s image, and even the lower part of the 3rd dislocation started to attach to the grain boundary after 27 s. Other lattice dislocations followed behind the 3rd dislocation appear to pile up within the bulk crystal regions. Finally, the specimen edge was fractured due to the stress concentration at the indenting point. After releasing the concentrated stress due to the fracture, the intervals of dislocations were extended. This may be due to the relaxation of the repulsive forces [41] exerted between the dislocations with the same Burgers vector. In addition, the dislocations attached on the grain boundaries during the nanoindentation experiment remain trapped even after the external stress was released. This indicates that the grain boundary core can be energetically stable site for lattice dislocations. Thus, the Σ5 grain boundary can act as a barrier for the lattice dislocation glide across it. To understand the mechanisms of the impediment process, we first consider the effect of geometric discontinuity across the grain boundary geometry. When the lattice dislocations are to cross the grain boundary, the Burgers vector has to be rotated. Since Burgers vectors should be conserved in crystals, the rotation of the Burgers vector inevitably leaves a residual dislocation on the grain boundary plane [282], whose Burgers vector is equal to the difference vector of the lattice Burgers vectors within the respective crystals. The formation of the residual dislocation on the grain boundary should increase the total energy by the self-energy of it. This should be one of the reasons why the grain boundary behaves as the barrier for the lattice dislocations. Secondly, we consider the effect of the dislocation stabilization at the grain boundary core. According to the previous reports, this effect may be caused by several possible reasons, such as dislocation dissociation on the boundary plane [283], elastic discontinuity (attractive image stress) and core structural stabilization. In any scenario, dislocations can be trapped by the grain boundary core, adding another component for the dislocation barrier across the grain boundary.

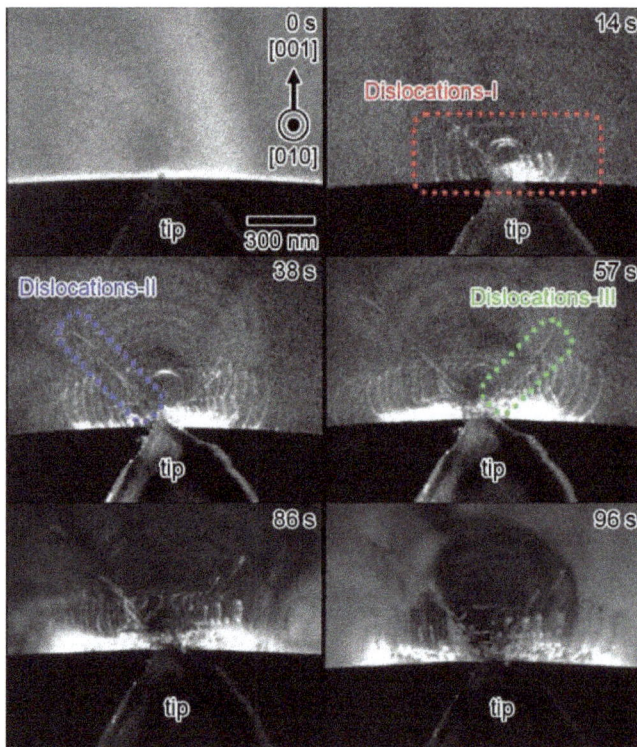

Figure 71. Sequential dark-field TEM images captured from the movie recorded during the nanoindentation experiment. The image was taken using **g** = 002. The broken lines indicate three different kinds of dislocations. Adapted with permission from Ref. [108].

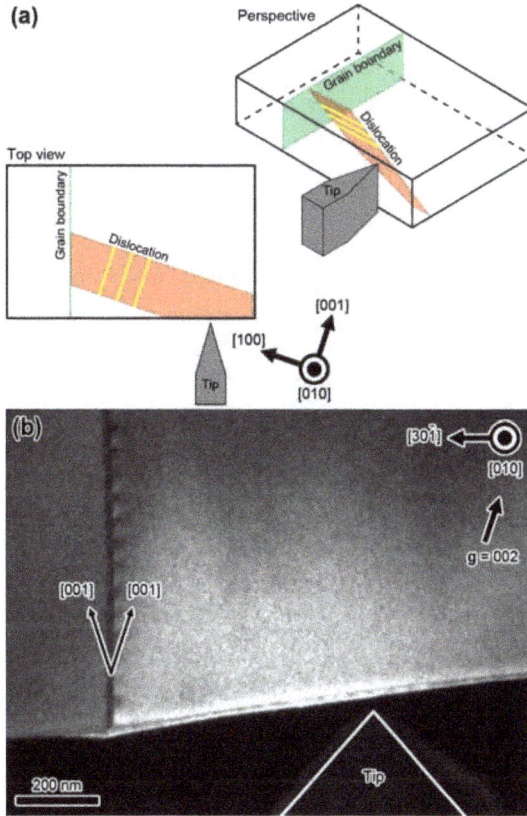

Figure 72. (a) Schematic illustration showing the geometric arrangement of the specimen, the [010](30$\bar{1}$) Σ5 grain boundary, the indenter tip, and the introduced lattice dislocation; (b) A dark-field TEM image just before the nanoindentation experiment. The weak triangular contrast at the lower right is the indenter tip and the vertical line contrast inside the specimen corresponds to the grain boundary. The sample thickness is estimated to be about 300 nm. Adapted with permission from Ref. [276].

In order to minimize the above-mentioned geometric discontinuity effects, we further performed in situ nanoindentation experiments for a low-angle grain boundary. In low-angle grain boundaries, grain boundary structures can be described by the periodic array of dislocations [284]. Here, (100) low-angle tilt grain boundary with the rotation angle of 1.2° around the [010] axis was selected. The actual structure of the low-angle tilt grain boundary is shown in Figure 74a. The grain boundary structure consists of the periodic array of edge dislocations with the Burgers vector of the [100] [11,40]. The interval of the edge dislocations is estimated to be about 19 nm. Figure 74b shows the geometric set-up of the in situ TEM nanoindentation experiments. Considering the Burgers vector of the introduced lattice dislocations, the moving dislocations interacting with the low-angle grain boundary should be screw dislocations because the Burgers vector is parallel to the grain boundary plane. Figure 74c shows a dark-field TEM image observed from the [010] direction just before the nanoindentation experiment. We then dynamically observe the interaction processes between the individual lattice dislocations and the low-angle tilt grain boundary. Figure 75 shows sequential TEM images captured from the movie of the nanoindentation experiment. The complete movie is available elsewhere [276]. The introduced lattice dislocations have the glide system of the [0$\bar{1}$1](011) by **g·b** contrast analyses after the nanoindentation experiment. By the continuous insertion of the indenter tip, the lattice dislocations were found to be able to traverse the grain boundary plane (see the 69 s image), contrary to the Σ5 grain boundary case. In addition, the dislocation motion was seen to be slightly impeded by the grain boundary core region. During the extraction of the indenter tip, the lattice dislocations moved backwards due to the stress relaxation, and some dislocations crossed the grain boundary again to go back into the right-hand side crystal. In this unloading process, we again observed the grain boundary impediment effect on the lattice dislocation. Finally, one dislocation indicated by the red arrow in Figure 75 appeared to be caught by the grain boundary plane.

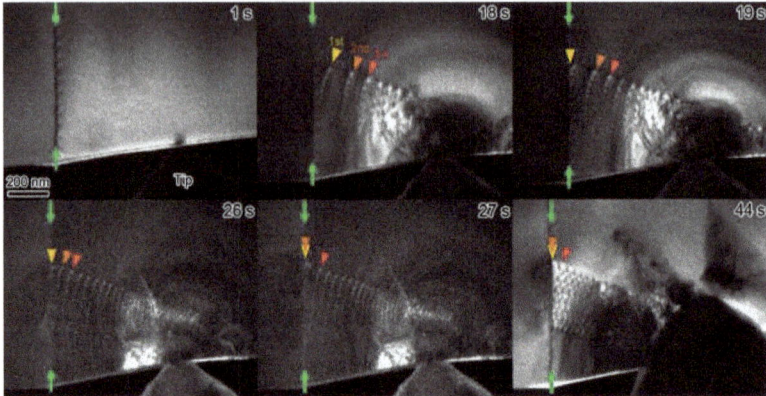

Figure 73. Sequential dark-field TEM images captured from the movie recorded during the nanoindentation experiment. The line contrasts indicated by the green arrows correspond to the Σ5 grain boundary. The positions of the leading three lattice dislocations are indicated by the triangles. The indenter tip was gradually inserted from 0 s to 42 s, and the specimen edge was fractured at 43 s. The dislocation motion was strongly impeded by the grain boundary, which resulted in the dislocation pile-up. In this experiment, the 1st and 2nd dislocations and the lower part of the 3rd dislocation were trapped on the grain boundary plane even after the external stress was removed. Adapted with permission from Ref. [276].

Figure 76a shows a dark-field image of the grain boundary edge dislocations after the crossing of the lattice screw dislocations. The dislocation lines of the grain boundary edge dislocations are seen to be shifted toward the $[0\bar{1}1]$ direction, and jogs are formed at the grain boundary dislocations. This indicates that the lattice screw dislocations intersected with the grain boundary edge dislocations during the crossing process [21]. A single intersection results in the formation of kinks on the lattice screw dislocations and jogs on the grain boundary edge dislocations (or making the jog length longer) as schematically shown in Figure 76b. Since the jog length estimated from the image is much longer than the unit cell of $SrTiO_3$, these jogs can be classified into "superjog" [41]. The estimated superjog length well corresponds with the product of the Burgers vector and the number of the screw dislocations crossing the grain boundary on the same glide plane. Thus, the superjogs are formed by the multiple intersections with the lattice screw dislocations on the same glide plane. Considering the dislocation reaction between the lattice dislocation (Burgers vector of $b_{Lattice}$: b = $[0\bar{1}1]$) and the grain boundary dislocation ($b_{G.B.}$: b = [100]), the Burgers vector of the superjog segments (b_{Jog}) should become as follows.

$$b_{Jog} = b_{Lattice} + b_{G.B.} = [0\bar{1}1] + [100] = [1\bar{1}1]$$

The detailed image contrast analysis showed that the lattice screw dislocations partially become the mixed dislocations just on the grain boundary plane by the dislocation reaction shown above [276].

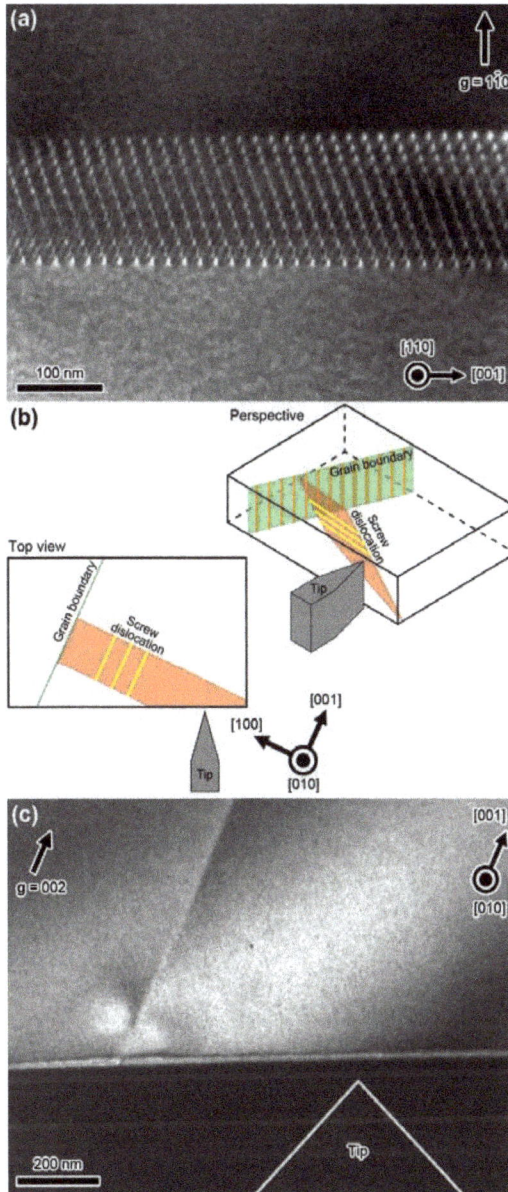

Figure 74. (a) A dark-field TEM image of the initial low-angle tilt grain boundary. The specimen was tilted from the edge-on condition to observe the grain boundary plane; (b) Schematic illustration of the geometric arrangement of the specimen, the grain boundary, the grain boundary edge dislocations, the indenter tip, and the introduced lattice screw dislocation; (c) A dark-field TEM image just before the nanoindentation experiment. The thickness of the specimen is about 150 nm. Adapted with permission from Ref. [276].

The low-angle tilt grain boundary slightly impeded the lattice screw dislocation motion not only when they crossed the grain boundary plane in the loading process, but also when they moved back into the initial crystal in the unloading process. In the loading process, the intersection of the lattice screw dislocation with the grain boundary edge dislocations forms the jogs or lengthens the superjogs. This process corresponds to the formation of residual grain boundary dislocation in the high-angle grain boundary case, although the residual dislocations in the low-angle grain boundary case is the discrete (super)jog row [285]. It means that the intersecting process during the loading process should lead to increase the self-energies of the grain boundary dislocations, consequently increasing the total energy. This may work as an effective energy barrier during the loading process to some extent. However, the impediment of the lattice screw dislocation motion was also observed in the unloading process, nevertheless the inverse intersection always shortens the length of the superjogs, and thus the total energy. Considering the fact that one screw dislocation was trapped on the grain boundary plane after removing the external stress, it can be concluded that the low-angle tilt grain boundary acts as a stable site for the lattice screw dislocations. As already mentioned, the screw dislocations partially become the mixed dislocations when they react with the grain boundary dislocations at the intermediate stage of the intersections. It should be noted that this reaction will not increase the total elastic energy of the system. Thus, the core relaxation induced by the dislocation reaction could be the main source of the stabilization at the grain boundary. This mechanism consistently explains the dislocation impediments in both loading and unloading processes.

In summary, recent in situ TEM nanoindentation experiments for microscopic plastic deformation processes of $SrTiO_3$ at room temperatures are reviewed. We suggest that the origin of the lattice dislocation impediment by grain boundaries is not simply the geometric effects but also the local structural stabilization effects at their cores. The present findings indicate that both the geometric effects and the stabilization effects should be simultaneously considered to quantitatively understand the dislocation interaction processes at grain boundaries in $SrTiO_3$. Combined with the well-controlled $SrTiO_3$ samples and careful TEM contrast analysis, in situ TEM nanoindentation can be extremely powerful tool for fundamentally understanding the dynamic and microscopic mechanisms of defect-defect interactions in $SrTiO_3$.

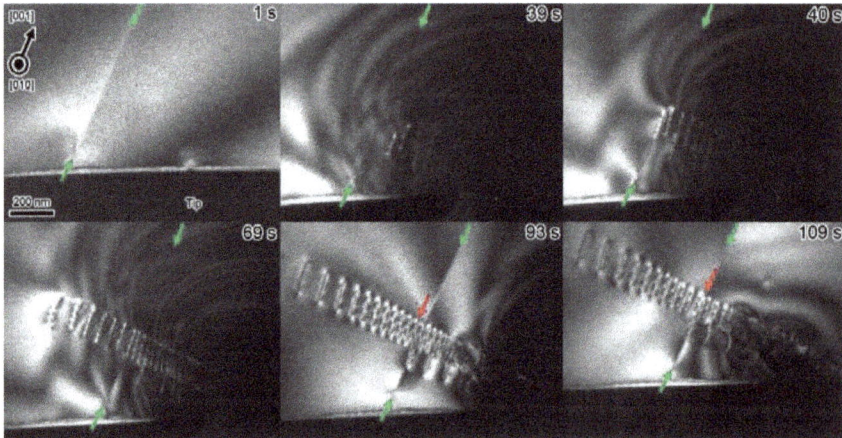

Figure 75. Sequential dark-field TEM images captured from the movie recorded during the nanoindentation experiment. The green arrows indicate the grain boundary position. The indenter tip was gradually inserted to the specimen edge from 0 s to 85 s and extracted from 86 s to 109 s. Adapted with permission from Ref. [276].

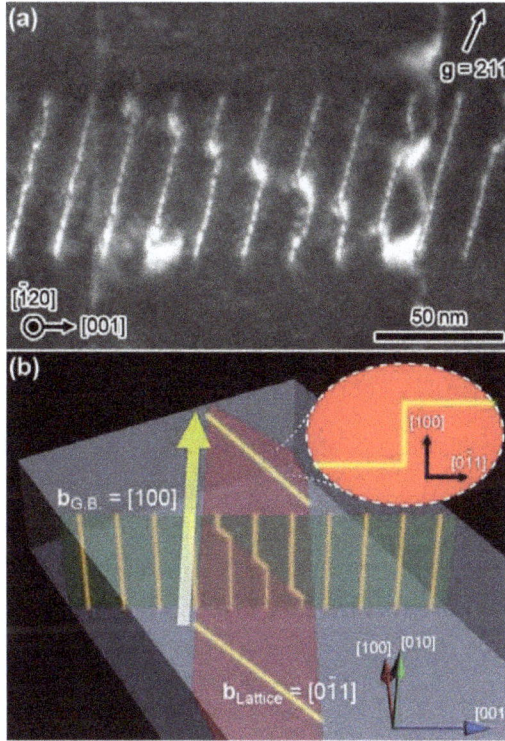

Figure 76. (**a**) A dark-field TEM image of the grain boundary edge dislocations after the nanoindentation experiment. It is clear that the grain boundary edge dislocation lines are shifted and the superjogs were formed on them; (**b**) Schematic illustration of the crossing process of the lattice screw dislocations. The intersection of the screw dislocation with the grain boundary edge dislocations forms the jogs on the grain boundary dislocations and the kinks on the lattice dislocation. Adapted with permission from Ref. [276].

11. Conclusions

The idea of Shockley [286], who has written back in 1983 that the dislocations hold potential technological promise and may be used in microelectronics when it becomes possible to control and manipulate their electronic properties, has been fulfilled in the case of band insulating oxides. The impressive role of dislocations as metallic nano-wires (a kind of short-circuit) has been for the first time presented for Al_2O_3 by incorporations of Ti metal atoms into the matrix. A similar effect, namely the transition of the core of dislocations from semiconducting to metallic state can also be generated via self-doping in the case of TiO_2 and $SrTiO_3$. This process can easily be controlled by reduction under low oxygen partial pressure, electrodegradation or nanoscale electrical-addressing of individual dislocations; in all cases the dislocations or dislocations network can be switched from a semiconducting state into a metallic state. This nano-manipulation of the chemical composition and electronic structure opens in our opinion a new era in the nano-engineering of oxides on the basis of the multinary transition metal oxides. The challenge for the future applications of these line defects with invariance is connected with the control of the dislocations in terms of their density, their distribution and, most important, an appropriate pinning at defined locations.

Author Contributions: The manuscript was written through contributions of all authors. All authors have given approval to the final version of the manuscript.

Acknowledgments: We thank P. Gao, S. Kondou, R. Sun, B. Feng, A. Kumamoto, E. Tochigi of University of Tokyo for their collaboration in a part of this paper. A part of this study was supported by Grant-in-Aid for Specially promoted Research (Grant No. JP17H06094) from Japan Society for the Promotion of Science (JSPS), Grant-in-Aid for Scientific Research on Innovative Areas "Nano Informatics" from JSPS, "Nanotechnology Platform" (Project No. 12024046) of MEXT. This work was supported in parts by the Deutsche Forschungsgemeinschaft (SFB 917 "Nanoswitches").

Conflicts of Interest: The authors declare no conflict of interest.

References

1. Veyssiere, P. Dislocation-Structure in Non-Stoichiometric Oxides. *Radiat. Eff.* **1983**, *74*, 1–15. [CrossRef]
2. Doukhan, N.; Doukhan, J.C. Dislocations in Perovskites $BaTiO_3$ and $CaTiO_3$. *Phys. Chem. Miner.* **1986**, *13*, 403–410.
3. Nakamura, A.; Matsunaga, K.; Tohma, J.; Yamamoto, T.; Ikuhara, Y. Conducting nanowires in insulating ceramics. *Nat. Mater.* **2003**, *2*, 453–456. [CrossRef] [PubMed]
4. Shibata, N.; Chisholm, M.F.; Nakamura, A.; Pennycook, S.J.; Yamamoto, T.; Ikuhara, Y. Nonstoichiometric dislocation cores in α-alumina. *Science* **2007**, *316*, 82–85. [CrossRef] [PubMed]
5. Arredondo, M.; Ramasse, Q.M.; Weyland, M.; Mahjoub, R.; Vrejoiu, I.; Hesse, D.; Browning, N.D.; Alexe, M.; Munroe, P.; Nagarajan, V. Direct Evidence for Cation Non-Stoichiometry and Cottrell Atmospheres Around Dislocation Cores in Functional Oxide Interfaces. *Adv. Mater.* **2010**, *22*, 2430–2434. [CrossRef] [PubMed]
6. Shimada, T.; Xu, T.; Araki, Y.; Wang, J.; Kitamura, T. Multiferroic Dislocations in Ferroelectric $PbTiO_3$. *Nano Lett.* **2017**, *17*, 2674–2680. [CrossRef] [PubMed]
7. Nagarajan, V.; Jia, C.L.; Kohlstedt, H.; Waser, R.; Misirlioglu, I.B.; Alpay, S.P.; Ramesh, R. Misfit dislocations in nanoscale ferroelectric heterostructures. *Appl. Phys. Lett.* **2005**, *86*, 192910. [CrossRef]
8. De Souza, R.A.; Fleig, J.; Maier, J.; Kienzle, O.; Zhang, Z.L.; Sigle, W.; Rühle, M. Electrical and structural characterization of a low-angle tilt grain boundary in iron-doped strontium titanate. *J. Am. Ceram. Soc.* **2003**, *86*, 922–928. [CrossRef]
9. Mao, Z.; Knowles, K.M. Dissociation of lattice dislocations in $SrTiO_3$. *Philos. Mag. A* **1996**, *73*, 699–708. [CrossRef]
10. Wang, R.H.; Zhu, Y.M.; Shapiro, S.M. Structural defects and the origin of the second length scale in $SrTiO_3$. *Phys. Rev. Lett.* **1998**, *80*, 2370–2373. [CrossRef]
11. Zhang, Z.L.; Sigle, W.; Kurtz, W.; Rühle, M. Electronic and atomic structure of a dissociated dislocation in $SrTiO_3$. *Phys. Rev. B* **2002**, *66*, 214112. [CrossRef]
12. Zhang, Z.L.; Sigle, W.; Rühle, M. Atomic and electronic characterization of the a[100] dislocation core in $SrTiO_3$. *Phys. Rev. B* **2002**, *66*, 094108. [CrossRef]
13. Chen, J.; Sekiguchi, T.; Li, J.Y.; Ito, S. Investigation of dislocations in Nb-doped (100) $SrTiO_3$ single crystals and their impacts on resistive switching. *Superlattices Microstruct.* **2016**, *99*, 182–185. [CrossRef]
14. Sun, H.P.; Tian, W.; Pan, X.Q.; Haeni, J.H.; Schlom, D.G. Evolution of dislocation arrays in epitaxial $BaTiO_3$ thin films grown on (100) $SrTiO_3$. *Appl. Phys. Lett.* **2004**, *84*, 3298–3300. [CrossRef]
15. Brunner, D.; Taeri-Baghbadrani, S.; Sigle, W.; Rühle, M. Surprising results of a study on the plasticity in strontium titanate. *J. Am. Ceram. Soc.* **2001**, *84*, 1161–1163. [CrossRef]
16. Gumbsch, P.; Taeri-Baghbadrani, S.; Brunner, D.; Sigle, W.; Rühle, A. Plasticity and an inverse brittle-to-ductile transition in strontium titanate. *Phys. Rev. Lett.* **2001**, *87*, 085505. [CrossRef] [PubMed]
17. Metlenko, V.; Ramadan, A.H.H.; Gunkel, F.; Du, H.C.; Schraknepper, H.; Hoffmann-Eifert, S.; Dittmann, R.; Waser, R.; De Souza, R.A. Do dislocations act as atomic autobahns for oxygen in the perovskite oxide $SrTiO_3$? *Nanoscale* **2014**, *6*, 12864–12876. [CrossRef] [PubMed]
18. Wang, Z.C.; Karato, S.; Fujino, K. High-Temperature Creep of Single-Crystal Strontium-Titanate ($SrTiO_3$)—A Contribution to Creep Systematics in Perovskites. *Phys. Earth Planet. Inter.* **1993**, *79*, 299–312. [CrossRef]
19. Adepalli, K.K.; Kelsch, M.; Merkle, R.; Maier, J. Influence of Line Defects on the Electrical Properties of Single Crystal TiO_2. *Adv. Funct. Mater.* **2013**, *23*, 1798–1806. [CrossRef]

20. Adepalli, K.K.; Kelsch, M.; Merkle, R.; Maier, J. Enhanced ionic conductivity in polycrystalline TiO_2 by "one-dimensional doping". *Phys. Chem. Chem. Phys.* **2014**, *16*, 4942–4951. [CrossRef] [PubMed]

21. Anderson, J.S.; Hyde, B.G. On Possible Role of Dislocations in Generating Ordered and Disordered Shear Structures. *J. Phys. Chem. Solids* **1967**, *28*, 1393–1408. [CrossRef]

22. Ashbee, K.H.G.; Smallman, R.E. Plastic Deformation of Titanium Dioxide Single Crystals. *Proc. R. Soc. Lond. Ser. A Math. Phys. Sci.* **1963**, *274*, 195–205. [CrossRef]

23. Ashbee, K.H.G.; Smallman, R.E. The Fracture of Titanium Dioxide Single Crystals with Particular Reference to Non-Stoichiometry. *Phys. Status Solidi* **1964**, *4*, 289–298. [CrossRef]

24. Barber, D.J.; Wenk, H.-R.; Hirth, G.; Kohlstedt, H.L. Chapter 95 Dislocations in Minerals. *Dislocat. Solids* **2010**, *16*, 171–232.

25. Blanchin, M.G.; Fontaine, G. Transmission Electron-Microscope Observations of Deformed Rutile (TiO_2). *Phys. Status Solidi A* **1975**, *29*, 491–501. [CrossRef]

26. Bursill, L.A.; Smith, D.J. Interaction of Small and Extended Defects in Nonstoichiometric Oxides. *Nature* **1984**, *309*, 319–321. [CrossRef]

27. Hirthe, W.M.; Brittain, J.O. Dislocations in Rutile as Revealed by the Etch-Pit Technique. *J. Am. Ceram. Soc.* **1962**, *45*, 546–554. [CrossRef]

28. Hirthe, W.M.; Brittain, J.O. High-Temperature Steady-State Creep in Rutile. *J. Am. Ceram. Soc.* **1963**, *46*, 411–417. [CrossRef]

29. Lenser, C.; Connell, Z.; Kovacs, A.; Dunin-Borkowski, R.; Kohl, A.; Waser, R.; Dittmann, R. Identification of screw dislocations as fast-forming sites in Fe-doped $SrTiO_3$. *Appl. Phys. Lett.* **2013**, *102*, 183504. [CrossRef]

30. De Souza, R.A.; Fleig, J.; Maier, J.; Zhang, Z.L.; Sigle, W.; Rühle, M. Electrical resistance of low-angle tilt grain boundaries in acceptor-doped $SrTiO_3$ as a function of misorientation angle. *J. Appl. Phys.* **2005**, *97*, 053502. [CrossRef]

31. Marrocchelli, D.; Sun, L.X.; Yildiz, B. Dislocations in $SrTiO_3$: Easy To Reduce but Not so Fast for Oxygen Transport. *J. Am. Chem. Soc.* **2015**, *137*, 4735–4748. [CrossRef] [PubMed]

32. Sun, R.; Wang, Z.C.; Shibata, N.; Ikuhara, Y. A dislocation core in titanium dioxide and its electronic structure. *RSC Adv.* **2015**, *5*, 18506–18510. [CrossRef]

33. Sigle, W.; Sarbu, C.; Brunner, D.; Rühle, M. Dislocations in plastically deformed $SrTiO_3$. *Philos. Mag.* **2006**, *86*, 4809–4821. [CrossRef]

34. Motohashi, Y.; Blanchin, M.G.; Vicario, E.; Fontaine, G.; Otake, S. Elastic Parameters, Elastic Energy, and Stress-Fields of Dislocations in TiO_2 Rutile Crystals. *Phys. Status Solidi A* **1979**, *54*, 355–364. [CrossRef]

35. Ohno, T.; Ii, S.; Shibata, N.; Matsunaga, K.; Ikuhara, Y.; Yamamoto, T. High resolution Microscopy study for [001] symmetric tilt boundary with a tilt angle of 66 degrees in rutile-type TiO_2 bicrystal. *Mater. Trans.* **2004**, *45*, 2117–2121. [CrossRef]

36. Stephens, D.L.; Alford, W.J. Dislocation Structures in Single-Crystal Al_2O_3. *J. Am. Ceram. Soc.* **1964**, *47*, 81–86. [CrossRef]

37. Suzuki, K.; Ichihara, M.; Takeuchi, S. High-Resolution Electron-Microscopy of Lattice-Defects in TiO_2 and SnO_2. *Philos. Mag. A* **1991**, *63*, 657–665. [CrossRef]

38. Wang, Z.C.; Tsukimoto, S.; Saito, M.; Ikuhara, Y. Individual charge-trapping dislocations in an ionic insulator. *Appl. Phys. Lett.* **2009**, *95*, 184101. [CrossRef]

39. Jia, C.L.; Thust, A.; Urban, K. Atomic-scale analysis of the oxygen configuration at a $SrTiO_3$ dislocation core. *Phys. Rev. Lett.* **2005**, *95*, 225506. [CrossRef] [PubMed]

40. Buban, J.P.; Chi, M.F.; Masiel, D.J.; Bradley, J.P.; Jiang, B.; Stahlberg, H.; Browning, N.D. Structural variability of edge dislocations in a $SrTiO_3$ low-angle [001] tilt grain boundary. *J. Mater. Res.* **2009**, *24*, 2191–2199. [CrossRef]

41. Hirth, J.P.; Lothe, J. *Theory of Dislocations*; Krieger Publishing Company: Malabar, FL, USA, 1992.

42. Jin, L.; Guo, X.; Jia, C.L. TEM study of -type 35.26 degrees dislocations specially induced by polishing of $SrTiO_3$ single crystals. *Ultramicroscopy* **2013**, *134*, 77–85. [CrossRef] [PubMed]

43. McGibbon, M.M.; Browning, N.D.; Chisholm, M.F.; Mcgibbon, A.J.; Pennycook, S.J.; Ravikumar, V.; Dravid, V.P. Direct Determination of Grain-Boundary Atomic-Structure in $SrTiO_3$. *Science* **1994**, *266*, 102–104. [CrossRef] [PubMed]

44. Paladino, A.E.; Rubin, L.G.; Waugh, J.S. Oxygen Ion Diffusion in Single Crystal $SrTiO_3$. *J. Phys. Chem. Solids* **1965**, *26*, 391–397. [CrossRef]

45. Szot, K.; Speier, W.; Carius, R.; Zastrow, U.; Beyer, W. Localized metallic conductivity and self-healing during thermal reduction of SrTiO₃. *Phys. Rev. Lett.* **2002**, *88*, 075508. [CrossRef] [PubMed]
46. Zhang, Z.L.; Sigle, W.; Kurtz, W. HRTEM and EELS study of screw dislocation cores in SrTiO₃. *Phys. Rev. B* **2004**, *69*, 144103. [CrossRef]
47. Kwon, J.; Sharma, A.A.; Bain, J.A.; Picard, Y.N.; Skowronski, M. Oxygen Vacancy Creation, Drift, and Aggregation in TiO₂-Based Resistive Switches at Low Temperature and Voltage. *Adv. Funct. Mater.* **2015**, *25*, 2876–2883. [CrossRef]
48. Rogala, M.; Bihlmayer, G.; Speier, W.; Klusek, Z.; Rodenbücher, C.; Szot, K. Resistive Switching of a Quasi-Homogeneous Distribution of Filaments Generated at Heat-Treated TiO₂ (110)-Surfaces. *Adv. Funct. Mater.* **2015**, *25*, 6382–6389. [CrossRef]
49. Strukov, D.B.; Snider, G.S.; Stewart, D.R.; Williams, R.S. The missing memristor found. *Nature* **2008**, *453*, 80–83. [CrossRef] [PubMed]
50. Waser, R.; Aono, M. Nanoionics-based resistive switching memories. *Nat. Mater.* **2007**, *6*, 833–840. [CrossRef] [PubMed]
51. Waser, R.; Dittmann, R.; Staikov, G.; Szot, K. Redox-Based Resistive Switching Memories—Nanoionic Mechanisms, Prospects, and Challenges. *Adv. Mater.* **2009**, *21*, 2632–2663. [CrossRef]
52. Kim, K.M.; Jeong, D.S.; Hwang, C.S. Nanofilamentary resistive switching in binary oxide system; a review on the present status and outlook. *Nanotechnology* **2011**, *22*, 254002. [CrossRef] [PubMed]
53. Szot, K.; Rogala, M.; Speier, W.; Klusek, Z.; Besmehn, A.; Waser, R. TiO₂-a prototypical memristive material. *Nanotechnology* **2011**, *22*, 254001. [CrossRef] [PubMed]
54. Jeong, D.S.; Thomas, R.; Katiyar, R.S.; Scott, J.F.; Kohlstedt, H.; Petraru, A.; Hwang, C.S. Emerging memories: Resistive switching mechanisms and current status. *Rep. Prog. Phys.* **2012**, *75*, 076502. [CrossRef] [PubMed]
55. Szot, K.; Speier, W.; Bihlmayer, G.; Waser, R. Switching the electrical resistance of individual dislocations in single-crystalline SrTiO₃. *Nat. Mater.* **2006**, *5*, 312–320. [CrossRef] [PubMed]
56. Watanabe, Y.; Bednorz, J.G.; Bietsch, A.; Gerber, C.; Widmer, D.; Beck, A.; Wind, S.J. Current-driven insulator-conductor transition and nonvolatile memory in chromium-doped SrTiO₃ single crystals. *Adv. Funct. Mater.* **2001**, *78*, 3738–3740. [CrossRef]
57. Kwon, D.H.; Kim, K.M.; Jang, J.H.; Jeon, J.M.; Lee, M.H.; Kim, G.H.; Li, X.S.; Park, G.S.; Lee, B.; Han, S.; et al. Atomic structure of conducting nanofilaments in TiO₂ resistive switching memory. *Nat. Nanotechnol.* **2010**, *5*, 148–153. [CrossRef] [PubMed]
58. Wojtyniak, M.; Szot, K.; Wrzalik, R.; Rodenbücher, C.; Roth, G.; Waser, R. Electro-degradation and resistive switching of Fe-doped SrTiO₃ single crystal. *J. Appl. Phys.* **2013**, *113*, 083713. [CrossRef]
59. Szot, K.; Speier, W.; Eberhardt, W. Microscopic Nature of the Metal to Insulator Phase-Transition Induced through Electroreduction in Single-Crystal KNbO₃. *Appl. Phys. Lett.* **1992**, *60*, 1190–1192. [CrossRef]
60. Urban, K.W. Studying atomic structures by aberration-corrected transmission electron microscopy. *Science* **2008**, *321*, 506–510. [CrossRef] [PubMed]
61. Matsunaga, T.; Saka, H. Transmission electron microscopy of dislocations in SrTiO₃. *Philos. Mag. Lett.* **2000**, *80*, 597–604. [CrossRef]
62. Jia, C.L.; Mi, S.B.; Urban, K.; Vrejoiu, I.; Alexe, M.; Hesse, D. Effect of a Single Dislocation in a Heterostructure Layer on the Local Polarization of a Ferroelectric Layer. *Phys. Rev. Lett.* **2009**, *102*, 117601. [CrossRef] [PubMed]
63. Nishigaki, J.; Kuroda, K.; Saka, H. Electron-Microscopy of Dislocation-Structures in SrTiO₃ Deformed at High-Temperatures. *Phys. Status Solidi A* **1991**, *128*, 319–336. [CrossRef]
64. Yang, G.Y.; Finder, J.M.; Wang, J.; Wang, Z.L.; Yu, Z.; Ramdani, J.; Droopad, R.; Eisenbeiser, K.W.; Ramesh, R. Study of microstructure in SrTiO₃/Si by high-resolution transmission electron microscopy. *J. Mater. Res.* **2002**, *17*, 204–213. [CrossRef]
65. Du, H.C.; Jia, C.L.; Houben, L.; Metlenko, V.; De Souza, R.A.; Waser, R.; Mayer, J. Atomic structure and chemistry of dislocation cores at low-angle tilt grain boundary in SrTiO₃ bicrystals. *Acta Mater.* **2015**, *89*, 344–351. [CrossRef]
66. Diebold, U. The surface science of titanium dioxide. *Surf. Sci. Rep.* **2003**, *48*, 53–229. [CrossRef]
67. Fukui, K.; Onishi, H.; Iwasawa, Y. Atom-resolved image of the TiO₂(110) surface by noncontact atomic force microscopy. *Phys. Rev. Lett.* **1997**, *79*, 4202–4205. [CrossRef]

68. Szot, K.; Bihlmayer, G.; Speier, W. Nature of the Resistive Switching Phenomena in TiO_2 and $SrTiO_3$: Origin of the Reversible Insulator-Metal Transition. *Solid State Phys.* **2014**, *65*, 353–559.

69. Jia, C.L.; Lentzen, M.; Urban, K. Atomic-resolution imaging of oxygen in perovskite ceramics. *Science* **2003**, *299*, 870–873. [CrossRef] [PubMed]

70. Bieger, T.; Maier, J.; Waser, R. Optical Investigation of Oxygen Incorporation in $SrTiO_3$. *Solid State Ion.* **1992**, *53*, 578–582. [CrossRef]

71. Bieger, T.; Maier, J.; Waser, R. Kinetics of Oxygen Incorporation in $SrTiO_3$ (Fe-Doped)—An Optical Investigation. *Sens. Actuators B Chem.* **1992**, *7*, 763–768. [CrossRef]

72. Ferre, D.; Carrez, P.; Cordier, P. Modeling dislocation cores in $SrTiO_3$ using the Peierls-Nabarro model. *Phys. Rev. B* **2008**, *77*, 014106. [CrossRef]

73. Clinard, F.W.; Hobbs, L.W. Radiation effects in non-metals. In *Physics of Radiation Effects in Crystals*; Johnson, R.A., Orlov, A.N., Eds.; North-Holland Physics Publishing: Amsterdam, The Nederlands, 1986.

74. Furushima, Y.; Arakawa, Y.; Nakamura, A.; Tochigi, E.; Matsunaga, K. Nonstoichiometric [012] dislocation in strontium titanate. *Acta Mater.* **2017**, *135*, 103–111. [CrossRef]

75. Weertman, J.; Weertman, J.R. *Elementary Dislocation Theory*; Oxford University Press Oxford: New York, NY, USA, 1992.

76. Kofstad, P. Note on Defect Structure of Rutile (TiO_2). *J. Less Common Met.* **1967**, *13*, 635–638. [CrossRef]

77. Kofstad, P. *Electrical Conductivity and Diffusion in Binary Metal Oxides*; Wiley: New York, NY, USA, 1972.

78. Marucco, J.F.; Gautron, J.; Lemasson, P. Thermogravimetric and Electrical Study of Nonstoichiometric Titanium-Dioxide TiO_{2-x} between 800 and 1100 °C. *J. Phys. Chem. Solids* **1981**, *42*, 363–367. [CrossRef]

79. Marucco, J.F.; Poumellec, B.; Gautron, J.; Lemasson, P. Thermodynamic Properties of Titanium-Dioxide, Niobium Dioxide and Their Solid-Solutions at High-Temperature. *J. Phys. Chem. Solids* **1985**, *46*, 709–717. [CrossRef]

80. Millot, F.; Blanchin, M.G.; Tetot, R.; Marucco, J.F.; Poumellec, B.; Picard, C.; Touzelin, B. High-Temperature Nonstoichiometric Rutile TiO_{2-x}. *Prog. Solid State Chem.* **1987**, *17*, 263–293. [CrossRef]

81. Balachandran, U.; Eror, N.G. Electrical-Conductivity in Non-Stoichiometric Titanium-Dioxide at Elevated-Temperatures. *J. Mater. Sci.* **1988**, *23*, 2676–2682. [CrossRef]

82. Bak, T.; Burg, T.; Kang, S.J.L.; Nowotny, J.; Rekas, M.; Sheppard, L.; Sorrell, C.C.; Vance, E.R.; Yoshida, Y.; Yamawaki, M. Charge transport in polycrystalline titanium dioxide. *J. Phys. Chem. Solids* **2003**, *64*, 1089–1095. [CrossRef]

83. Bak, T.; Nowotny, J.; Rekas, M.; Sorrell, C.C. Defect chemistry and semiconducting properties of titanium dioxide: III. Mobility of electronic charge carriers. *J. Phys. Chem. Solids* **2003**, *64*, 1069–1087. [CrossRef]

84. Yamada, H.; Miller, G.R. Point-Defects in Reduced Strontium-Titanate. *J. Solid State Chem.* **1973**, *6*, 169–177. [CrossRef]

85. Chan, N.H.; Sharma, R.K.; Smyth, D.M. Non-Stoichiometry in $SrTiO_3$. *J. Electrochem. Soc.* **1981**, *128*, 1762–1769. [CrossRef]

86. Smyth, D.M. Defects and Order in Perovskite-Related Oxides. *Annu. Rev. Mater. Sci.* **1985**, *15*, 329–357. [CrossRef]

87. Eror, N.G.; Balachandran, U. Electrical-Conductivity in Strontium-Titanate with Nonideal Cationic Ratio. *J. Solid State Chem.* **1982**, *42*, 227–241. [CrossRef]

88. Waser, R. Bulk Conductivity and Defect Chemistry of Acceptor-Doped Strontium-Titanate in the Quenched State. *J. Am. Ceram. Soc.* **1991**, *74*, 1934–1940. [CrossRef]

89. Smyth, D.M. Oxidative Nonstoichiometry in Perovskite Oxides. In *Properties and Application of Perovskite-Type Oxides*; Tejuca, L.G., Fierro, J.L.G., Eds.; CRC Press: New York, NY, USA, 1993; p. 63.

90. Ravikumar, V.; Rodrigues, R.P.; Dravid, V.P. Direct Imaging of Spatially Varying Potential and Charge across Internal Interfaces in Solids. *Phys. Rev. Lett.* **1995**, *75*, 4063–4066. [CrossRef] [PubMed]

91. Ravikumar, V.; Rodrigues, R.P.; Dravid, V.P. An investigation of acceptor-doped grain boundaries in $SrTiO_3$. *J. Phys. D Appl. Phys.* **1996**, *29*, 1799–1806. [CrossRef]

92. Von Alfthan, S.; Benedek, N.A.; Chen, L.; Chua, A.; Cockayne, D.; Dudeck, K.J.; Elsässer, C.; Finnis, M.W.; Koch, C.T.; Rahmati, B.; et al. The Structure of Grain Boundaries in Strontium Titanate: Theory, Simulation, and Electron Microscopy. *Annu. Rev. Mater. Res.* **2010**, *40*, 557–599. [CrossRef]

93. Wang, Y.G.; Dravid, V.P. Determination of electrostatic characteristics at a 24 degrees, [001] tilt grain boundary in a $SrTiO_3$ bicrystal by electron holography. *Philos. Mag. Lett.* **2002**, *82*, 425–432. [CrossRef]

94. Kalinin, S.V.; Bonnell, D.A. Surface potential at surface-interface junctions in SrTiO$_3$ bicrystals. *Phys. Rev. B* **2000**, *62*, 10419–10430. [CrossRef]

95. Gao, P.; Ishikawa, R.; Feng, B.; Kumamoto, A.; Shibata, N.; Ikuhara, Y. Atomic-scale structure relaxation, chemistry and charge distribution of dislocation cores in SrTiO$_3$. *Ultramicroscopy* **2018**, *184*, 217–224. [CrossRef] [PubMed]

96. Yang, S.F.; Xiong, L.M.; Deng, Q.; Chen, Y.P. Concurrent atomistic and continuum simulation of strontium titanate. *Acta Mater.* **2013**, *61*, 89–102. [CrossRef]

97. Yang, S.F.; Chen, Y.P. Concurrent atomistic and continuum simulation of bi-crystal strontium titanate with tilt grain boundary. *Proc. R. Soc. A Math. Phys. Eng. Sci.* **2015**, *471*, 20140758. [CrossRef] [PubMed]

98. Seltzer, M.S.; Wilcox, B.A.; Clauer, A.H. The Influence of Stoichiometric Defects on the Creep of Oxides with the Fluorite Structure. In *Defects and Transport in Oxides*; Jaffee, R.I., Ed.; Springer: Berlin, Germany, 1973; pp. 443–457.

99. Yamanaka, J.; Yoshimura, J.; Kimura, S. Characterization of lattice defects in strontium titanate single crystals by X-ray topography and transmission electron microscopy. *J. Electron Microsc.* **2000**, *49*, 89–92. [CrossRef]

100. Yoshimura, J.; Sakamoto, T.; Yamanaka, J. X-ray double-crystal diffractometry of Verneuil-grown SrTiO$_3$ crystals. *Jpn. J. Appl. Phys.* **2001**, *40*, 6536–6542. [CrossRef]

101. Yoshimura, J.; Yamanaka, J.; Iwamoto, T.; Kimura, S. X-ray topographic observation of lattice defects in heat-treated SrTiO$_3$ crystals. *Mater. Trans. JIM* **2000**, *41*, 559–562. [CrossRef]

102. Yamanaka, J. High-resolution electron microscopy of a subgrain boundary in strontium titanate single crystal. *Mater. Trans.* **2001**, *42*, 1131–1134. [CrossRef]

103. Yoshimura, J.; Sakamoto, T.; Usui, S.; Kimura, S. X-ray perfection study of Verneuil-grown SrTiO$_3$ crystals. *J. Cryst. Growth* **1998**, *191*, 483–491. [CrossRef]

104. Nabokin, P.I.; Souptel, D.; Balbashov, A.M. Floating zone growth of high-quality SrTiO$_3$ single crystals. *J. Cryst. Growth* **2003**, *250*, 397–404. [CrossRef]

105. Hirel, P.; Carrez, P.; Cordier, P. From glissile to sessile: Effect of temperature on <110> dislocations in perovskite materials. *Scr. Mater.* **2016**, *120*, 67–70. [CrossRef]

106. Hirel, P.; Carrez, P.; Clouet, E.; Cordier, P. The electric charge and climb of edge dislocations in perovskite oxides: The case of high-pressure MgSiO$_3$ bridgmanite. *Acta Mater.* **2016**, *106*, 313–321. [CrossRef]

107. Szot, K.; Speier, W. Verfahren zur Herstellung eines ABO$_3$ Substrates mit einer Stufe. Patent DE19808778 C2, 9 December 1999.

108. Kondo, S.; Shibata, N.; Mitsuma, T.; Tochigi, E.; Ikuhara, Y. Dynamic observations of dislocation behavior in SrTiO$_3$ by in situ nanoindentation in a transmission electron microscope. *Appl. Phys. Lett.* **2012**, *100*, 181906. [CrossRef]

109. Yang, K.H.; Ho, N.J.; Lu, H.Y. Kink-Pair Mechanism in <001> SrTiO$_3$ Single Crystal Compression-Deformed at Room Temperature. *Jpn. J. Appl. Phys.* **2011**, *50*, 105601.

110. Prabhumirashi, P.; Dravid, V.P.; Lupini, A.R.; Chisholm, M.F.; Pennycook, S.J. Atomic-scale manipulation of potential barriers at SrTiO$_3$ grain boundaries. *Appl. Phys. Lett.* **2005**, *87*, 121917. [CrossRef]

111. Mizoguchi, T.; Sato, Y.; Buban, J.P.; Matsunaga, K.; Yamamoto, T.; Ikuhara, Y. Sr vacancy segregation by heat treatment at SrTiO$_3$ grain boundary. *Appl. Phys. Lett.* **2005**, *87*, 241920. [CrossRef]

112. Zhang, Z.; Yates, J.T. Defects on TiO$_2$—Key Pathways to important surface processes. In *Defects at Oxide Surfaces*; Jupille, J., Thorton, G., Eds.; Springer: Berlin, Germany, 2015; p. 86.

113. Javaid, F.; Stukowski, A.; Durst, K. 3D Dislocation structure evolution in strontium titanate: Spherical indentation experiments and MD simulations. *J. Am. Ceram. Soc.* **2017**, *100*, 1134–1145. [CrossRef]

114. Hanzig, J.; Abendroth, B.; Hanzig, F.; Stocker, H.; Strohmeyer, R.; Meyer, D.C.; Lindner, S.; Grobosch, M.; Knupfer, M.; Himcinschi, C.; et al. Single crystal strontium titanate surface and bulk modifications due to vacuum annealing. *J. Appl. Phys.* **2011**, *110*, 064107. [CrossRef]

115. Guo, X.; Zhang, Z.L.; Sigle, W.; Wachsman, E.; Waser, R. Schottky barrier formed by network of screw dislocations in SrTiO$_3$. *Appl. Phys. Lett.* **2005**, *87*, 162105. [CrossRef]

116. Anderson, S.; Jahnberg, L. Crystal structure studies on the homologous series Ti$_n$O$_{2n-1}$, V$_n$O$_{2n-1}$ and Ti$_{n-2}$Cr$_2$O$_{2n-1}$. *Ark. Kemi* **1963**, *21*, 413.

117. Bursill, L.A.; Hyde, B.G. Crystallographic shear in the higher titanium oxides: Structure, texture, mechanisms and thermodynamics. *Prog. Solid State Chem.* **1972**, *7*, 177–253. [CrossRef]

118. Bursill, L.A.; Peng, J.; Fan, X. Structure and reactivity of atomic surfaces of barium-titanate under electron-irradiation. *Ferroelectrics* **1989**, *97*, 71–84. [CrossRef]

119. Szot, K.; Pawelczyk, M.; Herion, J.; Freiburg, C.; Albers, J.; Waser, R.; Hulliger, J.; Kwapulinski, J.; Dec, J. Nature of the surface layer in ABO$_3$-type Perovskites at elevated temperatures. *Appl. Phys. A* **1996**, *62*, 335–343. [CrossRef]

120. Szot, K.; Speier, W. Surfaces of reduced and oxidized SrTiO$_3$ from atomic force microscopy. *Phys. Rev. B* **1999**, *60*, 5909–5926. [CrossRef]

121. Lee, W.; Yoo, S.; Yoon, K.J.; Yeu, I.W.; Chang, H.J.; Choi, J.H.; Hoffmann-Eifert, S.; Waser, R.; Hwang, C.S. Resistance switching behavior of atomic layer deposited SrTiO$_3$ film through possible formation of Sr$_2$Ti$_6$O$_{13}$ or Sr$_1$Ti$_{11}$O$_{20}$ phases. *Sci. Rep.* **2016**, *6*, 20550. [CrossRef] [PubMed]

122. Meyer, D.C.; Levin, A.A.; Leisegang, T.; Gutmann, E.; Paufler, P.; Reibold, M.; Pompe, W. Reversible tuning of a series of intergrowth phases of the Ruddlesden-Popper type SrO(SrTiO$_3$)$_n$ in an (001) SrTiO$_3$ single-crystalline plate by an external electric field and its potential use for adaptive X-ray optics. *Appl. Phys. A* **2006**, *84*, 31–35. [CrossRef]

123. Szot, K.; Speier, W.; Breuer, U.; Meyer, R.; Szade, J.; Waser, R. Formation of micro-crystals on the (100) surface of SrTiO$_3$ at elevated temperatures. *Surf. Sci.* **2000**, *460*, 112–128. [CrossRef]

124. Van Landuyt, J. Shear structures and crystallographic shear propagation. *J. Phys. Colloq.* **1974**, *35*, C7–C35. [CrossRef]

125. Vogel, F.L.; Pfann, W.G.; Corey, H.E.; Thomas, E.E. Observations of Dislocations in Lineage Boundaries in Germanium. *Phys. Rev.* **1953**, *90*, 489–490. [CrossRef]

126. Barber, D.J.; Tighe, N.J. Observations of Dislocations and Surface Features in Corundum Crystals by Electron Transmission Microscopy. *J. Res. Natl. Bur. Stand. Sect. A Phys. Chem.* **1965**, *A69*, 271. [CrossRef]

127. Sangwal, K. *Etching of the Crystals*; North-Holland: Amsterdam, The Netherlands, 1987.

128. Ashbee, K.A.G.; Smallman, R.E.; Williamson, G.K. Stacking Faults and Dislocations in Titanium Dioxide, with Special Reference to Non-Stoichiometry. *Proc. R. Soc. Lond. Ser. A Math. Phys. Sci.* **1963**, *276*, 542–552. [CrossRef]

129. Barber, D.J.; Farabaug, E.N. Dislocations and Stacking Faults in Rutile Crystals Grown by Flame-Fusion Methods. *J. Appl. Phys.* **1965**, *36*, 2803–2806. [CrossRef]

130. Bell, H.; Jones, J.T.; Krishnam, V. Recovery of High-Temperature Creep-Resistant Substructure in Rutile. *J. Am. Ceram. Soc.* **1972**, *55*, 6–10. [CrossRef]

131. Fleischauer, P.D.; Chase, A.B. Coordination Chemistry and Kinetics of Preferential Etching on Surfaces of TiO$_2$ (Rutile). *J. Phys. Chem. Solids* **1974**, *35*, 1211–1219. [CrossRef]

132. Bright, E.; Readey, D.W. Dissolution Kinetics of TiO$_2$ in Hf-HCl Solutions. *J. Am. Ceram. Soc.* **1987**, *70*, 900–906. [CrossRef]

133. Gorokhovsky, A.V.; Escalante-García, J.I.; Sánchez-Monjarás, T.; Gutiérrez-Chavarría, C.A. Synthesis of potassium polytitanate precursors by treatment of TiO$_2$ with molten mixtures of KNO$_3$ and KOH. *J. Eur. Ceram. Soc.* **2004**, *24*, 3541–3546. [CrossRef]

134. Zaremba, T. Investigation on synthesis and microstructure of potassium tetratitanate. *J. Therm. Anal. Calorim.* **2008**, *91*, 911–913. [CrossRef]

135. Waugh, J.S.; Paladino, A.E.; Dibenedetto, B.; Wantman, R. Effect of Dislocations on Oxidation and Reduction of Single-Crystal SrTiO$_3$. *J. Am. Ceram. Soc.* **1963**, *46*, 60. [CrossRef]

136. Rhodes, W.H. Etching and Chemical Polishing of Single-Crystal SrTiO$_3$. *J. Am. Ceram. Soc.* **1966**, *49*, 110–112. [CrossRef]

137. Spalding, G.; Murphy, W.I.; Davidsmeier, T.M.; Elenewski, J. Faceting of single-crystal SrTiO$_3$ during wet chemical etching. In *Substrate Engineering: Paving the Way to Epitaxy*; Norton, D., Schlom, D.G., Newman, N., Matthiesen, D., Eds.; Materials Research Society: Warrendale, PA, USA, 1999.

138. Yamamoto, T.; Oba, F.; Ikuhara, Y.; Sakuma, T. Current-voltage characteristics across small angle symmetric tilt boundaries in Nb-doped SrTiO$_3$ bicrystals. *Mater. Trans.* **2002**, *43*, 1537–1541. [CrossRef]

139. Raghothamachar, B.; Dhanaraj, G.; Bai, J.; Dudley, M. Defect analysis in crystals using X-ray topography. *Microsc. Res. Tech.* **2006**, *69*, 343–358. [CrossRef] [PubMed]

140. Moore, M. White-beam X-ray topography. *Crystallogr. Rev.* **2012**, *18*, 205–233. [CrossRef]

141. Tanner, B.K. Contrast of defects in X-ray diffraction topographs. *X-Ray Neutron Dyn. Diffr.* **1996**, *357*, 147–166.

142. Rozhestrevenskaya, M.B.; Kazurov, K. The growth of SrTiO₃ single crystals using the melt method in a flame and X-ray graphic studies of the degree of their perfection. In *Kristallizatsiia i Fazovye Prevrashcheniia*; Sirota, N.N., Ed.; Akademii Nauk SSSR Nauchnyi Sov. Pol., Nauka i tekhnika: Minsk, Russia, 1971; p. 267.
143. Mochizuki, S.; Fujishiro, F.; Shibata, K.; Ogi, A.; Konya, T.; Inaba, K. Optical, electrical, and X-ray-structural studies on verneuil-grown SrTiO₃ single crystal: Annealing study. *Physica B* **2007**, *401*, 433–436. [CrossRef]
144. Haider, M.; Uhlemann, S.; Schwan, E.; Rose, H.; Kabius, B.; Urban, K. Electron microscopy image enhanced. *Nature* **1998**, *392*, 768–769. [CrossRef]
145. Krivanek, O.L.; Dellby, N.; Lupini, A.R. Towards sub-angstrom electron beams. *Ultramicroscopy* **1999**, *78*, 1–11. [CrossRef]
146. Morishita, S.; Ishikawa, R.; Kohno, Y.; Sawada, H.; Shibata, N.; Ikuhara, Y. Attainment of 40.5 pm spatial resolution using 300 kV scanning transmission electron microscope equipped with fifth-order aberration corrector. *Microsopy* **2017**, in press.
147. Kimoto, K.; Asaka, T.; Nagai, T.; Saito, M.; Matsui, Y.; Ishizuka, K. Element-selective imaging of atomic columns in a crystal using STEM and EELS. *Nature* **2007**, *450*, 702–704. [CrossRef] [PubMed]
148. Bosman, M.; Keast, V.J.; Garcia-Munoz, J.L.; D'Alfonso, A.J.; Findlay, S.D.; Allen, L.J. Two-dimensional mapping of chemical information at atomic resolution. *Phys. Rev. Lett.* **2007**, *99*, 086102. [CrossRef] [PubMed]
149. Muller, D.A.; Kourkoutis, L.F.; Murfitt, M.; Song, J.H.; Hwang, H.Y.; Silcox, J.; Dellby, N.; Krivanek, O.L. Atomic-scale chemical imaging of composition and bonding by aberration-corrected microscopy. *Science* **2008**, *319*, 1073–1076. [CrossRef] [PubMed]
150. D'Alfonso, A.J.; Freitag, B.; Klenov, D.; Allen, L.J. Atomic-resolution chemical mapping using energy-dispersive X-ray spectroscopy. *Phys. Rev. B* **2010**, *81*, 100101. [CrossRef]
151. Chu, M.W.; Liou, S.C.; Chang, C.P.; Choa, F.S.; Chen, C.H. Emergent Chemical Mapping at Atomic-Column Resolution by Energy-Dispersive X-Ray Spectroscopy in an Aberration-Corrected Electron Microscope. *Phys. Rev. Lett.* **2010**, *104*, 196101. [CrossRef] [PubMed]
152. Pennycook, S.J.; Nellist, P.D. *Scanning Transmission Electron Microscopy Imaging and Analysis*; Springer: New York, NY, USA, Dordrecht, The Netherlands; Heidelberg, Germany; 2011.
153. Crewe, A.V. Scanning Electron Microscopes—Is High Resolution Possible. *Science* **1966**, *154*, 729–738. [CrossRef] [PubMed]
154. Crewe, A.V.; Wall, J.; Langmore, J. Visibility of Single Atoms. *Science* **1970**, *168*, 1338–1340. [CrossRef] [PubMed]
155. Pennycook, S.J.; Boatner, L.A. Chemically Sensitive Structure-Imaging with a Scanning-Transmission Electron-Microscope. *Nature* **1988**, *336*, 565–567. [CrossRef]
156. Krivanek, O.L.; Chisholm, M.F.; Nicolosi, V.; Pennycook, T.J.; Corbin, G.J.; Dellby, N.; Murfitt, M.F.; Own, C.S.; Szilagyi, Z.S.; Oxley, M.P.; et al. Atom-by-atom structural and chemical analysis by annular dark-field electron microscopy. *Nature* **2010**, *464*, 571–574. [CrossRef] [PubMed]
157. Krivanek, O.L.; Lovejoy, T.C.; Dellby, N.; Aoki, T.; Carpenter, R.W.; Rez, P.; Soignard, E.; Zhu, J.T.; Batson, P.E.; Lagos, M.J.; et al. Vibrational spectroscopy in the electron microscope. *Nature* **2014**, *514*, 209. [CrossRef] [PubMed]
158. Mukai, M.; Okunishi, E.; Ashino, M.; Omoto, K.; Fukuda, T.; Ikeda, A.; Somehara, K.; Kaneyama, T.; Saitoh, T.; Hirayama, T.; et al. Development of a monochromator for aberration-corrected scanning transmission electron microscopy. *Microscopy* **2015**, *64*, 151–158. [CrossRef] [PubMed]
159. Ishikawa, R.; Lupini, A.R.; Findlay, S.D.; Pennycook, S.J. Quantitative Annular Dark Field Electron Microscopy Using Single Electron Signals. *Microsc. Microanal.* **2014**, *20*, 99–110. [CrossRef] [PubMed]
160. Shibata, N.; Kohno, Y.; Findlay, S.D.; Sawada, H.; Kondo, Y.; Ikuhara, Y. New area detector for atomic-resolution scanning transmission electron microscopy. *J. Electron Microsc.* **2010**, *59*, 473–479. [CrossRef] [PubMed]
161. Shibata, N.; Seki, T.; Sanchez-Santolino, G.; Findlay, S.D.; Kohno, Y.; Matsumoto, T.; Ishikawa, R.; Ikuhara, Y. Electric field imaging of single atoms. *Nat. Commun.* **2017**, *8*, 15631. [CrossRef] [PubMed]
162. Yang, H.; Rutte, R.N.; Jones, L.; Simson, M.; Sagawa, R.; Ryll, H.; Huth, M.; Pennycook, T.J.; Green, M.L.H.; Soltau, H.; et al. Simultaneous atomic-resolution electron ptychography and Z-contrast imaging of light and heavy elements in complex nanostructures. *Nat. Commun.* **2016**, *7*, 12532. [CrossRef] [PubMed]
163. Kirkland, E.J. *Advanced Computing in Electron Microscopy*; Springer: New York, NY, USA, 2010.

164. Pennycook, S.J.; Jesson, D.E. High-Resolution Incoherent Imaging of Crystals. *Phys. Rev. Lett.* **1990**, *64*, 938–941. [CrossRef] [PubMed]

165. Pennycook, S.J.; Jesson, D.E. High-Resolution Z-Contrast Imaging of Crystals. *Ultramicroscopy* **1991**, *37*, 14–38. [CrossRef]

166. Gao, P.; Kumamoto, A.; Ishikawa, R.; Lugg, N.; Shibata, N.; Ikuhara, Y. Picometer-scale atom position analysis in annular bright-field STEM imaging. *Ultramicroscopy* **2018**, *184*, 177–187. [CrossRef] [PubMed]

167. LeBeau, J.M.; Findlay, S.D.; Allen, L.J.; Stemmer, S. Standardless Atom Counting in Scanning Transmission Electron Microscopy. *Nano Lett.* **2010**, *10*, 4405–4408. [CrossRef] [PubMed]

168. Van Aert, S.; De Backer, A.; Martinez, G.T.; Goris, B.; Bals, S.; Van Tendeloo, G.; Rosenauer, A. Procedure to count atoms with trustworthy single-atom sensitivity. *Phys. Rev. B* **2013**, *87*, 064107. [CrossRef]

169. Jones, L.; MacArthur, K.E.; Fauske, V.T.; van Helvoort, A.T.J.; Nellist, P.D. Rapid Estimation of Catalyst Nanoparticle Morphology and Atomic-Coordination by High-Resolution Z-Contrast Electron Microscopy. *Nano Lett.* **2014**, *14*, 6336–6341. [CrossRef] [PubMed]

170. Ishikawa, R.; Lupini, A.R.; Findlay, S.D.; Taniguchi, T.; Pennycook, S.J. Three-Dimensional Location of a Single Dopant with Atomic Precision by Aberration-Corrected Scanning Transmission Electron Microscopy. *Nano Lett.* **2014**, *14*, 1903–1908. [CrossRef] [PubMed]

171. Findlay, S.D.; Shibata, N.; Sawada, H.; Okunishi, E.; Kondo, Y.; Yamamoto, T.; Ikuhara, Y. Robust atomic resolution imaging of light elements using scanning transmission electron microscopy. *Appl. Phys. Lett.* **2009**, *95*, 191913. [CrossRef]

172. Findlay, S.D.; Shibata, N.; Sawada, H.; Okunishi, E.; Kondo, Y.; Ikuhara, Y. Dynamics of annular bright field imaging in scanning transmission electron microscopy. *Ultramicroscopy* **2010**, *110*, 903–923. [CrossRef] [PubMed]

173. Findlay, S.D.; LeBeau, J.M. Detector non-uniformity in scanning transmission electron microscopy. *Ultramicroscopy* **2013**, *124*, 52–60. [CrossRef] [PubMed]

174. Findlay, S.D.; Huang, R.; Ishikawa, R.; Shibata, N.; Ikuhara, Y. Direct visualization of lithium via annular bright field scanning transmission electron microscopy: A review. *Microscopy* **2017**, *66*, 3–14. [CrossRef] [PubMed]

175. Huang, R.; Ikuhara, Y.H.; Mizoguchi, T.; Findlay, S.D.; Kuwabara, A.; Fisher, C.A.J.; Moriwake, H.; Oki, H.; Hirayama, T.; Ikuhara, Y. Oxygen-Vacancy Ordering at Surfaces of Lithium Manganese(III,IV) Oxide Spinel Nanoparticles. *Angew. Chem. Int. Ed.* **2011**, *50*, 3053–3057. [CrossRef] [PubMed]

176. Findlay, S.D.; Saito, T.; Shibata, N.; Sato, Y.; Matsuda, J.; Asano, K.; Akiba, E.; Hirayama, T.; Ikuhara, Y. Direct Imaging of Hydrogen within a Crystalline Environment. *Appl. Phys. Express* **2010**, *3*, 116603. [CrossRef]

177. Ishikawa, R.; Okunishi, E.; Sawada, H.; Kondo, Y.; Hosokawa, F.; Abe, E. Direct imaging of hydrogen-atom columns in a crystal by annular bright-field electron microscopy. *Nat. Mater.* **2011**, *10*, 278–281. [CrossRef] [PubMed]

178. Egerton, R.F. *Electron Energy-Loss Spectroscopy in the Electron Microscope*; Springer: New York, NY, USA, 2011.

179. Matsukawa, M.; Ishikawa, R.; Hisatomi, T.; Moriya, Y.; Shibata, N.; Kubota, J.; Ikuhara, Y.; Domen, K. Enhancing Photocatalytic Activity of LaTiO$_2$N by Removal of Surface Reconstruction Layer. *Nano Lett.* **2014**, *14*, 1038–1041. [CrossRef] [PubMed]

180. Ishikawa, R.; Shimbo, Y.; Sugiyama, I.; Lugg, N.R.; Shibata, N.; Ikuhara, Y. Room-temperature dilute ferromagnetic dislocations in Sr$_{1-x}$Mn$_x$TiO$_{3-\delta}$. *Phys. Rev. B* **2017**, *96*, 024440. [CrossRef]

181. Sugiyama, I.; Shibata, N.; Wang, Z.C.; Kobayashi, S.; Yamamoto, T.; Ikuhara, Y. Ferromagnetic dislocations in antiferromagnetic NiO. *Nat. Nanotechnol.* **2013**, *8*, 266–270. [CrossRef] [PubMed]

182. Chen, Z.; Weyland, M.; Sang, X.; Xu, W.; Dycus, J.H.; LeBeau, J.M.; D'Alfonso, A.J.; Allen, L.J.; Findlay, S.D. Quantitative atomic resolution elemental mapping via absolute-scale energy dispersive X-ray spectroscopy. *Ultramicroscopy* **2016**, *168*, 7–16. [CrossRef] [PubMed]

183. Chen, M.S.; Goodman, D.W. The structure of catalytically active gold on titania. *Science* **2004**, *306*, 252–255. [CrossRef] [PubMed]

184. Chen, X.B.; Liu, L.; Yu, P.Y.; Mao, S.S. Increasing Solar Absorption for Photocatalysis with Black Hydrogenated Titanium Dioxide Nanocrystals. *Science* **2011**, *331*, 746–750. [CrossRef] [PubMed]

185. Rodriguez, J.A.; Ma, S.; Liu, P.; Hrbek, J.; Evans, J.; Perez, M. Activity of CeO$_x$ and TiO$_x$ nanoparticles grown on Au(111) in the water-gas shift reaction. *Science* **2007**, *318*, 1757–1760. [CrossRef] [PubMed]

186. Diguna, L.J.; Shen, Q.; Kobayashi, J.; Toyoda, T. High efficiency of CdSe quantum-dot-sensitized TiO$_2$ inverse opal solar cells. *Appl. Phys. Lett.* **2007**, *91*, 023116. [CrossRef]

187. Gai, Y.Q.; Li, J.B.; Li, S.S.; Xia, J.B.; Wei, S.H. Design of Narrow-Gap TiO$_2$: A Passivated Codoping Approach for Enhanced Photoelectrochemical Activity. *Phys. Rev. Lett.* **2009**, *102*, 036402. [CrossRef] [PubMed]

188. Wendt, S.; Sprunger, P.T.; Lira, E.; Madsen, G.K.H.; Li, Z.S.; Hansen, J.O.; Matthiesen, J.; Blekinge-Rasmussen, A.; Laegsgaard, E.; Hammer, B.; et al. The role of interstitial sites in the Ti3d defect state in the band gap of Titania. *Science* **2008**, *320*, 1755–1759. [CrossRef] [PubMed]

189. Nakano, Y.; Morikawa, T.; Ohwaki, T.; Taga, Y. Electrical characterization of band gap states in C-doped TiO$_2$ films. *Appl. Phys. Lett.* **2005**, *87*, 052111. [CrossRef]

190. Ahn, H.S.; Han, S.; Hwang, C.S. Pairing of cation vacancies and gap-state creation in TiO$_2$ and HfO$_2$. *Appl. Phys. Lett.* **2007**, *90*, 252908. [CrossRef]

191. Li, H.; Bradt, R.C. Knoop Microhardness Anisotropy of Single-Crystal Rutile. *J. Am. Ceram. Soc.* **1990**, *73*, 1360–1364. [CrossRef]

192. Basu, S.; Elshrief, O.A.; Coward, R.; Anasori, B.; Barsoum, M.W. Microscale deformation of (001) and (100) rutile single crystals under spherical nanoindentation. *J. Mater. Res.* **2012**, *27*, 53–63. [CrossRef]

193. Ikuhara, Y.; Nishimura, H.; Nakamura, A.; Matsunaga, K.; Yamamoto, T.; Lagerlof, K.P.D. Dislocation structures of low-angle and near-Σ3 grain boundaries in alumina bicrystals. *J. Am. Ceram. Soc.* **2003**, *86*, 595–602. [CrossRef]

194. Tochigi, E.; Shibata, N.; Nakamura, A.; Mizoguchi, T.; Yamamoto, T.; Ikuhara, Y. Structures of dissociated <1 $\bar{1}$ 0 0> dislocations and {1 $\bar{1}$ 0 0} stacking faults of alumina (α-Al$_2$O$_3$). *Acta Mater.* **2010**, *58*, 208–215. [CrossRef]

195. Stoyanov, E.; Langenhorst, F.; Steinle-Neumann, G. The effect of valence state and site geometry on Ti L-3,L-2 and O K electron energy-loss spectra of Ti$_x$O$_y$ phases. *Am. Mineral.* **2007**, *92*, 577–586. [CrossRef]

196. Morin, P.; Pitaval, M.; Besnard, D.; Fontaine, G. Electron-channelling imaging in scanning electron-microscopy. *Philos. Mag. A* **1979**, *40*, 511–524. [CrossRef]

197. Czernuszka, J.T.; Long, N.J.; Boyes, E.D.; Hirsch, P.B. Imaging of dislocations using backscattered electrons in a scanning electron-microscope. *Philos. Mag. Lett.* **1990**, *62*, 227–232. [CrossRef]

198. Kamaladasa, R.J.; Liu, F.; Porter, L.M.; Davis, R.F.; Koleske, D.D.; Mulholland, G.; Jones, K.A.; Picard, Y.N. Identifying threading dislocations in GaN films and substrates by electron channelling. *J. Microsc.* **2011**, *244*, 311–319. [CrossRef] [PubMed]

199. Kuwano, N.; Itakura, M.; Nagatomo, Y.; Tachibana, S. Scanning electron microscope observation of dislocations in semiconductor and metal materials. *J. Electron Microsc.* **2010**, *59*, S175–S181. [CrossRef] [PubMed]

200. Kamaladasa, R.J.; Jiang, W.K.; Picard, Y.N. Imaging Dislocations in Single-Crystal SrTiO$_3$ Substrates by Electron Channeling. *J. Electron. Mater.* **2011**, *40*, 2222–2227. [CrossRef]

201. Lichte, H.; Lehmann, M. Electron holography—Basics and applications. *Rep. Prog. Phys.* **2008**, *71*, 016102. [CrossRef]

202. Jones, R.O. Density functional theory: Its origins, rise to prominence, and future. *Rev. Mod. Phys.* **2015**, *87*, 897–923. [CrossRef]

203. Shanthi, N.; Sarma, D.D. Electronic structure of electron doped SrTiO$_3$: SrTiO$_{3-\delta}$ and Sr$_{1-x}$La$_x$TiO$_3$. *Phys. Rev. B* **1998**, *57*, 2153–2158. [CrossRef]

204. Ricci, D.; Bano, G.; Pacchioni, G.; Illas, F. Electronic structure of a neutral oxygen vacancy in SrTiO$_3$. *Phys. Rev. B* **2003**, *68*, 224105. [CrossRef]

205. He, J.; Behera, R.K.; Finnis, M.W.; Li, X.; Dickey, E.C.; Phillpot, S.R.; Sinnott, S.B. Prediction of high-temperature point defect formation in TiO$_2$ from combined ab initio and thermodynamic calculations. *Acta Mater.* **2007**, *55*, 4325–4337. [CrossRef]

206. Carrasco, J.; Illas, F.; Lopez, N.; Kotomin, E.A.; Zhukovskii, Y.F.; Piskunov, S.; Maier, J.; Hermansson, K. First-principles calculations of the atomic and electronic structure of F centers in the bulk and on the (001) surface of SrTiO$_3$. *Phys. Status Solidi* **2005**, *2*, 153–158. [CrossRef]

207. Alexandrov, V.E.; Kotomin, E.A.; Maier, J.; Evarestov, R.A. First-principles study of bulk and surface oxygen vacancies in SrTiO$_3$ crystal. *Eur. Phys. J. B* **2009**, *72*, 53–57. [CrossRef]

208. Lin, C.; Demkov, A.A. Electron Correlation in Oxygen Vacancy in SrTiO$_3$. *Phys. Rev. Lett.* **2013**, *111*, 217601. [CrossRef] [PubMed]

209. Park, S.G.; Magyari-Köpe, B.; Nishi, Y. Electronic correlation effects in reduced rutile TiO_2 within the LDA+*U* method. *Phys. Rev. B* **2010**, *82*, 115109. [CrossRef]
210. Cuong, D.D.; Lee, B.; Choi, K.M.; Ahn, H.S.; Han, S.; Lee, J. Oxygen vacancy clustering and electron localization in oxygen-deficient SrTiO3: LDA+*U* study. *Phys. Rev. Lett.* **2007**, *98*, 115503. [CrossRef] [PubMed]
211. Park, S.G.; Magyari-Köpe, B.; Nishi, Y. Impact of Oxygen Vacancy Ordering on the Formation of a Conductive Filament in TiO_2 for Resistive Switching Memory. *IEEE Electr. Device Lett.* **2011**, *32*, 197–199. [CrossRef]
212. Le Bacq, O.; Salinas, E.; Pisch, A.; Bernard, C.; Pasturel, A. First-principles structural stability in the strontium-titanium-oxygen system. *Philos. Mag.* **2005**, *86*, 2283. [CrossRef]
213. Jia, C.L.; Mi, S.B.; Urban, K.; Vrejoiu, I.; Alexe, M.; Hesse, D. Atomic-scale study of electric dipoles near charged and uncharged domain walls in ferroelectric films. *Nat. Mater.* **2008**, *7*, 57–61. [CrossRef] [PubMed]
214. Rahmanizadeh, K.; Wortmann, D.; Bihlmayer, G.; Blugel, S. Charge and orbital order at head-to-head domain walls in $PbTiO_3$. *Phys. Rev. B* **2014**, *90*, 115104. [CrossRef]
215. Basletic, M.; Maurice, J.L.; Carretero, C.; Herranz, G.; Copie, O.; Bibes, M.; Jacquet, E.; Bouzehouane, K.; Fusil, S.; Barthelemy, A. Mapping the spatial distribution of charge carriers in $LaAlO_3/SrTiO_3$ heterostructures. *Nat. Mater.* **2008**, *7*, 621–625. [CrossRef] [PubMed]
216. Hirel, P.; Mrovec, M.; Elsässer, C. Atomistic simulation study of <1 1 0> dislocations in strontium titanate. *Acta Mater.* **2012**, *60*, 329–338. [CrossRef]
217. Vollman, M.; Waser, R. Grain-Boundary Defect Chemistry of Acceptor-Doped Titanates—Space-Charge Layer Width. *J. Am. Ceram. Soc.* **1994**, *77*, 235–243. [CrossRef]
218. Vollmann, M.; Hagenbeck, R.; Waser, R. Grain-boundary defect chemistry of acceptor-doped titanates: Inversion layer and low-field conduction. *J. Am. Ceram. Soc.* **1997**, *80*, 2301–2314. [CrossRef]
219. Hagenbeck, R.; Waser, R. Influence of temperature and interface charge on the grain-boundary conductivity in acceptor-doped $SrTiO_3$ ceramics. *J. Appl. Phys.* **1998**, *83*, 2083–2092. [CrossRef]
220. Dudler, R.; Albers, J.; Muser, H.E. Dielectric Behavior of Pure $BaTiO_3$ at Ultralow Frequencies. *Ferroelectrics* **1978**, *21*, 381–383. [CrossRef]
221. Waser, R. (Forschungszentrum Jülich, Jülich, Germany). Personal communication, 2018.
222. Rodenbücher, C.; Bihlmayer, G.; Speier, W.; Kubacki, J.; Wojtyniak, M.; Rogala, M.; Wrana, D.; Krok, F.; Szot, K. Detection of confined paths on oxide surfaces by local-conductivity atomic force microscopy with atomic resolution. *arXiv* **2016**, arXiv:1611.07773.
223. Psiuk, B.; Szade, J.; Schroeder, H.; Haselier, H.; Mlynarczyk, M.; Waser, R.; Szot, K. Photoemission study of $SrTiO_3$ surface layers instability upon metal deposition. *Appl. Phys. A Mater.* **2007**, *89*, 451–455. [CrossRef]
224. Szot, K.; Otto, R.; Herion, J. Strom-Spannungswandler zur Erfassung eines Tunnelstroms eines Rastertunnelmikroskops. Patent DE4438960, 31 October 1994.
225. Szot, K.; Dittmann, R.; Speier, W.; Waser, R. Nanoscale resistive switching in $SrTiO_3$ thin films. *Phys. Status Solidi Rapid Res. Lett.* **2007**, *1*, R86–R88. [CrossRef]
226. Wrana, D.; Rodenbücher, C.; Belza, W.; Szot, K.; Krok, F. In situ study of redox processes on the surface of $SrTiO_3$ single crystals. *Appl. Surf. Sci.* **2018**, *432*, 46–52. [CrossRef]
227. Rodenbücher, C.; Speier, W.; Bihlmayer, G.; Breuer, U.; Waser, R.; Szot, K. Cluster-like resistive switching of $SrTiO_3$: Nb surface layers. *New J. Phys.* **2013**, *15*, 103017. [CrossRef]
228. Pilch, M. Role of the La Doping in SrTiO3 Crystals. Ph.D. Thesis, University of Silesia, Kattowice, Poland, 2010.
229. Reichenberg, B.; Szot, K.; Schneller, T.; Breuer, U.; Tiedke, S.; Waser, R. Inhomogeneous local conductivity induced by thermal reduction in $BaTiO_3$ thin films and single crystals. *Integr. Ferroelectr.* **2004**, *61*, 43–49. [CrossRef]
230. Szot, K.; Reichenberg, B.; Peter, F.; Waser, R.; Tiedke, S. Electrical Characterization of Perovskite Nanostructures. In *Scanning Probe Microscopy*; Kalinin, S., Gruverman, A., Eds.; Springer: New York, NY, USA, 2007.
231. Jankowska-Sumara, I.; Szot, K.; Majchrowski, A.; Roleder, K. Effect of resistive switching and electrically driven insulator-conductor transition in $PbZrO_3$ single crystals. *Phys. Status Solidi A Appl. Mater. Sci.* **2013**, *210*, 507–512. [CrossRef]

232. Kubacki, J.; Molak, A.; Rogala, M.; Rodenbücher, C.; Szot, K. Metal-insulator transition induced by non-stoichiometry of surface layer and molecular reactions on single crystal $KTaO_3$. *Surf. Sci.* **2012**, *606*, 1252–1262. [CrossRef]

233. Molak, A.; Szot, K.; Kania, A.; Friedrich, J.; Penkalla, H.J. Insulator-metal transition in Mn-doped $NaNbO_3$ induced by chemical and thermal treatment. *Phase Transit.* **2008**, *81*, 977–986. [CrossRef]

234. Molak, A.; Szot, K. Insulator-semiconductor-metallic state transition induced by electric fields in Mn-doped $NaNbO_3$. *Phys. Status Solidi Rapid Res. Lett.* **2009**, *3*, 127–129. [CrossRef]

235. Markiewicz, E.; Szot, K.; Hilczer, B.; Pietraszko, A.A. $BiFeO_3$ single crystal as resistive switching element for application in microelectronic devices. *Phase Transit.* **2013**, *86*, 284–289. [CrossRef]

236. Choi, B.J.; Jeong, D.S.; Kim, S.K.; Rohde, C.; Choi, S.; Oh, J.H.; Kim, H.J.; Hwang, C.S.; Szot, K.; Waser, R.; et al. Resistive switching mechanism of TiO_2 thin films grown by atomic-layer deposition. *J. Appl. Phys.* **2005**, *98*, 033715. [CrossRef]

237. Menke, T.; Meuffels, P.; Dittmann, R.; Szot, K.; Waser, R. Separation of bulk and interface contributions to electroforming and resistive switching behavior of epitaxial Fe-doped $SrTiO_3$. *J. Appl. Phys.* **2009**, *105*, 066104. [CrossRef]

238. Muenstermann, R.; Dittmann, R.; Szot, K.; Mi, S.B.; Jia, C.L.; Meuffels, P.; Waser, R. Realization of regular arrays of nanoscale resistive switching blocks in thin films of Nb-doped $SrTiO_3$. *Appl. Phys. Lett.* **2008**, *93*, 023110. [CrossRef]

239. Menke, T.; Dittmann, R.; Meuffels, P.; Szot, K.; Waser, R. Impact of the electroforming process on the device stability of epitaxial Fe-doped $SrTiO_3$ resistive switching cells. *J. Appl. Phys.* **2009**, *106*, 114507. [CrossRef]

240. Peter, F.; Rudiger, A.; Dittmann, R.; Waser, R.; Szot, K.; Reichenberg, B.; Prume, K. Analysis of shape effects on the piezoresponse in ferroelectric nanograins with and without adsorbates. *Appl. Phys. Lett.* **2005**, *87*, 082901. [CrossRef]

241. Chen, J.; Sekiguchi, T.; Li, J.Y.; Ito, S.; Yi, W.; Ogura, A. Investigation of dislocations in Nb-doped $SrTiO_3$ by electron-beam-induced current and transmission electron microscopy. *Appl. Phys. Lett.* **2015**, *106*, 102109. [CrossRef]

242. Jiang, W.; Evans, D.; Bain, J.A.; Skowronski, M.; Salvador, P.A. Electron beam induced current investigations of interfaces exposed to chemical and electrical stresses. *Appl. Phys. Lett.* **2010**, *96*, 092102. [CrossRef]

243. Kim, D.C.; Seo, S.; Ahn, S.E.; Suh, D.S.; Lee, M.J.; Park, B.H.; Yoo, I.K.; Baek, I.G.; Kim, H.J.; Yim, E.K.; et al. Electrical observations of filamentary conductions for the resistive memory switching in NiO films. *Appl. Phys. Lett.* **2006**, *88*, 202102. [CrossRef]

244. Zhang, K.; Lanza, M.; Shen, Z.Y.; Fu, Q.; Hou, S.M.; Porti, M.; Nafria, M. Analysis of Factors in the Nanoscale Physical and Electrical Characterization of High-K Materials by Conductive Atomic Force Microscope. *Integr. Ferroelectr.* **2014**, *153*, 1–8. [CrossRef]

245. Son, J.Y.; Shin, Y.H. Direct observation of conducting filaments on resistive switching of NiO thin films. *Appl. Phys. Lett.* **2008**, *92*, 222106. [CrossRef]

246. Rodenbücher, C.; Luysberg, M.; Schwedt, A.; Havel, V.; Gunkel, F.; Mayer, J.; Waser, R. Homogeneity and variation of donor doping in Verneuil-grown $SrTiO_3$:Nb single crystals. *Sci. Rep.* **2016**, *6*, 32250. [CrossRef] [PubMed]

247. Kim, Y.; Disa, A.S.; Babakol, T.E.; Fang, X.Y.; Brock, J.D. Strain and oxygen vacancy ordering in $SrTiO_3$: Diffuse X-ray scattering studies. *Phys. Rev. B* **2015**, *92*, 064105. [CrossRef]

248. Mott, N.F. Metal-Insulator Transition. *Rev. Mod. Phys.* **1968**, *40*, 677. [CrossRef]

249. Frederikse, H.P.R.; Thurber, W.R.; Hosler, W.R. Electronic Transport in Strontium Titanate. *Phys. Rev.* **1964**, *134*, A442. [CrossRef]

250. Calvani, P.; Capizzi, M.; Donato, F.; Lupi, S.; Maselli, P.; Peschiaroli, D. Observation of a Midinfrared Band in $SrTiO_{3-y}$. *Phys. Rev. B* **1993**, *47*, 8917–8922. [CrossRef]

251. Spinelli, A.; Torija, M.A.; Liu, C.; Jan, C.; Leighton, C. Electronic transport in doped $SrTiO_3$: Conduction mechanisms and potential applications. *Phys. Rev. B* **2010**, *81*, 155110. [CrossRef]

252. Lin, X.; Rischau, C.W.; Buchauer, L.; Jaoui, A.; Fauque, B.; Behnia, K. Metallicity without quasi-particles in room-temperature strontium titanate. *npj Quantum Mater.* **2017**, *2*, 41. [CrossRef]

253. Pergament, A.; Stefanovich, G.; Markova, N. The Mott criterion: So simple and yet so complex. *arXiv* **2014**, arXiv:1411.4372.

254. Lehovec, K.; Shirn, G.A. Conductivity Injection and Extraction in Polycrystalline Barium Titanate. *J. Appl. Phys.* **1962**, *33*, 2036–2044. [CrossRef]

255. Lin, Y.H.; Chen, Y.; Goldman, A.M. Indications of superconductivity at somewhat elevated temperatures in strontium titanate subjected to high electric fields. *Phys. Rev. B* **2010**, *82*, 172507. [CrossRef]

256. Pilch, M.; Szot, K. Resistive switching in $Sr_{1-0.05}La_{0.05}TiO_3$. In Proceedings of the 2012 International Symposium on Applications of Ferroelectrics Held Jointly with 11th IEEE ECAPD and IEEE PFM (ISAF/ECAPD/PFM), Aveiro, Portugal, 9–13 July 2012.

257. Szot, K.; Rytz, D.; Lazar, I.; Kajewki, D.; Roleder, K. The macro- and nanoscale phenomena in $BaTiO_3$ single crystals (inv. talk). In Proceedings of the International Meeting on Ferroelectricity (IFM), San Antonio, TX, USA, 4–8 September 2017.

258. Van Landuyt, J.; Amelinckx, S. Generation Mechanism for Shear Planes in Shear Structures. *J. Solid State Chem.* **1973**, *6*, 222–229. [CrossRef]

259. Sakaguchi, I.; Komastu, M.; Watanabe, A.; Haneda, H. Oxygen diffusion along the short-circuit paths in bicrystal $SrTiO_3$. *J. Mater. Res.* **2000**, *15*, 2598–2601. [CrossRef]

260. Szot, K.; Rodenbücher, C. Insulator-Metal Transition Associated with Resistive Switching in Real $SrTiO_3$ and TiO_2 Crystals. In Proceedings of the 2015 Joint IEEE International Symposium on the Applications of Ferroelectric, International Symposium on Integrated Functionalities and Piezoelectric Force Microscopy Workshop (ISAF/ISIF/PFM), Singapore, 24–27 May 2015; pp. 143–146.

261. Menzel, S.; Waters, M.; Marchewka, A.; Bottger, U.; Dittmann, R.; Waser, R. Origin of the Ultra-nonlinear Switching Kinetics in Oxide-Based Resistive Switches. *Adv. Funct. Mater.* **2011**, *21*, 4487–4492. [CrossRef]

262. Jiang, W.; Kamaladasa, R.J.; Lu, Y.M.; Vicari, A.; Berechman, R.; Salvador, P.A.; Bain, J.A.; Picard, Y.N.; Skowronski, M. Local heating-induced plastic deformation in resistive switching devices. *J. Appl. Phys.* **2011**, *110*, 054514. [CrossRef]

263. Kamaladasa, R.J.; Noman, M.; Chen, W.; Salvador, P.A.; Bain, J.A.; Skowronski, M.; Picard, Y.N. Dislocation impact on resistive switching in single-crystal $SrTiO_3$. *J. Appl. Phys.* **2013**, *113*, 234510. [CrossRef]

264. Kim, G.H.; Lee, J.H.; Seok, J.Y.; Song, S.J.; Yoon, J.H.; Yoon, K.J.; Lee, M.H.; Kim, K.M.; Lee, H.D.; Ryu, S.W.; et al. Improved endurance of resistive switching TiO_2 thin film by hourglass shaped Magneli filaments. *Appl. Phys. Lett.* **2011**, *98*, 262901.

265. Bobeth, M.; Farag, N.; Levin, A.A.; Meyer, D.C.; Pompe, W.; Romanov, A.E. Reversible electric field-induced structure changes in the near-surface region of strontium titanate. *J. Ceram. Soc. Jpn.* **2006**, *114*, 1029–1037. [CrossRef]

266. Kamaladasa, R.J.; Sharma, A.A.; Lai, Y.T.; Chen, W.H.; Salvador, P.A.; Bain, J.A.; Skowronski, M.; Picard, Y.N. In Situ TEM Imaging of Defect Dynamics under Electrical Bias in Resistive Switching Rutile-TiO_2. *Microsc. Microanal.* **2015**, *21*, 140–153. [CrossRef] [PubMed]

267. Cooper, D.; Baeumer, C.; Bernier, N.; Marchewka, A.; La Torre, C.; Dunin-Borkowski, R.E.; Menzel, S.; Waser, R.; Dittmann, R. Anomalous Resistance Hysteresis in Oxide ReRAM: Oxygen Evolution and Reincorporation Revealed by In Situ TEM. *Adv. Mater.* **2017**, *29*, 1700212. [CrossRef] [PubMed]

268. Meuffels, P.; Soni, R. Fundamental Issues and Problems in the Realization of Memristors. *arXiv* **2012**, arXiv:1207.7319.

269. Baeumer, C.; Schmitz, C.; Marchewka, A.; Mueller, D.N.; Valenta, R.; Hackl, J.; Raab, N.; Rogers, S.P.; Khan, M.I.; Nemsak, S.; et al. Quantifying redox-induced Schottky barrier variations in memristive devices via in operando spectromicroscopy with graphene electrodes. *Nat. Commun.* **2016**, *7*, 12398. [CrossRef] [PubMed]

270. Menzel, S.; Bottger, U.; Wimmer, M.; Salinga, M. Physics of the Switching Kinetics in Resistive Memories. *Adv. Funct. Mater.* **2015**, *25*, 6306–6325. [CrossRef]

271. Ikuhara, Y.; Suzuki, T.; Kubo, Y. Transmission Electron-Microscopy Insitu Observation of Crack-Propagation in Sintered Alumina. *Philos. Mag. Lett.* **1992**, *66*, 323–327. [CrossRef]

272. Oh, S.H.; Legros, M.; Kiener, D.; Dehm, G. In situ observation of dislocation nucleation and escape in a submicrometre aluminium single crystal. *Nat. Mater.* **2009**, *8*, 95–100. [CrossRef] [PubMed]

273. Kiener, D.; Hosemann, P.; Maloy, S.A.; Minor, A.M. In situ nanocompression testing of irradiated copper. *Nat. Mater.* **2011**, *10*, 608–613. [CrossRef] [PubMed]

274. Minor, A.M.; Asif, S.A.S.; Shan, Z.W.; Stach, E.A.; Cyrankowski, E.; Wyrobek, T.J.; Warren, O.L. A new view of the onset of plasticity during the nanoindentation of aluminium. *Nat. Mater.* **2006**, *5*, 697–702. [CrossRef] [PubMed]

275. De Hosson, J.T.M.; Soer, W.A.; Minor, A.M.; Shan, Z.W.; Stach, E.A.; Asif, S.A.S.; Warren, O.L. In situ TEM nanoindentation and dislocation-grain boundary interactions: A tribute to David Brandon. *J. Mater. Sci.* **2006**, *41*, 7704–7719. [CrossRef]

276. Kondo, S.; Mitsuma, T.; Shibata, N.; Ikuhara, Y. Direct observation of individual dislocation interaction processes with grain boundaries. *Sci. Adv.* **2016**, *2*, e1501926. [CrossRef] [PubMed]

277. Hirsch, P.B.; Howie, A.; Nicholson, R.B.; Pashley, D.W.; Whelan, M.J. *Electron Microscopy of Thin Crystals*, 2nd ed.; Krieger: New York, NY, USA, 1977.

278. Yang, K.H.; Ho, N.J.; Lu, H.Y. Deformation Microstructure in (001) Single Crystal Strontium Titanate by Vickers Indentation. *J. Am. Ceram. Soc.* **2009**, *92*, 2345–2353. [CrossRef]

279. Yang, K.H.; Ho, N.J.; Lu, H.Y. Plastic Deformation of <001> Single-Crystal SrTiO$_3$ by Compression at Room Temperature. *J. Am. Ceram. Soc.* **2011**, *94*, 3104–3111. [CrossRef]

280. Ravikumar, V.; Dravid, V.P. Atomic-Structure of Undoped Σ5 Symmetrical Tilt Grain-Boundary in Strontium-Titanate. *Ultramicroscopy* **1993**, *52*, 557–563. [CrossRef]

281. Imaeda, M.; Mizoguchi, T.; Sato, Y.; Lee, H.S.; Findlay, S.D.; Shibata, N.; Yamamoto, T.; Ikuhara, Y. Atomic structure, electronic structure, and defect energetics in [001](310) Σ5 grain boundaries of SrTiO$_3$ and BaTiO$_3$. *Phys. Rev. B* **2008**, *78*, 245320. [CrossRef]

282. Lim, L.C.; Raj, R. Continuity of Slip Screw and Mixed-Crystal Dislocations across Bicrystals of Nickel at 573-K. *Acta Metall.* **1985**, *33*, 1577–1583. [CrossRef]

283. Priester, L. On the accommodation of extrinsic dislocations in grain boundaries. *Interface Sci.* **1997**, *4*, 2005–2219. [CrossRef]

284. Sutton, A.P.; Balluffi, R.W. *Interfaces in Crystalline Materials*; Oxford University Press: New York, NY, USA, 1995.

285. Hirth, J.P.; Pond, R.C.; Lothe, J. Spacing defects and disconnections in grain boundaries. *Acta Mater.* **2007**, *55*, 5428–5437. [CrossRef]

286. Shockley, W. Do Dislocations Hold Technological Promise. *Solid State Technol.* **1983**, *26*, 75–78.

crystals

MDPI

Review

Dislocations and Plastic Deformation in MgO Crystals: A Review

Jonathan Amodeo [1,*] , Sébastien Merkel [2,3] , Christophe Tromas [4] , Philippe Carrez [2] , Sandra Korte-Kerzel [5], Patrick Cordier [2] and Jérôme Chevalier [1]

[1]   Université de Lyon, INSA-Lyon, CNRS, MATEIS UMR5510, F-69621 Villeurbanne, France; jerome.chevalier@insa-lyon.fr

[2]   Université de Lille, CNRS, INRA, ENSCL, UMR 8207-UMET-Unité Matériaux et Transformations, F-59000 Lille, France; sebastien.merkel@univ-lille1.fr (S.M.); philippe.carrez@univ-lille1.fr (P.C.); patrick.cordier@univ-lille1.fr (P.C.)

[3]   Institut Universitaire de France, F-75005 Paris, France

[4]   Institut Pprime, UPR 3346 CNRS, Université de Poitiers, ENSMA, Département de Physique et Mécanique des Matériaux, SP2MI Bd Marie et Pierre Curie BP 30179, F-86962 Futuroscope Chasseneuil CEDEX, France; christophe.tromas@univ-poitiers.fr

[5]   Institute of Physical Metallurgy and Metal Physics, RWTH Aachen University, Kopernikusstr. 14, 52074 Aachen, Germany; Korte-Kerzel@imm.rwth-aachen.de

*   Correspondence: jonathan.amodeo@insa-lyon.fr; Tel.:+33-(0)4-7243-8235

Received: 29 March 2018; Accepted: 10 May 2018; Published: 31 May 2018

Abstract: This review paper focuses on dislocations and plastic deformation in magnesium oxide crystals. MgO is an archetype ionic ceramic with refractory properties which is of interest in several fields of applications such as ceramic materials fabrication, nano-scale engineering and Earth sciences. In its bulk single crystal shape, MgO can deform up to few percent plastic strain due to dislocation plasticity processes that strongly depend on external parameters such as pressure, temperature, strain rate, or crystal size. This review describes how a combined approach of macro-mechanical tests, multi-scale modeling, nano-mechanical tests, and high pressure experiments and simulations have progressively helped to improve our understanding of MgO mechanical behavior and elementary dislocation-based processes under stress.

Keywords: MgO; dislocations; mechanical properties; nano-mechanics; multi-scale modeling; extreme conditions

1. Introduction

MgO is a well-known crystalline ceramic, maybe the simplest one, and its mechanical properties were widely investigated during the second half of the 20th century, mainly at the macroscopic scale. In this review, we emphasize how more modern methods, e.g., multi-scale modeling, nano-indentation and high-pressure experiments, lead to significant advances in our understanding of the plasticity of MgO. For example, atomic scale simulations in MgO allow characterizing pressure, temperature and strain-rate controlled transitions in plasticity processes responsible for viscosity variations in the deep Earth. At the same time, recent experimental developments such as testing at the nano-scale allowed investigating dislocation nucleation processes in MgO, opening new routes toward ultra-hard ceramics fabrication. These recent and exciting technical developments allow theorists and experimentalists to work even more closely together.

The article presents past and current progresses on dislocation and plasticity understanding in MgO (and related ceramics) based on several technical points of view. An effort was made to parse various fields of applications such as mechanical engineering, theoretical physics, materials science as well as geosciences, equally represented through experimental and theoretical case studies. The article is organized as follows:

- In Section 2, basic knowledge on MgO plasticity is described based on an exhaustive literature review of macro-mechanical testing. Lattice structure and slip systems are introduced as well as the main mechanical properties of MgO based on constant strain-rate and creep experiments.
- In Section 3, we focus on multi-scale modeling of MgO plasticity. This section relies on atomic scale simulations (dislocation core modeling) and their implications for large scale simulations such as dislocation dynamics and crystal plasticity.
- In Section 4, recent improvements in nano-mechanical testing, i.e., nano-indentation, micro- and nano-compression, are introduced. Special attention will be paid on elementary dislocation processes occurring in the small interrogated volumes.
- In Section 5, the role of high pressure on dislocations and MgO plasticity is detailed. Based on both modeling and experimental approaches, we describe how pressure affects basic deformation processes in MgO with implications in both materials sciences and geosciences.

The knowledge on MgO mechanical properties recently reached a focal point in which dislocations properties at different scales play a key role. From the atoms to the polycrystal, experimental and numerical investigations enable multi-scale crossed-connections that drastically multiply the field of applications of this new knowledge to the entire class of ceramic materials.

2. Mechanical Properties of MgO Single Crystal: The Contribution of Macro-Mechanical Tests

MgO is an ionic ceramic which is mainly used for its refractory properties in furnaces and flame retardants as well as a compound for some technical ceramics for the construction industry (e.g., cements). Moreover, MgO is also a model material for the investigation of plastic deformation and dislocation mobility in ionic ceramics. The very first studies focusing on dislocation traces close to a micro-indent were performed in the 1950s [1]. These years marked the opening of a new field of research with the refined analysis of dislocation-based deformation processes in ionic crystals, especially those with relatively simple crystalline structure.

2.1. Lattice, Dislocations and Slip Systems in MgO

MgO atomic stacking is characterized by the B1 cubic crystalline structure (space group Fm$\bar{3}$m) that is also called the rock-salt (NaCl) structure. Its lattice parameter is about 4.21 Å at ambient conditions and its atomic structure is made of oxygen and magnesium atoms spread by half the diagonal of the cube. The B1 crystal structure can be described as two entangled anionic and cationic fcc sub lattices (see Figure 1a).

Unlike the widespread belief that ceramics, including MgO, are too brittle to deform by anything other than cracking, they may indeed also deform plastically by dislocation glide or climb. This controversial affirmation is particularly true in single crystals assuming both limited plastic and fracture strains. Furthermore, the *dense planes and directions* paradigm that usually rules slip system definition does not apply for oxides and especially not for the MgO B1 structure that is characterized by an alternative sequence of ionic and cationic {111} planes. In ionic crystals, an additional prerequisite is deduced from the impossibility of shearing planes that bring same charge sign ions close together. This explains why dislocation glide in MgO is localized in less dense {110} or {100} slip planes with ½<110> Burgers vectors, as shown Figure 1 [2–4].

Whatever the temperature range, there is a vast amount of valuable data about <100>-oriented single crystals available in the literature. This is due to {100} cleavage planes that facilitate such sample preparation. Hence, compression tests and critical resolved shear stresses (CRSS) focusing on {110} slip are often shown for a wide range of temperatures. It remains challenging, however, to obtain single-crystal deformation data for {100} slip for which sample preparation is more difficult.

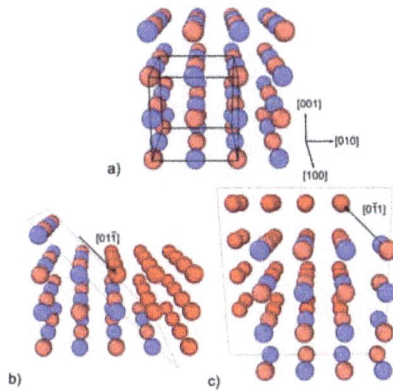

Figure 1. (a) Perspective view of the MgO crystalline structure. The unit cell is drawn with black lines and the crystal is oriented along the cubic orientation. (b,c) Illustrations of the shearing process for the (b) ½<110>{110} and (c) ½<110>{100} slip systems. The same orientation is used and the atoms displaced during the shearing process are labeled in red. Black arrows refer to typical Burgers directions.

Dislocations slip systems in MgO were investigated using ex situ atomic force and transmission microscopy. They are referenced in Table 1. In the study of Hulse and collaborators, the authors showed that CRSS are larger in ½<110>{100} than in ½<110>{110} slip systems for temperatures ranging from 300 K to 1500 K [5]. This study emphasizes that only four out of six ½<110>{110} slip systems are effective under <100> and <110> compression, with Schmid factors of about 0.5 and 0.25, respectively. Investigating the influence of orientation and anisotropy properties, the authors underline that the compressive yield stress is twice as high for orientations with 0.25 Schmid factors verifying the validity of the Schmid law for MgO. This study does not show any evidence of dislocation slip in ½<110>{111} slip systems. Schmid factors for {110}, {100} and {111} modes and <100>, <110> and <111> compression axis (CA) are shown Table 2.

Table 1. ½<110>{110} and ½<110>{100} slip systems in MgO.

Index	Plane	$\vec{b}$	Index	Plane	$\vec{b}$
1	$(\bar{1}10)$	$\frac{1}{2}[110]$	7	(001)	$\frac{1}{2}[110]$
2	(110)	$\frac{1}{2}[1\bar{1}0]$	8	(001)	$\frac{1}{2}[1\bar{1}0]$
3	$(0\bar{1}1)$	$\frac{1}{2}[011]$	9	(100)	$\frac{1}{2}[011]$
4	(011)	$\frac{1}{2}[01\bar{1}]$	10	(100)	$\frac{1}{2}[01\bar{1}]$
5	$(\bar{1}01)$	$\frac{1}{2}[101]$	11	(010)	$\frac{1}{2}[101]$
6	(101)	$\frac{1}{2}[10\bar{1}]$	12	(010)	$\frac{1}{2}[10\bar{1}]$

Table 2. Schmid factors in main ½<110>{110} and ½<110>{100} slip modes and additional ½<110>{111} for compression axis CA = <100>, <110> and <111>.

CA	½<110>{110}	½<110>{100}	½<110>{111}
<100>	0.5 for 4/6 systems	0 for all systems	$\sqrt{6}/6$ (~0.41) for 8/12 systems
<110>	0.25 for 4/6 systems	$\sqrt{2}/4$ (~0.35) for 4/6 systems	$\sqrt{6}/6$ (~0.41) for 4/12 systems
<111>	0 for all systems	$\sqrt{2}/3$ (~0.47) for 3/6 systems	$\sqrt{6}/9$ (~0.27) for 6/12 systems

Etched surface analyses were also performed to characterize slip localization [6–8]. In the study of Day et al. [7], dislocation slip is confined in a single slip system between room temperature and 1000 K. From 1000 K to 1400 K, two conjugate ½<110>{110} slip systems (with orthogonal slip planes) interpenetrate and produce the strain. This configuration is also mentioned in the work of Copley [8]. In both temperature regimes, the initial distribution of dislocations plays a key role on the primary slip plane activity. At higher temperatures, the authors mentioned the activity of both 90° and 60°/120° slip system configurations. In all configurations, TEM analyses show an increase of the dislocation density along the <111> directions, typical of {110} slip plane intersections [9].

Contrary to metallic fcc structures, very few studies mention ½<110>{111} dislocation slip in MgO single crystals [10,11]. On the other hand, dislocation cross-slip was observed in {112} and {122} planes during in situ TEM tests [12]. These observations were confirmed in nano-indentation tests [13] considering confined deformation at particularly high-stress under the indenter (see Section 4).

2.2. Critical Resolved Shear Stress (CRSS)

Hulse, Copley and Pask performed a set of compression tests on single crystals investigating the CRSS and deformation processes in both slip modes [5,8,14]. Temperatures investigated range from 100 K up to 1400 K for the ½<110>{110} slip systems. In their analysis, the authors discuss the role of thermally-activated processes (e.g., lattice friction, impurities) on CRSS variations. The reader can refer to Ref. [8] for a detailed overview of this work.

The study of Appel and Wielke focuses on CRSS measurements in the ½<110>{110} slip systems down to 4 K [15]. The authors emphasize that, even if impurities (especially iron) interact with defects restricting dislocation glide, they cannot explain alone the non-linear variations of CRSS with temperature and thus assume a Peierls, lattice-friction, mechanism to proceed. The lowest temperatures investigated lead to stresses of about 118 MPa (T = 4.2 K) and 132 MPa (T = 5.6 K). Extrapolating Appel and Wielke data, one can evaluate a Peierls stress of about 150 MPa in ½<110>{110}. While the uncertainty is of the order of a few tens of MPa close to 0 K, this study confirms this value as an approximate proxy of the Peierls stress for ½<110>{110} slip in MgO, as already proposed in previous studies [16,17].

Several additional studies addressed the effect of temperature on the CRSS for the ½<110>{110} slip systems [18–22] while only few datasets focus on ½<110>{100} [5,8,22]. These data are shown in Figure 2. Several interpretations can be made from this master plot. First, one can confirm Appel and Wielke [15] hypothesis stating that deformation is controlled by lattice friction at low temperature (T < 100 K) as most of the experiments show significant non-linear variations of the CRSS over few tens of MPa, which is the signature of a Peierls mechanism. Nevertheless, the data scatter observed at intermediate temperatures (100 K < T < 600 K) might be the sign of other thermally activated processes such as dislocation versus impurities interactions or precipitate hardening. This point will be discussed later. At high temperature (T > 600 K), forest dislocation interactions lead to a stress plateau of about 15 MPa at least up to 1900 K. Concerning the ½<110>{100} hard mode, besides wider stress variations, no conclusion can be made about ½<110>{100} Peierls stress from macro-mechanical tests. A simple guess larger than 1 GPa can be inferred from Barthel's data extrapolation.

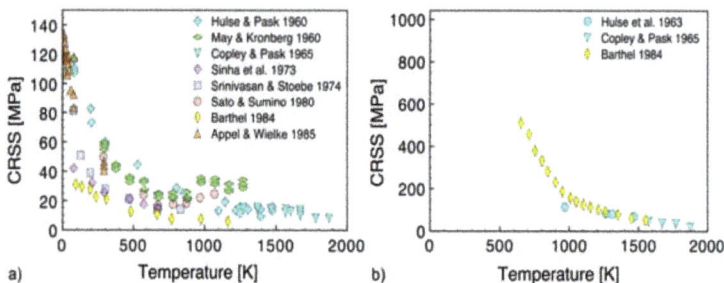

Figure 2. CRSS for (**a**) ½<110>{110} and (**b**) ½<110>{100} slip systems in MgO single crystal under compression. Details about impurity contents can be found in original articles and in Ref. [23].

2.3. Temperature-Dependent Dislocation Microstructures

Based on the CRSS master plot in Figure 2, one can define two main deformation regimes for MgO single crystal. At low to intermediate temperatures, MgO deformation is in a thermally-activated, "bcc-like", regime characterized by wide CRSS variations for which both lattice and solute frictions influence the deformation. At high-temperature, the mechanical response of MgO is described by a "fcc-like" regime, where dislocation interactions lead to a stress steady-state (or weak variations). Based on the metallic alloys terminology, an athermal transition temperature T_a that emphasizes the transition between the thermally-activated regime ($T < T_a$) and the so-called athermal regime ($T > T_a$) is defined. For ½<110>{110}, T_a ranges from 600 K to 800 K depending on the experimental dataset. CRSS vary less rapidly for ½<110>{100} slip systems and $T_a > 1300$ K.

Dislocation microstructures developed during single crystal compression tests were characterized using in situ TEM [24,25]. At room temperature, straight and long screw dislocations are observed in both the {110} and the {100} slip planes [24,26]. Their velocity is assumed to follow a stress-exponent [16,27] or an Arrhenius-like dependency [28,29]. This dislocation microstructure is typical of high lattice friction materials. Edge dislocations are shorter and glide at higher velocity when compared to screw. In MgO, they are generally observed during the first stage of deformation similarly to α-iron [30], except in the case of dipoles formation [25]. Beyond the transition temperature, dislocations exhibit a viscous behavior typical of fcc metals as shown, e.g., in the work of Clauer and Wilcox [10]. Both dislocation microstructures are illustrated on Figure 3.

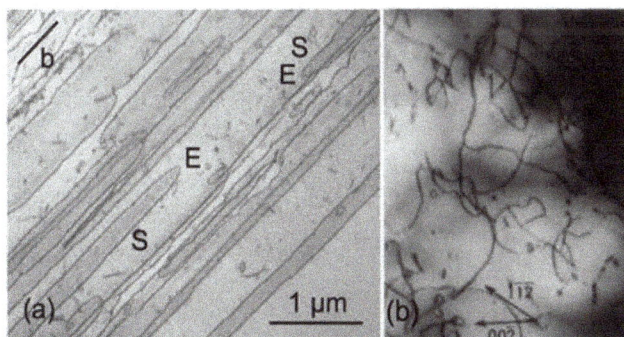

Figure 3. TEM micrographs of (**a**) room temperature dislocations extracted from a 45° shear band using the High-Voltage Electron Microscope (HVEM). Dislocations are elongated along the screw direction. S, E and b refers to Screw, Edge and to the Burgers vector respectively. Image reproduced from Ref. [26] with the permission of Springer Nature; (**b**) [110] thin section, 1400 °C. The dislocation curvature is more isotropic when compared to room temperature. Image reproduced from Ref. [10] with the permission of John Wiley and Sons.

2.4. On the Role of Impurities and Heat Treatments on CRSS and Dislocation Mobility

Several studies refer to the influence of point defects on CRSS in MgO [8,14,18,20,21]. Generally, the CRSS increase in the intermediate range of temperature is justified by the impurities oxidation state [21,31,32] and their spatial distribution [1,33,34], both parameters being intrinsically linked to the heat treatment processed before deformation.

In the study of Gorum and collaborators [33], the authors investigate the role of iron on ½<110>{110} CRSS. Three main results can be emphasized: (i) For an as cast sample including 10 ppm of iron, the CRSS is of about 40 MPa at room temperature. For a 30 ppm sample, it is 140 MPa and increases up to 180 MPa for a concentration of about 3000 ppm. Therefore, we can conclude that, without heat treatment, MgO CRSS depend strongly on the impurities content (at room temperature); (ii) A heat treatment at 2000 °C on a sample containing 3000 ppm of iron deformed at room temperature leads to lower CRSS when compared to an as cast sample that contains only 30 ppm of iron. Furthermore, it increases its ductility; (iii) A faster cooling rate after heat treatment decreases the CRSS while increasing the sample ductility.

Obviously, heat treatment and cooling rate control the spatial distribution of impurities in crystalline materials. A slow cooling rate allows the impurities to drift toward dislocation and pin them. With a faster cooling rate, impurities are maintained in solid solution which limits the impurity hardening effect. For ceramics, this process was discussed by Johnston and Gilman [1]. In addition, several studies emphasize the role of heating atmosphere during heat treatment [21,32]. Indeed, a reductive atmosphere (e.g., argon gas) will favor divalent ions that are known to have a weaker strengthening effect than trivalent impurities, merely obtained when the heat treatment is performed in air.

In the early 1970s, the group of Messerschmidt (Halle, Germany) did an impressive work quantifying the interaction between dislocations and point defects, especially to interpret observations made in the High-Voltage Electron Microscope (HVEM) [12,24,35–42]. Among others, a model for point obstacle crossing activation energy is proposed [37] as well as a comparison between TEM observations and calculation of point defect distance distribution [43]. For an exhaustive review of this work, the reader can refer to the book by Messerschmid [26].

Several sets of experiments were proposed by Singh and Coble to investigate stress and temperature dependency of dislocation velocity in MgO [16,27]. One concerns *pure* samples with about 100 ppm impurities (including 20 ppm of iron). *Pure* samples were all subjected to a preliminary 40 h heat treatment in air at 1300 °C. Then, some of the samples were again heated at 1200 °C for one week under a reductive atmosphere to decrease the content of trivalent cations. These last are referred to as "*Pre*" while others are called "*Pox*". Finally, successive micro-indentations tests were performed to measure dislocation velocity at room, 100 °C and 150 °C temperatures. Results can be resumed as follows. First, the velocity of edge dislocations is always larger than the one of screw dislocations which confirms the aforementioned TEM studies. At room temperature, "*Pox*" samples are characterized by a lower dislocation velocity than "*Pre*" samples, even for comparable trivalent Fe ions concentrations (respectively, 19 and 6 ppm). These results show that "*Pox*" samples already contain a sufficient amount of Fe^{3+} ions for dislocation pinning confirming Gorum's hypothesis about the non-linear dependency of the dislocation velocity regarding the concentration of trivalent ions [33].

2.5. Strain Hardening in MgO Single Crystal

Dislocation-point defects interactions, dislocation dipoles and more generally dislocation reactions can lead to strain hardening in MgO single crystals. Hulse and Pask have shown that MgO single crystal can deform plastically down to very low temperatures, measuring a fracture strain of about 6% at 77 K [14]. Surface analyses emphasize 90° shear bands that confirm the joint activity of two conjugate ½<110>{110} slip systems as already observed in Day and Stokes' work [7]. While the hardening coefficient $\frac{\theta}{\mu}$ (where θ and μ are, respectively, the stress strain derivative and the shear modulus) ranges from ~200 to 1100 when increasing temperature up to 1273 K (see Figure 4), it remains constant (for a given temperature) up to the fracture strain. Similar compression tests were performed by Copley and Pask between 1273 K and 1873 K, with <100> and <111> compression axes. <100> tests show a decrease of θ attributed to high-temperature restoring processes such as cross-slip and dislocation climb.

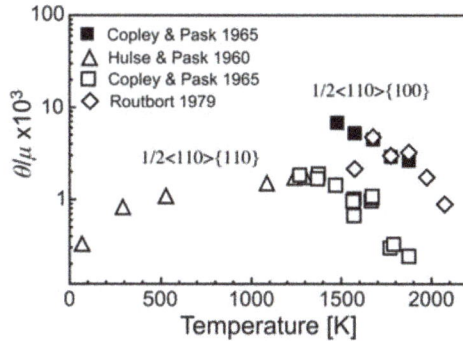

Figure 4. Strain hardening coefficient in MgO. White and black symbols refer to experimental data for the ½<110>{110} and ½<110>{100} slip modes respectively. Adapted from [8,14,44,45].

Up to highest temperatures, two conjugate ½<110>{110} slip planes control the deformation while four of the six slip systems have the same (maximum) Schmid factors. Therefore, it is assumed that two slip systems are surprisingly inhibited. On the other hand, dislocation interactions between 60°/120° slip systems are stronger than 90° reactions. The authors also refer to Kear's calculation about dislocation junction formation [46], i.e., two ½<110>{110} dislocations with 60°/120° oriented Burgers vectors can react and make a junction following, e.g., Equation (1).

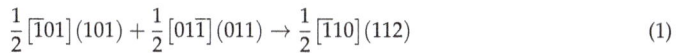

$$\frac{1}{2}\left[\bar{1}01\right](101) + \frac{1}{2}\left[01\bar{1}\right](011) \rightarrow \frac{1}{2}\left[\bar{1}10\right](112) \tag{1}$$

The resulting dislocation is of edge character, aligned along one of the <111> directions and can eventually glide in {112} slip planes which were observed only under very high-stress conditions [13]. As dislocation junctions are obstacles to dislocation glide, shear bands observations can be interpreted by a joint process of a pair of conjugated slip systems producing deformation in the same time than making junctions, with remaining two least activated slip systems. While the decrease of θ at high temperature can be attributed to restoring processes, the similar trend at low temperature can only be justified by the weakening of the hardening processes.

<111> compression tests allow investigating strain hardening in ½<110>{100} slip systems. Results show higher value of θ when compared to {110} (5–10 times larger depending on temperature) but a similar trend against temperature variations (see Figure 4).

2.6. High-Temperature Creep of MgO

At high temperatures, solids can be deformed by creep, i.e., under a constant load, following the evolution of strain with time. The range of *high-temperatures* is conventionally considered to start at $0.5 \times T_m$ (T_m being the melting temperature). With a melting temperature close to 3100 K, MgO is a refractory material and some applications (material for crucibles) are a consequence of its stability (physical and chemical) at high temperature. Most creep tests were, as expected, performed at temperatures above $0.5 \times T_m$. However, very few were performed above $0.7 \times T_m$. Ruano and colleagues reviewed the creep behavior of MgO single crystals compressed along <100> [47]. Creep of MgO can be described by a power law with a stress exponent close to 5. A comparable behavior ($3.8 < n < 4.5$) was found for tensile creep tests of <110> single crystals between 1200 and 1500 °C [10]. Three-point bending tests on MgO single crystals lead to similar results with a slightly wider range ($4 < n < 7$, see [48]). It is to be noted that similar tests performed in Ref. [49] showed that the application of an electric field gives a transient acceleration of the creep rate. However, very few studies have considered explicitly the implication of charged species on the creep behavior. Recently, Mariani and collaborators [11] have shown using EBSD that dynamic recovery by sub-grain rotation is the main recrystallization mechanism in MgO single crystals deformed at high strains. The activation enthalpy is usually reported to be in the range 320–400 kJ/mol. Under those conditions, the dislocation density ρ is found to scale with the applied stress σ following the equation $\rho \sim \sigma^m$ with $m = 1.4$ [50], or $m = 2.15$ [51]. At low stress (below 10 MPa) and usually above $0.6 \times T_m$, the creep rate of MgO single crystal may exhibit a stress dependence close to 1 [52]. Studies on polycrystals lead to comparable results with however a stress dependence closer to 3 [53–55]. In one case [54], a much smaller activation enthalpy was reported (213 kJ/mol) with a polycrystal characterized by a rather small grain size (12 μm). Given the importance of iron-bearing MgO (ferropericlase) in Earth sciences, several studies have also investigated the influence of this alloying element which does not seem to have a strong influence on creep properties in this temperature range, at least up to 20% content [56,57].

These parameters suggest that single crystal MgO deforms at high temperature through a dislocation creep mechanism. As described by Weertman, this creep regime involves dislocation glide, which controls the strain-rate by inducing recovery and strain mechanisms [58]. The recovery mechanism is usually controlled by climb of edge dislocations although Poirier pointed out that cross-slip of screw dislocations could also control recovery and creep [59].

MgO is one of the ceramic material for which mechanical properties were the more intensively studied in the mid-20th century. In the Section 3, we see how multi-scale simulations and modeling were able to recently propose theoretically-based clarifications for most of the mechanical behavior highlighted in experiments.

3. Multi-Scale Modeling of MgO Deformation

The numerical and multi-scale investigation of plastic deformation in MgO reflects a renewed interest in the field of both materials (nano-mechanics, ceramics applications) and applications in the Earth's sciences. These studies all rely on the basic MgO dislocation core properties, down at the atomic scale, which started to be described several decades ago, in the 1970s. In this section, we first outline dislocation core modeling, from cluster-based embedded models up to the more recent Peierls–Nabarro–Galerkin (PNG) approach. Then, a short section will focus on dislocation mobility and kink-pair modeling in MgO. The sensitivity of dislocation velocity to stress and temperature are discussed. To conclude, a final section focuses on grain-scale plasticity modeling using dislocation dynamics simulations, a meso-scale tool for the modeling of the collective behavior of dislocations.

3.1. Dislocation Core Modeling: Methods and Results

Over the last fifty years, only few studies focused on the dislocation (core) properties at the atomic scale in MgO. One of the issues lies in the choice of an accurate interatomic potential to describe the core configuration. Rigorous determinations often require the use of first-principles calculations. However, this choice was not always numerically tractable for studies of dislocations due to the long-range stress and strain field associated to these defects. Empirical potentials are thus often used to circumvent this limitation. In ionic materials, classically, the pair potential involves coulombic interactions (at long range), non-coulombic short-range interactions plus eventually polarization effects. In case of MgO, short range interactions are usually described through a Buckingham term (a semi-empirical description based on a Born-Mayer and a Van Der Waals term) [60]. In addition, as MgO is known to exhibit a strong deviation from the Cauchy conditions, $C_{12} = C_{44}$, most accurate potentials go beyond the description of central forces, using for instance a core-shell or a breathing shell model to allow a rigorous treatment of polarization effects (e.g., [61]). A review of the different methodologies used to model dislocation cores [62] shows that the three most common approaches were applied to MgO: cluster-based embedded models, fully periodic dipole models and the semi-continuum Peierls–Nabarro model. In the following, we review the literature according to these methods.

3.1.1. Cluster-Based Embedded Models

The earliest atomistic models of dislocation cores in MgO were proposed by Woo and co-workers at the end of the 1970s [17,63–65]. Woo and colleagues performed simulations on isolated edge dislocations in the ½<110>{110} slip systems. In their simulations, a straight edge dislocation is placed at a center of a cylindrical central region where atoms are free to relax during energy minimization. To ensure that the simulation corresponds to a single dislocation in an infinite crystal, an outer cylindrical region is added to the boundary. In this boundary region, the atomic positions are well defined and kept fixed to the solution of the dislocation displacement field given by linear anisotropic elasticity. Despite the various drawbacks of the cluster approach with rigid boundary, Puls and Norgett [64] succeeded in stabilizing an edge dislocation into two sets of symmetrical configurations corresponding to a core position in the minimum of the Peierls potential (as shown Figure 5) and to a metastable core located on the top of the Peierls potential. They were thus able to predict a Peierls energy barrier for a ½<110>{110} edge dislocation between 0.01 and 0.04 eV/b (depending on the choice of the potential used to perform the simulation) which corresponds to a reasonable Peierls stress estimation of a few hundreds of MPa.

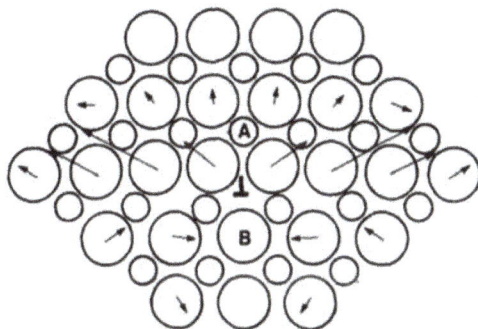

Figure 5. Atomic configuration of a 1/2<110> edge dislocation. Reproduced from [64] with the permission of AIP publishing.

Later, a similar cluster-based methodology was adopted by Watson [66] to simulate the atomic arrangement around different screw dislocations with <100> and ½<110> Burgers vectors. Motivated by the importance of dislocations in crystal growth, their study represents the first simulation of screw dislocations in ionic systems.

3.1.2. Periodic Dipole Model

In a periodic dipole model, a dipole of dislocations with opposite Burgers vectors is placed according to a periodically quadrupolar arrangement to minimize the residual strains in the simulation volume [67]. The fully periodic dipole model was used in MgO in the recent study of Carrez and colleagues [68] to take advantage of first-principles calculations. The ½<110> screw core structure is shown in Figure 6 using a so-called differential displacement map [69] to reveal the details of the atomic arrangement inside the screw core. As expected, due to the low critical shear stresses on ½<110>{110}, the screw core structure is found almost entirely spread in {110}. A close inspection of the atomic arrangement in the vicinity of the dislocation core shows that the spreading allows bringing oppositely charged ions on top of each other (see Figure 6). To further accommodate the O-Mg bonds, the screw core lying in a $(1\bar{1}0)$ glide plane is associated with a dilation state with an edge displacement component restricted to $(1\bar{1}0)$ (see Figure 7).

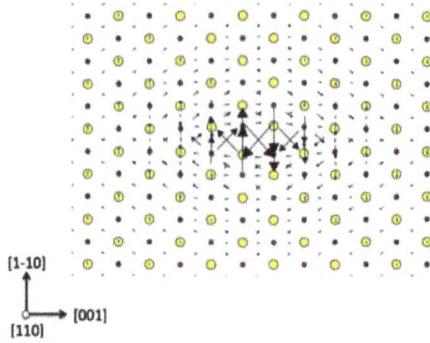

Figure 6. Differential displacement plot of ½<110> screw core at ambient pressure. As for a screw dislocation, most of the displacements are along the direction of the Burgers vector (i.e., along the [110] direction of the dislocation line), the dislocation core spreading can only be visualized by plotting arrows between pairs of neighboring atoms with lengths proportional to the magnitudes of their relative displacements along [110] as compared to ideal positions in the perfect crystal. Adapted from Ref. [68].

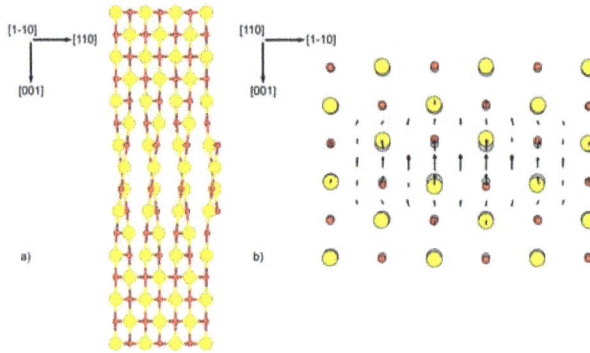

Figure 7. (a) Normal view to a screw dislocation line lying on a {110} glide plane. Only ions below and above the glide plane are shown; (b) Edge displacement field at the center of ½<110> screw dislocation core. Adapted from [68]).

3.1.3. Semi-Continuum Peierls–Nabarro Model

The Peierls–Nabarro (PN) model [70,71] is an efficient approach to describe dislocation cores incorporating the equivalent of an inelastic atomic energy instead of explicitly treating the atomic scale details of the core. In this model, a dislocation is viewed as a continuous distribution of infinitesimal dislocations ρ of elementary Burgers vectors db with $db = \rho(x)dx = (df/dx')xdx$, where f corresponds to the disregistry of the dislocation. In an elastically isotropic system, each infinitesimal dislocation creates a stress field which tends to extend the core. The equilibrium shape of the dislocation results then from the balance between non-elastic restoring force and the elastic ones described above. Whereas, in the original model, the restoring force used was an analytical sine function, Christian [69] demonstrated that it can be approximated to the gradient of the so-called generalized stacking fault (GSF) energy. Thus, atomistic effects are introduced in the PN model according to GSF calculations. The concept of the GSF energy (also called γ-surface) was initially introduced by Vitek [72] to systematically explore the stacking faults in bcc metals. The idea is to map out atomistic interaction energies across a shear plane as a function of any displacement. Corresponding to a rigid shear of two half crystals, one over the other, this type of calculation has a low CPU cost and is therefore feasible using first-principles simulations. It is worth noticing that repeating the same calculations with any empirical potential is also an interesting option to quantify the transferability of any empirical potential to the proper description of atomic displacements within dislocation cores.

GSF energies for {100} and {110} in MgO were computed using either first-principles calculations or several pairwise potentials [23,73,74]. In both planes, shear along ½<110> corresponds to the lowest energy path (Figure 8). According to first-principles calculations [23], the maximum unstable stacking fault energy along <110> ranges from 1.05 J/m^2 in {110} to 2.46 J/m^2 in {111}, with 2.18 J/m^2 in {100} in between.

In the PN model, the dislocation core width can be calculated from a balance between restoring forces derived from γ-surfaces and the elastic properties of the crystal. Thus, according to the respective γ-surface energies for {110} and {100}, the 1/2<110>{100} edge dislocation is characterized by a smaller core width than the one found for 1/2<110>{110} [23,74]. For the 1/2<110>{110} edge dislocation, solutions of the PN model lead to a wide spreading of the dislocation core [73,74] in agreement with the early atomic configuration suggested by Woo and collaborators [17,63–65]. This shows that, despite the various assumptions, the PN model leads to a good description of the dislocation core geometry in case of planar core configurations. To accurately model screw dislocation cores using the PN model, some of the intrinsic limitations like the need to focus *a priori* on a particular slip plane have to be overcome [75]. A modified version of the PN model, the Peierls–Nabarro–Galerkin (PNG) method [76], was applied in MgO to model the core structure of the screw dislocation with ½<110> Burgers vector [23]. In this model, core spreading in several planes can be considered simultaneously. Given this improvement, the screw core computed according to the PNG approach perfectly matches the atomistic configurations depicted above [68].

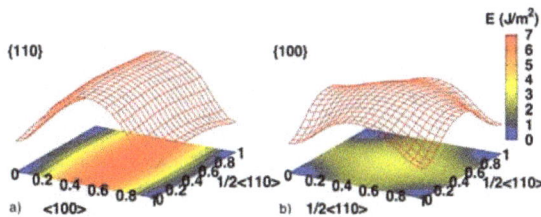

Figure 8. MgO γ-surface for (**a**) {110} and (**b**) {100} slip modes computed with first-principles. Adapted from [23].

One of the main achievements of the PN model is to provide reasonable estimates of dislocation core sizes (e.g., Ref. [77]). However, equally importantly, the PN approaches allow one to evaluate the Peierls stress. Classically in the PN model, the evaluation of the Peierls potential is performed by a discrete summation of the γ-energy corresponding to the disregistry function of the dislocation core solution of the PN equation (for instance [78]). The evaluation of the Peierls stress is however influenced by the numerical scheme used for the discrete summation. Consequently, the various stresses computed according to PN approaches in MgO agree with the order of magnitude but may show some discrepancies [15,18]. According to the most recent evaluation [23], the Peierls stresses for ½<110>{110} are 150 MPa for screw and 80 MPa for edge components. For ½<100>{100}, the Peierls stresses are 1600 MPa and 300 MPa for screw and edge components respectively. For the two slip systems, the estimation of Peierls stresses are consistent with the experimental determination discussed above and shown in Figure 2. Finally, as reviewed in Section 5, the introduction of γ-surfaces calculated from first-principles at high pressures enable investigations of the effect of pressure on the dislocation properties in MgO.

3.2. Dislocation Mobility in MgO

MgO is a material with reasonably high lattice friction, like for example in bcc metals (e.g., iron and tungsten). Therefore, below the athermal temperature T_a, dislocation glide is controlled by thermally-activated jumps of dislocation portions called kink-pairs, while phonon drag constrains dislocation velocity at higher temperatures.

3.2.1. The Kink-Pair Mechanism

In the thermally-activated regime, the dislocation motion over the Peierls potential is assisted by the conjugate effect of stress and thermal activation (see, e.g., [79,80] for an extended description). During this process, the dislocation does not move as a straight line but through the nucleation of kink-pairs. The further propagation of the kinks along the dislocation line is responsible for the glide of the whole dislocation. The theoretical description of dislocation motion involving the kink-pair mechanism was successfully applied to the understanding of elemental deformation processes (see, for instance, [81]) in several materials, including bcc metals [82–89], covalent materials such as silicon [90–92] and ionic materials [93–95], including MgO [23].

The kink-pair nucleation process is usually associated with a critical change in enthalpy that has to be supplied by thermal activation under a given stress. The change in enthalpy ΔH_k, consisting of three contributions, follows the general equation:

$$\Delta H_k = \Delta E_{elast} + \Delta W_p - W_\tau \qquad (2)$$

The two positive contributions, ΔE_{elast} and ΔW_p correspond respectively to the increase of the elastic energy of the kinked dislocation line and to the increase of the Peierls energy due to the part of the line lying on the Peierls barrier. These quantities are balanced by the work of the applied stress τ. Several theoretical models, only differing by the treatment of the change in elastic energy, were proposed to evaluate the critical kink pair nucleation energy Δ_k^{crit}. Among these are the line tension (LT) model [29,96], the kink-kink interaction model [97] and the elastic interaction (EI) model [98,99]. Atomistic simulations can also be used to compute the critical nucleation enthalpy (e.g., [91]) or to evaluate the different energy components involved in the LT model (e.g., [89]). Nevertheless, such approaches have not been applied to MgO up to now. Instead, in Ref. [23], the EI model was used to numerically estimate the evolution of Δ_k^{crit} as a function of the applied stress.

The evolution of Δ_k^{crit} for screw components belonging to 1/2<110>{110} and 1/2<110>{100} is shown in Figure 9. The nucleation of a double-kink on a 1/2<110>{110} dislocation requires a lower energy when compared to 1/2<110>{100} which is consistent with the analysis of Takeuchi [4] based on experimental mechanical data. For slip on {110}, the isolated kink energy U_k (corresponding here to $0.5 \times \Delta_k^{crit}(\tau = 0)$) is 0.55 eV, in a rather good quantitative agreement with the activation energies of 0.2–1.0 eV, derived by the measurement of the Bordoni peaks and accounting for the possible scatter induced by impurities [100].

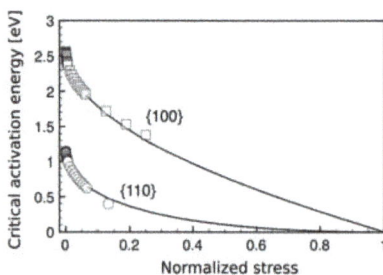

Figure 9. Kink-pair activation energy (EI model) for screw dislocation in MgO as function of stress (normalized by the Peierls stress). Adapted from [23].

More generally, the activation enthalpy for the kink-pair mechanism involves both nucleation and migration of kinks. However, only one of the two processes generally act as a limiting factor. In the case of MgO, according to the EI model, the kink Pairs widths are found to be larger than 50b. Such a large width is not compatible with sharp rectangular kink pairs but more probably accounts for both the kink pair width configuration plus the spreading of the two individual kinks [23]. It was then concluded that the individual kinks should be widely spread, bearing a low migration enthalpy barrier. Based on the Frenkel–Kontorova model (e.g., Ref. [101]), for an individual kink width larger than a few b, one can show that the migration energy would be at least of one order of magnitude lower than U_k. The kink-pair mechanism in MgO is then assumed to be controlled exclusively by the nucleation of the kink-pair. Nevertheless, there is still scope to improvement. With the rapid increase of atomistic simulation capabilities, the shape and energy of kinks could be determined by atomistic simulations following the recent scheme proposed in bcc metals or covalent materials (see [92,102–104] for instance).

3.2.2. Dislocation Glide Velocity

Based on the Guyot and Dorn formalism [29], the jump frequency of a dislocation in the thermally-activated regime ($T < T_a$) is defined by Equation (3).

$$\nu(\tau, T) = \frac{\nu_D b}{w^*(\tau)} \frac{L}{2w^*(\tau)} exp^{-\Delta H^*(\tau)/kT} \tag{3}$$

where $\nu_D b/w^*(\tau)$ is the oscillation frequency of a dislocation segment of length $w^*(\tau)$ and ν_D is the Debye frequency. $L/2w^*(\tau)$ refers to the number of nucleation sites for kink-pairs and the exponential term accounts for the kink-pair nucleation probability which can be either deduced from experimental data following an inverse engineering process or computed using atomistic simulation as described in the Section 3.2.1. Please note that this formalism is only valid when kink-pair nucleation is the limiting process when compared to kinks propagation along the line.

Equation (3) only refers to forward kink-pair jumps. However, at low stress (e.g., close to T_a), the kink-pair nucleation process can proceed either along the forward displacement direction or along the opposite one. In addition, it is obvious that Equation (3) does not go to zero at the zero-stress state. To solve both inconsistencies, Nabarro [105] proposed to integrate kink-pair backward jumps using a double-exponential equation leading to a new frequency definition:

$$\nu(\tau, T) = \frac{\nu_D b}{w^*(\tau)} \frac{L}{2w^*(\tau)} \left(exp^{-\Delta H^+(\tau)/kT} - exp^{-\Delta H^-(\tau)/kT} \right) \tag{4}$$

where $\Delta H^+(\tau)$ and $\Delta H^-(\tau)$ refer to forward and backward kink-pair activation energy, respectively.

Finally, based on Nabarro's work [105], the dislocation mobility in the thermally-activated regime is defined by reintroducing the Peierls potential periodicity a' (Equation (5)). This formulation applies specifically for each dislocation character and its validity for screw dislocations in MgO was confirmed by experimental measurements in the case of ½<110>{110} dislocations [16,23].

$$\nu(\tau, T) = a' \frac{\nu_D b}{w^*(\tau)} \frac{L}{w^*(\tau)} exp^{-\Delta H_0/kT} \sinh\left(\frac{\Delta H_0 - \Delta H^*(\tau)}{kT}\right) \tag{5}$$

On the other hand, beyond the athermal transition ($T > T_a$), lattice friction vanishes and the resulting dislocation structure is characterized by more isotropically shaped defects, as shown by TEM micrographs (see, e.g., Figure 3b). In this "fcc-like" regime, the velocity of dislocations is constrained by a viscous drag that is attributed to the interactions between dislocations, phonons and electrons (Equation (6)).

$$\nu(\tau, T) = \frac{b\tau}{B(T)} \tag{6}$$

where $B(T)$ is a viscous drag coefficient generally in the range of 10^{-5} Pa·s in ionic crystals.

3.3. Grain-Scale Plasticity Modeling Using Dislocation Dynamics: On the Collective Behavior of Dislocations

3.3.1. Methods: 3D and 2.5D Dislocation Dynamics

Dislocation dynamics (DD) simulations are at the interface between atomic scale and continuum approaches. They provide the link between physics-based plasticity laws (continuum elasticity theory) and macroscopic deformation.

In 3D simulations, the elementary unit is a dislocation segment or a node that bridges two dislocation portions, depending on the chosen approach [106,107]. Therefore, while atomistic simulations commonly refer to the nm-scale, DD simulations allow modeling volumes up to a few hundreds of μm³ including large populations of dislocations. Many DD applications exist for the cases of nano-indentation [108], precipitation hardening [109–111], fatigue [112] or micropillar compression [113–115], all applied whatever the crystalline materials. Mainly, all DD codes follow similar calculation steps. After the definition of a starting configuration (including sample design, dislocation density distribution and mechanical setup), a stress or strain is applied. Then, forces between dislocations are computed (following iso- or anisotropic elasticity) and an effective stress is defined on each dislocation portion as the sum of each contributions, e.g., applied stress, line tension, lattice friction and neighboring stress fields. Finally, dislocations displacements are solved iteratively using a dislocation mobility law (see e.g., Equation (5) or Equation (6)) integrated using a simple time step that scales from 10^{-10} s up to approximately 10^{-6} s depending on the problem under study. As inputs, DD simulations generally refer to crystallography, temperature, elastic constants as well as dislocation core properties. The testing conditions are also obviously required. In the evolution of the dislocation microstructure under stress, data such as stress–strain curves, dislocation density evolution, plastic shear per slip mode as well as local stress maps are then easily obtained.

On the other hand, 2.5D simulations were developed in the early 90s [116–118] but are still used to investigate particular questions, such as dislocation organization, mobility or diffusion-assisted plasticity. In 2.5D DD simulations, dislocations are introduced as straight segments of constant length, all perpendicular to a reference plane giving rise to a quasi-2D projection. When compared to 3D simulations, local rules are introduced to describe dislocation multiplication, short-range interactions and junction formation to reproduce at best the 3D dislocation evolution but using a simplified and "cheaper" (in terms of CPU costs) framework. More details about 2.5D simulations can be found, e.g., in the recent study of Curtin and collaborators [119].

3.3.2. Dislocation Interactions Mapping

Dislocation contact reactions are one of the key processes to understand strain hardening. They can be computed using 3D-DD simulations based on Kroupa's force equation (see, e.g., Ref. [79]) that characterizes the attractive/repulsive character of the interaction $\vec{F}_{12}$ between two dislocations and the line tension equilibration equation calculated at the triple nodes which defines the (φ_1, φ_2) ensemble favorable to the junction formation (Equations (7) and (8)).

$$\vec{F}_{12} = \frac{\mu}{\left|\vec{l}_1 \times \vec{l}_2\right|} \frac{\vec{R}_{12}}{R_{12}} \left\{ \frac{1}{2}\left(\vec{b}_1 \cdot \vec{l}_1\right)\left(\vec{b}_2 \cdot \vec{l}_2\right) - \left(\vec{b}_1 \times \vec{b}_2\right) \cdot \left(\vec{l}_1 \times \vec{l}_2\right) + \frac{1}{1-\nu}\left[\left(\vec{b}_1 \times \vec{l}_1\right) \cdot \frac{\vec{R}_{12}}{R_{12}}\right]\left[\left(\vec{b}_2 \times \vec{l}_2\right) \cdot \frac{\vec{R}_{12}}{R_{12}}\right] \right\} \qquad (7)$$

where $\vec{l}_1$ and $\vec{l}_2$ refer to unit line vectors of dislocations 1 and 2, respectively; $\vec{b}_1$ and $\vec{b}_2$ are Burgers vectors; $\frac{\vec{R}_{12}}{R_{12}}$ is a unity vector along the force direction (i.e., along the shorter distance between the two dislocations); and ν is the Poisson ratio.

$$f(\beta_j)b_j^2 = f(\beta_1)b_1^2\cos(\varphi_1^*) - f'(\beta_1)b_1^2\sin(\varphi_1^*) + f(\beta_2)b_2^2\cos(\varphi_2^*) - f'(\beta_2)b_2^2\sin(\varphi_2^*) \qquad (8)$$

where $f(\beta_j) = 1 - \nu\cos^2(\beta)$ and β is the dislocation angle character.

These two equations lead to the definition of three configurations, i.e., the repulsive state, the crossed attractive state and the junction state. In addition, attractive dislocations with the same Burgers vector can lead to collinear annihilations. These simulations are driven by elastic energy minimization and no applied stress is imposed. More technical details on dislocation reactions can be found in [79]. The results of dislocation interaction mapping are independent of parent dislocations lengths and the junction region (made of φ_1 and φ_2 favorable configurations) is increased for energetically favorable configurations (Franck criterion). Finally, the junction strength τ_j scales with $\mu b/l_j$ (with l_j the junction arm length) depending on (φ_1, φ_2). Interactions maps were computed for a wide range of materials as, e.g., in fcc and bcc metals [120–122], hcp structures [123,124], ice [125] and minerals like olivine [126].

Figure 10, based on MgO slip system analysis, ½<110>{110} and ½<110>{100} dislocations lead to eight distinct types of interactions (see [45] for more details):

- Three maps for the {110} mode (self-interaction + {110} junction + {110} crossed/repulsive states): These reactions were investigated in Ref. [127] and show the possible formation of edge ½<110> junctions between dislocations in 60°/120° oriented slip systems. Junctions are oriented along <111> and belong to {112} slip planes. These dislocations could possibly justify {112} dislocation slip observed during nano-indentation tests (see Section 4). Ninety-degree oriented slip systems only lead to crossed or repulsive states.

- Two maps (plus self-interactions) for the {100} mode ({100} junctions + coplanar interactions): Coplanar interaction can lead to the formation of dislocation dipoles and only 90° oriented {100} slip systems can make ½<110>{110} edge junctions oriented along the <100> directions.

- Three *crossed* maps that involve one slip system from each {110} and {100} mode (collinear interactions + *crossed* junctions + crossed/repulsive states): Collinear annihilations can occur if dislocations with identical Burgers vectors interact what is the case for six combinations of slip systems. Collinear annihilation is the reaction promoting the strongest of all possible forest strengthening mechanisms [128,129]. Mixed junctions with ½<110> Burgers vectors can be made out of 60°/120° oriented slip systems.

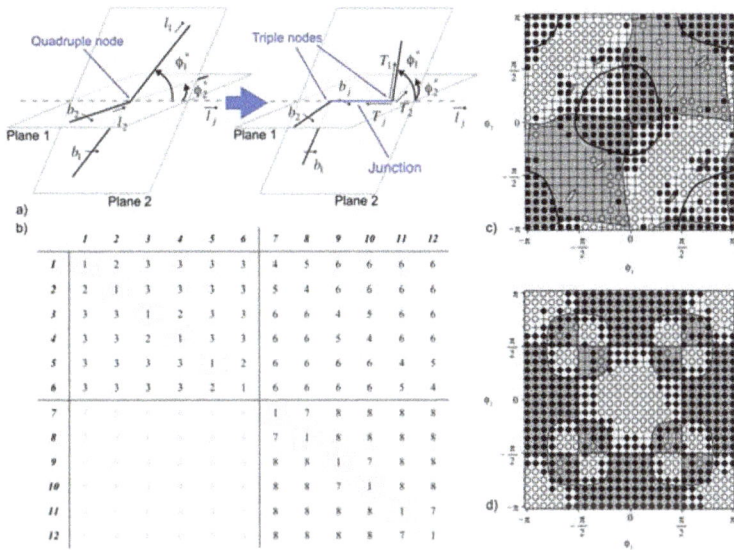

Figure 10. Interaction between dislocations in MgO modeled using DD simulations: (**a**) dislocation reaction scheme leading to a junction; and (**b**) interaction matrix between ½<110>{110} and ½<110>{100}. Indices refer to slip systems as described Table 1. Numbers inside the matrix refer to each type of reaction i.e., 1 is self-interaction; 2 and 3 are, respectively, {110} junctions and {110} crossed/repulsive states; 7 and 8 are, respectively, {100} coplanar interactions and junctions; 4 refers to collinear interactions; and 5 and 6 refer to {110} + {100} junctions and crossed/repulsive states that involve both slip modes. (**c,d**) Example of dislocation reaction maps ({110} junction and collinear interactions). Black lines refer to the elastic solution for dislocations interaction based on Equations (7) and (8). White and grey regions describe repulsive and attractive areas. Symbols are DD simulation results where crosses are attractive crossed-states, circles refer to repulsive states, filled circles are junctions and filled diamond are annihilations.

3.3.3. Grain-Scale Deformation: On the Role of Temperature on MgO Single Crystal Flow

Only few DD studies were performed to investigate oxides' deformation. In addition to the work of Cordier's group for mineral physics applications [23,45,126,127,130–132], one can refer to the work of Chang and coworkers who investigated the plastic deformation of <111>-oriented LiF micropillars from room temperature up to 600 K [113]. In this study, DD simulations predict a size-dependent flow stress in the micrometer-size regime ("*smaller is stronger*") in good agreement with experimental results [133], see also Section 4.5.

In MgO, Amodeo and collaborators investigated the deformation of bulk MgO single crystal using DD simulations [23,45]. In Ref. [23], the authors used the multi-scale approach described in Sections 3.1–3.3 to build a model able to mimic single crystalline compression tests for a large range of temperatures. 3D-periodic micrometer-sized sample were modeled including two dislocation distributions. In the thermally activated-regime, as the CRSS is believed to be intrinsically controlled by the velocity of long screw dislocations (Equation (5)), the simulation cells include only a single infinite (i.e., periodic) screw dislocation, the cell volume being tailored according to a reasonable initial dislocation density of 10^{12} m^{-2}. Only single slip configurations were tested under compression at $\dot{\varepsilon} = 10^{-4}$ s^{-1} for $T < T_a$. Within the athermal regime $(T > T_a)$, simulations are performed in larger simulation cells including an initial dislocation microstructure made of several dislocation sources $(\rho = 10^{12}$ m$^{-2})$ equally distributed between each slip system. Two sets of orientations were tested to activate glide on the {110} and {100} planes independently.

As shown Figure 11, the conclusions of the study show that this atomistically-informed DD approach can be used to describe bulk incipient plasticity in MgO single crystal, especially when compared to pure (or heat-treated) samples. Modeled CRSS vs. temperature curves are in good agreement with experimental data for the two slip modes, including the both thermal and athermal deformation regimes. While this study appears as one of the first *full-numerical* model of single crystal plasticity (no experimental parameterization), one can miss: (i) the lack of solute hardening processes in the model which are known to strongly influence MgO plasticity at intermediate temperatures; and (ii) the distinction made between {110} and {100} modes in bulk simulations (no multi-mode configuration tested).

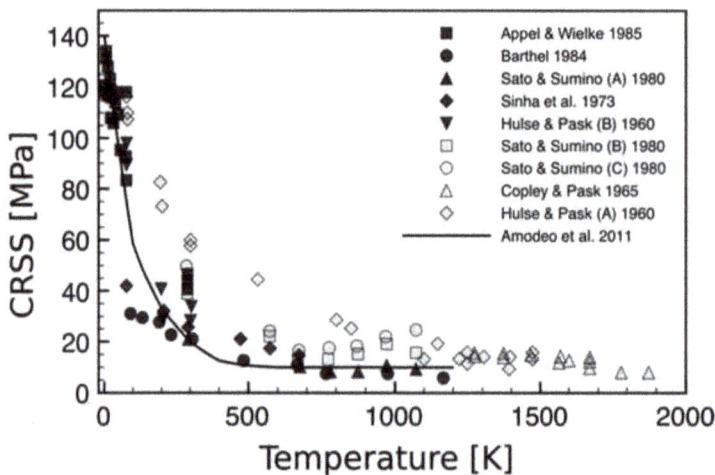

Figure 11. CRSS for 1/2<110>{110} slip systems. Comparison of the multi-scale model of Amodeo et al., 2011 (black line) [23] and experiments (symbols). The model is in better agreement with data found in pure or heat-treated MgO (black symbols). Adapted from Ref. [23].

Finally, strain-hardening of MgO single crystals at high-temperature was investigated using a similar DD simulation setup [45]. As in the case of fcc metals, MgO forest hardening is ruled by a Taylor type equation (Equation (9)) with moderated hardening coefficients α of about 0.24 and 0.28 respectively for {110} and {100} slip modes. Such values are attributed to harder reactions as the 60°/120° {110} junction and the {100} junction (see Figure 10).

$$\tau_f = \alpha \mu b \sqrt{\rho} \qquad (9)$$

Strain-hardening rates were also computed and show a satisfactory agreement with experiments [8,14,44] at 1000 K and 2000 K for both slip modes. Again, no multi-mode configurations between slip modes were investigated.

3.3.4. Creep Modeling

Weertman creep based on dislocation glide and climb was recently modeled using DD. Due to the additional computational cost associated with modeling diffusion, the models used are usually based on 2.5D simulations [131,134]. An illustration is presented in Figure 12. Reali and collaborators [132] used this approach to model creep in MgO. The chosen reference plane was (111) that contains the glide and climb directions of the two investigated slip systems (with Burgers vectors $\frac{1}{2}[01\bar{1}]$ and $\frac{1}{2}[\bar{1}01]$, gliding in (011) and (101) respectively) [132].

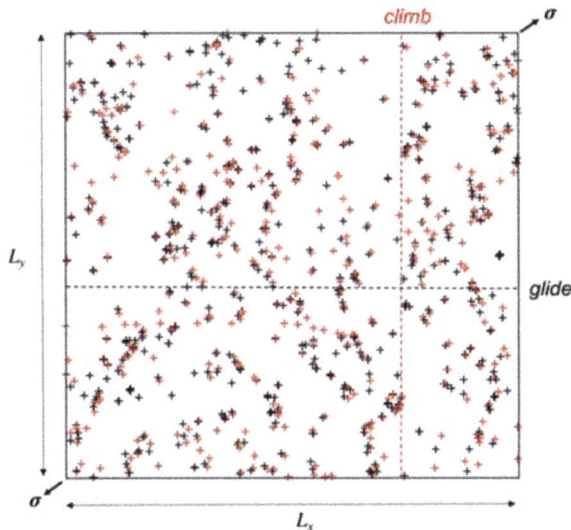

Figure 12. 2.5D-DD simulations. Black and red crosses refer to dislocations with opposite Burgers vectors.

The mobility of dislocations is determined by ascribing them relevant velocity laws. In MgO 2.5-DD simulations, Equation (6) (viscous drag equation) is used to describe the glide mobility of dislocations in the athermal regime while the climb velocity is described by the following equation:

$$v_{climb} = \eta \frac{D}{b} \left[\exp\left(\frac{\tau_c \Omega}{k_B T} \right) - \frac{c_\infty}{c_0} \right] \qquad (10)$$

where η is a factor which depends on the geometry of the flux field; D is the diffusion coefficient of the species which controls the process; Ω is the activation volume for self-diffusion; τ_c is the effective stress resolved in the climb direction; c_∞ is the vacancy concentration far from the dislocation lines; and c_0 is the intrinsic vacancy concentration at the temperature T considered. For more technical details, the reader is invited to refer to Boioli and Reali's studies [131,132].

A major difficulty of these simulations is to handle two very different kinetics: the one of glide and the one of climb. Both are associated with a characteristic time step. First, stress is applied and only glide is allowed, all glide events being resolved using the time step for glide (which is the smallest). The corresponding plastic strain is calculated. However, with glide only, the microstructure rapidly reaches a quasi-equilibrium situation and plastic strain cannot proceed any further. Then, the mobility is set to climb using the longer climb time step until at least one dislocation has climbed over a distance b. The simulation is then switched back to glide and the process is repeated. Activation of climb allows recovery processes such as annihilation, which in turn allow some dislocations to be freed and glide again until a new jammed configuration is reached. The repetition of these processes leads to a steady-state process, as shown by a constant rate of plastic strain evolution and a constant dislocation density. As usually assumed in Weertman creep, these results demonstrate that the contribution of climb to the total plastic strain is negligible, although this process controls the creep rate.

The creep rate being controlled by climb, it depends critically on the diffusion of the slowest species. In MgO, it is usually assumed that oxygen diffusion is the slowest. The simulations presented here were performed with the oxygen self-diffusion coefficients of Yoo and collaborators (i.e., 5.2×10^{-20} m^2/s at 1700 K) [135]. The creep results are presented in Figure 13 where the plastic strain rate is plotted versus strain and reciprocal temperature. These results show that the 2.5D-DD model predicts a creep behavior which follows a power law of the type:

$$\dot{\varepsilon} = \dot{\varepsilon}_0 \sigma^n \exp\left(-\frac{E}{k_B T}\right) \tag{11}$$

with $n = 2.9$, and $E = 300$ kJ/mol which is in excellent agreement with experimental results.

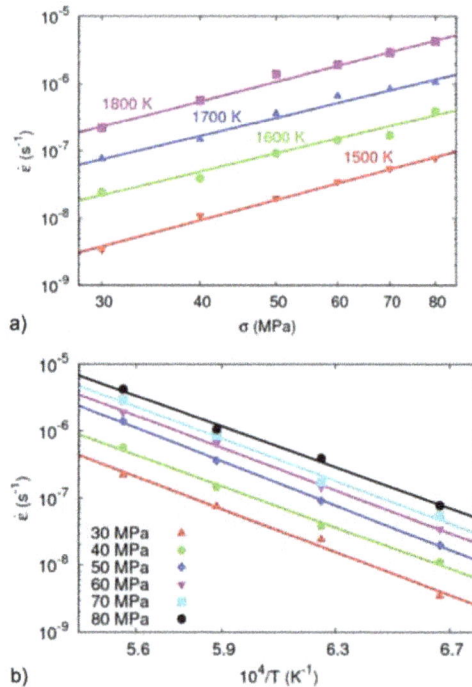

Figure 13. Results of 2.5D-DD simulation of creep in MgO: (**a**) creep strain-rates as a function of stress; and (**b**) creep strain-rates as a function of reciprocal temperature. Adapted from [132].

MgO is one of the top most advanced ceramics in terms of multi-scale modeling of mechanical properties and dislocation-based processes. Specific attributes that are crucial to explain plastic deformation such as dislocation core, mobility and interactions were investigated using the more modern modeling approaches (atomistic simulations, dislocation dynamics, and FEM), for a wide set of PT conditions. While simulation and modeling approaches highlighted prior experimental observations, it allows today opening new routes describing the mechanical behavior of MgO under extreme conditions, at the limit of experimental capabilities. In the Section 4, we see how miniaturizing mechanical tests recently brought additional perspectives on the elementary deformation processes of MgO.

4. Nano-Mechanical Testing of MgO

Nano-indentation testing is widely used for the determination of local mechanical properties such as hardness and elastic modulus [136–139]. Thanks to the accurate control of the force and displacement of the indenter tip, this technique is also used to induce local plastic deformation and obtain load–displacement curves. For nanometer sized indents, the plastic deformation mechanisms can be analyzed in terms of individual dislocations. Nano-indentation tests also allow to study dislocation nucleation, since the volume involved during a nano-indentation tests is often small enough to be free of pre-existing defects and thus of potential dislocation sources. MgO can be considered as a model material for this kind of approach because of its reduced number of slip systems, and because single crystals can be easily cleaved along a (001) plane (as mentioned in Section 1), providing well oriented and very flat surfaces suitable for nano-indentation tests. There is also an extensive literature on plastic dislocation structures generated during conventional micro-indentation tests in MgO.

In the following, a short review on micro-indentation in MgO is first presented to describe the general dislocation organization around an indent. A second part describes the elementary dislocation mechanisms in MgO around a nano-indentation imprint through slip line or etch pits analysis at the nanometer scale. In the third part, dislocation nucleation is considered thanks to spherical nano-indentation. The fourth part describes how the analysis and modeling of simple dislocation pile ups nucleated during a spherical nano-indentation can be used to determine mechanical properties of the material such as the lattice friction stress. Finally, the last part is dedicated to micro-compression tests performed with a nano-indenter equipped with a flat punch on micropillars and nanocubes.

4.1. Micro-Indentation in MgO

Since MgO and relatives such as LiF or KCl are brittle at room temperature, micro-indentation test was used early in the 1950s to introduce local plasticity [140,141]. The first analyses of the dislocation structure around the indent were carried out by Keh [142] and Newey [143]. They used chemical etching to produce etch pits on the surface at the emergence points of the dislocations. These micrometer size etch pits were then observed by optical microscopy (Figure 14). These initial studies were then followed by TEM studies [144] to identify the Burgers vector of the dislocation, as well as their organization in the volume around the indent. The dislocations are organized along a height rosette arms pattern as shown in Figure 14. The four arms oriented along the <110> directions correspond to dislocation half loops lying in {110}$_{90}$ planes perpendicular to the surface. Since these dislocations emerge with a pure edge character and a Burgers vector parallel to the surface, their propagation does not produce any visible surface deformation around the indent. The four other arms oriented along the <100> directions correspond to dislocation half loops lying in {110}$_{45}$ planes inclined at 45° from the surface. For these ones, the dislocations emerge with a screw character, so that their propagation gives rise to slip lines at the surface associated to a downward material displacement. The surface deformation associated with these slip lines was analyzed using SEM by Armstrong and Wu [145] and by Chaudhri [146]. At the intersection of different rosette arms, dislocations reaction, according to Equation (1), are reported by Keh [142,147]. These reactions lead to the formation of a Lomer lock supposed to be sessile for the aforementioned reasons. This obstacle to further dislocation propagation was associated with crack opening in {110}$_{45}$ planes by Keh [147] and Hammond and Armstrong [148].

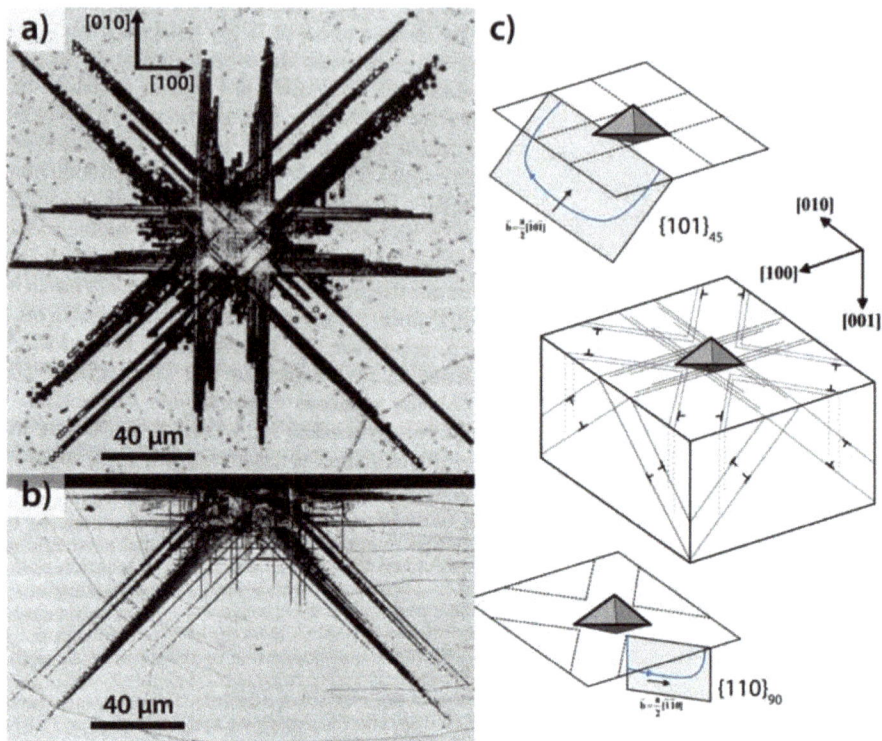

Figure 14. Dislocation structure around a Vickers micro-indentation in a (001) MgO surface. (**a**) Optical microscopy image of the etch pits pattern. The dislocations are organized in a rosette arms pattern (from Keh [142]). (**b**) Etch pits pattern in cross section in a (010) plane (from Keh [142]). (**c**) Schematic representation of the rosette arms dislocation structure (from Gaillard [149]). Images (**a**,**b**) are reprinted with permission from AIP Publishing.

4.2. Nano-Indentation: Analysis of Elementary Deformation Mechanisms

The analysis of dislocation structures around an indent has gained a renewed interest since the development of the nano-indentation technique in the 1990s. The increased sensitivity of load and displacement control of the indenter tip allowed inducing very small amount of dislocations in the material. Tromas et al. [150] used Atomic Force Microscopy (AFM) to observe the rosette arms around a nano-indentation imprint. The accuracy of the AFM allowed to observe individual slip lines in the <100> direction. Furthermore, its ability to measure heights at the sub-nanometer scale was used to determine the exact number of dislocations involved in the slip lines, since for a single dislocation the slip line height corresponds to the projection of the Burgers vector along the normal to the surface. Similar slip line analyses around indents were performed in other materials with the rock-salt crystal structure. In particular, Bennewitz and collaborators [151,152] produced and characterized atomic-scale plastic deformation by means of atomic force microscopy in ultra-high vacuum (UHV) on (001) surfaces of Potassium Bromide (KBr) single crystals.

Figure 15 presents a comparison between the slip lines at the surface around a Berkovich indent (triangular shape pyramid) observed by AFM and the dislocation structure in the volume observed by TEM [153].

These two observations are complementary since the slip lines only show the track left at the surface by the emergence and the propagation of the dislocations. Interestingly, this feature provides an access to the history of the dislocations, because the slip lines reveal the path that was followed by the dislocations.

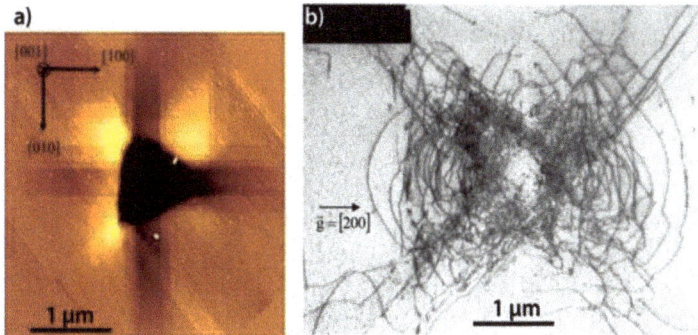

Figure 15. (**a**) AFM image; and (**b**) TEM image of two 3 mN Berkovich nano-indentation imprints in a (001) MgO surface. The scale and the orientations are the same for both images. The AFM image reveals the slip lines due to the emergence and propagation of dislocations while TEM reveals the dislocation structure in the volume.

Following this property, Tromas and collaborators observed an isolated slip line with a three-stages behavior around a Berkovich indent [13]. This line, which originates from a <100> rosette arm, is first deflected at 26° along a <120> direction over a short extent, before deviating again to propagate along a <010> direction perpendicular to its initial rosette arm (Figure 16). This threefold path was explained by the following mechanism. A screw dislocation gliding in a {101}$_{45}$ plane along the <100> direction reacts with a second dislocation gliding in a {110}$_{90}$ slip plane along a <110> directions. These two dislocations form a junction along a <111> direction according to Equation (1). Due to the high stress induced by the indenter, this segment glides in a {211} plane, leaving a slip line along a <120> direction at the surface, before cross-slipping into its natural (011)$_{45}$ slip plane. The dislocation can then easily glide in this new slip plane, leaving at the surface a slip line along a <010> direction perpendicular to the initial rosette arm (see Figure 16). Thus, this observation shows that the junction resulting from Equation (1) is not truly sessile and that the <110>{211} slip systems can be activated at room temperature in MgO. Furthermore, after the cross slip, while the dislocation segment reached its easy slip plane, its extremity is still pinned at a triple node formed by the two parent dislocations. Tromas and colleagues showed that this configuration leads to the activation of a spiral source ensuring the plastic deformation required for the indenter penetration [13].

Slip line analysis can thus provide detailed information about complex dislocation mechanisms that would be difficult to reveal by TEM. The drawback of this approach is that only four of the eight rosette arms can be observed, since the dislocations in the <110> directions do not produce any slip line. Furthermore, the slip lines do not provide direct information of the 3D organization of the underlying dislocation structure. Gaillard and collaborators [154] adapted the etch pits method of Gilman and colleagues [155] and Shaw and collaborators [156] at the nanometer-scale. By an accurate control of the etchant concentration and of the etching time, they produced nanometer-scale etch pits at the emergence points of the dislocations and observed them by AFM. This technique was repeated sequentially in a same area after successive chemo-mechanical polishing (CMP) stages. This so called nano-etching technique consists thus in a kind of dislocation tomography, allowing to reconstruct, layer by layer, the 3D dislocation structure. Figure 17 shows the etching patterns obtained at the surface and after successive CMP stages around a 80 mN Berkovich indent. The eight rosette arms are observed and each individual dislocation can be resolved. The dislocation reactions at the intersection of the rosette arms are also identified by following two etch pits along the successive CMP stages. These two etch pits progressively combine to form a single one, which is then always localized along the same <111> direction. The presence of Lomer junctions is also characterized by high etch pits concentration in the adjacent rosette arms, due to dislocation pile ups induced by these junctions.

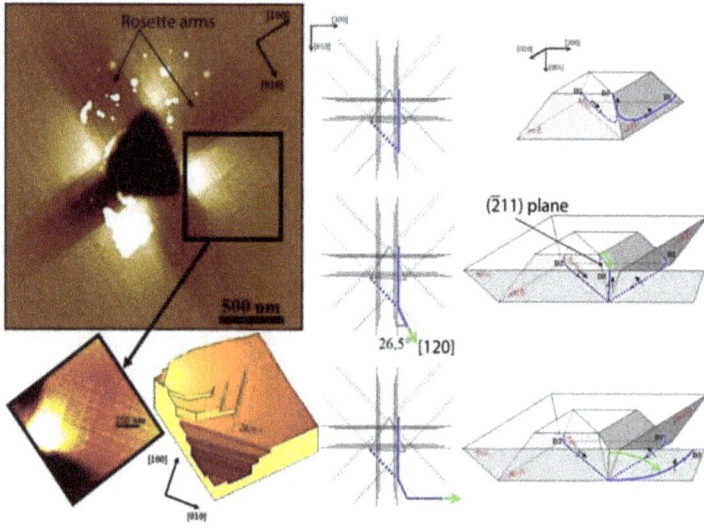

Figure 16. On the left side: AFM image of a 2.6 mN load Berkovich indent in a (001) MgO surface. A three-stages slip line is observed (enlarged view and schematic at the bottom). This trace shows the complex path followed by a dislocation segment gliding in a (211) plane and then cross slipping in a (011) plane (right side).

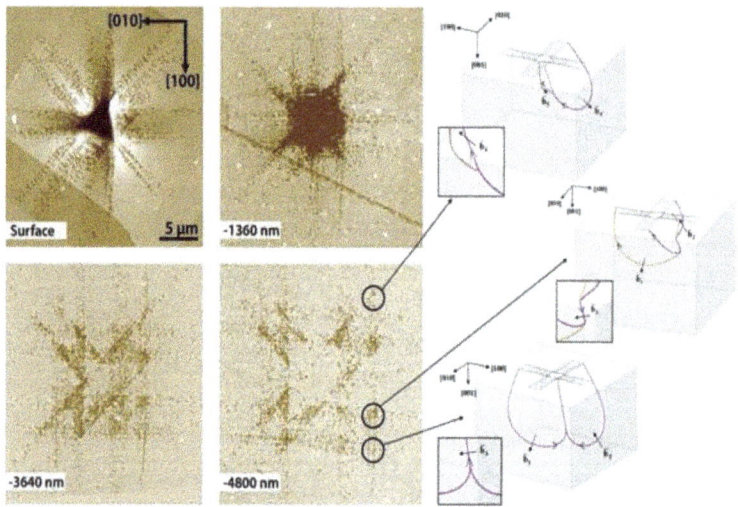

Figure 17. Left side: Successive AFM observations of the nano-etching pattern produced around an 80 mN nano-indentation in MgO at the surface and after chemo mechanical polishing at −1360 nm, −3640 nm and −4800 nm below the initial surface. Right side: Different junctions resulting from interaction between dislocations from different rosette arms (from Gaillard [149]).

In MgO, as in many other materials, an indentation size effect (ISE) is observed. This significant increase of the measured hardness for small indentations is often explained based on the Nix and Gao (NG) model [157] that relies on the concept of geometrically necessary dislocations (GND). In the NG model, the GNDs are contained into a hemispherical plastic zone with a radius equal to the contact radius. However,

Feng and collaborators [158] observed that for a penetration depth lower than 200 nm, the NG model overestimates the hardness in MgO. Following the work of Gaillard and colleagues [154], they applied the nano-etching method to reveal dislocations around the indent and determine the size of an effective plastic zone. Then, they have shown that the plastic zone radius was higher than the contact radius for low penetration depth. They thus proposed a correction for the NG model to take into account the non self-similarity of the plastic zone for low penetration depth. This result was strengthened by the work of Huang and collaborators [159], even if they conclude that, for penetration depth as low as few nanometers, the model is no longer valid since hardness measurements are scattered because of discrete dislocations events. In 2009, Sadrabadi and coworkers [160] proposed a similar study of the ISE by nano-etching in CaF$_2$, another ionic crystal. Indentation size effect in MgO was also studied as a function of the temperature [161] and after ion irradiation [162].

4.3. Incipient Plasticity: Pop-In and Dislocation Nucleation

In the nano-indentation load–displacement curves, the first plastic event in load-controlled experiment is characterized by a pop-in phenomenon, i.e., an abrupt increase in the penetration depth [163–168]. In MgO, if the pop-in appears at different loads, it always makes the transition from an elastic behavior to an elasto-plastic regime characterized by a master curve. This is clearly illustrated by Figure 18 where different nano-indentation curves obtained with a Berkovich indenter in a (001) oriented MgO single crystal are superimposed. This phenomenon, which is observed in many different materials, has aroused a great interest since it is commonly associated to homogeneous nucleation of dislocations. It also must be pointed out that the nanometer scale volumes of material involved in these nano-indentation experiments are very well suited for mesoscopic and atomistic simulations [169–171].

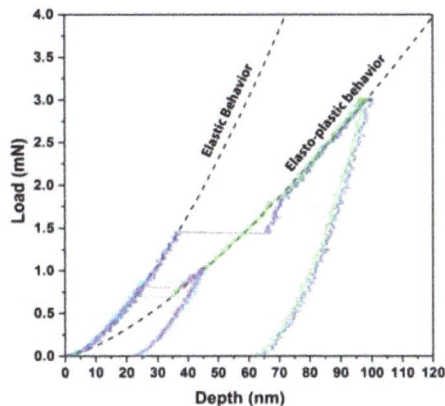

Figure 18. Superposition of nano-indentation curves in MgO (Berkovich indenter). A pop-in, which can appear at different loads, characterizes the transition from an elastic behavior to an elasto-plastic behavior.

Gaillard et al. [172] applied the nano-etching technique to study the dislocation structures nucleated during nano-indentation tests that are stopped just after the pop-in in both MgO and LiF. They first used a Berkovich indenter. In this case, they report that the dislocations are concentrated in a small volume below the contact area between the indenter and the surface. They concluded that, since a high number of dislocations are nucleated at the same time during the pop-in, dislocation interactions leading to Lomer locks should be enhanced, which impedes the dislocation propagation in the conventional rosette arm distribution. However, even if few dislocations are nucleated during the pop-in, their concentration is still high when using a Berkovich indenter because of the sharpness of the pyramidal tip. For a more accurate analysis of the dislocation nucleation phenomenon, Tromas and colleagues [153] used spherical indenters with radius of curvature of few micrometers.

In that case, the pop-in appears at very low penetration depth compared to the radius of the indenter. The ratio of the contact area over the penetration depth is thus higher than in the case of a Berkovich indenter. The dislocations structures nucleated during the pop-in with a spherical indenter are thus extremely simple since they are limited to only one or two pile ups of few dislocations. In this configuration, there are no (or few) dislocation reactions. Furthermore, the elastic contact between a sphere and a plane is well described by the Hertz law [173,174], and the stress field distribution can be determined analytically [175]. The dislocation structure can thus be compared to the stress field generated by the indenter at the beginning of the pop-in, when they were nucleated (Figure 19). From this comparison between the dislocation organization below the contact area and the mapping of the resolved shear stress acting on the activated slip systems, Tromas and colleagues concluded that these initial dislocation pile ups are nucleated at the point of maximum resolved shear stress, that is along the indentation axis at a depth of $0.44 \times a_c$, where a_c is the contact radius for the $1/2<110>\{101\}_{45}$ slip systems, and at a radial distance of $1.1 \times a_c$ from the indentation axis and a depth of $0.65 \times a_c$ for the $1/2<110>\{101\}_{45}$ slip systems [153].

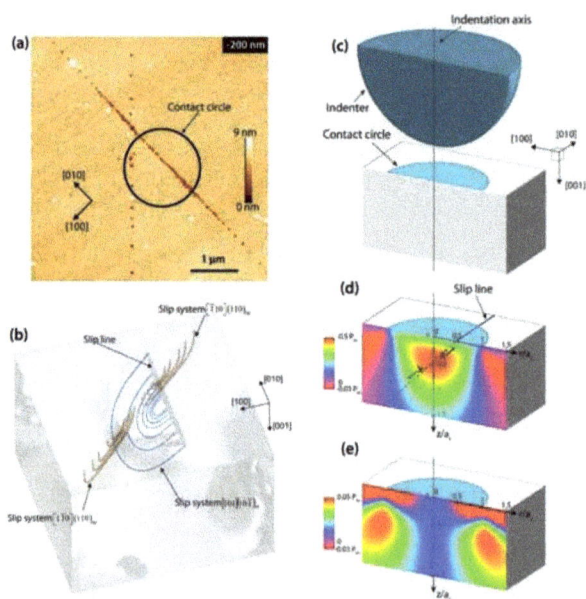

Figure 19. (a) Nano-etching pattern, after removing a layer of 200 nm by CMP, of a spherical nano-indentation stopped after the pop-in; (b) schematic of the corresponding dislocation structure; and (c–e) schematic representation of the resolved stress field (Equation (12)) in the slip systems $<110>\{110\}_{45}$ (d) and $<110>\{110\}_{90}$ (e). The stress field is presented in a plane of cross section (010) containing the indentation axis. The contact circle is represented in blue.

While it is admitted that the pop-in phenomenon corresponds in many materials to dislocation nucleation, there is still a debate on its nature: homogeneous or heterogeneous? The question is thus to answer if the nucleation process is favored by pre-existing defects or not. The experimental difficulty is thus to find a model material where the pre-existing defects can be accurately controlled and characterized prior to indentation. As developed before, MgO single crystals can be easily cleaved, with a simple hammer and a fine blade, providing flat and well oriented (001) surfaces. Under certain conditions, cleavage can also generate a high dislocation density in a very thin surface layer [176–181]. According to the previous work of Gilman and colleagues in LiF and MgO [179,180], Montagne and collaborators [182] used "slow" cleavage in MgO to generate dislocations in the subsurface, the dislocation density being measured by nano-etching. The surface was then carefully polished by CMP to reach regions with different dislocation density. With this method, it was thus possible to prepare flat (001) MgO surfaces with a controlled density of (pre-existing) dislocations. By performing spherical nano-indentation, they observed that the existence of pre-existing dislocations in the contact area between the indenter and the surface decreases the pop-in load. Dislocation nucleation is thus promoted by pre-existing dislocations. However, after a careful analysis of the dislocation structures by nano-etching, they showed that the effect of pre-existing dislocation was more complex than expected (see Figure 20). Indeed, the dislocations nucleated during the pop-in do not result from dislocation source activation or reactions with pre-existing dislocations. Furthermore, the organization of the dislocation structures during the pop-in is still consistent with a nucleation event at the point of maximum resolved shear stress, independently of the pre-exiting dislocations positions. They suggest that the influence of pre-existing dislocations should be linked to their displacement in the indenter stress field during the nano-indentation test.

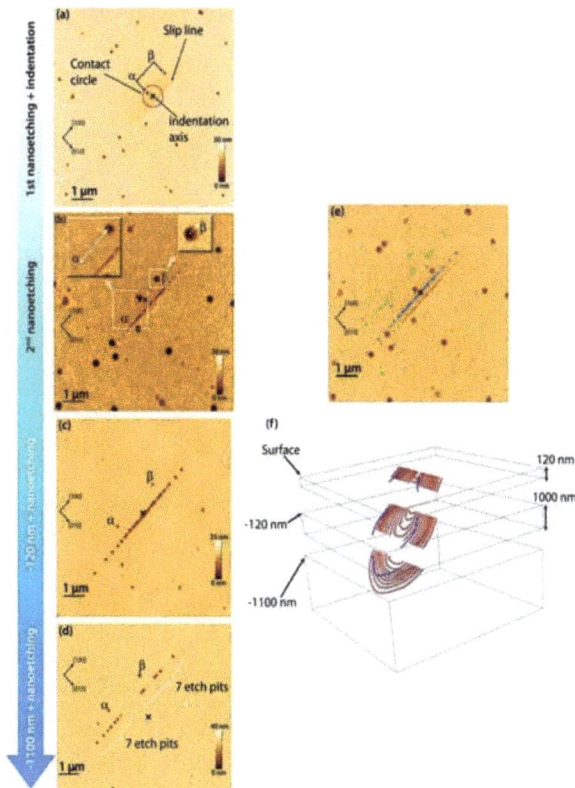

Figure 20. (**a**) Initial surface first etched to reveal pre-existing dislocations and then indented with a spherical indenter. The indentation was stopped after the pop-in; (**b**) Second etching on the initial surface to reveal the dislocations nucleated during the indentation. The pre-existing dislocation are clearly identified by bigger etch pits; (**c**) New nano-etching obtained after removing a layer of 120 nm by CMP; (**d**) New nano-etching after removal of a 1100 nm total layer. The pre-existing dislocation half loop αβ does not interact with the dislocations nucleated during the pop-in. However, this initial half loop extends in the indenter stress field; (**e**) Superposition of the successive polishing and nano-ecthing stages; (**f**) Schematic representation of the dislocation structure. Images are adapted and reproduced from Ref. [182] with permission from Elsevier.

4.4. Determination of the Lattice Friction Stress through the Dislocation Structure around a Spherical Indent

The complete analysis of individual dislocation positions in the indenter stress field can be used to measure specific mechanical properties of MgO. Gaillard and collaborators [183] performed spherical nano-indentation test in MgO, stopped just after the pop-in. They obtained very simple dislocation structure limited to two pile ups of few dislocations: a first one oriented along a <110> direction corresponding to dislocations emerging with an edge character, and a second one oriented along a <100> direction corresponding to dislocations emerging with a screw character. The structure consisted in dislocation half loops pile-up, so that the etch pits could be paired for each pile up. Then, similar pile-ups were analytically modeled considering that the stress acting on each dislocation is the sum of the external applied stress τ_a, the stress due to other dislocations in the pile-up τ_i, the image forces stress due to the surface τ_{im} and the lattice friction stress τ_f.

The external applied stress τ_a is the spherical indenter stress $\overset{=ind}{\sigma}$ resolved on the dislocation slip system, $\overset{=ind}{\sigma}$ being calculated by Hanson [175], Equation (12).

$$\tau_a = \sigma\left(\vec{b}, \vec{l}\right) = \frac{\vec{b}}{\|\vec{b}\|} \cdot \overset{=ind}{\sigma}(x, y, z) \cdot \vec{n} \tag{12}$$

The stress due to other dislocations in the pile-up τ_i is defined by

$$\tau_i = \sum_{j \neq i}^{n} \frac{A}{x_i - x_j} \tag{13}$$

where x_i is the position of the dislocation i and A is defined by Equation (14):

$$A = \frac{\mu}{2\pi}\left(\cos^2\theta + \frac{\sin^2\theta}{1 - \nu}\right) \tag{14}$$

with θ the angle between the Burgers vector $\vec{b}$ and the line direction $\vec{l}$ of the dislocation [184].

The Lattice friction stress τ_f is the parameter to be determined. This is thus a free parameter which is adjusted by comparing the analytical calculation and the experimental observations of the dislocation pile-up. At the equilibrium:

$$\tau_a(x_i) + \tau_{im}(x_i) - \tau_f + \sum_{j \neq i}^{n} \frac{A}{x_i - x_j} = 0 \tag{15}$$

In their calculation, Gaillard and coworkers introduced successively pairs of dislocation with opposite Burgers vectors to model dislocation half loop, at the point of maximum resolved shear stress, i.e., where they are supposed to nucleate according to the work of Tromas et al. [153]. Once all the dislocations were introduced in the calculation, the indenter stress field is progressively decreased to model the indenter unloading, and the pile-up relaxation is calculated. Finally, the calculated dislocation positions in the pile-ups are compared to dislocation etch pits (see Figure 21). This calculation was performed for different lattice friction stress values, until the calculated pile up length fits with the experimental observations. The lattice friction stress in MgO at room temperature is 65 MPa for edge dislocations and 86 MPa for screw dislocations i.e., values in the same range than {110} experimental CRSS measured under compression [14,18,21].

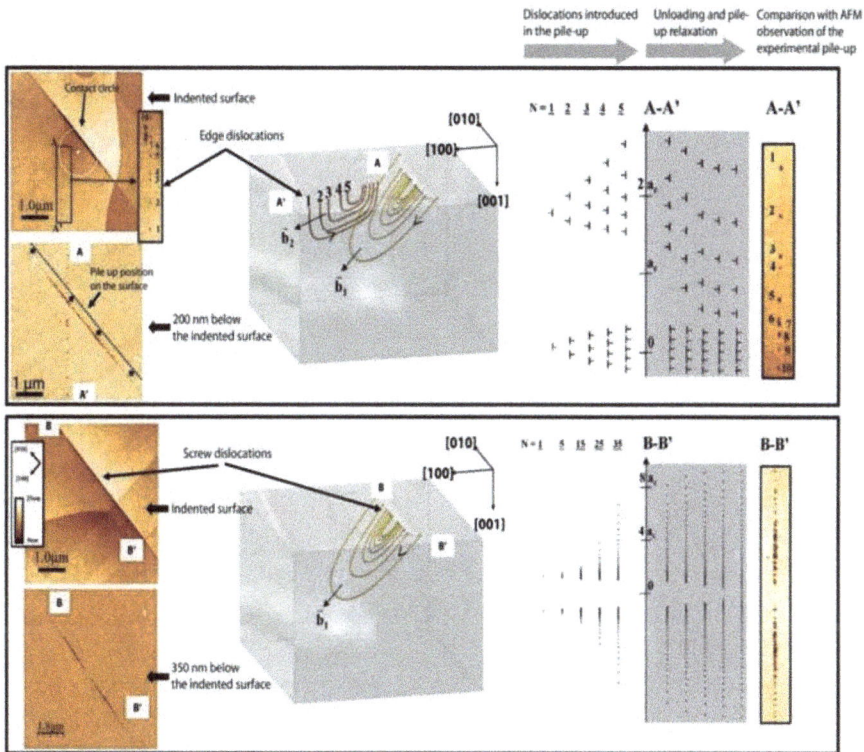

Figure 21. Dislocation structure induced by spherical nano-indentation in MgO stopped just after the pop-in: (**top**) analysis and modeling of a pile up of edge dislocations; and (**bottom**) analysis and modeling of a pile up of screw dislocations. Images are adapted and reproduced from Ref. [183] with permission from Elsevier.

4.5. Plasticity in MgO and Other Ionic Crystals Studied by Micro-Compression

Indentation of MgO was used extensively to study its deformation mechanisms. However, although brittle fracture can be suppressed owing to the confining pressure [185,186], there are certain drawbacks. The most important one is that it is not possible to directly distinguish the contributions of the soft ½<110>{110} slip systems and the hard ½<110>{100} slip systems unless a modeling approach is used in conjunction with experiments on different crystal orientations [187]. The reason for this is that the three-dimensional deformation underneath the indenter requires deformation on several independent slip systems. The magnitude of the hardness often corresponds most closely to the estimate of the room temperature hardness of the hard slip system, justified as this system may be the limiting component in achieving the required deformation [188]. In contrast, the slip traces observed at the surface belong to the soft slip system [145], which appears to accommodate the majority of the deformation. Therefore, the hard slip system may only be identified by TEM, where the sample preparation and large dislocation densities in indentation render such observations difficult [188]. Deformation after indentation in the hexagonal and therefore intrinsically anisotropic GaN as well as in MgO are shown in Figure 22 to illustrate this effect.

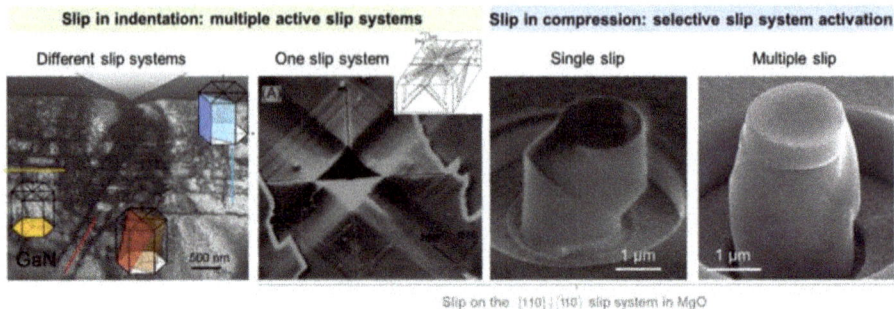

Figure 22. Geometry of slip in indentation and micro-compression. In indentation, multiple slip systems are always required and may result in the activation of several equivalent (e.g., in MgO [145] or very different systems, GaN [189]). In micro-compression, as in any single crystalline compression test, both single (here MgO near [001]) and multiple slip orientations (here MgO aligned with [110], see Ref. [190]) may be interrogated. Images are adapted and reproduced with permission from K. McLaughlin [189], John Wiley and Sons, Inc. [145] and Cambridge University Press [190].

The simplest approach to avoid this issue is to change the test geometry and use uniaxial stresses on single crystals instead. In uniaxial compression, CRSS on the individual slip mode may then be obtained, as collated in Figure 2 from macroscopic tests. Unfortunately, this approach is limited to conditions in which plastic deformation by dislocation glide is favored over cracking. For compression along soft directions, in which the $\frac{1}{2}$<110>{110} slip systems are activated, this is nearly always the case. In contrast, in the hard direction near a [111] compression axis, where the Schmid factor on the soft system diminishes, the uniaxial stress has to be increased until the CRSS of the $\frac{1}{2}$ <110>{100} system is exceeded. At low temperatures, this is normally preceded by the occurrence of fracture, such that no CRSS for dislocation glide can be obtained. In Figure 2, this is evident from the lack of data without cracking below 650 K [3,5]. In recent years, a new approach in mechanical testing has emerged, which overcomes this challenge of suppressing brittle fracture at low temperatures: the down-scaling of samples to the micrometer range in the so-called micro-compression technique [191,192]. As a general trend, the idea is to test samples with a size lower than that of the plastic zone to favor ductility before failure. More precisely, in uniaxial tests at small scales, cracking is suppressed due to two reasons [193]: (i) small samples are statistically less prone to contain critical pre-existing flaws or these would be more easily visible prior to testing; and (ii) where no pre-existing cracks exist, these have to be nucleated and extended in order for fracture to occur. Analyses of the second point have shown that indeed, fracture is often suppressed in small specimens and that if it does occur it is where slip bands intersect (Figure 23) [190,193–195]. This is also seen in macroscopic observations of deformed MgO [196] (Figure 23) and the geometry of slip then determines the nucleation of cracks. It can be shown that a size effect exists where a crack splitting a pillar through the middle and that this may be exploited to suppress unstable crack growth [193,194,197]. The most favorable condition for this is single slip [193]. However, even where this cannot be achieved or multiple slip conditions are reached, a size effect is nevertheless observed, resulting effectively in a substantial reduction of the brittle-to-ductile transition temperature with size, i.e., the competition between deformation by cracking and dislocation glide is shifted in favor of the latter. In testing micropillars at different temperatures, this was shown on $MgAl_2O_4$ spinel [198], where plastic deformation was achieved more than 1000 °C below the macroscopic brittle to ductile transition temperature. The yield stress required to initiate substantial dislocation glide in small volumes increases as the size decreases towards and below the μm regime [199]. However, this increase is relatively small in intrinsically hard materials [200–202] and the critical stress for cracking increases more sharply [193,197].

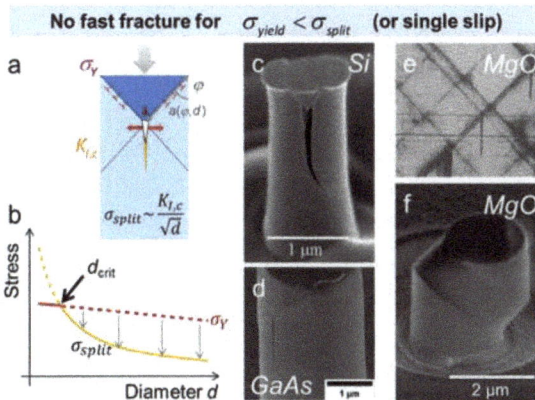

Figure 23. Suppression of unstable fracture in micropillars: (**a**) schematic drawing of the model relating size, slip geometry and the splitting stress; (**b**) the size effects on critical stresses for a typical hard crystal showing a transition from failure to plastic deformation; and an experimental geometry of this type in (**c**) silicon and (**d**) GaAs as well as (**e**) at slip band intersections in macroscopic experiments in MgO. In single slip orientation fracture can then be suppressed to the largest diameters, shown here for MgO in (**f**). Figures in (**c**) [195], (**d**) [193], and (**e**) [196] reprinted with permission from Wiley, Taylor & Francis and Springer.

The micro-compression technique is therefore exceedingly powerful in resolving the deformation mechanisms at temperatures below the macroscopic brittle to ductile transition temperature in brittle materials, such as MgO, and also those which are difficult to obtain in bulk form, such as coatings, manipulated surface layers, complex materials or phases which cannot be solidified as single crystals from the melt [203–205].

A schematic illustration of the application of this method is given in Figure 24 and a review covering its use on a much wider range of brittle crystals may be found in reference [200].

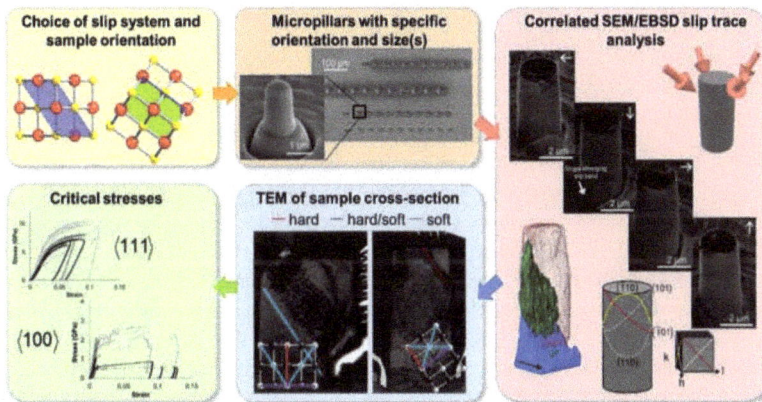

Figure 24. Schematic representation of a study of deformation mechanisms by micro-compression and electron microscopy. After the choice of crystal orientation, pillars or specific size(s) are milled and typically analyzed by SEM and EBSD. Where fracture is successfully suppressed TEM can further elucidate the operative deformation mechanisms to correlate these with the stress–strain curves collected during the experiment. SEM [206] and TEM [201] images as well 3D-EBSD results [206] and stress–strain curves [201] reprinted with permission from Elsevier and Taylor & Francis.

This section is divided into three subsections: measurements of CRSS in MgO and related crystals at room temperature using micro-compression, studies of the thermal activation of dislocation motion in micro-compression and the extension of the technique to compression of nanocubes, where dislocation nucleation governs deformation.

4.5.1. Measurements of CRSS in MgO and Other Rock-Salt Crystals at Room Temperature

Micro-compression experiments on single crystals allow the individual activation of the soft and hard slip systems. The <111> orientation requiring the hard slip mode to operate is of particular interest due to the difficulties outlined above in testing macroscopically at low temperatures. The critical resolved shear stress of MgO for both slip modes, using crystals compressed in <100>, <110>, and <111> direction, was determined by Korte and collaborators as shown in Figure 25. Similar results were reported by Zou and Spolenak for compression along <100> in MgO [207]. As expected, the CRSS on the soft mode is strongly overestimated owing to the size effect on dislocation glide. Several studies covered the current understanding of this effect that is generally thought to be governed by the effective dislocation line length and its statistical variation in a small volume [192,208], the exhaustion of dislocation due to escape at the surface [209], image forces [210], the nucleation of partial dislocations without the need to activate the trailing partial for slip distances in the pillar in the order of the dissociation length [211] and finally the effect of Focused Ion Beam (FIB) induced defects on dislocation nucleation and confinement [197]. In contrast to the soft slip system, the CRSS on the hard slip system shows excellent agreement with an extrapolation of the macroscopic values towards room temperature. This is consistent with findings in other hard materials [200] and with the idea that on the hard slip system, the lattice resistance governs the magnitude of the yield stress and the effect of changes in line length is relatively weak [133,200,201] as the relevant dimensions are of the order of a kink pair. This is shown for a selection of materials ranging from metals over semiconductors to oxides.

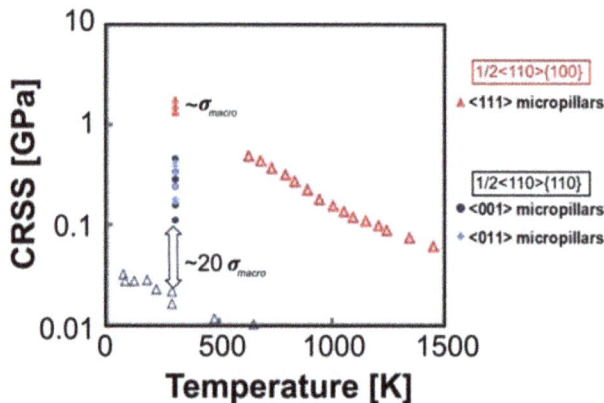

Figure 25. The critical resolved shear stress of the soft 1/2<110>{110} and the hard 1/2<110>{100} slip systems in MgO measured by micro-compression at room temperature along the <100>, <110> and <111> orientations compared with measurements on macroscopic single crystals in <100> and <111> orientation. A size effect on CRSS is observed on the soft slip system, while the previously unavailable data on the hard slip system correspond well to the extrapolation from higher temperatures due to its higher intrinsic stress level. Data from references [3,5,201].

This difference in the effect of size can be rationalized rather simply: most models of plasticity size effect for a size regime of a few micrometers, i.e., where dislocation motion and multiplication rather than nucleation are thought to be the dominant mechanism, consider the importance of only a limited number and truncated single armed sources. Using such a model, a first estimate of the expected effect of size may be made using the statistical source length suggested by Parthasarathy and colleagues [212] with the average source length λ taken as 0.2 times the pillar diameter. This leads to a stress contribution, most simply assumed to be additive to the bulk strength, of the order of $Gb/2\lambda$, where G is the shear modulus. Taking a commonly used pillar diameter of 2 μm, a shear modulus of $G_{\{110\}}$ = 101 GPa and $G_{\{100\}}$ = 155 GPa for the two slip systems this amounts to 38 and 58 MPa on the soft and hard slip system, respectively. This is of the same order of the expected CRSS in macroscopic crystals in the case of the soft slip mode, while in the case of the hard slip systems, the contribution would be approximately 5% of the room temperature value extrapolated from macroscopic high temperature data. Of course, the effects of ion milling point and line defect densities and types, contributions of dislocation nucleation stresses, the uncertainties in the pillar dimensions at small scales, their taper and the resulting inhomogeneous stress state (which may give much higher local strain rates) are neglected in this simple estimate. Note that, in the data shown in Figure 25, the effect of FIB damage should be approximately constant as the same material is milled and simply compressed in different crystal directions. Despite these simplifications, this estimate does, however, illustrate in an easy manner why a much stronger size effect is to be expected on the soft slip system, even if the exact size of the effect is in fact still under debate in the literature. The importance of the lattice resistance, governing the bulk strength above which a size effect becomes significant or not, was also shown using tests at different temperatures in LiF [133] and other materials, such as bcc metals [213,214]. Owing to the large difference in CRSS it is of course important to verify the operation of the hard slip system (normally achieved by TEM, as in reference [201], see Figure 24) and address potential artefact encountered due to imperfect misalignment. In MgO, 3D-EBSD was used to track deformation on secondary systems not detectable by surface trace analysis [206]. Taking LiF as a model material, Soler and coworkers [215] assessed the effect of misalignment in experiment and finite element models and Kiener and collaborators [216] collated a general review on experimental constraints specific to micro-compression. Across these high strength materials, a challenge remains to clearly distinguish the contributions of dislocation glide and nucleation regarding the measured stresses. Indeed, the underlying activation volumes and critical stresses are much closer than in the metals and surface alteration, e.g., by oxidation or FIB milling, is known to influence the measured stress level and its scatter [197].

4.5.2. Rate and Temperature Dependence in Other Ionic Rock Salt Crystals

Following these initial tests initiated on MgO, there was a wider effort to characterize the ionic rock salt crystals with LiF in the focus [133,215,217] using the same approach and including variable temperatures [206,218,219] and rates [220,221] with a combination with simulation [133]. An excerpt from these works is shown in Figure 26, highlighting the characteristics of the micro-compression tests described above for MgO: an increase in flow stress with size, which becomes more pronounced as the temperature is increased and, hence, the lattice resistance reduced [113,133]. A drop in stress level consistent with thermal activation of lattice resistance controlled flow is confirmed in elevated temperature tests [133] and variable rate tests yield an appropriate activation volume of several b^3 [133]. A fit to a theoretical model describing the bulk strength, lattice resistance and size distributions separately indeed also explains these results quantitatively (Figure 26b) [133].

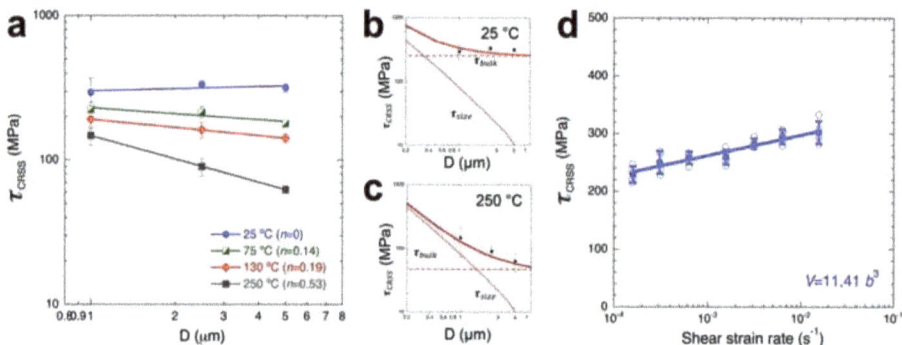

Figure 26. The effects of temperature, strain rate and size in LiF. (**a**) Experiments and discrete dislocation dynamics simulations showing the drop-in flow stress with temperature and increasing size effects towards lower overall stresses. These results are consistent with a size (source controlled) term and a temperature dependent bulk term (governed by the lattice resistance) contributing to the measured stresses at both (**b**) room and (**c**) elevated temperature. (**d**) Strain rate jump tests offer a consistent view with an activation volume of the order of 10 b^3. Figures reproduced from Ref. [133] with permission from Elsevier.

4.5.3. Dislocation Nucleation in Small MgO Volumes

Although the samples tested by micro-compression are very small, the volume is normally not dislocation free, particularly where it was milled by FIB, or possesses a roughened surface from etching [133] or lithography [222]. However, in some cases, it is possible to produce pristine samples which allow the investigation of this rarely accessible regime [223–225]. Issa and colleagues [224] were able to obtain pristine MgO nanocubes from burning Mg chips in air, tested them in-situ in the TEM and compared these results with MD simulations [224,226] (Figure 27). In both simulations and experiments, the expected dislocations on the softer {110} planes are observed and nucleation occurs at the expected locations, i.e., the edges and corners of the cubes. In contrast to other tests, the modeled nucleation stress is very high and a substantial curvature of the dislocation lines is observed (especially in the simulation). Thus, the dislocation microstructures observed at the nanoscale, under a compression stress in the GPa range, may therefore appear more similar those found at high temperatures [224].

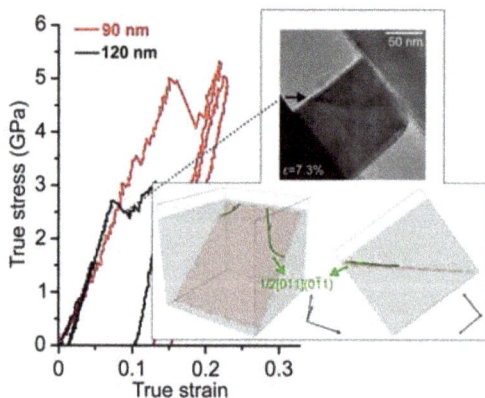

Figure 27. Experiments in initially dislocation free MgO nanocubes inside the TEM. Stress–strain curves and a corresponding TEM image at 7.3% strain with two MD simulation snapshots are showing ½<110>{110} dislocations, which are in green. Figure reproduced from Ref. [224] with permission from Elsevier.

In this section, we outline how various nano-mechanical methods in conjunction with electron and atomic force microscopy were used to unravel the dominant plastic deformation mechanisms in MgO. The local deformation, when limited to its very early stages, allows individual dislocation analysis. A particular strength of these methods is the ability to extend the measurable range to room temperature even in the hard directions. In micro-compression, a size effect on fracture facilitates the study of plasticity, while, in indentation, it is the confining stress field provided by the surrounding material that suppresses cracking. In this context, and in other common high-pressure situations for MgO, such as in the Earth's mantle, it is important to also understand the effect of hydrostatic pressure on the crystal and the resulting deformation mechanisms. These is reviewed in Section 5.

5. High-Pressure Plasticity in MgO

Hydrostatic pressures in the GPa range induce drastic changes in the electronic, structural, and mechanical properties of materials. A simple ionic salt such as NaCl, for instance, transforms from the classical rock-salt structure (B1) into a cesium chloride structure (B2) at 20 to 30 GPa [227]. CsI, another simple ionic salt at ambient conditions, transforms to a hexagonal-closed-packed structure at ca. 50 GPa [228]. In addition to this structural transition, CsI transforms from an insulator to a semi-conductor at pressure of about 55 GPa and becomes metallic above ~100 GPa [229,230]. As such, hydrostatic pressure can have a profound influence on the properties of matter.

MgO remains in the B1 structure over a wide pressure and temperature range. Recently, it was shown that the B1 to B2 phase transition in MgO occurs at pressures of the order of 400–600 GPa [231,232] and that it becomes metallic at even higher pressures [231]. Nevertheless, pressure already influences the properties of MgO even below these extreme values. For instance, the elastic anisotropy of MgO reverses close to 15 GPa [233]: below 15 GPa, longitudinal and shear-wave propagations are faster along [111] and [100], respectively, whereas above 15 GPa, the corresponding fast directions are [100] and [110]. Based on considerations on elasticity and electronic polarizability, it was suggested that the primary slip plane of MgO would change from {110} to {100} at increasing pressures [234].

The effect of pressure on materials properties can be counter-intuitive. It can also serve as a test of current modeling capabilities and has implications for Earth and planetary sciences. As such, plastic properties are being studied at pressures in the 10s of GPa range and above. This section reviews current methods for such experiments, their capabilities/limitations and application to MgO. Recent outcomes from simulations are be discussed.

5.1. Methods for HP Experiments and Simulations

5.1.1. Experimental Devices

The deformation-DIA (D-DIA) [235], Rotational Drickamer Apparatus (RDA) [236], and newly developed Deformation T-Cup (DT-Cup) [237] are the three main large volume presses (LVP) used for high pressure plasticity studies. All allow the controlled deformation of millimeter size samples. The D-DIA is designed for deformation at constant pressure, both compressional and extensional, at strain rates between 10^{-3} and 10^{-7}/s and, currently, up to 18 GPa and 1900 K [238]. The DT-Cup allows for controlled deformation in axial geometry with experiments at about 20 GPa at 300 K and 10 GPa at 1100 K [237]. Finally, the RDA allows a large rotational shear deformation of the sample up to pressures and temperatures over 27 GPa at 2150 K [239]. In the D-DIA and DT-Cup, the stress boundary acting on the sample is axial [240] while, in the RDA, it may be considered as a combination of simple shear and axial compression [241]. Depending on experiments, strains are measured directly using X-ray radiography or can be reconstructed from changes in sample pressure or based on markers placed inside the sample.

LVP allow for controlled deformation experiments but their P-T range does not yet reach over 30 GPa. The DAC, on the other hand, in combination with heating methods reach the terapascal at ambient temperature [242] and 350 GPa at 6000 K [243]. DAC may not only impose pressure but also a compressive stress that produces elastic and plastic deformation. A limitation of the DAC is that pressure and deformation cannot be decoupled. Despite these limitations, the DAC is a useful tool for investigating plastic properties up to 300 GPa at ambient temperature [244] with developments under way to allow for performing high-T experiments, either using resistive heating elements [245] or lasers [246]. Stress in DAC experiments is typically assumed to be axial with cylindrical symmetry around the compression direction [247] and with low pressure gradients across the sample. Note that this assumption relies on the use of small samples and gasket confinement, unlike older experiments that did not confine the sample within gaskets [248]. Sample strain along the diamond direction is compressive. Radial strain is dependent on sample, gasket, and sample loading.

5.1.2. Data Collection

Because of structural and physical changes induced by P and T, analytical methods that can be used in-situ as the sample is being deformed are often preferred. These in-situ analyses rely on synchrotron radiation, either in the form of X-ray diffraction (XRD) or X-ray radiography (Figure 28).

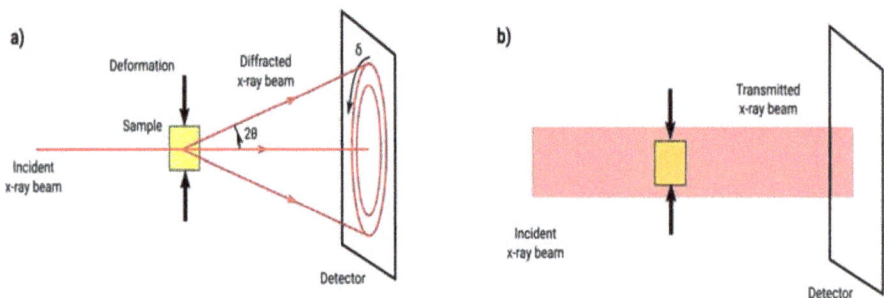

Figure 28. Geometry for in-situ X-ray diffraction and radiography in high pressure deformation experiments. The sample is deformed inside a high-pressure device (not shown) with the direction of deformation perpendicular to the incident X-ray beam. In diffraction mode (**a**), the incident beam is smaller than the sample and diffraction is collected on a detector perpendicular to the incident beam. Features of the diffraction image are analyzed as a function of the azimuth angle δ. In imaging mode (**b**), the incident beam is larger than the sample and its shape and dimensions can be measured on the detector.

Typical experiments tend to investigate average properties inside a polycrystal, including strain and strain rate, average stress, and texture. Others focus on single-crystals. In this case, experiments measure the sample strain, strain rate, and average stress. In the D-DIA, RDA, and DT-Cup, sample strain and strain rates are measured using X-ray radiograph images of the sample [249] (Figure 28). Similar measurements were performed in the DAC in the radial diffraction geometry [250] with limited use, however, since sample deformation is not controlled.

Deviatoric stress and texture are keys to quantify the plasticity of materials such as MgO. The stress itself provides information about strength and can yield flow laws when combined with information on strain rate. In-situ measurements of stress rely on methods derived from XRD residual stress analysis [251]. The basic principle is that the macroscopic sample stress induces a distortion of the Debye diffraction rings, resulting from elastic strains within the sample. The elastic strains vary with orientation relative to the applied stress direction and can be used to invert the macroscopic stress (Figure 29).

When a material deforms plastically, grains rotate under the action of dislocation glide on a given plane. This gives rise to lattice preferred orientations (LPO) or texture. In a textured material, diffraction intensities vary with orientation (Figure 29) and this can be used to invert the sample LPO [252].

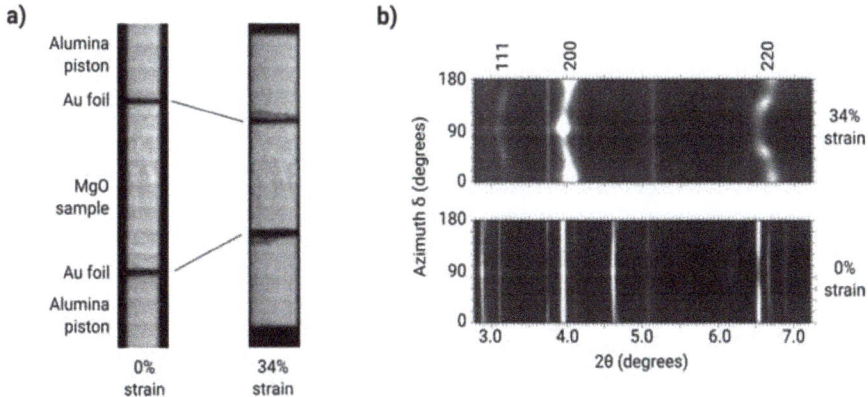

Figure 29. Typical in-situ X-ray data, collected on polycrystalline MgO at 2–5 GPa and 300 K at the ID06 beamline of the ESRF [253]: (**a**) X-ray radiograph for measuring the sample size and macroscopic strain; and (**b**) X-ray diffraction after 0% and 34% axial strain. Diffraction peaks of the MgO sample are labeled. Other peaks are from the sample assembly. Variations of diffraction intensity with azimuth are related to LPO. Variations of peak positions with orientation are a measure of the local state of strain.

5.1.3. Data Analysis and Polycrystalline Simulation Methods

Stress estimation from the diffraction data are often performed using equations derived from linear elasticity [241,252,254]. However, there are limitations, as these equations do not account for plastic deformation. Plastic deformation induces a local relaxation of stress, leading to inconsistent stress interpretations with purely elastic models [255]. Plastic deformation can be modeled, e.g., by comparing the strains measured using XRD to results of elasto-plastic or elasto-visco-plastic self-consistent (EPSC or EVPSC) calculations that account for both elastic and plastic relaxation [253,255,256] (see below). Assuming that the experimental textures arise from plastic deformation, they can also be compared with the same polycrystalline plasticity simulations to obtain information about the deformation modes operating in the sample [252]. Atomistically-informed crystal plasticity finite-element modeling (CPFEM) can also be used to investigate the polycrystalline mechanical response as well as the texture evolution [257].

E(V)PSC calculations rely on effective medium self-consistent methods, which treat each grain of the polycrystal as an inclusion in a homogeneous but anisotropic medium (Figure 30). The properties of the medium are determined by the average of all the inclusions. At each deformation step, the inclusion and medium interact and the macroscopic elasto-plastic properties are updated iteratively until the average strain and stress of all the inclusions equals the macroscopic strain and stress.

CPFEM produces the full-field response of the heterogeneous deformation inside a representative volume element (RVE) of the polycrystalline aggregate [258]. Each grain is discretized into a large number of finite elements and the stress response is computed at each integration point while accounting for the local crystal orientation and constitutive equations. Consequently, each grain deforms heterogeneously and the average stress and strain per grain depend on both the local lattice orientation and the interaction with neighboring grains.

Parameters for both CPFEM and E(V)PSC are the sample starting texture, elasticity, applied deformation, and plastic deformation modes. In their coupled XRD-EVPSC approach, Lin and collaborators [253] optimize the agreement between experiments and simulations adjusting the list of active deformation modes and their physical properties, i.e., their CRSS, including (sometimes) effects of hardening. Comparisons between experimental and simulated textures (using VPSC, for instance) allow constraining the relative CRSS of the deformation modes. Comparison between experimental and simulated microscopic strains (as measured from diffraction) using E(V)PSC allows constraining the absolute values of the CRSS.

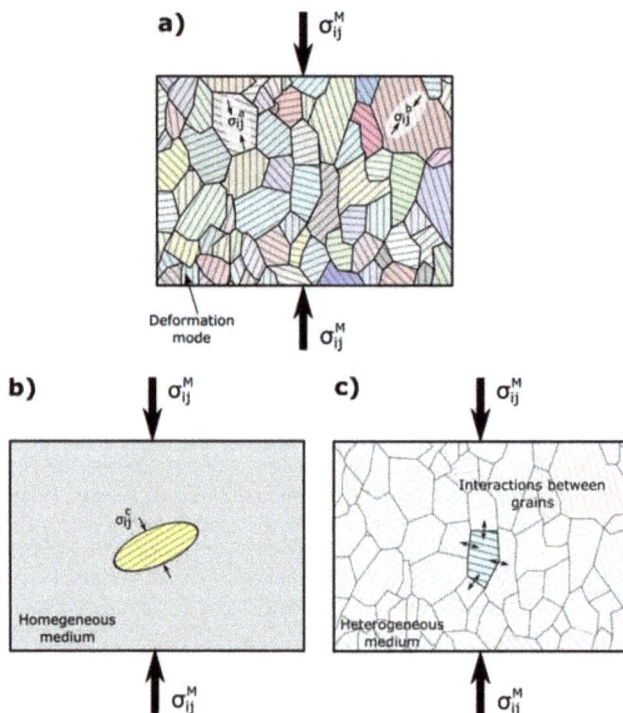

Figure 30. Polycrystalline simulation. (**a**) In a polycrystal, stress and strain distributions are heterogeneous. The behavior of each grain depends on its local environment, elastic, and plastic properties; (**b**) In self-consistent models, each grain is treated as an elliptical inclusion inside a homogeneous medium; (**c**) In CPFEM, while the whole polycrystal deforms according to boundary conditions, each grain deforms due to a full-field approach where all grains are considered.

Finally, TEM is the method of choice for characterizing individual deformation mechanisms such as dislocation types and microstructures for materials that are stable or can be brought back to room pressure [259]. It can be used to analyze the types of deformation modes present in the sample but also their arrangements and interactions [260]. As shown Sections 1 and 4, microscopy in the TEM or in the SEM can also be used to characterize the grain sizes, shapes, and orientation distributions [253].

5.1.4. Effect of Pressure on Rheological Laws

The effect of pressure on materials rheology is typically analyzed using rheological laws such as Equation (16).

$$\dot{\varepsilon} = A\tau^n exp\left(-\frac{Q+PV^*}{RT}\right) \qquad (16)$$

where $\dot{\varepsilon}$ is the strain rate, A is a constant, τ is the differential stress, n is the stress exponent, Q is the activation energy, P is the hydrostatic pressure, V^* is the activation volume, R is the gas constant, and T is the temperature. In experiments, $\dot{\varepsilon}$, P, T, and τ are measured. A, n, Q, and V^* are obtained using fits of Equation (16). In such formulation, the effect of pressure on rheology is captured by the value of the activation volume.

5.2. High-Pressure Plasticity of MgO and Iron Bearing (Mg,Fe)O

First, note that all experimental results presented below were performed at the strain rates attainable in a static deformation experiment, between 10^{-6} and 10^{-4}/s. All quoted results are hence given for these ranges of strain rates. Furthermore, the review is extended here to polycrystalline MgO that is more adapted for high pressure plasticity and $(Mg_x,Fe_{1-x})O$ as the effect of Fe is relevant for deep planetary interiors.

5.2.1. Effect of P and T on MgO and (Mg,Fe)O Deformation Textures

Polycrystalline MgO compressed at ambient temperature up to ~50 GPa shows textures with a concentration at 001 (Figure 31). A comparison between these observations and simulations using VPSC models indicate that such textures are compatible with slip on ½<110>{110} [252,253].

In the study of Amodeo and colleagues [261], the authors investigate the pressure sensitivity of dislocation cores in MgO from ambient up to 100 GPa hydrostatic P using the multi-scale modeling approach described Section 2. Due to the increase of P, the ½<110> screw dislocation (the main character for plasticity in MgO) reconstructs to further spread in the {100} slip planes at pressures larger than 30 GPa. This pressure-induced atomic reconstruction induces a relative hardening of the ½<110>{110} slip mode when compared to ½<110>{100}, i.e., at the grain scale, ½<110>{100} slip becomes easier than ½<110>{110} at high-pressure (see Figures 32 and 33).

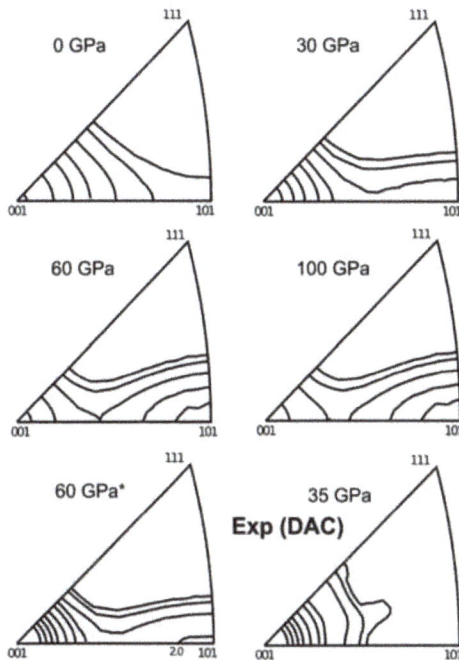

Figure 31. Illustration of polycrystalline MgO compression textures (IPF of the compression axis) at various pressure modeled using the CPFEM approach, starting from a random distribution of crystal orientations. Adapted from Ref. [257]. Black lines refer to grain orientation intensity; hydrostatic pressure is indicated at the top left of each subfigure; and "60 GPa*" refers to a test for which the sample starting texture was focused at 001 rather than random. The Exp (DAC) subfigure is adapted from the experimental work of Merkel and collaborators [252].

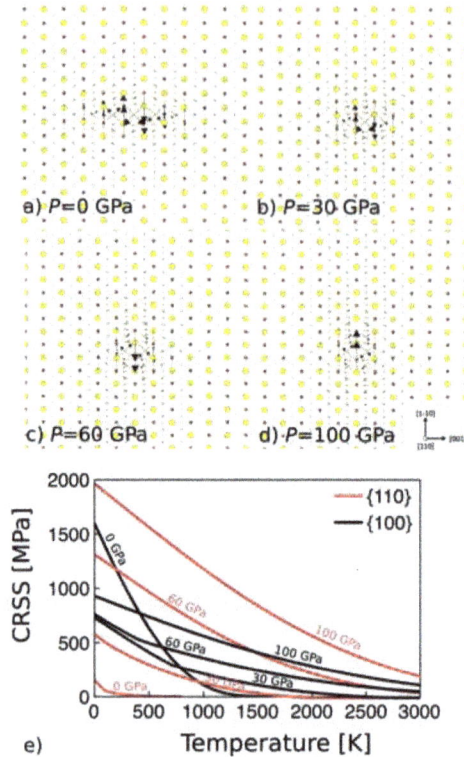

Figure 32. Screw dislocation core (relative displacement mapping) and CRSS evolution under hydrostatic pressure computed using the multiscale modeling approach of Amodeo and collaborators [23,261]. (**a–d**) Screw dislocation core structure from ambient up to 100 GPa pressure computed using the PNG model (see Section 2). Size of arrows refer to the relative displacement amplitude of core atoms. Yellow regions emphasize larger atomic displacements. They show the core spread variations i.e., from {110} spread at low-P to {100} at high-P. (**e**) CRSS for ½<110>{110} and ½<110>{100} slip modes computed for various pressure conditions. Results adapted from [23,261].

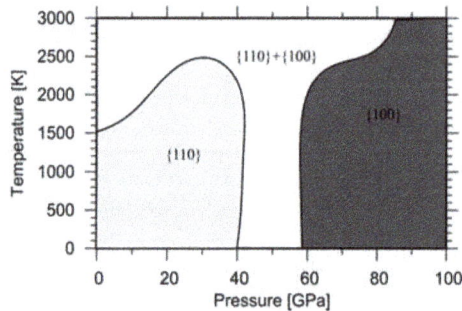

Figure 33. Illustration of the predicted slip system activity in MgO single crystal vs. pressure and temperature based on the multi-scale modeling approach of Amodeo and collaborators. Light and dark grey areas refer to ½<110>{110} and ½<110>{100} as dominant modes, respectively, while the white region draws a PT region where both slip modes are effective. Adapted from [23,261].

Nevertheless, this pressure-induced transition remains difficult to observe experimentally, with confirmations in single crystal experiments [262] and in (Mg,Fe)O polycrystals at high temperature [263] (see below). At 300 K, the reported texture maxima tend to remain at 001 [252,264]. In fact, recent atomistically-informed CPFEM simulations including a constitutive model parameterization based on Figure 32 data [261] show that polycrystalline textures are not as sensitive to P as expected. Indeed, similar to the discussion in Section 4.5, the plastic activity in MgO polycrystals cannot be restricted to a single mode, neither of the suggested slip modes provides enough freedom degrees to accommodate the polycrystal macroscopic deformation. In CPFEM simulations, the transition from dominant {110} to dominant {100} slip induces only slight texture variations from 001 compression texture at low-P to 001 + 011 texture at high-P rather than a strict mode transition (Figure 31). It should also be noted that the strong 001 compression texture induced by slip on {110} early in the compression in DAC experiments could inhibit further texture evolution in experiments, even if slip on {100} becomes dominant at higher pressures as confirmed by the simulation [257].

The addition of Fe into (Mg,Fe)O lowers the strength of the 001 compression textures with the apparition of a secondary maximum at 011 for large Fe contents [265]. This could be associated to a decrease of activity on ½<110>{110}, but also the fact that, as (Mg,Fe)O is more compressible than pure MgO, a larger portion of the strain is accommodated elastically. Between 40 and 80 GPa, Fe in (Mg,Fe)O undergoes electronic spin-pairing transition from a high-spin to a low-spin state [266]. The pressure at which this transition occurs depends on stress state and exact Fe content of the material. It was shown, however, that the 001 compression texture maximum remains in $(Mg_{0.83}Fe_{0.17})O$ compressed through the spin transition up to 81 GPa [267], indicating that the spin transition in Fe has a limited effect on the plasticity of (Mg,Fe)O.

Deformation textures were also investigated at higher temperatures but lower pressures. At 300 MPa and up to 1400 K, $(Mg_{0.80}Fe_{0.20})O$ shows 001 and 011 compression textures [57]. This suggests that temperature plays a relevant role on deformation mechanisms of this material as this type of textures could be induced by a switch of dominant ½<110>{110} to dominant ½<110>{100} slip at higher temperatures. Shear deformation of (Mg,Fe)O also shows a change of textures with increasing Fe content with a clear transition when the Fe content reaches 50% [264,268]. This indicates that indeed, Fe-content does affect the plastic behavior of (Mg,Fe)O although the microscopic interpretation of this effect of Fe on the fundamentals of plasticity in (Mg,Fe)O and dislocation structure remains to be understood.

High-strain shear deformation of $(Mg_{0.80}Fe_{0.20})O$ at 300 MPa and temperatures between 1300 and 1400 K indicate textures compatible with slip on ½<110>{110} at low strain, with a change dominant ½<110>{100} and ½<110>{111} at intermediate strains and, finally, strong recrystallization at the highest strains [269]. Recrystallization aligns {112} with the shear plane and <110> with the shear direction, which was interpreted again as the action of dominant slip on ½<110>{100} and potentially ½<110>{111}.

Finally, the recent study of Immoor and coworkers [270] shows the compression textures of (Mg,Fe)O up to combined pressures and temperatures of 80 GPa and 1400 K. These experiments indicate a shift from a 001 to a combined (001 + 011) compression texture maximum as pressure and temperature increase. The texture transition, which could be assigned to a change from dominant {110} to dominant {100} slip is observed at ~1400 K at low pressure and for pressures above 50 GPa at ~1200 K.

5.2.2. Effect of P, T, and Fe on the Mechanical Properties of Polycrystalline MgO and $(Mg_x,Fe_{1-x})O$

The effect of hydrostatic pressure on MgO plasticity was first investigated up to 1 GPa and 750 °C [271,272]. These studies showed that: (i) under ambient temperature MgO becomes ductile between 100 and 200 MPa; (ii) that the ambient temperature strength beyond yielding increases with confining pressure; and (iii) that the pressure sensitivity of the stress–strain curve becomes less pronounced above 300 °C. In all cases, the plasticity of MgO was shown to be dominated by slip in ½<110>{110}.

The effect of hydrostatic pressure on the polycrystalline rheological law was quantified by experiments up to 10 GPa at temperatures up to 1600 K in the D-DIA [255,263,273–277]. Most of these studies focused on establishing the stress–strain behavior and rheological laws for MgO. Under ambient temperature and a pressure of 8 GPa [274] the measured strength is 4 GPa, with a sharp decrease above 1200 K, and down to a strength of ~200 MPa at 1400 K (Figure 34b).

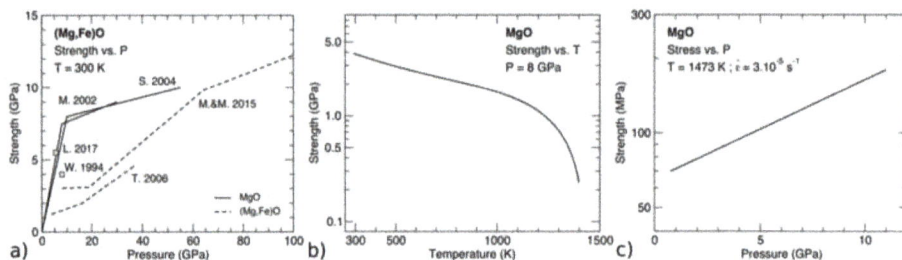

Figure 34. Effect of P, T, and Fe content on the strength of polycrystalline (Mg,Fe)O. (**a**) Ambient T strength of (Mg,Fe)O vs. P and Fe content. Solid lines are the diamond anvil cell experiments for pure MgO of Merkel and colleagues [252] and Singh and collaborators [278]. Squares are the multi-anvil press data for pure MgO of Weidner et al. [263] and Lin and collaborators [253]. Dashed lines are the diamond anvil data for Fe-bearing (Mg,Fe)O of Tommaseo et al. [265] and Marquardt and Miyagi [279]; (**b**) Strength of pure MgO vs. temperature measured at 8 GPa in the multi-anvil press by Weidner and collaborators [263]; (**c**) Flow stress of pure MgO vs. pressure at 1473 K and 3×10^{-5}/s from the multi-anvil press results of Mei and coworkers [273].

More recently, the creep behavior of MgO was investigated up to 10 GPa [273]. Based on their data and stresses evaluated on the 200 diffraction line (see discussion above), the flow strength of MgO at 1473 K and at a strain rate of 3×10^{-5}/s increases from 80 MPa at 2 GPa to 200 MPa at 10 GPa, indicating an activation volume $V^* = 2.4 \pm 0.9 \times 10^{-6}$ m^3/mol (Equation (16), Figure 34c). Other parameters for their rheological law are a stress exponent $n = 3$ and an activation energy $Q = 72 \pm 50$ kJ/mol. Do note, however, that the choice of the 200 diffraction line for stress evaluation is arbitrary and that stresses evaluated using the 111 reflection are about two times higher [273].

Under ambient pressure, pure MgO was studied multiple times up to pressure of 60 GPa [248,252,278,280,281]. From these studies, the strength of polycrystalline MgO was shown to increase from 1 to ca. 8 GPa as the pressure is raised from ambient to about 9 GPa. The increase in strength is much less pronounced at higher pressures, reaching 10 GPa at a pressure of 60 GPa (Figure 34a). Earlier works of [248] lead to lower values of MgO strength but this deduction of stress is probably based on improper assumptions regarding the sample stress state [282].

Interestingly, the addition of Fe into (Mg,Fe)O seems to change this behavior [265,267,279]. Between 40 and 80 GPa, Fe in (Mg,Fe)O undergoes electronic spin-pairing transition from a high-spin to a low-spin state [266]. This transition leads to a reduction of the unit-cell volume which can induce a reduction of the sample stress [267], although it is not always observed [279]. Measurements up to 100 GPa, however, seem to indicate that the strength of (Mg,Fe)O does not follow the same behavior than that of pure MgO (Figure 34a), with a slow increase of strength up 3 GPa at a pressure of 20 GPa, a sharp increase up to 10 GPa at 60 GPa, and a further increase up to 12 GPa at 100 GPa [279].

5.2.3. Experimental Identification of Individual Deformation Mechanisms

Comparisons between deformation textures in polycrystals and simulations such as VPSC are often used to infer dominant deformation mechanisms at the single-crystal scale. There are limitations, however, as: (i) deformation textures are not always sensitive to a change in dominant deformation mechanisms; (ii) the solution from modeling such as VPSC can be non-unique; (iii) textures cannot be used to evaluate the absolute values of the slip system's CRSS; and (iv) polycrystal deformation includes effects of grain-to-grain interactions, microstructures, and hardening and, hence, can differ from that of a single crystal. The identification of dominant deformation mechanisms in MgO at high pressure can be further addressed using two methods: experiments on single-crystals, and more advanced comparison between experimental results on polycrystals and polycrystalline plasticity simulations.

Recent works in the D-DIA [262] focused on the effect of pressure on slip and dislocation in MgO single crystals with the aim of testing the pressure-induced dominant slip system inversion predicted by numerical models [261]. In these experiments, MgO single crystals are deformed in orientations chosen to activate either slip on {110} or {100}, at temperatures between 1273 K and 1473 rac12<110>{110}, the activation volume is $V^* = 1.0 \times 10^{-6}$ m^3/mol. For single crystals oriented to favor slip on ½<110>{100}, the activation volume is close to zero. This implies that, as pressure increases, the strength of slip on ½<110>{100} does not change while ½<110>{110} slip becomes harder. The predicted inversion of dominant slip system in MgO based on these measurements is at ~23 GPa.

Uncertainties remain, however, on the relative plastic behavior of single-crystal and polycrystalline MgO. Grain-grain interactions in polycrystals can fundamentally change their behavior. Moreover, dislocation interactions between the different slip systems may change the behavior relative to perfectly oriented single crystals. To address this issue, recent studies focused on the behavior of polycrystalline MgO at ambient temperature, at strain rates of ~3 × 10^{-5} s^{-1} and pressures up to 5.4 GPa [253]. In this case, the experimental data is fully compared to EVPSC models to match not only the measured textures but also the microscopic strains inferred from individual lattice planes. Such a procedure provides much stronger constrains on the active deformation mechanisms and their physical properties, i.e., CRSS and hardening. This study leads to a steady state CRSS of 1.2 and 3.2 GPa for ½<110>{110} and ½<110>{100}, respectively. However, their results are significantly larger than results of numerical models on single crystals [261] and demonstrate the importance of strain hardening or grain size effects in polycrystals.

As highlighted above, there are still many questions regarding the combined effects of pressure, temperature on the plasticity of MgO. At present, works are under way to extend the measurements into several directions: (i) larger P, T, and composition ranges [245,270]; (ii) finer interpretation of the experimental data with polycrystal plasticity models [253]; and (iii) complex deformation history, such as sinusoidal oscillations [283]. These works will allow, in the future, clarifying the combined effects of P and T on MgO and (Mg,Fe)O plasticity.

6. Conclusions and Prospects

This review paper focused on dislocations and plastic deformation in MgO crystals and here is an overview of the key-points covered:

- Standard compression tests helped to describe dislocation-based plasticity processes in MgO in the 1950s. Plasticity is controlled by screw dislocations gliding in soft ½<110>{110} and hard ½<110>{100} slip systems and MgO is referred as a material with high lattice friction.
- Atomistic simulations based on dislocation core calculation confirmed the key role of screw dislocation for MgO plasticity. Multi-scale modeling approaches allow describing MgO plasticity on a wide range of temperatures and pressures, up to the polycrystal.
- The development of existing and novel experimental techniques since the 1990s, especially with the development of tests at the micro- and nano-scales, allows investigating elementary deformation processes providing a better description of dislocation mechanisms.
- Experiments and modeling show that MgO is sensitive to pressure increases. Dislocation cores and elementary deformation processes evolve under pressure inducing a softening of the ½<110>{100} mode when compared to ½<110>{110}. This change has significant consequences on macroscopic mechanical properties such as polycrystalline textures.

The general understanding of MgO plasticity from both experimental tests at different scales and multi-scale modeling opens new avenues to describe its behavior under stress and, in particular, in the case of high hydrostatic pressure. While this is of particular importance for the prediction of deep-Earth rheology, this may impact our view of ceramic plasticity, since it can occur for particular loading conditions and/or small sample sizes in inorganic materials. This is now no longer restricted to MgO, and plasticity of other inorganic materials (e.g., Alumina) in the form of nanoparticles was shown to be possible. New paradigms in the process of ceramics can then be forecast as it is the case of isostatic compression and shaping of ceramic green bodies at low or moderate temperature or the use of external electric field to accelerate dislocation motion and thus improve plasticity. Playing with the plasticity of inorganic crystals may thus to a certain extent open the door towards "plastic ceramics".

Author Contributions: J.A. designed the project. J.A. and J.C. were in charge of the macro-testing section. P.C. (Philippe Carrez), P.C. (Patrick Cordier) and J.A. designed the multi-scale modeling section. C.T. and S.K. were in charge of the micro- and nano-mechanics section. S.M. designed the high-pressure section with J.A. All authors discussed the study and wrote the article.

Funding: Patrick Cordier and Philippe Carrez are supported by funding from the European Research Council under the Seventh Framework Programme (FP7), ERC Grant No. 290424–RheoMan.

Conflicts of Interest: The authors declare no conflict of interest.

References

1. Johnston, W.G.; Gilman, J.J. Dislocation Velocities, Dislocation Densities, and Plastic Flow in Lithium Fluoride Crystals. *J. Appl. Phys.* **1959**, *30*, 1–16. [CrossRef]
2. Stokes, R.J. *Microstructure and Mechanical Properties of Ceramics*; Accession Number: AD0407614; Defense Technical Information Center: Fort Belvoir, VA, USA, 1963.
3. Haasen, P.; Barthel, C.; Suzuki, T. Choice of slip system and Peierls stresses in the NaCl structure. In *Dislocations in solids*; Suzuki, H., Ninomiya, T., Sumino, K., Takeuchi, S., Eds.; University of Tokyo Press: Tokyo, Japan, 1985; pp. 455–462.
4. Takeuchi, S.; Koizumi, H.; Suzuki, T. Peierls stress and kink pair energy in NaCl type crystals. *Mater. Sci. Eng. A* **2009**, *521*, 90–93. [CrossRef]
5. Hulse, C.; Copley, S.; Pask, J. Effect of crystal orientation on plastic deformation of magnesium oxide. *J. Am. Ceram. Soc.* **1963**, *46*, 317–323. [CrossRef]
6. Groves, G.W.; Kelly, A. Independent slip systems in crystals. *Philos. Mag. A* **1963**, *8*, 877–887. [CrossRef]
7. Day, R.; Stokes, R. Mechanical behavior of Magnesium Oxide at high temperatures. *J. Am. Ceram. Soc.* **1964**, *47*, 493–503. [CrossRef]
8. Copley, S.; Pask, J. Plastic deformation of MgO single crystals up to 1600 C. *J. Am. Ceram. Soc.* **1965**, *48*, 139–146. [CrossRef]
9. Washburn, J.; Groves, G.; Kelly, A.; Williamson, G. Electron microscope observations of deformed magnesium oxide. *Philos. Mag. A* **1960**, *5*, 991–999. [CrossRef]
10. Clauer, A.; Wilcox, B. High temperature tensile creep of magnesium oxide single crystals. *J. Am. Ceram. Soc.* **1976**, *59*, 89–96. [CrossRef]
11. Mariani, E.; Mecklenburgh, J.; Wheeler, J.; Prior, D.J.; Heidelbach, F. Microstructure evolution and recrystallization during creep of MgO single crystals. *Acta Mater.* **2009**, *57*, 1886–1898. [CrossRef]
12. Messerschmidt, U.; Appel, F. Dislocation motion and formation of dislocation structures during in situ deformation in a high voltage electron microscope. *Mater. Sci. Eng. A* **1989**, *113*, 409–414. [CrossRef]
13. Tromas, C.; Girard, J.; Woirgard, J. Study by atomic force microscopy of elementary deformation mechanisms involved in low load indentations in MgO crystals. *Phil. Mag. A* **2000**, *80*, 2325–2335. [CrossRef]
14. Hulse, C.; Pask, J. Mechanical Properties of Magnesia Single Crystals Compression. *J. Am. Ceram. Soc.* **1960**, *43*, 373–378. [CrossRef]
15. Appel, F.; Wielke, B. Low temperature deformation of impure MgO single crystals. *Mater. Sci. Eng.* **1985**, *73*, 97–103. [CrossRef]
16. Singh, R.; Coble, R. Dynamic dislocation behavior in "pure" magnesium oxide single crystals. *J. Appl. Phys.* **1974**, *45*, 981–989. [CrossRef]

17. Woo, C.H.; Puls, M.P. The Peierls mechanism in MgO. *Phil. Mag. A J. Theor. Expe. Appl. Phys.* **1977**, *35*, 1641–1652. [CrossRef]

18. May, J.; Kronberg, M. Temperature Dependence of Plastic Yield Stress of Single Crystals of Magnesium Oxide. *J. Am. Ceram. Soc.* **1960**, *43*, 525–530. [CrossRef]

19. Sinha, M.; Lloyd, D.; Tangri, K. Dislocation dynamics and thermally-activated deformation of MgO single crystals. *Philos. Mag. A* **1973**, *28*, 1341–1352. [CrossRef]

20. Srinivasan, M.; Stoebe, T. Temperature dependence of yielding and work-hardening rates in magnesium oxide single crystals. *J. Mat. Sci.* **1974**, *9*, 121–128. [CrossRef]

21. Sato, F.; Sumino, K. The yield strength and dynamic behaviour of dislocations in MgO crystals at high temperatures. *J. Mat. Sci.* **1980**, *15*, 1625–1634. [CrossRef]

22. Barthel, C. Plastiche Anisotropie von Bleisulfid und Magnesiumoxid. Diploma Thesis, University of Gottingen, Göttingen, Germany, 1984.

23. Amodeo, J.; Carrez, P.; Devincre, B.; Cordier, P. Multiscale modelling of MgO plasticity. *Acta Mater.* **2011**, *59*, 2291–2301. [CrossRef]

24. Appel, F.; Bethge, H.; Messerschmidt, U. Dislocation motion and multiplication at the deformation of MgO single crystals in the high voltage electron microscope. *Phys. Status Solidi A* **1977**, *42*, 61–71. [CrossRef]

25. Foitzik, A.; Skrotzki, W.; Haasen, P. Correlation between microstructure, dislocation dissociation and plastic anisotropy in ionic crystals. *Mater. Sci. Eng. A* **1989**, *113*, 399–407. [CrossRef]

26. Messerschmidt, U. *Dislocation Dynamics during Plastic Deformation*; Springer Series in Materials Science: New York, NY, USA, 2010; Volume 129, pp. 1–509.

27. Singh, R.; Coble, R. Dynamic dislocation behavior in iron-doped magnesium oxide crystals. *J. Appl. Phys.* **1974**, *45*, 990–995. [CrossRef]

28. Dorn, J.; Rajnak, S. Nucleation of kink pairs and the Peierls mechanism of plastic deformation. *Trans. Metall. Soc. AIME* **1964**, *230*, 1052–1064.

29. Guyot, P.; Dorn, J. A critical review of the Peierls mechanism. *Can. J. Phys.* **1967**, *45*, 983–1016. [CrossRef]

30. Boudet, A.; Kubin, L.P. Exhaustion mechanisms in the Preyield domain of niobium single crystals at low temperatures. *J. Phys.* **1975**, *36*, 823–833. [CrossRef]

31. Groves, G.; Fine, M. Solid Solution and Precipitation Hardening in Mg-Fe-O Alloys. *J. Appl. Phys.* **1964**, *35*, 3587–3593. [CrossRef]

32. Srinivasan, M.; Stoebe, T. Effect of Impurities on the Mechanical Behavior of MgO Single Crystals. *J. Appl. Phys.* **1970**, *41*, 3726–3730. [CrossRef]

33. Gorum, A.; Luhman, W.; Pask, J. Effect of Impurities and Heat-Treatment on Ductility of MgO. *J. Am. Ceram. Soc.* **1960**, *43*, 241–245. [CrossRef]

34. Davidge, R. The distribution of iron impurity in single-crystal magnesium oxide and some effects on mechanical properties. *J. Mater. Sci.* **1967**, *2*, 339–346. [CrossRef]

35. Messerschmidt, U.; Appel, F. Characterization of precipitation hardening mechanisms by investigating dislocation dynamics in the HVEM. *Czechoslov. J. Phys.* **1985**, *35*, 217–220. [CrossRef]

36. Appel, F.; Messerschmidt, U. Dislocation Processes in MgO Single Crystals Observed by In-situ Deformation in the HVEM. *Krist. Tech.* **1979**, *14*, 1329. [CrossRef]

37. Messerschmidt, U. Line tension model of the interaction between dislocations and extended obstacles to glide. *Mater. Sci. Eng. A* **1988**, *100*, 101–108. [CrossRef]

38. Messerschmidt, U.; Appel, F.; Schmid, H. The radius of curvature of dislocation segments in MgO crystals stressed in the high-voltage electron microscope. *Philos. Mag.* **1985**, *51*, 781–796. [CrossRef]

39. Kardashev, B.K.; Kustov, S.B.; Lebedev, A.B.; Berezhkova, G.V.; Perstnev, P.P.; Appel, F.; Messerschmidt, U. Acoustic and Electron Microscopy Study of the Dislocation Structure in MgO Crystals. *Phys. Status Solidi A* **1985**, *91*, 79–87. [CrossRef]

40. Appel, F.; Bartsch, M.; Messerschmidt, U.; Nadgornyi, E.; Valkovskii, S. Dislocation motion and plasticity in MgO single crystals. *Phys. Status Solidi A* **1984**, *83*, 179–194. [CrossRef]

41. Appel, F.; Berezhkova, G.V.; Messerschmidt, U.; Perstnev, P.P.; Rozhanskii, V.N. HVEM In-Situ Straining Experiments on MgO Specimens Predeformed at High Temperature. *Cryst. Res. Technol.* **1981**, *16*, 1309–1314. [CrossRef]

42. Appel, F.; Messerschmidt, U.; Nadgornyi, E.M.; Zaitsev, S.I. The interaction between dislocations and point obstacles: A comparison of the interaction parameter distributions obtained from computer simulation and from In situ high voltage electron microscopy straining experiments. *Mater. Sci. Eng.* **1982**, *52*, 69–74. [CrossRef]

43. Appel, F.; Bethge, H.; Messerschmidt, U. Distribution of point-obstacle distance during the motion of screw dislocations in MgO single crystals. *Phys. Status Solidi A* **1976**, *38*, 103–108. [CrossRef]

44. Routbort, J.L. Work-Hardening and Creep of MgO. *Acta Metall.* **1979**, *27*, 649–661. [CrossRef]

45. Amodeo, J.; Devincre, B.; Carrez, P.; Cordier, P. Dislocation reactions, plastic anisotropy and forest strengthening in MgO at high temperature. *Mech. Mater.* **2014**, *71*, 62–73. [CrossRef]

46. Kear, B.; Taylor, A.; Pratt, P. Some dislocation interactions in simple ionic crystals. *Philos. Mag. A* **1959**, *4*, 665–680. [CrossRef]

47. Ruano, O.A.; Wolfenstine, J.; Wadsworth, J.; Sherby, O.D. Harper-Dorn and power law creep in uranium dioxide. *Acta Metallurgica et Materialia* **1991**, *39*, 661–668. [CrossRef]

48. Cummerow, R. High-Temperature Steady-State Creep Rate in Single-Crystal MgO. *J. Appl. Phys.* **1963**, *34*, 1724–1729. [CrossRef]

49. Rothwell, W.S.; Neiman, A.S. Creep in Vacuum of MgO Single Crystals and the Electric Field Effect. *J. Appl. Phys.* **1965**, *36*, 2309–2316. [CrossRef]

50. Hüther, W.; Reppich, B. Dislocation structure during creep of MgO single crystals. *Philos. Mag. A* **1973**, *28*, 363–371. [CrossRef]

51. Ramesh, K.S.; Yasuda, E.; Kimura, S.; Urabe, K. High-temperature creep and dislocation structure of MgO single crystals at low stresses. *J. Mater. Sci.* **1986**, *21*, 4015–4018. [CrossRef]

52. Ramesh, K.S.; Yasuda, E.; Kimura, S. Negative creep and recovery during high-temperature creep of MgO single crystals at low stresses. *J. Mater. Sci.* **1986**, *21*, 3147–3152. [CrossRef]

53. Hensler, J.H.; Cullen, G.V. Stress, Temperature, and Strain Rate in Creep of Magnesium Oxide. *J. Am. Ceram. Soc.* **1968**, *51*, 557–559. [CrossRef]

54. Langdon, T.G.; Pask, J.A. The mechanism of creep in polycrystalline magnesium oxide. *Acta Metall.* **1970**, *18*, 505–510. [CrossRef]

55. Bilde-Sörensen, J.B. Dislocation Structures in Creep-Deformed Polycrystalline MgO. *J. Am. Ceram. Soc.* **1972**, *55*, 606–610. [CrossRef]

56. Wolfenstine, J.; Kohlstedt, D.L. Creep of (Mg, Fe)O single crystals. *J. Mater. Sci.* **1988**, *23*, 3550–3557. [CrossRef]

57. Stretton, I.; Heidelbach, F.; Mackwell, S.; Langenhorst, F. Dislocation creep of magnesiowüstite ($Mg_{0.8}Fe_{0.2}$). *Earth Planet. Sci. Lett.* **2001**, *194*, 229–240. [CrossRef]

58. Weertman, J. Steady-state creep through dislocation climb. *J. Appl. Phys.* **1957**, *28*, 362–364. [CrossRef]

59. Poirier, J.P. On the symmetrical role of cross-slip of screw dislocations and climb of edge dislocations as recovery processes controlling high-temperature creep. *Revue de Physique Appliquée* **1976**, *11*, 731–738. [CrossRef]

60. Lewis, G.V.; Catlow, C.R.A. Potential models for ionic oxides. *J. Phys. C Solid State Phys.* **1985**, *18*, 1149–1161. [CrossRef]

61. Sangster, M.J.L. Interionic potentials and force constant models for rocksalt structure crystals—II. *J. Phys. Chem. Solids* **1974**, *35*, 195–200. [CrossRef]

62. Woodward, C. First-principles simulations of dislocation cores. *Mat. Sci. Eng. A* **2005**, *400–401*, 59–67. [CrossRef]

63. Woo, C.H.; Puls, M.P. An improved method of calculating the lattice friction stress using an atomistic model. *J. Phys. C Solid State Phys.* **1976**, *9*, L27. [CrossRef]

64. Puls, M.; Norgett, M. Atomistic calculation of the core structure and Peierls energy of an (a/2)[110] edge dislocation in MgO. *J. Appl. Phys.* **1976**, *47*, 466–477. [CrossRef]

65. Woo, C.H.; Puls, M.P. Atomistic breathing shell model calculations of dislocation core configurations in ionic crystals. *Philos. Mag. A* **1977**, *35*, 727–756. [CrossRef]

66. Watson, G.W.; Kelsey, E.T.; de Leeuw, N.H.; Harris, D.J.; Parker, S.C. Atomistic simulation of dislocations, surfaces and interfaces in MgO. *Faraday Trans.* **1996**, *92*, 433–438. [CrossRef]

67. Lehto, N.; Öberg, S. Effects of Dislocation Interactions: Application to the Period-Doubled Core of the 90° Partial in Silicon. *Phys. Rev. Lett.* **1998**, *80*, 5568–5571. [CrossRef]

68. Carrez, P.; Godet, J.; Cordier, P. Atomistic simulations of ½<110> screw dislocation core in magnesium oxide. *Comput. Mater. Sci.* **2015**, *103*, 250–255. [CrossRef]
69. Christian, J.W.; Vítek, V. Dislocations and stacking faults. *Rep. Prog. Phys.* **1970**, *33*, 307–411. [CrossRef]
70. Peierls, R. The size of a dislocation. *Proc. Phys. Soc.* **1940**, *52*, 34–37. [CrossRef]
71. Nabarro, F. Dislocations in a simple cubic lattice. *Proc. Phys. Soc.* **1947**, *59*, 256–272. [CrossRef]
72. Vítek, V. Intrinsic stacking faults in body-centred cubic crystals. *Philos. Mag. A* **1968**, *18*, 773–786. [CrossRef]
73. Miranda, C.; Scandolo, S. Computational materials science meets geophysics: Dislocations and slip planes of MgO. *Comput. Phys. Commun.* **2005**, *169*, 24–27. [CrossRef]
74. Carrez, P.; Ferré, D.; Cordier, P. Peierls-Nabarro modelling of dislocations in MgO from ambient pressure to 100 GPa. *Model. Sim. Mater. Sci. Eng.* **2009**, *17*, 035010. [CrossRef]
75. Schoeck, G. The Peierls model: Progress and limitations. *Mater. Sci. Eng. A* **2005**, *400*, 7–17. [CrossRef]
76. Denoual, C. Dynamic dislocation modeling by combining Peierls Nabarro and Galerkin methods. *Phys. Rev. B* **2004**, *70*, 024106. [CrossRef]
77. Lu, G. The Peierls–Nabarro model of dislocations: a venerable theory and its current development. In *Handbook of Materials Modeling*; Springer: Dordrecht, The Netherlands, 2005; pp. 773–792.
78. Joos, B.; Ren, Q.; Duesbery, M.S. Peierls-Nabarro model of dislocations in silicon with generalized stacking-fault restoring forces. *Phys. Rev. B* **1994**, *50*, 5890–5898. [CrossRef]
79. Hirth, J.P.; Lothe, J. *Theory of Dislocations*; John Willey and Sons: Hoboken, NJ, USA, 1982.
80. Caillard, D.; Martin, J.L. *Thermally Activated Mechanisms in Crystal Plasticity*; Elsevier: New York, NY, USA, 2003.
81. Kamimura, Y.; Edagawa, K.; Takeuchi, S. Experimental evaluation of the Peierls stresses in a variety of crystals and their relation to the crystal structure. *Acta Mater.* **2013**, *61*, 294–309. [CrossRef]
82. Vitek, V.; Yamaguchi, M. Core structure of nonscrew 1/2 (111) dislocations on (110) planes in bcc crystals. II. Peierls stress and the effect of an external shear stress on the cores. *J. Phys. F Metal Phys.* **1973**, *3*, 537. [CrossRef]
83. Duesbery, M.; Vitek, V. Plastic anisotropy in bcc transition metals. *Acta Mater.* **1998**, *46*, 1481–1492. [CrossRef]
84. Xu, W.; Moriarty, J.A. Accurate atomistic simulations of the Peierls barrier and kink-pair formation energy for ⟨111⟩ screw dislocations in bcc Mo. *Comput. Mater. Sci.* **1998**, *9*, 348–356. [CrossRef]
85. Woodward, C.; Rao, S.I. Flexible Ab Initio Boundary Conditions: Simulating Isolated Dislocations in bcc Mo and Ta. *Phys. Rev. Lett.* **2002**, *88*, 216402. [CrossRef] [PubMed]
86. Wang, G.; Strachan, A.; Çağın, T.; Goddard, W.A. Atomistic simulations of kinks in 1/2a⟨111⟩screw dislocations in bcc tantalum. *Phys. Rev. B Condens. Matter* **2003**, *68*, 558. [CrossRef]
87. Gröger, R.; Bailey, A.; Vitek, V. Multiscale modeling of plastic deformation of molybdenum and tungsten: I. Atomistic studies of the core structure and glide of 1/2<111> screw dislocations at 0 K. *Acta Mater.* **2008**, *56*, 5401–5411. [CrossRef]
88. Rodney, D.; Proville, L. Stress-dependent Peierls potential: Influence on kink-pair activation. *Phys. Rev. B* **2009**, *79*, 1–9. [CrossRef]
89. Proville, L.; Ventelon, L.; Rodney, D. Prediction of the kink-pair formation enthalpy on screw dislocations in α-iron by a line tension model parametrized on empirical potentials and first-principles calculations. *Phys. Rev. B* **2013**, *87*, 244–248. [CrossRef]
90. Edagawa, K.; Koizumi, H.; Kamimura, Y.; Suzuki, T. Temperature dependence of the flow stress of III–V compounds. *Philos. Mag. A* **2000**, *80*, 2591–2608. [CrossRef]
91. Pizzagalli, L.; Pedersen, A.; Arnaldsson, A.; Jonsson, H.; Beauchamp, P. Theoretical study of kinks on screw dislocation in silicon. *Phys. Rev. B* **2008**, *77*, 064106. [CrossRef]
92. Pedersen, A.; Pizzagalli, L.; Jónsson, H. Finding mechanism of transitions in complex systems: Formation and migration of dislocation kinks in a silicon crystal. *J. Phys. Condens. Matter* **2009**, *21*, 084210. [CrossRef] [PubMed]
93. Mitchell, T.E.; Anderson, P.M.; Baskes, M.I.; Chen, S.P.; Hoagland, R.G.; Misra, A. Nucleation of kink pairs on partial dislocations: A new model for solution hardening and softening. *Philos. Mag. A* **2003**, *83*, 1329–1346. [CrossRef]
94. Carrez, P.; Ferré, D.; Denoual, C.; Cordier, P. Modelling thermal activation of <110>{110} slip at low temperature in SrTiO3. *Scr. Mater.* **2010**, *63*, 434–437. [CrossRef]
95. Castillo-Rodríguez, M.; Sigle, W. The kink-pair mechanism and low-temperature flow-stress behaviour of strontium titanate single crystals. *Scr. Mater.* **2011**, *64*, 241–244. [CrossRef]

96. Celli, V.; Kabler, M.; Ninomiya, T.; Thomson, R. Theory of Dislocation Mobility in Semiconductors. *Phys. Rev.* **1963**, *131*, 58–72. [CrossRef]

97. Seeger, A.; Schiller, P. Bildung und diffusion von kinken als grundprozess der versetzungsbewegung bei der messung der inneren reibung. *Acta Metall.* **1962**, *10*, 348–357. [CrossRef]

98. Koizumi, H.; Kirchner, H.; Suzuki, T. Kink pair nucleation and critical shear stress. *Acta Metallurgica et Materialia* **1993**, *41*, 3483–3493. [CrossRef]

99. Koizumi, H.; Kirchner, H.O.K.; Suzuki, T. Nucleation of trapezoidal kink pairs on a Peierls potential. *Philos. Mag. A* **1994**, *69*, 805–820. [CrossRef]

100. Southgate, P.D.; Mendelson, K.S. Kilocycle-Range Dislocation Damping in Magnesium Oxide. *J. Appl. Phys.* **1966**, *37*, 206–215. [CrossRef]

101. Joos, B.; Zhou, J. The Peierls-Nabarro model and the mobility of the dislocation line. *Philos. Mag. A* **2001**, *81*, 1329–1340. [CrossRef]

102. Ventelon, L.; Willaime, F.; Leyronnas, P. Atomistic simulation of single kinks of screw dislocations in α-Fe. *J. Nucl. Mater.* **2009**, *386–388*, 26–29. [CrossRef]

103. Proville, L.; Rodney, D.; Marinica, M.-C. Quantum effect on thermally activated glide of dislocations. *Nat. Mater.* **2012**, *11*, 845–849. [CrossRef] [PubMed]

104. Fitzgerald, S.P. Kink pair production and dislocation motion. *Sci. Rep.* **2016**, *6*, 1–7. [CrossRef] [PubMed]

105. Nabarro, F. One-dimensional models of thermal activation under shear stress. *Philos. Mag. A* **2003**, *83*, 3047–3054. [CrossRef]

106. Devincre, B.; Madec, R.; Monnet, G.; Queyreau, S.; Gatti, R.; Kubin, L. Modeling crystal plasticity with dislocation dynamics simulations: The "microMegas" code. In *Mechanics of Nano-Objects*; Presse des Mines: Paris, France, 2011; pp. 81–100.

107. Arsenlis, A.; Cai, W.; Tang, M.; Rhee, M.; Oppelstrup, T.; Hommes, G.; Pierce, T.; Bulatov, V. Enabling strain hardening simulations with dislocation dynamics. *Model. Sim. Mater. Sci. Eng.* **2007**, *15*, 553–595. [CrossRef]

108. Chang, H.-J.; Fivel, M.; Rodney, D.; Verdier, M. Multiscale modelling of indentation in FCC metals: From atomic to continuum. *Comptes Rendus Physique* **2010**, *11*, 285–292. [CrossRef]

109. Queyreau, S.; Monnet, G.; Devincre, B. Orowan strengthening and forest hardening superposition examined by dislocation dynamics simulations. *Acta Mater.* **2010**, *58*, 5586–5595. [CrossRef]

110. Mohles, V. Simulations of dislocation glide in overaged precipitation-hardened crystals. *Philos. Mag. A* **2001**, *81*, 971–990. [CrossRef]

111. Monnet, G. Investigation of precipitation hardening by dislocation dynamics simulations. *Philos. Mag. A* **2006**, *86*, 5927–5941. [CrossRef]

112. Shin, C.; Fivel, M.; Verdier, M.; Robertson, C. Dislocation dynamics simulations of fatigue of precipitation-hardened materials. *Mater. Sci. Eng. A* **2005**, *400*, 166–169. [CrossRef]

113. Chang, H.-J.; Segurado, J.; Molina-Aldareguía, J.M.; Soler, R.; LLorca, J. A 3D dislocation dynamics analysis of the size effect on the strength of [1 1 1] LiF micropillars at 300K and 600K. *Model. Sim. Mater. Sci. Eng.* **2016**, *24*, 035009. [CrossRef]

114. Cui, Y.; Po, G.; Ghoniem, N. Controlling Strain Bursts and Avalanches at the Nano- to Micrometer Scale. *Phys. Rev. Lett.* **2016**, *117*, 155502. [CrossRef] [PubMed]

115. El-Awady, J.; Woodward, C.; Dimiduk, D.; Ghoniem, N. Effects of focused ion beam induced damage on the plasticity of micropillars. *Phys. Rev. B* **2009**, *80*, 104104. [CrossRef]

116. Amodeo, R.J.; Ghoniem, N.M. Dislocation Dynamics. 1. A Proposed Methodology for Deformation Micromechanics. *Phys. Rev. B Condens. Matter* **1990**, *41*, 6958–6967. [CrossRef] [PubMed]

117. Amodeo, R.J.; Ghoniem, N.M. Dislocation Dynamics. 2. Applications to the Formation of Persistent Slip Bands, Planar Arrays and Dislocation Cells. *Phys. Rev. B Condens. Matter* **1990**, *41*, 6968–6976. [CrossRef] [PubMed]

118. Van der Giessen, E.; Needleman, A. Discrete dislocation plasticity: a simple planar model. *Model. Sim. Mater. Sci. Eng.* **1995**, *3*, 689–735. [CrossRef]

119. Keralavarma, S.M.; Curtin, W.A. Strain hardening in 2D discrete dislocation dynamics simulations—A new "2.5D" algorithm. *J. Mech. Phys. Solids* **2016**, *95*, 132–146. [CrossRef]

120. Wickham, L.; Schwarz, K.; Stölken, J. Rules for forest interactions between dislocations. *Phys. Rev. Lett.* **1999**, *83*, 4574–4577. [CrossRef]

121. Madec, R.; Kubin, L.P. Dislocations interactions and symmetries in BCC crystals. In *IUTAM Symposium on Mesoscopic Dynamics of Fracture Process and Materials Strength*; Springer: Dordrecht, The Netherlands, 2004; pp. 69–79.
122. Madec, R.; Devincre, B.; Kubin, L.P. On the nature of attractive dislocation crossed states. *Comput. Mater. Sci.* **2002**, *23*, 219–224. [CrossRef]
123. Monnet, G.; Devincre, B.; Kubin, L.P. Dislocation study of prismatic slip systems and their interactions in hexagonal close packed metals: application to zirconium. *Acta Mater.* **2004**, *52*, 4317–4328. [CrossRef]
124. Capolungo, L. Dislocation junction formation and strength in magnesium. *Acta Mater.* **2011**, *59*, 2909–2917. [CrossRef]
125. Devincre, B. Dislocation dynamics simulations of slip systems interactions and forest strengthening in ice single crystal. *Philos. Mag. A* **2012**, *93*, 235–246. [CrossRef]
126. Durinck, J.; Devincre, B.; Kubin, L.P.; Cordier, P. Modeling the plastic deformation of olivine by dislocation dynamics simulations. *Am. Mineral.* **2007**, *92*, 1346–1357. [CrossRef]
127. Carrez, P.; Cordier, P.; Devincre, B.; Kubin, L.P. Dislocation reactions and junctions in MgO. *Mater. Sci. Eng. A* **2005**, *400*, 325–328. [CrossRef]
128. Devincre, B.; Hoc, T.; Kubin, L.P. Collinear interactions of dislocations and slip systems. *Mater. Sci. Eng. A* **2005**, *400*, 182–185. [CrossRef]
129. Madec, R.; Devincre, B.; Kubin, L.P.; Hoc, T.; Rodney, D. The role of collinear interaction in dislocation-induced hardening. *Science* **2003**, *301*, 1879–1882. [CrossRef] [PubMed]
130. Cordier, P.; Amodeo, J.; Carrez, P. Modelling the rheology of MgO under Earth's mantle pressure, temperature and strain rates. *Nature* **2012**, *481*, 177–180. [CrossRef] [PubMed]
131. Boioli, F.; Carrez, P.; Cordier, P.; Devincre, B.; Marquille, M. Modeling the creep properties of olivine by 2.5-dimensional dislocation dynamics simulations. *Phys. Rev. B* **2015**, *92*, 014115-1–014115-12. [CrossRef]
132. Reali, R.; Boioli, F.; Gouriet, K.; Carrez, P.; Devincre, B.; Cordier, P. Modeling plasticity of MgO by 2.5D dislocation dynamics simulations. *Mat. Sci. Eng. A* **2017**, *690*, 52–61. [CrossRef]
133. Soler, R.; Wheeler, J.M.; Chang, H.-J.; Segurado, J.; Michler, J.; LLorca, J.; Molina-Aldareguía, J.M. Understanding size effects on the strength of single crystals through high-temperature micropillar compression. *Acta Mater.* **2014**, *81*, 50–57. [CrossRef]
134. Keralavarma, S.M.; Benzerga, A.A. High-temperature discrete dislocation plasticity. *J. Mech. Phys. Solids* **2015**, *82*, 1–22. [CrossRef]
135. Yoo, H. Oxygen self-diffusion in single-crystal MgO: Secondary-ion mass spectrometric analysis with comparison of results from gas–solid and solid–solid exchange. *Solid State Ion.* **2002**, *150*, 207–221. [CrossRef]
136. Fischer-Cripps, A.C. *Nanoindentation*; Mechanical Engineering Series; Springer: New York, NY, USA, 2004.
137. Oliver, W.C.; Pharr, G.M. An improved technique for determining hardness and elastic modulus using load and displacement sensing indentation experiments. *J. Mater. Res.* **1992**, *7*, 1564–1583. [CrossRef]
138. Tromas, C.; Stinville, J.C.; Templier, C.; Villechaise, P. Hardness and elastic modulus gradients in plasma nitrided 316l polycrystalline stainless steel investigated by nanoindentation tomography. *Acta Mater.* **2012**, *60*, 1965–1973. [CrossRef]
139. Woirgard, J.; Tromas, C.; Girard, J.C.; Audurier, V. Study of the mechanical properties of ceramic materials by the nanoindentation technique. *J. Eur. Ceram. Soc.* **1998**, *18*, 2297–2305. [CrossRef]
140. Gilman, J.J.; Johnston, W.G. Dislocations in lithium fluoride crystals. *Solid State Phys.* **1961**, *13*, 147–222.
141. Stokes, R.J.; Johnston, T.L.; Li, C.H. Effect of surface condition on the initiation of plastic flow in MgO. *Trans. AIME* **1959**, *215*, 437.
142. Keh, A.S. Dislocations in indented magnesium oxide crystals. *J. Appl. Phys.* **1960**, *31*, 1538–1545. [CrossRef]
143. Newey, C.; Davidge, R.W. *Dislocations in Lithium Fluoride*; Metallurgical Services: Betchworth, UK, 1965.
144. Velednitskaya, M.A.; Rozhanskii, V.N.; Comolova, L.F.; Saparin, G.V.; Schreiber, J.; Brümmer, O. Investigation of the deformation mechanism of MgO crystals affected by concentrated load. *Phys. Status Solidi A* **1975**, *32*, 123–132. [CrossRef]
145. Armstrong, R.W.; Wu, C.C. Lattice Misorientation and Displaced Volume for Microhardness Indentations in MgO Crystals. *J. Am. Ceram. Soc.* **1978**, *61*, 102–106. [CrossRef]
146. Chaudhri, M.M. The displacement of material and the formation of {110}90, cracks around spherical indentations in MgO crystals. *Philos. Mag. A* **1986**, *53*, L55–L63. [CrossRef]

147. Keh, A.S. Cracks due to the piling-up of dislocations on two intersecting slip planes in MgO crystals. *Acta Metall.* **1959**, *7*, 694–696. [CrossRef]

148. Hammond, B.L.; Armstrong, R.W. Recovered elastic and plastic strains at residual micro-indentations in an MgO crystal. *Philos. Mag. Lett.* **1988**, *57*, 41–47. [CrossRef]

149. Gaillard, Y. Initiation de la Plasticité sous Nanoindentation dans MgO et LiF: Etude de L'organisation des Dislocations et des Contraintes Associées. PhD Thesis, Université de Poitiers, Poitiers, France, 2004.

150. Tromas, C.; Girard, J.; Audurier, V.; Woirgard, J. Study of the low stress plasticity in single-crystal MgO by nanoindentation and atomic force microscopy. *J. Mater. Sci.* **1999**, *34*, 5337–5342. [CrossRef]

151. Egberts, P.; Bennewitz, R. Atomic-scale nanoindentation: detection and identification of single glide events in three dimensions by force microscopy. *Nanotechnology* **2011**, *22*, 425703. [CrossRef] [PubMed]

152. Filleter, T.; Maier, S.; Bennewitz, R. Atomic-scale yield and dislocation nucleation in KBr. *Phys. Rev. B Condens. Matter Mater. Phys.* **2006**, *73*, 155433. [CrossRef]

153. Tromas, C.; Gaillard, Y.; Woirgard, J. Nucleation of dislocations during nanoindentation in MgO. *Philos. Mag. A* **2006**, *86*, 5595–5606. [CrossRef]

154. Gaillard, Y.; Tromas, C.; Woirgard, J. Study of the dislocation structure involved in a nanoindentation test by atomic force microscopy and controlled chemical etching. *Acta Mater.* **2003**, *51*, 1059–1065. [CrossRef]

155. Gilman, J.J.; Johnston, W.G.; Sears, G.W. Dislocation etch pit formation in lithium fluoride. *J. Appl. Phys.* **1958**, *29*, 747–754. [CrossRef]

156. Shaw, M.; Brookes, C. Dislocations produced in magnesium oxide crystals due to contact pressures developed by softer cones. *J. Mater. Sci.* **1989**, *24*, 2727–2734. [CrossRef]

157. Nix, W.D.; Gao, H. Indentation size effects in crystalline materials: A law for strain gradient plasticity. *J. Mech. Phys. Solids* **1998**, *46*, 411–425. [CrossRef]

158. Feng, G.; Nix, W.D. Indentation size effect in MgO. *Scr. Mater.* **2004**, *51*, 599–603. [CrossRef]

159. Huang, Y.; Zhang, F.; Hwang, K.C.; Nix, W.D.; Pharr, G.M.; Feng, G. A model for size effect in nano-indentation. *J. Mech. Phys. Sol.* **2006**, *54*, 1668–1986.

160. Sadrabadi, P.; Durst, K.; Göken, M. Study on the indentation size effect in CaF2: Dislocation structure and hardness. *Acta Mater.* **2009**, *57*, 1281–1289. [CrossRef]

161. Ren, X.J.; Hooper, R.M.; Griffiths, C.; Henshall, J.L. Indentation-size effects in single-crystal MgO. *Philos. Mag. A* **2002**, *82*, 2113–2120. [CrossRef]

162. Richter, A.; Wolf, B.; Nowicki, M.; Usov, I.O.; Valdez, J.A.; Sickafus, K. Multi-cycling nanoindentation in MgO single crystal before and after ion irradiation. *J. Appl. Phys. D Appl. Phys.* **2006**, *39*, 3342–3349. [CrossRef]

163. Bahr, D.F.; Kramer, D.E.; Gerberich, W.W. Non-linear deformation mechanisms during nanoindentation. *Acta Mater.* **1998**, *46*, 3605–3617. [CrossRef]

164. Chiu, Y.L.; Ngan, A.H.W. Time-dependent characteristics of incipient plasticity in nanoindentation of a Ni$_3$Al single crystal. *Acta Mater.* **2002**, *50*, 1599–1611. [CrossRef]

165. Gerberich, W.W.; Nelson, J.C.; Lilleodden, E.T.; Anderson, P.; Wyrobek, J.T. Indentation induced dislocation nucleation: The initial yield point. *Acta Mater.* **1996**, *44*, 3585–3598. [CrossRef]

166. Page, T.F.; Oliver, W.C.; McHargue, C.J. Deformation behavior of ceramic crystals subjected to very low load (nano)indentations. *J. Mater. Res.* **1992**, *7*, 450–473. [CrossRef]

167. Schuh, C.A.; Mason, J.K.; Lund, A.C. Quantitative insight into dislocation nucleation from high-temperature nanoindentation experiments. *Nat. Mater.* **2005**, *4*, 617–621. [CrossRef] [PubMed]

168. Tymiak, N.I.; Daugela, A.; Wyrobek, T.J.; Warren, O.L. Acoustic emission monitoring of the earliest stages of contact-induced plasticity in sapphire. *Acta Mater.* **2004**, *52*, 553–563. [CrossRef]

169. Kelchner, C.L.; Plimpton, S.J.; Hamilton, J.C. Dislocation nucleation and defect structure during surface indentation. *Phys. Rev. B* **1998**, *58*, 11085–11088. [CrossRef]

170. Van Vliet, K.; Li, J.; Zhu, T.; Yip, S.; Suresh, S. Quantifying the early stages of plasticity through nanoscale experiments and simulations. *Phys. Rev. B* **2003**, *67*, 104105. [CrossRef]

171. Lilleodden, E.T.; Zimmerman, J.A.; Foiles, S.M.; Nix, W.D. Atomistic simulations of elastic deformation and dislocation nucleation during nanoindentation. *J. Mech. Phys. Solids* **2003**, *51*, 901–920. [CrossRef]

172. Gaillard, Y.; Tromas, C.; Woirgard, J. Pop-in phenomenon in MgO and LiF: Observation of dislocation structures. *Philos. Mag. Lett.* **2003**, *83*, 553–561. [CrossRef]

173. Hertz, H. Über die Berührung fester elastischer Körper. *J. Reine Angew. Mat.* **1882**, *92*, 156–171.

174. Johnson, K.L. *Contact Mechanics*; Cambridge University Press: Cambridge, UK, 1987.

175. Hanson, M.T.; Johnson, T. Elastic field for spherical Hertzian contact of isotropic bodies revisited. Some alternative expressions. *J. Tribol.* **1993**, *115*, 327–332. [CrossRef]

176. Booth, A.S.; Ellis, M.; Roberts, S.G.; Hirsch, P.B. Dislocation-controlled stable crack growth in Mo and MgO. *Mater. Sci. Eng. A* **1993**, *164*, 270–274. [CrossRef]

177. Burns, S.J.; Webb, W.W. Fracture surface energies and dislocation processes during dynamical cleavage of LiF. I. Theory. *J. Appl. Phys.* **1970**, *41*, 2078–2085. [CrossRef]

178. Burns, S.J.; Webb, W.W. Fracture surface energies and dislocation processes during dynamical cleavage of LiF. II. Experiments. *J. Appl. Phys.* **1970**, *41*, 2086–2095. [CrossRef]

179. Gilman, J.J. Nucleation of dislocation loops by cracks in crystals. *J. Metall.* **1957**, *9*, 449–454. [CrossRef]

180. Gilman, J.J.; Knudsen, C.; Walsh, W.P. Cleavage cracks and dislocations in LiF crystals. *J. Appl. Phys.* **1958**, *29*, 601–607. [CrossRef]

181. Robins, J.L.; Rhodin, T.N.; Gerlach, R.L. Dislocation structures in cleaved magnesium oxide. *J. Appl. Phys.* **1966**, *37*, 3893–3903. [CrossRef]

182. Montagne, A.; Audurier, V.; Tromas, C. Influence of pre-existing dislocations on the pop-in phenomenon during nanoindentation in MgO. *Acta Mater.* **2013**, *61*, 4778–4786. [CrossRef]

183. Gaillard, Y.; Tromas, C.; Woirgard, J. Quantitative analysis of dislocation pile-ups nucleated during nanoindentation in MgO. *Acta Mater.* **2006**, *54*, 1409–1417. [CrossRef]

184. Jouiad, M.; Pettinari, F.; Coujou, A.; Clément, N. Evaluation of friction stresses in the γ phase of a nickel-base superalloy "In situ" deformation experiments. *Mater. Sci. Eng. A* **1997**, *234–236*, 1041–1044. [CrossRef]

185. Bridgman, P.W.; Simon, I. Effects of Very High Pressures on Glass. *J. Appl. Phys.* **1953**, *24*, 405–413. [CrossRef]

186. Paterson, M.S.; Wong, T. *Experimental Rock Deformation-the Brittle Field*; Springer: Berlin, Germany, 2005.

187. Zambaldi, C.; Zehnder, C.; Raabe, D. Orientation dependent deformation by slip and twinning in magnesium during single crystal indentation. *Acta Mater.* **2015**, *91*, 267–288. [CrossRef]

188. Lloyd, S.J.; Molina-Aldareguia, J.M.; Clegg, W.J. Deformation under nanoindents in sapphire, spinel and magnesia examined using transmission electron microscopy. *Philos. Mag. A* **2002**, *82*, 1963–1969. [CrossRef]

189. McLaughlin, K.K. *TEM Diffraction Analysis of the Deformation Underneath Low Load Indentations*; University of Cambridge: Cambridge, UK, 2007.

190. Howie, P.R.; Korte, S.; Clegg, W.J. Fracture modes in micropillar compression of brittle crystals. *J. Mater. Res.* **2011**, *27*, 141–151. [CrossRef]

191. Uchic, M.D.; Dimiduk, D.M.; Florando, J.N.; Nix, W.D. Sample Dimensions Influence Strength and Crystal Plasticity. *Science* **2004**, *305*, 986–989. [CrossRef] [PubMed]

192. Uchic, M.D.; Shade, P.A.; Dimiduk, D.M. Plasticity of Micrometer-Scale Single Crystals in Compression. *Annu. Rev. Mater. Res.* **2009**, *39*, 361–386. [CrossRef]

193. Östlund, F.; Howie, P.R.; Ghisleni, R.; Korte, S.; Leifer, K.; Clegg, W.J.; Michler, J. Ductile–brittle transition in micropillar compression of GaAs at room temperature. *Philos. Mag. A* **2011**, *91*, 1190–1199. [CrossRef]

194. Kendall, K. Impossibility of comminuting small particles by compression. *Nature* **1978**, *272*, 710–711. [CrossRef]

195. Östlund, F.; Rzepiejewska-Malyska, K.; Leifer, K.; Hale, L.M.; Tang, Y.; Ballarini, R.; Gerberich, W.W.; Michler, J. Brittle-to-Ductile Transition in Uniaxial Compression of Silicon Pillars at Room Temperature. *Adv. Funct. Mater.* **2009**, *19*, 2439–2444. [CrossRef]

196. Cottrell, P.A.H. Lectures. 1961. Available online: https://link.springer.com/article/10.1007%2FBF00812985 (accessed on 29 March 2018).

197. Korte, S.; Barnard, J.S.; Stearn, R.J.; Clegg, W.J. Deformation of silicon–insights from microcompression testing at 25–500 C. *Int. J. Plasticity* **2011**, *27*, 1853–1866. [CrossRef]

198. Korte, S.; Clegg, W.J. Micropillar compression of ceramics at elevated temperatures. *Scr. Mater.* **2009**, *60*, 807–810. [CrossRef]

199. Uchic, M.; Shade, P.; Dimiduk, D. Micro-compression testing of fcc metals: A selected overview of experiments and simulations. *J. Miner. Metals Mater. Soc.* **2009**, *61*, 36–41. [CrossRef]

200. Korte-Kerzel, S. Microcompression of brittle and anisotropic crystals: Recent advances and current challenges in studying plasticity in hard materials. *MRS Commun.* **2017**, *7*, 109–120. [CrossRef]

201. Korte, S.; Clegg, W. Discussion of the dependence of the effect of size on the yield stress in hard materials studied by microcompression of MgO. *Philos. Mag. A* **2011**, *91*, 1150–1162. [CrossRef]

202. Zou, Y. Materials selection in micro- or nano-mechanical design: Towards new Ashby plots for small-sized materials. *Mat. Sci. Eng. A* **2017**, *680*, 421–425. [CrossRef]
203. Gibson, J.S.K.L.; Rezaei, S.; Rueß, H.; Hans, M.; Music, D.; Wulfinghoff, S.; Schneider, J.M.; Reese, S.; Korte-Kerzel, S. From quantum to continuum mechanics: Studying the fracture toughness of transition metal nitrides and oxynitrides. *Mater. Res. Lett.* **2017**, *6*, 142–151. [CrossRef]
204. Armstrong, D.E.J.; Hardie, C.D.; Gibson, J.S.K.L.; Bushby, A.J.; Edmondson, P.D.; Roberts, S.G. Small-scale characterisation of irradiated nuclear materials: Part II nanoindentation and micro-cantilever testing of ion irradiated nuclear materials. *J. Nucl. Mater.* **2015**, *462*, 374–381. [CrossRef]
205. Korte-Kerzel, S.; Schnabel, V.; Clegg, W.J.; Heggen, M. Room temperature plasticity in m-Al$_{13}$Co$_4$ studied by microcompression and high resolution scanning transmission electron microscopy. *Scr. Mater.* **2018**, *146*, 327–330. [CrossRef]
206. Korte, S.; Ritter, M.; Jiao, C.; Midgley, P.; Clegg, W. Three-dimensional electron backscattered diffraction analysis of deformation in MgO micropillars. *Acta Mater.* **2011**, *59*, 7241–7254. [CrossRef]
207. Zou, Y.; Spolenak, R. Size-dependent plasticity in micron- and submicron-sized ionic crystals. *Philos. Mag. Lett.* **2013**, *93*, 431–438. [CrossRef]
208. Kraft, O.; Gruber, P.A.; Mönig, R. Plasticity in confined dimensions. *Annu. Rev. Mater. Res.* **2010**, *40*, 293–317. [CrossRef]
209. Kiener, D.; Minor, A.M. Source Truncation and Exhaustion: Insights from Quantitative in situ TEM Tensile Testing. *Nano Lett.* **2011**, *11*, 3816–3820. [CrossRef] [PubMed]
210. Kaufmann, D.; Mönig, R.; Volkert, C.A.; Kraft, O. Size dependent mechanical behaviour of tantalum. *Int. J. Plasticity* **2011**, *27*, 470–478. [CrossRef]
211. Michler, J.; Wasmer, K.; Meier, S.; Östlund, F.; Leifer, K. Plastic deformation of gallium arsenide micropillars under uniaxial compression at room temperature. *Appl. Phys. Lett.* **2007**, *90*, 043123. [CrossRef]
212. Parthasarathy, T.A.; Rao, S.I.; Dimiduk, D.M.; Uchic, M.D.; Trinkle, D.R. Contribution to size effect of yield strength from the stochastics of dislocation source lengths in finite samples. *Scr. Mater.* **2007**, *56*, 313–316. [CrossRef]
213. Schneider, A.S.; Frick, C.P.; Arzt, E.; Clegg, W.J.; Korte, S. Influence of test temperature on the size effect in molybdenum small-scale compression pillars. *Philos. Mag. Lett.* **2013**, *93*, 331–338. [CrossRef]
214. Torrents Abad, O.; Wheeler, J.M.; Michler, J.; Schneider, A.S.; Arzt, E. Temperature-dependent size effects on the strength of Ta and W micropillars. *Acta Mater.* **2016**, *103*, 483–494. [CrossRef]
215. Soler, R.; Molina-Aldareguia, J.M.; Segurado, J.; LLorca, J.; Merino, R.I.; Orera, V.M. Micropillar compression of LiF [112] single crystals: Effect of size, ion irradiation and misorientation. *Int. J. Plasticity* **2012**, *36*, 50–63. [CrossRef]
216. Kiener, D.; Motz, C.; Dehm, G. Micro-compression testing: A critical discussion of experimental constraints. *Mater. Sci. Eng. A* **2009**, *505*, 79–87. [CrossRef]
217. Zou, Y.; Spolenak, R. Size-dependent plasticity in KCl and LiF single crystals: influence of orientation, temperature, pre-straining and doping. *Philos. Mag. A* **2015**, 1–19. [CrossRef]
218. Gibson, J.S.K.L.; Schröders, S.; Zehnder, C.; Korte-Kerzel, S. On extracting mechanical properties from nanoindentation at temperatures up to 1000°C. *Extreme Mech. Lett.* **2017**, *17*, 43–49. [CrossRef]
219. Kang, W.; Merrill, M.; Wheeler, J.M. In situ thermomechanical testing methods for micro/nano-scale materials. *Nano* **2017**, *9*, 2666–2688. [CrossRef] [PubMed]
220. Durst, K.; Maier, V. Dynamic nanoindentation testing for studying thermally activated processes from single to nanocrystalline metals. *Curr. Opin. Solid State Mater. Sci.* **2015**, *19*, 340–353. [CrossRef]
221. Maier, V.; Durst, K.; Mueller, J.; Backes, B.; Höppel, H.W.; Göken, M. Nanoindentation strain-rate jump tests for determining the local strain-rate sensitivity in nanocrystalline Ni and ultrafine-grained Al. *J. Mater. Res.* **2011**, *26*, 1421–1430. [CrossRef]
222. Moser, B.; Wasmer, K.; Barbieri, L.; Michler, J. Strength and fracture of Si micropillars: A new scanning electron microscopy-based micro-compression test. *J. Mater. Res.* **2007**, *22*, 1004–1011. [CrossRef]
223. Bei, H.; Shim, S.; George, E.P.; Miller, M.K.; Herbert, E.G.; Pharr, G.M. Compressive strengths of molybdenum alloy micro-pillars prepared using a new technique. *Scr. Mater.* **2007**, *57*, 397–400. [CrossRef]
224. Issa, I.; Amodeo, J.; Réthoré, J.; Joly-Pottuz, L.; Esnouf, C.; Morthomas, J.; Perez, M.; Chevalier, J.; Masenelli-Varlot, K. In situ investigation of MgO nanocube deformation at room temperature. *Acta Mater.* **2015**, *86*, 295–304. [CrossRef]

225. Maaß, R.; Meza, L.; Gan, B.; Tin, S.; Greer, J.R. Ultrahigh strength of dislocation-free Ni3Al nanocubes. *Small* **2012**, *8*, 1869–1875. [CrossRef] [PubMed]

226. Amodeo, J.; Lizoul, K. Mechanical properties and dislocation nucleation in nanocrystals with blunt edges. *Mater. Des.* **2017**, *135*, 223–231. [CrossRef]

227. Nishiyama, N.; Katsura, T.; Funakoshi, K.-I.; Kubo, A.; Kubo, T.; Tange, Y.; Sueda, Y.-I.; Yokoshi, S. Determination of the phase boundary between the B1 and B2 phases in NaCl by in situ x-ray diffraction. *Phys. Rev. B Condens. Matter* **2003**, *68*, 319. [CrossRef]

228. Mao, H.K.; Wu, Y.; Hemley, R.J.; Chen, L.C.; Shu, J.F.; Finger, L.W.; Cox, D.E. High-pressure phase transition and equation of state of CsI. *Phys. Rev. Lett.* **1990**, *64*, 1749–1752. [CrossRef] [PubMed]

229. Williams, Q.; Jeanloz, R. Measurements of CsI band-gap closure to 93 GPa. *Phys. Rev. Lett.* **1986**, *56*, 163–164. [CrossRef] [PubMed]

230. Jeanloz, R. Physical Chemistry at Ultrahigh Pressures and Temperatures. *Ann. Rev. Phys. Chem.* **1989**, *40*, 237–259. [CrossRef]

231. McWilliams, R.S.; Spaulding, D.K.; Eggert, J.H.; Celliers, P.M.; Hicks, D.G.; Smith, R.F.; Collins, G.W.; Jeanloz, R. Phase Transformations and Metallization of Magnesium Oxide at High Pressure and Temperature. *Science* **2012**, *338*, 1330–1333. [CrossRef] [PubMed]

232. Coppari, F.; Smith, R.F.; Eggert, J.H.; Wang, J.; Rygg, J.R.; Lazicki, A.; Hawreliak, J.A.; Collins, G.W.; Duffy, T.S. Experimental evidence for a phase transition in magnesium oxide at exoplanet pressures. *Nat. Geosci.* **2013**, *6*, 926–929. [CrossRef]

233. Karki, B.; Stixrude, L.; Clark, S.; Warren, M.; Ackland, G.; Crain, J. Structure and elasticity of MgO at high pressure. *Am. Mineral.* **1997**, *82*, 51–60. [CrossRef]

234. Karato, S.-I. Some remarks on the origin of seismic anisotropy in the D″ layer. *Earth Planets Space* **1998**, *50*, 1019–1028. [CrossRef]

235. Wang, Y.; Durham, W.; Getting, I.; Weidner, D. The deformation-DIA: A new apparatus for high temperature triaxial deformation to pressures up to 15 GPa. *Rev. Sci. Instrum.* **2003**, *74*, 3002–3011. [CrossRef]

236. Yamazaki, D.; Karato, S. High-pressure rotational deformation apparatus to 15 GPa. *Rev. Sci. Instrum.* **2001**, *72*, 4207–4211. [CrossRef]

237. Hunt, S.A.; Weidner, D.J.; McCormack, R.J.; Whitaker, M.L.; Bailey, E.; Li, L.; Vaughan, M.T.; Dobson, D.P. Deformation T-Cup: A new multi-anvil apparatus for controlled strain-rate deformation experiments at pressures above 18 GPa. *Rev. Sci. Instrum.* **2014**, *85*, 085103. [CrossRef] [PubMed]

238. Kawazoe, T.; Ohuchi, T.; Nishihara, Y.; Nishiyama, N.; Fujino, K.; Irifune, T. Seismic anisotropy in the mantle transition zone induced by shear deformation of wadsleyite. *Phys. Earth Planet. Int.* **2013**, *216*, 91–98. [CrossRef]

239. Girard, J.; Amulele, G.; Farla, R.; Mohiuddin, A.; Karato, S.I. Shear deformation of bridgmanite and magnesiowustite aggregates at lower mantle conditions. *Science* **2016**, *351*, 144–147. [CrossRef] [PubMed]

240. Raterron, P.; Merkel, S.; Holyoke, C.W. Axial temperature gradient and stress measurements in the deformation-DIA cell using alumina pistons. *Rev. Sci. Instrum.* **2013**, *84*, 043906. [CrossRef] [PubMed]

241. Xu, Y.; Nishihara, Y.; Karato, S.-I. Development of a rotational Drickamer apparatus for large-strain deformation experiments at deep Earth conditions. In *Advances in High-Pressure Technology for Geophysical Applications*; Chen, J., Wang, Y., Duffy, T.S., Shen, G., Dobrzhinetskaya, L.F., Eds.; Elsevier: Amsterdam, The Netherlands, 2005; pp. 167–182.

242. Dubrovinskaia, N.; Dubrovinsky, L.; Solopova, N.A.; Abakumov, A.; Turner, S.; Hanfland, M.; Bykova, E.; Bykov, M.; Prescher, C.; Prakapenka, V.B.; et al. Terapascal static pressure generation with ultrahigh yield strength nanodiamond. *Sci. Adv.* **2016**, *2*, e1600341. [CrossRef] [PubMed]

243. Tateno, S.; Hirose, K.; Ohishi, Y.; Tatsumi, Y. The Structure of Iron in Earth's Inner Core. *Science* **2010**, *330*, 359–361. [CrossRef] [PubMed]

244. Hemley, R.J.; Mao, H.K.; Shen, G.; Badro, J.; Gillet, P.; Hanfland, M.; Häusermann, D. X-ray imaging of stress and strain of diamond, iron, and tungsten at megabar pressures. *Science* **1997**, *276*, 1242–1245. [CrossRef]

245. Liermann, H.P.; Merkel, S.; Miyagi, L.; Wenk, H.R.; Shen, G.; Cynn, H.; Evans, W.J. New Experimental Method for In Situ Determination of Material Textures at Simultaneous High-Pressure and Temperature by Means of Radial Diffraction in the Diamond Anvil Cell. *Rev. Sci. Instrum.* **2009**, *80*, 104501. [CrossRef] [PubMed]

246. Hirose, K.; Nagaya, Y.; Merkel, S.; Ohishi, Y. Deformation of MnGeO₃ post-perovskite at lower mantle pressure and temperature. *Geophys. Res. Lett.* **2010**, *37*, L20302. [CrossRef]

247. Merkel, S. X-ray diffraction evaluation of stress in high pressure deformation experiments. *J. Phys. Condens. Matter* **2006**, *18*, S949–S962. [CrossRef] [PubMed]

248. Meade, C.; Jeanloz, R. Yield strength of MgO to 40 GPa. *J. Geophys. Res.* **1988**, *93*, 3261–3269. [CrossRef]

249. Vaughan, M.; Chen, J.; Li, L.; Weidner, D.; Li, B. *Use of X-ray Imaging Techniques at High Pressure and Temperature for Strain Measurements*; Manghnani, M.H., Nellis, W.J., Nicol, M.F., Eds.; Universities Press (India) Limited: Hyderabad, India, 2000; pp. 1097–1098.

250. Merkel, S.; Yagi, T. X-ray transparent gasket for diamond anvil cell high pressure experiments. *Rev. Sci. Instrum.* **2005**, *76*, 046109. [CrossRef]

251. Noyan, I.C.; Cohen, J.B. *Residual Stress: Measurements by Diffraction and Interpretation*; Springer-Verlag: New York, NY, USA, 1987.

252. Merkel, S.; Wenk, H.-R.; Shu, J.; Shen, G.; Gillet, P.; Mao, H.K.; Hemley, R.J. Deformation of polycrystalline MgO at pressures of the lower mantle. *J. Geophys. Res.* **2002**, *107*, 2271. [CrossRef]

253. Lin, F.; Hilairet, N.; Raterron, P.; Addad, A.; Immoor, J.; Marquardt, H.; Tomé, C.N.; Miyagi, L.; Merkel, S. Elasto-viscoplastic self consistent modeling of the ambient temperature plastic behavior of periclase deformed up to 5.4 GPa. *J. Appl. Phys.* **2017**, *122*, 205902. [CrossRef]

254. Singh, A.K.; Balasingh, C.; Mao, H.K.; Hemley, R.J.; Shu, J. Analysis of lattice strains measured under non-hydrostatic pressure. *J. Appl. Phys.* **1998**, *83*, 7567–7575. [CrossRef]

255. Li, L. X-ray strain analysis at high pressure: Effect of plastic deformation in MgO. *J. Appl. Phys.* **2004**, *95*, 8357–8365. [CrossRef]

256. Merkel, S.; Tomé, C.N.; Wenk, H.R. A modeling analysis of the influence of plasticity on high pressure deformation of hcp-Co. *Phys. Rev. B* **2009**, *79*, 064110. [CrossRef]

257. Amodeo, J.; Dancette, S.; Delannay, L. Atomistically-informed crystal plasticity in MgO polycrystals under pressure. *Int. J. Plasticity* **2016**, *82*, 177–191. [CrossRef]

258. Roters, F.; Eisenlohr, P.; Hantcherli, L.; Tjahjanto, D.D.; Bieler, T.R.; Raabe, D. Overview of constitutive laws, kinematics, homogenization and multiscale methods in crystal plasticity finite-element modeling: Theory, experiments, applications. *Acta Mater.* **2010**, *58*, 1152–1211. [CrossRef]

259. Cordier, P. Dislocations and slip systems of mantle minerals. In *Pastic Deformation of Minerals and Rocks*; Karato, S., Wenk, H.-R., Eds.; Reviews in Mineralogy and Geochemistry; Mineralogical Society of America: Chantilly, VA, USA, 2002; Volume 51, pp. 137–179.

260. Mussi, A.; Cordier, P.; Demouchy, S.; Hue, B. Hardening mechanisms in olivine single crystal deformed at 1090 °C: an electron tomography study. *Philos. Mag.* **2017**, *97*, 3172–3185. [CrossRef]

261. Amodeo, J.; Carrez, P.; Cordier, P. Modelling the effect of pressure on the critical shear stress of MgO single crystals. *Philos. Mag. A* **2012**, *92*, 1523–1541. [CrossRef]

262. Girard, J.; Chen, J.; Raterron, P. Deformation of periclase single crystals at high pressure and temperature: Quantification of the effect of pressure on slip-system activities. *J. Appl. Phys.* **2012**, *111*, 112607. [CrossRef]

263. Weidner, D.J.; Li, L.; Davis, M.; Chen, J. Effect of plasticity on elastic modulus measurements. *Geophys. Res. Lett.* **2004**, *31*, L06621. [CrossRef]

264. Yamazaki, D.; Karato, S. Fabric development in (Mg, Fe) O during large strain, shear deformation: implications for seismic anisotropy in Earth's lower mantle. *Phys. Earth Planet. Inter.* **2002**, *131*, 251–267. [CrossRef]

265. Tommaseo, C.E.; Devine, J.; Merkel, S.; Speziale, S.; Wenk, H.R. Texture development and elastic stresses in magnesiowustite at high pressure. *Phys. Chem. Miner.* **2006**, *33*, 84–97. [CrossRef]

266. Badro, J.; Fiquet, G.; Guyot, F.; Rueff, J.-P.; Struzhkin, V.V.; Vankó, G.; Monaco, G. Iron partitioning in Earth's mantle: toward a deep lower mantle discontinuity. *Science* **2003**, *300*, 789–791. [CrossRef] [PubMed]

267. Lin, J.F.; Wenk, H.-R.; Voltolini, M.; Speziale, S.; Shu, J.; Duffy, T.S. Deformation of lower-mantle ferropericlase (Mg,Fe)O across the electronic spin transition. *Phys. Chem. Min.* **2009**, *36*, 585–592. [CrossRef]

268. Long, M.D.; Xiao, X.; Jiang, Z.; Evansa, B.; Karato, S. Lattice preferred orientation in deformed polycrystalline (Mg,Fe)O and implications for seismic anisotropy in D". *Phys. Earth Planet. Inter.* **2006**, *156*, 75–88. [CrossRef]

269. Heidelbach, F.; Stretton, I.; Langenhorst, F.; Mackwell, S. Fabric evolution during high shear strain deformation of magnesiowüstite (Mg₀.₈Fe₀.₂O). *J. Geophys. Res.* **2003**, *108*, 2154. [CrossRef]

270. Immoor, J.; Marquardt, H.; Miyagi, L.; Lin, F.; Speziale, S.; Merkel, S.; Buchen, J.; Kurnosov, A.; Liermann, H.P. Evidence for {100}<011> slip in ferropericlase in Earth's lower mantle from high-pressure/high-temperature experiments. *Earth. Planet. Sci. Lett.* **2018**, *489*, 251–257. [CrossRef]

271. Weaver, C.W.; Paterson, M.S. Deformation of cube-oriented MgO crystals under pressure. *J. Am. Ceram. Soc.* **1969**, *52*, 293–302. [CrossRef]

272. Paterson, M.; Weaver, C. Deformation of polycrystalline MgO under pressure. *J. Am. Ceram. Soc.* **1970**, *53*, 463–471. [CrossRef]

273. Mei, S.; Kohlstedt, D.L.; Durham, W.B.; Wang, L. Experimental investigation of the creep behavior of MgO at high pressures. *Phys. Earth Planet. Inter.* **2008**, *170*, 170–175. [CrossRef]

274. Weidner, D.J.; Wang, Y.B.; Vaughan, M.T. Yield Strength at High-Pressure and Temperature. *Geophys. Res. Lett.* **1994**, *21*, 753–756. [CrossRef]

275. Uchida, T.; Funamori, N.; Ohtani, T.; Yagi, T. *Differential Stress of MgO and Mg2SiO4 under Uniaxial Stress Field: Variation with Pressure, Temperature, and Phase Transition*; Trzeciatowski, W.A., Ed.; World Scientific Publishing: Singapore, 1996; pp. 183–185.

276. Uchida, T.; Wang, Y.; Rivers, M.L.; Sutton, S.R. Yield strength and strain hardening of MgO up to 8 GPa measured in the deformation-DIA with monochromatic X-ray diffraction. *Earth. Planet. Sci. Lett.* **2004**, *226*, 117–126. [CrossRef]

277. Weidner, D.J.; Li, L. Measurement of stress using synchrotron X-rays. *J. Phys. Condens. Matter* **2006**, *18*, S1061–S1067. [CrossRef] [PubMed]

278. Singh, A.K.; Liermann, H.P.; Saxena, S.K. Strength of magnesium oxide under high pressure: Evidence for the grain-size dependence. *Solid State Commun.* **2004**, *132*, 795–798. [CrossRef]

279. Marquardt, H.; Miyagi, L. Slab stagnation in the shallow lower mantle linked to an increase in mantle viscosity. *Nat. Geosci.* **2015**, *8*, 311–314. [CrossRef]

280. Kinsland, G.; Bassett, W. Strength of MgO and NaCl polycrystals to confining pressures of 250 kbar at 25 C. *J. Appl. Phys.* **1977**, *48*, 978–985. [CrossRef]

281. Duffy, T.; Hemley, R.; Mao, H. Equation of state and shear strength at multimegabar pressures: Magnesium oxide to 227 GPa. *Phys. Rev. Lett.* **1995**, *74*, 1371–1374. [CrossRef] [PubMed]

282. Reynard, B.; Caracas, R.; Cardon, H.; Montagnac, G.; Merkel, S. High-pressure yield strength of rocksalt structures using quartz Raman piezometry. *C. R. Geosci.*. in press. [CrossRef]

283. Li, L.; Weidner, D.J. In situ analysis of texture development from sinusoidal stress at high pressure and temperature. *Rev. Sci. Instrum.* **2015**, *86*, 125106. [CrossRef] [PubMed]

![crystals logo] *crystals*

MDPI

Review

Peak Broadening Anisotropy and the Contrast Factor in Metal Alloys

Thomas Hadfield Simm

Materials Research Centre, Swansea University, Wales SA1 8EN, UK; thomas.simm@dmata.co.uk

Received: 14 March 2018; Accepted: 8 May 2018; Published: 13 May 2018

Abstract: Diffraction peak profile analysis (DPPA) is a valuable method to understand the microstructure and defects present in a crystalline material. Peak broadening anisotropy, where broadening of a diffraction peak doesn't change smoothly with 2θ or *d*-spacing, is an important aspect of these methods. There are numerous approaches to take to deal with this anisotropy in metal alloys, which can be used to gain information about the dislocation types present in a sample and the amount of planar faults. However, there are problems in determining which method to use and the potential errors that can result. This is particularly the case for hexagonal close packed (HCP) alloys. There is though a distinct advantage of broadening anisotropy in that it provides a unique and potentially valuable way to develop crystal plasticity and work-hardening models. In this work we use several practical examples of the use of DPPA to highlight the issues of broadening anisotropy.

Keywords: diffraction peak profile analysis (DPPA); contrast factor; dislocations; twinning; crystal plasticity; planar faults; powder diffraction

---

1. Introduction

The ability to quantify the shape of diffraction peak profiles is an important aspect of crystallography and materials science [1–3]. The technique can be used to quantify the dislocation density, crystal/dislocation-cell size, planar fault percentage and the dislocation slip systems present in samples in a statistically significant manner that is either practically very difficult or not possible using other techniques. However, its more widespread use is limited due to practical and mathematical limitations of the technique [4,5]. One of the main limitations of the technique is due to peak broadening anisotropy, which has implications on the results of the broadening methods and its use with other techniques. But conversely, this limitation may also provide a way to extend polycrystal plasticity models to incorporate details at the nano scale such as the activity and arrangement of dislocations, and hence provide a way to combine dislocation and work-hardening models with polycrystal plasticity models.

Although the original formulism from Rietveld [6] assumed that the broadening of a diffraction peak varied smoothly with its *d*-spacing, it has long been found that peaks close together in *d*-spacing can vary significantly in their broadening [7,8]. Various models have been constructed to account for this peak broadening anisotropy in Rietveld refinements due to the presence of crystallite and microstrain effects [7,9,10]. One of the earlier works on the use of diffraction peak profiles to study deformed metal alloys was by Stokes and Wilson [8]. Based on a model that broadening was due to distortion within a crystal, they derived a formula for strain broadening that could explain this observed broadening anisotropy. If it is assumed that the stress distribution is statistically isotropic, they showed that elastically anisotropic crystals (as even cubic crystal are) would lead to a strain that varied with crystallographic direction. The term is the same as that used today [11], although now this strain anisotropy is derived in terms of the presence of dislocations.

In metal alloys the most important cause of broadening is believed to be dislocations [12], with planar faults, such as stacking faults or twin faults, being the second most important contributor.

Various models exist for how peak broadening anisotropy in a metal alloy can be accounted for, based mainly on the changes in dislocations and planar faults, these include: (a) the average character of dislocations in a sample, that is whether they are mainly edge or screw; (b) the average Burgers vector of dislocations in the sample; (c) the slip systems of the dislocation in different orientations found using a crystal plasticity model; (d) the dislocation density in different orientations; (e) the average amount of different planar fault types in a sample. In most cases broadening anisotropy is not directly addressed, but instead a method is applied only to improve the data analysis and reduce errors. Due to the number of different models to account for anisotropy it can be difficult to evaluate the best approach, and what the implications are of their use. This often means that the cause of the broadening anisotropy is just another fitting parameter which is used only to improve the data analysis and reduce errors.

Diffraction peak profile analysis (DPPA) is potentially a very powerful tool to probe the nano-structure of a metal alloy. However, a lack of understanding of the implications of how to address the broadening anisotropy can lead to errors and incorrect interpretation when applying DPPA methods. To address this issue, this paper seeks to answer the following questions: (1) is there a physical basis for this anisotropy in deformed metal alloys, or is it just a method to improve data analysis by using more diffraction peaks? (2) what are the implications of this anisotropy for diffraction peak profile analysis methods?

Supplementary data and analysis of this work can be found at Open Science Framework [13]. New analysis of data using DPPA methods in this review was done using the MATLAB graphical user interfaces BIGdippa and dippaFC (dippa_v3) [14].

2. Theory

2.1. The Contrast Factor

The displacement field around a dislocation is anisotropic (Figure 1). This is readily observed in transmission electron microscopy (TEM) imaging of dislocations, whereby rotating the sample so that a dislocation's Burgers vector (*b*) is perpendicular to the diffraction vector, *g*, gives a vanishing contrast. The modelling of a dislocation's displacement field is an important method used within TEM to be able to identify details of dislocations [15,16], especially in cases when **g.b** = 0 does not lead to vanishing contrast or due to practicalities of rotating a sample. In the same manner, the diffraction broadening caused by a dislocation varies depending on the diffraction vector, and can be calculated by modelling the dislocation's displacement field. The contribution of a dislocation's displacement field to the broadening of different diffraction profiles is through what is known as the contrast (or orientation) factor of dislocations. The contrast factor of an individual dislocation is dependent on the angles between the diffraction vector and the vectors that define the dislocation: it's Burgers vector (*b*), slip plane normal (*n*), and dislocation line (*s*). The contrast factor of an individual dislocation can be calculated using the computer program ANIZC [17]. For example, the edge dislocation in Figure 2 has a contrast factor of 0.46 when *g* is [110] and parallel to the dislocations Burgers vector. The value is 0.00 when *g* is parallel to the slip line ($[1\bar{1}2]$) and 0.05 when *g* is parallel to the slip plane normal ($[\bar{1}11]$).

The way in which the contrast factor is incorporated into diffraction peak profile analysis (DPPA) techniques can be illustrated with the Williamson-Hall method [19]. Incorporation within other methods, such as the Warren-Averbach method [1] uses the same approach. The Williamson-Hall method [19] is a common method to define the broadening of a diffraction peak; here the full-width at half maximum intensity (full-width, or FW) of a diffraction peak is defined as being due to micro-strain and size components. In terms of broadening by dislocations this equation can be written as [4,20]:

$$FW_{hkl} = \frac{K_{Sch}}{D} + f_m g \sqrt{(\rho \overline{C_{hkl}})} \qquad (1)$$

where, D is the crystal size, K_{Sch} is the Scherrer constant (often taken as 0.9 for spherical crystals), g is the reciprocal of the d-spacing of a peak, f_m is a function related to the arrangement of dislocations (and represents how the arrangement of groups of dislocations influence their total strain), ρ is the dislocation density and $\overline{C_{hkl}}$ is the average contrast factor of dislocations in grains that contribute to the hkl diffraction peak. In Equation 1 and other DPPA approaches, $\overline{C_{hkl}}$ is assumed to be the only term that accounts for broadening anisotropy and its value is required to be able to obtain the dislocation density.

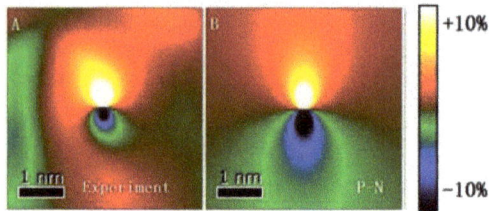

Figure 1. Experimental (**A**) and theoretical (**B**) strain fields around the edge component from a 60° dislocation core on a Ge/Si interface. Adapted from Liu et al. [15].

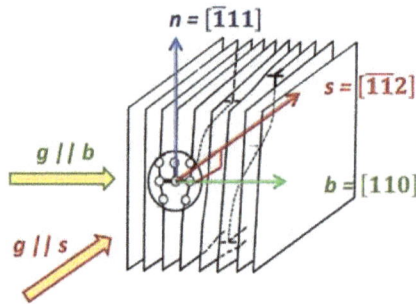

Figure 2. Schematic of an edge dislocation in a face-centred cubic crystal. The dislocation will cause maximum broadening when the diffraction vector is paralell to [110] and a minimum when it is parallel to $[1\bar{1}2]$. For the dislocation shown b is [110], n is $[\bar{1}11]$, and s is $[1\bar{1}2]$. Adapted from Armstrong [18].

As discussed by Scardi, Leoni and Delhez [21], the use of Williamson-Hall methods can be problematic for quantitative peak profile analysis, such as in determining details about broadening anisotropy and contrast factors, which is particularly relevant here. In addition, the use of the full-width is limited because it does not account for the whole of a peak profile. This can be especially important when broadening is due to dislocations, as shown by Wilkens [22] the same full-width can be caused by very different dislocation arrangements. A better parameter than the full-width, because it accounts for all the peak is the integral breadth, which is the area under a diffraction peak divided by the peaks height. The integral breadth can be replaced by the full-width in Equation (1) and in other Williamson-Hall equations. Due to their common use, both the full-width and the Williamson-Hall method are used throughout to display broadening anisotropy. The implications of this approximation are discussed by comparing full-width data, with integral breadths and Fourier coefficients of a titanium alloy in a later section (Section 3.2.1).

2.2. Contrast Factor, Homogeneous Approach

2.2.1. Cubic Alloys

For a powder diffraction pattern, a particular diffraction peak consists of contributions from many different crystals, of different orientations and containing a variety of different dislocation types.

Hence, to calculate the contrast factor for different diffraction peaks assumptions about the dislocations present are required. If it is assumed that, the sample has a random texture (no preferred orientation) and that slip system types, with a given Burgers vector and slip plane such as [110] (111) in face-centred cubic (FCC) crystals, are equally populated. This leads to a description for the contrast factor for a cubic material given as [11]:

$$\overline{C} = C_{h00}\left(1 - qH^2\right),\qquad(2)$$

where

$$H^2 = \frac{h^2k^2 + h^2l^2 + k^2l^2}{(h^2 + k^2 + l^2)^2}$$

C_{h00} and q are constants that are dependent on the elastic constants of the material and the dislocations present and h, k, l are the indices of the diffraction peak.

The value of q is used as a fitting parameter in DPPA methods, such as the modified-WH and modified-WA methods [19,20,23]. The value obtained from the fitting can be used to quantify the amount of different dislocation types since they also have different q values. For example, in stainless steel the q values for edge and screw dislocations are 1.72 and 2.47 [3,17], hence a fitted q value of 2.10 would be interpreted as the sample consisting of 50% screw and 50% edge dislocations. The homogeneous contrast factor approach has been widely used to quantify cubic alloys. Including, the amount of edge or screw dislocations [24–28], or the relative quantities of different slip types [29] in cubic crystals.

2.2.2. Hexagonal Close Packed Alloys

In a similar manner to that seen in Equation (2) for cubic crystals, Dragomir and Ungar [30] provided a formula for the contrast factor for hexagonal close packed (HCP) metals:

$$\overline{C} = \overline{C_{hk.0}}\left(1 + q_1x + q_2x^2\right),\qquad(3)$$

where

$$x = \frac{2l^2}{3(ga)^2}$$

$\overline{C_{hk.0}}$, q_1 and q_2 are constants that are dependent on the elastic constants of the material and the dislocations present, $g = 1/d$, l is the 3rd hkl index, and 'a' one of the lattice parameters (the HCP notation {$hkil$} is used but ignoring the 'i' index). As with the cubic formula the values of q are used as fitting parameters, which have a physical meaning. The method, described in more detail in [30], assumes that for a particular slip system type (such as <a> slip on the basal plane, or basal <a>), dislocations are equally populated (i.e., (0001) $[1\overline{2}10]$ has the same activity as (0001) $[\overline{2}110]$ and the other basal <a> slip systems). However, different slip system types are allowed to have different activities. Each slip system type has its own set of q values and values of $\overline{C_{hk.0}}$. The method then works by comparing the measured q values to those of the different slip system types, the amount of <a>, <c+a> and <c>, to determine their activity. The slip systems are grouped (Figure 3 and Table 1) into edge dislocations of the main slip systems in HCPs, and screw dislocations of the three Burgers vector types: <a>, <c+a> and <c>. From the q values of the different slip systems there is a greater tendency for <c+a> dislocations to have $q_1>0$ and for <a> dislocations to have $q_1<0$; in addition, prismatic <a> and pyramidal <a> are the only dislocations with $q_1<0$ and $q_2>0$. Hence, if the measured q_1 is less than 0 the method will provide <a> dislocations as the dominant dislocation type, and if the measure q_2 is more than 0 the value of <a> dislocations will be predicted to be close to 100%. Modifications to the approach have been made by some researchers to only include expected dislocation types; such as done by Seymour et al. for neutron irradiated zirconium alloys [31] using calculated contrast factor values for common dislocations found in the alloy, but the approach thereafter is essentially the same.

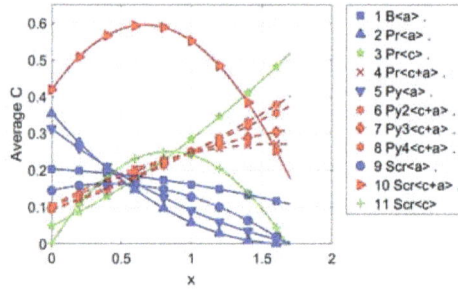

Figure 3. The change in the average contrast factor for the most common slip systems of titanium at different *x* values. The *q* values of the slip systems are detailed in Table 1. Adapted from Dragomir and Ungar [30].

Table 1. The *q* values and $\overline{C}_{hk.0}$ values for the most common slip system types of titanium. Data from Dragomir and Ungar [30].

Slip System Type	Slip Plane	Burgers Vector	$\overline{C}_{hk.0}$	q_1	q_2
1. Basal <a>	(0002)	[11$\overline{2}$0]	0.20227	−0.101142	−0.102625
2. Prismatic <a>	(1$\overline{1}$00)	[11$\overline{2}$0]	0.35387	−1.19272	0.355623
3. Prismatic <c>	(1$\overline{1}$00)	[0001]	0.04853	3.6161928	1.2264112
4. Prismatic <c+a>	(1$\overline{1}$00)	[11$\overline{2}$3]	0.10247	2.017177	−0.616631
5. Pyramidal <a>	(10$\overline{1}$1)	[11$\overline{2}$0]	0.3118	−0.894009	0.1833109
6. Pyramidal2 <c+a>	(11$\overline{2}$2)	[11$\overline{2}$3]	0.09227	1.299046	0.3972469
7. Pyramidal3 <c+a>	(11$\overline{2}$1)	[11$\overline{2}$3]	0.09813	1.89412	−0.365739
8. Pyramidal4 <c+a>	(10$\overline{1}$1)	[11$\overline{2}$3]	0.09323	1.5270212	0.14615
9. Screw <a>	multiple	[11$\overline{2}$0]	0.1444	0.59492	−0.710368
10. Screw <c+a>	multiple	[11$\overline{2}$3]	0.41873	1.25714	−0.94015
11. Screw <c>	multiple	[0001]	3.61×10^{-6}	165366	−98611

Whereas, in cubic alloys the value of the constant C_{h00} is very similar for edge and screw dislocations, in HCP alloys the corresponding constant $\overline{C}_{hk.0}$ can vary significantly for different slip systems. Therefore, in HCP alloys the contrast factor cannot just be a fitting variable, because it is needed to evaluate the dislocation density. This can be seen from Equation (1), which shows a feature of DPPA methods that they can only obtain the product $\rho \overline{C}_{hkl}$ and not the values individually. The homogeneous contrast factor approach has been widely used in HCP alloys to quantify the fraction of dislocations with different Burgers vectors [3,22,29–33]., or relative amounts of slip systems with different slip planes [3,30].

2.3. Planar Faults

Planar faults are defects in the crystal structure that are caused when the regular stacking of atoms is interrupted. Planar faults can be separated into two main types: stacking faults and twin faults. The presence of planar faults on diffraction peaks has been investigated in several works [23,32–35], with the work of Warren [23] being the most widely used for metal alloys.

In the approach of Warren, the contribution to broadening is given by adding an additional element to the size broadening. Hence, $1/D$ in Equation (1) is replaced with $1/Deff$ [23]:

$$\frac{1}{Deff_{hkl}} = \frac{1}{D} + \frac{(1.5\alpha + \beta)}{a}\omega_{hkl},$$

(4)

where a is the unit cell size, ω_{hkl} is a constant for a particular diffraction plane and given in Table 2, α the stacking (intrinsic stacking fault or deformation) fault probability and β the twin probability. The value of α and β are the fraction of crystal's layers that are of the planar fault types.

Warren [1] also showed that planar faults would also cause shifts in the position of a diffraction peak and a change in a peak's asymmetry. The magnitude of the shit in a peak was given by the following formula:

$$\Delta 2\theta = \frac{\sqrt{3}\alpha \tan\theta \chi_{hkl}}{2\pi}, \tag{5}$$

which can be converted to

$$\Delta g_{hkl} = \frac{\sqrt{3}}{4\pi}\alpha g_{hkl}\chi_{hkl}$$

where, χ_{hkl} is a constant for a given hkl plane and given in Table 2.

Balogh, Ribarik and Ungar [34] developed an analytical description of planar faults in FCC crystals to incorporate within the MWP diffraction peak profile analysis fitting program [36]. The procedure used the same formulism as Warren but was calculated with the computer program DIFFAX [32]. The approaches of Warren and Balogh et al. can both be used to estimate the broadening, peak profile asymmetry and shifts in the peaks. The main difference is that in the Balogh et al. approach the planar faults are separated into three types of faults: intrinsic and extrinsic stacking faults, and twin faults. Several authors have used the formulae of Warren [28,37] and that of Balogh et al. [38,39] to describe planar faults.

A formulism for twinning in HCP alloys was introduced by Balogh, Tichy and Ungar [40], in a similar manner to the work on planar faults in cubic alloys [34], using DIFFAX [32] and based on the boundaries of twins. The formulism has been used to quantify twinning in a magnesium alloy [41].

In the same manner as the contrast factor, there are multiple planes that planar faults can reside on; depending on the plane and the diffraction vector the contribution to broadening will be different. Hence, like the approach used with the contrast factor, it is often assumed that all planes are equally populated. Both planar fault broadening and dislocation broadening will give rise to anisotropic broadening, however there are differences between the two. Firstly, planar fault broadening is a type of 'size' broadening because it is doesn't have a dependence with g unlike 'strain' broadening. So, for example the 200 and 400 peaks are both broadened the same amount by planar faults, but not by the strain from dislocations. Secondly, planar faults can cause peak asymmetry and the shifting of the position of diffraction peaks, which does not occur in standard dislocation broadening formulae.

Table 2. The values of parameters ω_{hkl} used in Equation (4) and χ_{hkl} used in Equation (5), for quantifying the planar fault fraction to broadening in FCC alloys [23].

hkl	111	200	220	311	222	400
ω_{hkl}	$\frac{\sqrt{3}}{4}$	1	$\frac{1}{\sqrt{2}}$	$\frac{3}{2\sqrt{11}}$	$\frac{\sqrt{3}}{4}$	1
χ_{hkl}	$\frac{1}{4}$	$\frac{-1}{2}$	$\frac{1}{4}$	$\frac{-1}{11}$	$\frac{-1}{8}$	$\frac{1}{4}$

2.4. Contrast Factor, Plasticity Approach

In the plasticity approach to calculate the contrast factor, it is assumed that different grains have different dislocation populations that are calculated using a polycrystal plasticity model [42–44]. Polycrystal plasticity models are mathematical models of how polycrystals respond to an applied load. They can be used to predict various changes in a material due to plastic deformation: (1) changes in orientations [45,46], (2) the spreading of orientations [47], (3) intergranular strains [48], (4) dislocation and twin types present [42]. The simplest and most widely used models are the Taylor model [49] and the Schmid factor / Sachs model [50], but the use of more advanced models, such as visco-plastic self-consistent (VPSC) and crystal plasticity finite element models (CPFEM) [51,52], has increased significantly due to increased computational power and the accompanying development and availability of these models to researchers. These models do not provide dislocation density values, but instead provide the amount of slip on different slip systems. Therefore, to convert slip to dislocation density we need to make two main assumptions. Firstly, an assumption is needed about the relationship between slip activity and dislocation density of a particular slip system. One way to do this is by using the Orowan equation [53], as used in a few works to calculate the contrast factor [42–44], which relates an increment of shear (γ_i) on a slip system i to the increase in the mobile dislocation density on that system (ρ_m^i), based on the magnitude of the dislocation's Burger vector (b^i) and the dislocations speed (V^i).

$$\dot{\gamma}^i = \rho_m^i b^i V^i, \tag{6}$$

Two further assumptions are then taken. Firstly, about the screw or edge nature of the dislocations, which for simplicity it is often assumed that they have the same probability. Secondly, an assumption is needed about dislocation reactions, such as annihilation and cross-slip; the simplest ways to do this is by either assuming there are no dislocation reactions or that dislocations get distributed randomly to all slip systems with increments of strain. The actual situation is far more complicated than given by this equation and the two assumptions presented. Plastic deformation represents a complex and chaotic system, and as expressed succinctly by Alan Cottrell in 2002 [54] is the most challenging problem faced within classical physics. It is hoped that the development of dislocation models (such as dislocation dynamic models [55,56]) will aid with providing better formulation between slip activity from polycrystal models and details of dislocations such as their density, arrangement and the planes and Burgers vectors they have.

Each slip system in each grain will have a corresponding contrast factor that will differ depending on the diffraction plane being measured. Hence, to determine the average contrast factor for a diffraction peak involves finding the contrast factor of each dislocations that contributes to a peak and taking the average of these values. The individual contrast factors can be found using a program such as ANIZC [17] and the averaging can be done based on the texture. From this averaging, contrast factor values are obtained that depend on both the diffraction plane and the angle of that diffraction plane relative to the imposed load.

3. Practical Examples

3.1. Cubic Alloys

3.1.1. Homogeneous Approach

In the work of Simm et al. [4] two face-centred cubic (FCC) alloys, stainless steel and a commercially pure nickel, were deformed by compression to a range of applied strains. The samples were then measured using a laboratory X-ray diffractometer and subsequently analysed using different DPPA methods (more details are provided in [4]). The changes in the q values from three different modified Williamson-Hall analyses is shown in Figure 4. The data is relatively noisy and different q values are found depending on the Williamson-Hall equation used. Given these drawback, two main trends are observed.

Firstly, both alloys have falling q values with applied strain. This may be expected in stainless steel because at low strains there can be more screw dislocations due to the greater mobility of edge dislocations, whilst at higher strains dislocation interactions and activation of secondary slip systems can increase the relative amount of edge dislocations [57]. Partial dislocations are a feature of the dislocations in many alloys particularly those with a low stacking fault energy (SFE) in FCC alloys, such as stainless steel. Since partial dislocations have a lower q value than the full dislocation, the fall may also be due to a transition to partial dislocations. In nickel alloys recovery during plastic deformation results in the annihilation of screw dislocations and the formation of low-angle boundaries consisting of edge dislocations [57–59].

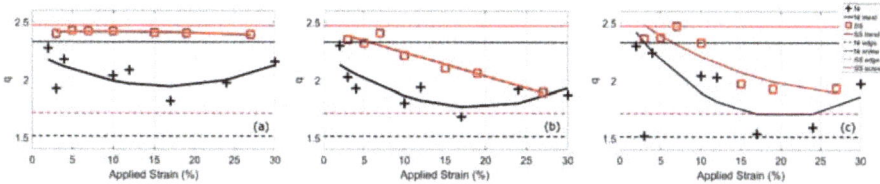

Figure 4. The change in *q* with applied strain of compression tested nickel and stainless-steel alloys. The lines indicate the predicted values for 100% edge and 100% screw dislocations for the two alloys using Equation (2) and ANIZC [17]. The different figures represent results using different modified Williamson-Hall methods; in the terminology of Simm et al. (**a**) is mWH-1, (**b**) mWH-2, and (**c**) mWH-3.

Secondly, the measured *q* values for nickel are closer to their calculated edge *q* values than that seen with the steel. This could be due to the dislocation structures that develop in high SFE alloys, consisting predominantly of edge dislocations [57].

Therefore, although most researchers do not use the *q* values to quantify the microstructure, there is reasonable justification from Figure 4 that *q* can be used to qualitatively evaluate the dislocation types in a sample, as also found by other researchers [24,60,61].

There is an additional benefit to the use of the contrast factor in this way, along with the details it provides of the microstructure, such as the edge to screw ratio. This is that it means that more diffraction peaks can be used during analysis. In FCC alloys, using a standard laboratory X-ray diffractometer, five diffraction peaks can be measured with two that are different orders of the same reflection, 111 and 222. Following the traditional method of using peaks of the same reflection can be problematic because of the relatively low intensity of the 222 peak, as well as the use of less diffraction peaks. An example of the difference between using the traditional and modified (with the contrast factor) approaches is shown in Figure 5, which shows the change in dislocation density using the Warren-Averbach method [4]. The figure shows that using the modified approach reduces the scatter, as given by the deviation from the best fit line. In addition, because different peaks are used that represent different orientations, the results should be more characteristic of the sample as a whole. The slightly different values obtained in some cases shown in the figure may be a result of this greater sampling of the sample.

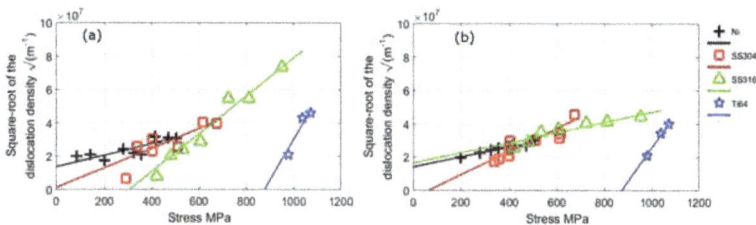

Figure 5. Plots of the change in dislocation density found by Warren-Averbach methods with applied deformation of compression and tensile samples, the stress the samples were at before unloading is shown. A commercially pure nickel alloy, two stainless steel alloys (304 and 316), and a titanium alloy (Ti-6Al-4V) are shown. In (**a**) the traditional approach is used, and in (**b**) the modified approach using the contrast factor is shown. Modified from Simm et al. [4].

3.1.2. Planar Faults

Dislocations are the most important defect contributing to the broadening of peaks in metal alloys, because of their importance in plastic deformation and how they contribute to broadening. The second most important defects are planar faults, as they are believed to be the only other defects in metals that cause significant peak broadening [12]. Other defects such as point defects or second phase particles have strain fields that fall off with distance at a higher rate than the reciprocal of the distance squared, and will not in general cause significant peak profile broadening. Planar faults can be separated into three different types: twin faults, and extrinsic and intrinsic stacking faults. Deformation twinning [62] is an important mode of deformation in HCP alloys [63–65] and cubic alloys with a low stacking fault energy (SFE) such as stainless steel alloys [66,67], whilst in martensitic alloys the lath boundaries can be a twin boundary rather than a dislocation one [68]. Stacking faults also play an important part in the plastic deformation of an alloy [69]. During plastic deformation dislocations can dissociate into two partial dislocations connected by an intrinsic stacking fault, as the partial dislocations can sometimes move more easily than the full dislocation [57]; dissociation occurs more readily in alloys with a low stacking fault energy (SFE). Hence, the ability to quantify planar faults is an important tool for studying plastic deformation. However, because of the small size of planar faults (of the order of the lattice spacing) they are difficult to characterise and quantify.

DPPA techniques are a valuable tool for quantify planar faults in a statically significant manner. As shown in Figure 6, planar faults influence a number of aspects of a diffraction peak including, (a) the peak's broadening, (b) the asymmetry of a peak, and (c) the shift of a peak's position. The figure also shows that the different types of planar faults change the peak in different ways. The values in the figure were calculated using the formulae of Warren [23] (Equation (4)) and derived from the work of Balogh et al. [34] using a fault probability of 5% (the calculations are shown in [13]). For the planar fault calculations using the work of Balogh et al., the planar broadening is convoluted with a theoretical instrumental broadened peak, the full-width of this instrumental peak is deducted from the convoluted peak to give only broadening from planar faults in the figure. The change in the peak profiles of the first six FCC diffraction peaks using the formulism of Balogh et al. are shown in Figure 7, for intrinsic stacking faults and twin faults.

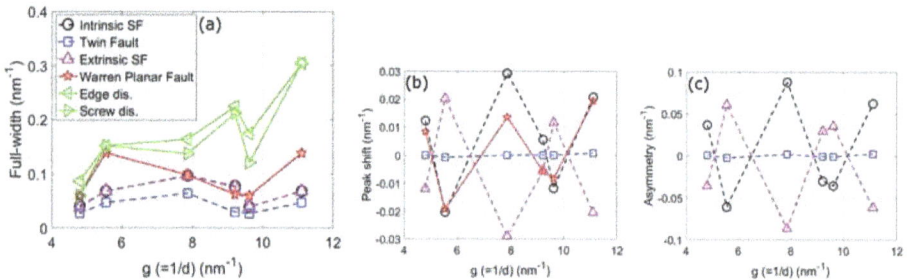

Figure 6. The effect of planar faults on the full-width (**a**), peak shift (**b**), and asymmetry (**c**) for the first six diffraction peaks of a FCC sample (111, 200, 220, 311, 222, 400). The intrinsic stacking faults and twin faults are determined using [34] convoluted to an instrumental broadened peak, and the instrumental FW is subtracted from the final FW. The Warren planar faults use formula from [23]. Values are for a fault probability of 5%. In (**a**) the broadening due to edge and screw dislocations using formula in [11] with a strain value to match the other values.

Reformulated values of ω_{hkl} used in Equation (4) and X_{hkl} used in Equation (5) using Balogh's approach are shown in Table 3. These values along with the accompanying equations provide an additional way to use the approach of Balogh et al., in addition to its part of the MWP procedure [36]. From this table and Figure 6, it can be seen that the approaches of Warren and Balogh et al. provide similar but markedly different results. Note that Warren's formula for broadening (Equation (4)) uses both intrinsic and twin faults, whereas the peak shift formula (Equation (5)) only uses intrinsic faults. The broadening from stacking fault and twin fault using the approach of Balogh et al. are similar, but stacking faults cause more broadening. The magnitude difference of twin and stacking faults is consistent with Warren's formula in Equation (4) where stacking faults contribute 1.5 times the broadening of twin faults. Warren's formula gives a greater broadening of the 200 peak than with Balogh's approach, but the broadening of other peaks is closer. The peak shift changes are relatively close for the two approaches, particularly for 111 and 200 peaks. The biggest difference in the approaches is for asymmetry. Whereas, the work of Warren suggested that only twin faults should cause asymmetric peaks, this is the opposite of that found using Balogh's approach. Warren only defined asymmetry in terms of the 200 peak and commented that it is not possible to determine fault probability accurately from peak asymmetry, and is therefore not included in the figure.

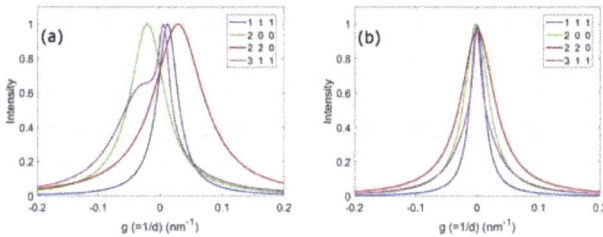

Figure 7. The effect of planar faults on the peak profile of the first four diffraction peaks of a FCC sample (111, 200, 220, 311). The figures are for intrinsic stacking faults (**a**) and twin faults (**b**) using the formulism of Balogh et al. [34] and convoluted to a theoretical instrumental broadened peak.

Table 3. The values of parameters ω_{hkl} used in Equation (4) and X_{hkl} used in Equation (5), for quantifying the planar fault fraction to broadening in FCC alloys [23]. The values ae calculated based on the formulism given by Balogh et al. [34], those by Warren [1] are shown for comparison.

Constant	Planar Fault Type	111	200	220	311	222	400
ω_{hkl}	Intrinsic fault	0.291	0.491	0.701	0.558	0.283	0.496
	Twin	0.195	0.342	0.465	0.217	0.195	0.341
	Extrinsic fault	0.274	0.495	0.696	0.552	0.284	0.491
	Warren	0.433	1.000	0.707	0.452	0.433	1.000
X_{hkl}	Intrinsic fault	0.368	−0.527	0.542	0.087	−0.178	0.271
	Twin	0	−0.019	0	0	0	0.010
	Extrinsic fault	−0.354	0.530	−0.532	−0.085	0.177	−0.265
	Warren	0.250	−0.500	0.250	−0.091	−0.125	0.250

There are several practical problems in being able to use DPPA techniques to quantify planar faults in cubic crystals. The way in which twin and stacking faults contribute to broadening (as shown in Figure 6) means it is very difficult to calculate values for both without making additional assumptions (such as the relative amount of each). However, a bigger problem is separating broadening anisotropy due to planar faults and dislocations. This is particularly a problem when fitting multiple peaks as demonstrated in Figure 8. The figure shows a way to visualise broadening anisotropy caused by dislocations with different q-values (using Equation (2)) and the presence of planar faults (using Equation (4)). The modified Williamson-Hall plots in this figure are for nickel and stainless-steel samples deformed by compression to 10% applied strain. The data is taken from a neutron diffraction experiment at HRPD, ISIS, Oxfordshire detailed in [3,42], but unlike the figures in those works the full-width values shown are averaged over a range of angles between the tensile and diffraction vector. Different q-values causes the value of $gC^{0.5}$ of a peak to shift by different amounts relative to the position of the 200 (and 400) peak, which stays in the same position. Whilst, in the figure, the broadening from planar faults causes reductions in the broadening depending on the peak (and value of ω_{hkl}, see Table 3). Hence, the q values cause movement of data points in the x-direction ($gC^{0.5}$), whereas planar faults cause movement of the data points in the y-direction (integral breadth). The success with which a definition describes the broadening anisotropy, can be given by the nearness to a fitted straight-line in Figure 8, root-mean-squared-errors of the data to the fitted lines are given in Table 4. The figure and accompanying table illustrates the difficulty that is encountered when trying to describe broadening anisotropy in terms of both q values and planar faults. For example, the root-mean-squared-errors can be very similar for very different descriptions of broadening anisotropy, such as those between edge and screw dislocations of nickel in the table. This problem is exacerbated because the description of broadening anisotropy by dislocations or planar faults are only an approximation, different descriptions of broadening anisotropy would be given by using different Williamson-Hall equations, or Warren-Averbach equations as shown in Figure 4. These difficulties are why it is common to only fit to either the q-value or only for planar faults.

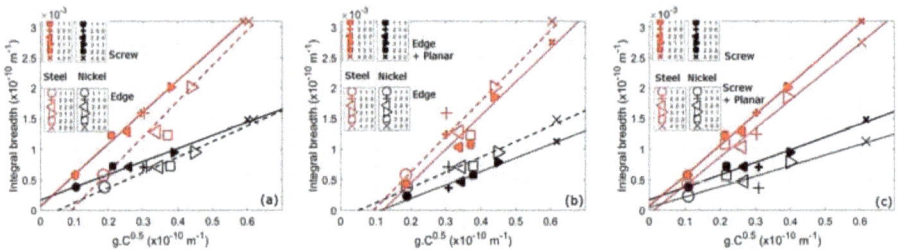

Figure 8. Modified Williamson-Hall plots of integral breadth against $gC^{0.5}$ for nickel and stainless-steel samples deformed to 10% applied strain. In (**a**) plots for different values of q of 1.5 (approximately edge) and 2.5 (approximately screw) using Equation (2). In (**b,c**) shows the effect of planar faults on the values in (**a**). The integral breadths are reduced by broadening from planar faults from Equation (4) of $\frac{(1.5\alpha+\beta)}{a}\varnothing_{hkl}$, using a nominal value of $1.5\alpha + \beta = 5.10^{-4}$. Straight lines have been fitted to the integral breadth against $gC^{0.5}$ for the different sets of values. The root-mean-squared-error (rmse) values for each of these are shown in Table 4. The data is for neutron diffraction measurements taken at HRPD, ISIS, Oxfordshire taken from [3]. The values are averaged over angles between the compression and diffraction vectors, between 0 and 90°.

Table 4. The root-mean-squared-error (rmse) values for the straight-line fits shown in Figure 8. For edge dislocation calculations q is taken as 1.5, for screw dislocations taken as 2.5, and for mixed as 2.0.

Alloy	Edge	Screw	Mixed	Edge + Planar Faults	Screw + Planar Faults	Mixed + Planar Faults
Steel	28.3×10^{-5}	6.8×10^{-5}	14.1×10^{-5}	22.1×10^{-5}	17.8×10^{-5}	10.3×10^{-5}
Nickel	7.1×10^{-5}	7.7×10^{-5}	2.9×10^{-5}	5.8×10^{-5}	13.6×10^{-5}	8.7×10^{-5}

Instead of using a whole powder diffraction profile approach (as shown in Figure 8), the planar faults can instead be determined from different orders of the same peak. These approaches are less prone to the problems presented above of incorporating planar faults as another fitting variable within a multiple peak fitting algorithm. In addition, they sample grains of the same orientation with the same strain distribution relative to the diffraction vector, which helps to reduce the problems with broadening anisotropy discussed in the next section, which are due to the heterogeneity of plastic deformation.

The crystal size broadening from different hkl reflections can be used to quantify the amount of planar faults using Equation (4). Since the crystal size (D) is not known, this means from Equation (4) that the size values from two different reflections are required (e.g., 111/222 and 200/400). The value of the planar fault density can be found by rearranging Equation (4), and assuming crystal size broadening is a constant for all peaks, to give:

$$(1.5\alpha + \beta) = a\frac{(1/Deff_{200}) - (1/Deff_{111})}{(\varpi_{200} - \varpi_{111})}, \tag{7}$$

In Figure 9 are Williamson-Hall plots of the same nickel and stainless-steel samples, compression tested to 10% applied strain, used above. The lines in the plot are Williamson-Hall fit lines (using g^2 against FW) for the 200/400 and 111/222 peaks of the alloys. The values of $(1/Deff_{200})$ and $(1/Deff_{111})$ are the intercepts of the lines in this figure.

In a similar manner Equation (5) needs to be rearranged to provide to provide an equation to quantify planar faults in deformed alloys. The intergranular strains, or the position of a diffraction peak, of a sample change due to plastic deformation, and can be predicted by crystal plasticity models [48]. The modelling is normally done on changes in intergranular strains during an in situ test (i.e., when there is an applied load), although these strains are less after unloading they are still present as shown in Figure 10. Since the intergranular strains due to plastic deformation are not generally known, it is not sufficient to use Equation (5) with a single diffraction peak and the lattice spacing measurement from an un-deformed sample. Instead the equation can be rearranged to give the intrinsic stacking fault fraction in terms of the difference in intergranular strains of two orders, as follows:

$$\alpha = \frac{4\pi}{\sqrt{3}(\chi_{111} - \chi_{222})}(\Delta g_{111}/g_{111} - \Delta g_{222}/g_{222}), \tag{8}$$

The quantity of planar faults in the nickel and stainless samples were found using Equation (7) (combined with Equation (1), i.e., using g not g^2 as shown in Figure 9) and Equation (8), and are shown in Table 5. The calculations were done using the average integral breadth, and average change in $\Delta g/g$ across the angles measured (between 0 and 90°). The values of planar faults in stainless steel are ~0.15%, whereas those for nickel are close to zero and often negative indicating the expected changes given by planar faults is not met. These results are consistent with the expectation that the quantity of planar faults is higher in stainless steel than in nickel. In addition, although not quantified the asymmetry changes shown in Figure 10 which show a greater difference in asymmetry between 111 and 222 peaks, or 200 and 400, in steel than in nickel is also consistent with these results. The change in intergranular strains of 200 and 400 peaks of stainless steel are unusual as they change in the opposite way than expected, i.e., g/g is higher for 200 the opposite of what is expected from the values in Table 3). This leads to a negative value for the fault percentage, although the absolute magnitude of the value is close to that found from the 111/222 peaks.

Table 5. The values of stacking fault (α) and twin faults (β) of stainless steel deformed to 10% applied strain using the data in Figures 9 and 10. The values ae calculated based on Equations (7) and (8) using values in Table 3 found from work of Balogh et al. [34], and Warren [1]. The values are given as percentages.

Alloy	Peaks	Warren Broadening (1.5α+β)	Balogh Broadening (1.5α+β)	Warren Intergranular strains (α)	Balogh Intergranular strains (α)
Steel	111/222	0.091	0.298	0.187	0.129
	200/400			−0.145	−0.137
Nickel	111/222	−0.058	−0.191	−0.059	−0.040
	200/400			−0.008	−0.007

There is an additional advantage of using the approach shown here to calculate planar faults. This is that the use of both the peak position and peak broadening can in theory offer a way to separate stacking and twin faults. Further investigation on alloys with known changes in these two could be used to establish whether this is possible, whether peak asymmetry is due to twin or stacking faults, and which constants are best to use in Equations (7) and (8).

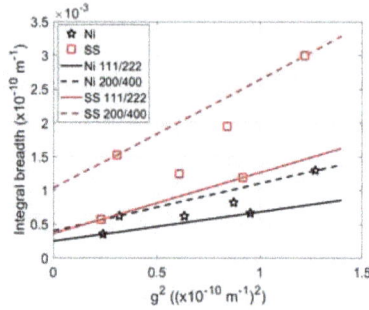

Figure 9. The full-width of the first six diffraction peaks (111, 200, 220, 311, 222, 400) of nickel and stainless-steel samples deformed to 10%. The data is for neutron diffraction measurements taken at HRPD, ISIS, Oxfordshire taken from [3]. The values are averaged over angles between the compression and diffraction vectors, between 0 and 90°.

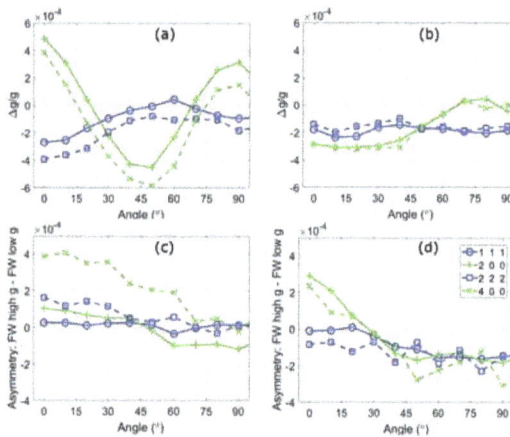

Figure 10. The measured peak positions and peak assymetry of a stainless steel (**a,c**) and nickel (**b,d**) alloy after compression testing to 10% applied strain, at different angles between the compression direction and the diffraction vector [3,42]. The change in peak position (**a,b**) are relative to the peak positions of an undeformed sample. The assymetry of a peak (**c,d**) is found from a pseudo-Voigt which has different full-width values either side of it's centre, the assymetry parameter is the full-width of the high g side minus the full-width of the low g side.

3.1.3. Plasticity Approach

In Figure 11 are full-width plots of compression deformed nickel and stainless-steel alloys at different applied strains. The figure is plotted at different angles between the compression direction and the diffraction vector, and shows that the full-width of a diffraction peak can vary significantly at the different angles. For example, for stainless steel after 10% strain the full-width of the 111 peak falls by ~20%, the 200 peak by ~20% and the 220 peak increases by ~40% from the values at 0°. Whereas for nickel after 10% strain, the full-width of the 111 peak falls by ~20%, the 200 peak by ~30% and the 220 peak increases by ~25% from the values at 0° These variations exist at all strains measured, and for both the fatigue and compression samples shown in the figure. Furthermore, the same broad changes in the diffraction peak broadening with angle is observed for all samples. Since, the magnitude of the full-width values at 2% strain are lower the relative error in the full-widths will be higher, which may explain the slight difference to the higher applied strains. The broadening anisotropy with angle are observed for both the low stacking fault energy stainless-steel and the high stacking fault energy nickel; including a fall in the full-width of most peaks from 0° to 90°, minima for 200 and 111 peaks near 45°, and maxima for the 220 peak near 45°.

There are a number of possible causes for these changes in broadening with angle: (1) variations in dislocations density in different orientations; (2) variations in the arrangement of dislocations in different orientations which can contribute to broadening through both crystal size and dislocation arrangement [4]; (3) variations in the dislocation slip systems present (or slip anisotropy); (4) variations in the character of dislocations, i.e., whether they are edge or screw, in different orientations; (4) variations in other defects, such as planar faults, in different orientations; (5) variations in the spread in the intergranular strains in different orientations. The plasticity approach predictions of the contrast factor shown in Figure 12 is based on No. 3, variations in the dislocation slip systems present. The prediction shares many of the features of the measured full-widths of Figure 11, hence the changes of broadening with viewing angle are likely to be mainly due to variations of slip systems in different orientations. These measurements were also done on rolled samples [3], and it was shown that the change in broadening also showed a good correlation with predictions based on the activity of different slip systems.

The use of the plasticity approach to predict contrast factors, based only on slip anisotropy, works better for low SFE alloys like steel, as shown in [42] or by comparing Figure 11 with Figure 12. But what is interesting is that the approach works for nickel even after large amounts of deformation [3,42]. In alloys with high SFE there is considerable amounts of cross-slip and recovery, which leads to dislocation structures like those shown in Figure 13 to develop. Although, it may be expected that this recovery process would remove most of the mobile dislocations that are predicted by the crystal plasticity models to accommodate the imposed loads, these plots of broadening with angle suggest that this is not the case.

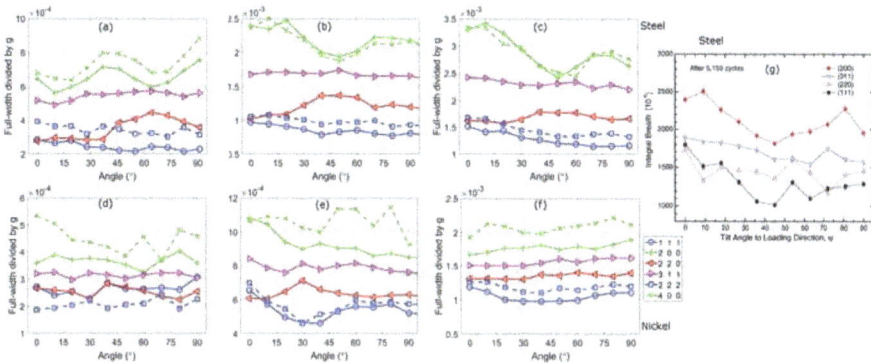

Figure 11. The measured full-width divided by *g* of a stainless steel (**a–c**) and nickel alloy (**d–f**) after compression testing, at different angles between the compression direction and the diffraction vector [3,42]. The figures are at increasing applied strain. For stainless steel the figures are at (**a**) 2%, (**b**) 10%, and (**c**) 16% applied strain. For nickel, the figures are at (**a**) 2%, (**b**) 10%, and (**c**) 30% applied strain. In (**g**) the integral breadth divided by *g* of a stainless steel alloy after fatigue testing, at angles between the tensile and diffraction vector [27]. The instrumental broadening has been subtracted from the measured values. In (**g**) are changes in the integral breadth (a function of FW and the shape of the tail of the peak) of a stainless steel after fatigue testing taken from Wang et al. [27].

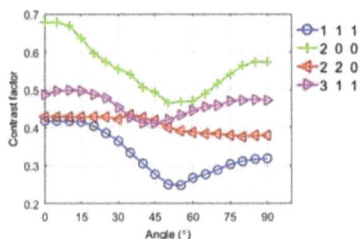

Figure 12. The predicted change in contrast factor for angles between compression and diffraction vectors using a Taylor model. Adapted from [42].

It has been observed by detailed TEM that the dislocation structure can vary significantly with orientations [70,71]. However, to be able to separate the slip anisotropy effect from other changes in DPPA methods, such as the dislocation structure, is difficult. This is partly because it is the product of the dislocation density and the contrast factor that is obtained by DPPA methods. But also due to practical and mathematical limitations of DPPA methods [4,5], such as: each method can give a slightly different results, limitations of separating size and strain components, along with the ambiguity of the parameters that are obtained, such as crystal size. A popular approach to take is to obtain the contrast factor from a crystal plasticity model and then calculate the dislocation density, and other parameters such as crystal size, using this value of the contrast factor. This was the approach taken by Guiglionda et al. [44], who obtained values of the dislocation density for different texture components which changed by a factor of 2 in an aluminium alloy. In their approach the edge to screw ratio was assumed to be 50:50 and no account was taken of dislocations interactions, i.e., all dislocations were assumed to be given by the Orowan equation (Equation (6)). But these assumptions need not be the case, and as shown by Simm et al. [42] can have a large influence on the calculated contrast factors. In their work Simm et al., using the data shown in Figure 11 showed that the difference between the broadening anisotropy of stainless steel and nickel can be described by nickel having: (1) a greater amount of edge dislocations, (2) less mobile dislocations or more cross-slip, and (3) a variation in dislocation density based on the Taylor factor, of up to ~10% in different orientations, which is not found for steel samples. These differences are consistent with differences that may be expected between the alloys. However, the approach does not fully address the cause of the broadening anisotropy because of its simplicity. A more comprehensive approach would instead incorporate more detailed models of work-hardening, along with more detailed descriptions of how different dislocation arrangements contribute to peak profiles, such as is being developed by Bertin and Cai [55].

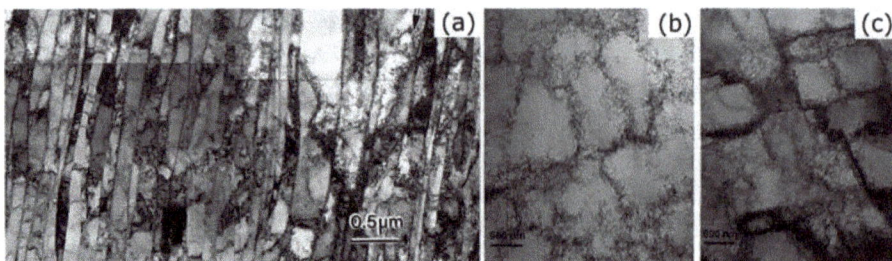

Figure 13. High purity polycrystalline nickel cold rolled to a cold reduction of 95% (**a**). Adapted from Hughes and Hansen [72]. High purity polycrystalline nickel deformed by uniaxial tension to (**b**) 10% and (**c**) 18% applied strain. Adapted from Keller et al. [59].

There are two important implication of broadening anisotropy in different orientations due to slip anisotropy. They can introduce errors in the results of DPPA and they limit the use of DPPA with other techniques.

If the correct value of the contrast factor is not used, because for example slip anisotropy is not accounted for, this can introduce errors in the results of DPPA methods. The potential contribution of this error was assessed in [42] by performing a Warren-Averbach analysis on diffraction patterns obtained at different angles between the tensile and diffraction vectors from the deformed nickel and stainless steel samples discussed previously. The analysis was done using different orders of the same diffraction peak for the 111/222 and 200/400 peaks separately. It was found that the dislocation density fell by ~50%, for both 111/22 and 200/400 analyses, from the values found when the diffraction vector and compression direction were parallel to each other to those found when they were perpendicular. The magnitude of these changes in dislocation density, and the overall trends, were consistent with the predicted changes in the contrast factor calculated by the plasticity approach, due only to slip anisotropy. In addition, the crystal size increased by ~40% with the fall in dislocation density, which may be due to the difficulty in separating size and strain components in the Warren-Averbach approach [5]. Therefore, the possible errors to the results of DPPA by incorrect calculation of the contrast factor are significant. The problem is that in many cases the active slip systems are not known. Even when the imposed deformation is known, limitations in crystal plasticity models and difficulties in relating the model to dislocations, means there will always be an uncertainty in knowing the actual average contrast factor.

A second implication of broadening anisotropy, is it limits the use of DPPA with other techniques. The average value of the broadening of a sample, or dislocation density and crystal size, can be related to other techniques. For example, TEM measurements [73] and positron annihilation spectroscopy [74], can provide dislocation density values to compare with those by DPPA [4,75]. However, broadening anisotropy does limit the use of DPPA alongside other techniques in quantifying differences in the microstructure in different orientations. These differences are crucial to be able to understand plastic deformation in general, and to verify or otherwise plasticity models, as well as to understand the kinetics of recrystallization [76–78] and failure [79,80] of a material.

A diffraction peak represents a fraction of the crystals within a sample based on their orientations. Hence, different diffraction peaks give information about different groups of orientations. This can be seen from the inverse pole figure plots in Figure 14 for the transverse and axial peaks of an untextured FCC alloy. Broadening anisotropy with g, and the difficulty in quantifying it, means that different diffraction peaks (such as 111 and 200) cannot be compared directly to provide the quantity of different defects in different orientations. In addition, because of the changes in the full-width of a given diffraction peak (e.g., 111) with viewing angle (such as shown in Figure 11), are largely due to changes in the contrast factor, it is difficult to even compare the same diffraction peak measured in different directions relative to the sample. This was highlighted by Das et al. [81] (Figure 15) who considered the change in shear strain by digital image correlation, grain orientation spread by EBSD, and the width of diffraction peaks from neutron diffraction, of a metastable austenitic stainless steel. The orientation spread and shear strain values for different grains families with 111, 200, 220, 311 plane normals parallel to the axial and transverse direction can be compared with each other. But the same comparison cannot be done for the integral breadth. For example, it can be said that the shear strain is on average higher in grains with 220 plane normals parallel to the tensile direction, than the other orientation groups. But the broadening of 111, 200 and 220 peaks cannot be compared directly because of broadening anisotropy with g. In addition, the broadening anisotropy with viewing angle, complicates comparison of assessing the broadening of a particular peak in the axial and transverse directions. That is the lower expected contrast factor for transverse peaks compared to axial peaks (i.e., 0° and 90° in Figure 12), mean that more broadening in the axial direction are not necessarily due to differences in dislocation density, but could be due to differences in contrast factor.

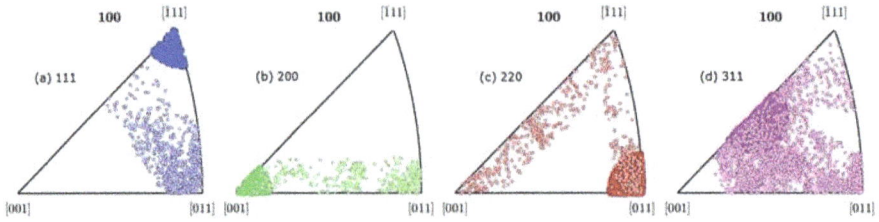

Figure 14. Inverse pole figure (IPF) plots to represent the different orientations given by diffraction peaks in the axial (full-circles) and transverse (empty circles) directions. (**a**) 111 peak, (**b**) 200 peak, (**c**) 220 peak, and (**d**) 311 peak. Where the IPF plots are with reference to the axial direction ([100]) and use a random orientation distribution of 14,000 grains. Note that if the refence was the transverse direction ([001]), the figure would look the same but the icons would be reversed (empty circles would become full circles and vice versa). Calculated using MTEX [82].

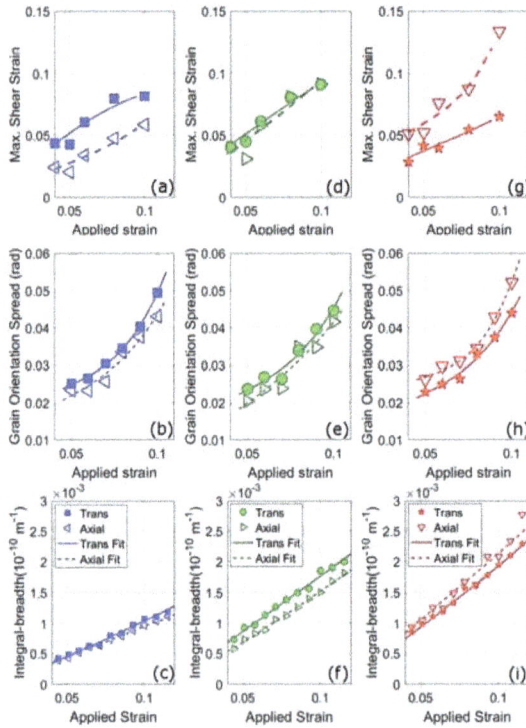

Figure 15. In situ measurements of the maximum shear strain, measured by digital image correlation in an SEM (**a,d,g**), grain orientation spread, measured by EBSD, (**b,e,h**), and integral breadth of diffraction peaks, measured using neutron diffraction at ENGIN-X, ISIS, Oxfordshire (**c,f,i**). The quantities are measured during tensile tests and plotted against applied strain. In (**a–c**) are orientations that contribute to the 111 diffraction peak, in (**d–f**) the 200 diffraction peak, and in (**g–i**) the 220 diffraction peak. Adapted from Das et al. [81].

3.2. Hexagonal Close Packed Alloys

3.2.1. Homogeneous Approach

The issue of peak broadening anisotropy is further complicated in hexagonal close packed (HCP) alloys (such as titanium, zirconium, magnesium), relative to cubic alloys. This is partly a practical issue due to the increased number of different slip systems, but also a result of the variations in the ease in the activity of these different slip systems [57,83,84]. For example, whereas in FCC alloys there is one slip system type {111} [110], in HCP alloys there are several active slip system types. In HCP alloys there are slip systems with both different Burgers vectors (<a> and <c+a>), and different slip planes (basal, prismatic and pyramidal). The homogeneous approach, as outlined by Dragomir and Ungar [30] and based on Equation (3), has been used in a number of works to evaluate the slip activity of different Burgers vector types [85–92]. However, because of the difficulty in quantifying the dislocation types in HCP alloys, corroborating characterisation test results to verify the approach is problematic.

The homogeneous approach in HCP alloys is based on their being a relationship between x (where x is 0 for the prismatic plane and a maximum for the basal plane) and the full-width. By combining Equation (1) with Equation (3), and assuming that crystal size broadening can be ignored and rearranging, the following equation is found:

$$\left(\frac{FW_{hkl}}{g}\right)^2 \approx f_m \rho \left\{\overline{C_{hk.0}}\left(1 + q_1 x + q_2 x^2\right)\right\}, \tag{9}$$

where $x = \frac{2l^2}{3(ga)^2}$ and $f_M^2 = b^2\left\{a_0 \ln(M+1) + b_0(\ln(M+1))^2 + c_0(\ln(M+1))^3 + d_0(\ln(M+1))^4\right\}$.

A similar approximation can be done for the Warren-Averbach method [23], a DPPA method that uses the Fourier coefficients (A_L) of a peak at different values of the Fourier length (L). The Warren-Averbach equation can be expressed in terms [4] of dislocations, with a density ρ, by the following equation:

$$\ln A_L = \ln A_L^S - \rho g^2 B \overline{C} L^2 f_{Wilk},$$

if size broadening is negligible $\ln A_L^S$ will be close to zero, hence:

$$-\frac{\ln A_L^S}{g^2} \approx \rho B L^2 f_{Wilk}\left(\overline{C}\right), \tag{10}$$

where $B = \pi b^2/2$, b being the magnitude of the Burgers vector of dislocations, f_{Wilk} the Wilkens function [93], and $\overline{C}$ the average contrast factor.

Hence from Equations (9) and (10) it may be expected that $(FW_{hkl}/g)^2$, $(\beta_{hkl}/g)^2$ (since full-width and integral breadth can be interchanged in the Williamson-Hall equation), and $-\ln A_L^S/g^2$ should change smoothly with x. For this to be the case we are assuming that: the Williamson-Hall and Warren-Averbach equations used are valid, and there is minimal size broadening. This is in general what is found for Ti-6Al-4V deformed by uni-axial tension to different strains, using the full-widths (Figure 16a), the integral breadths (Figure 16b), and the Fourier coefficients (Figure 17). Hence, this gives justification for the use of these equations and the assumption of minimal size broadening. Another feature that is evident from comparing the different figures is that for a given sample, the same changes are observed in $(FW_{hkl}/g)^2$, $(\beta_{hkl}/g)^2$, and $-\ln A_L^S/g^2$. For example, at low strains both $(FW_{hkl}/g)^2$, $(\beta_{hkl}/g)^2$ fall gradually with x, but at higher strains fall then increase, and for a given strain change in a similar manner. Although, the changes of $-\ln A_L^S/g^2$ vary somewhat with the Fourier length, they are comparable to the changes of the full-width or integral breadth of the 5% sample. Therefore, even though, as noted in Section 2.1, the Williamson-Hall method can be problematic for quantitative analysis of broadening anisotropy and the full-width is not the best measure for dislocation broadening [21], the use of the full-width through the Williamson-Hall method provides a similar description of broadening anisotropy to using integral breadths or Fourier coefficients for this alloy. This gives a justification for the common use of the full-width to describe broadening anisotropy.

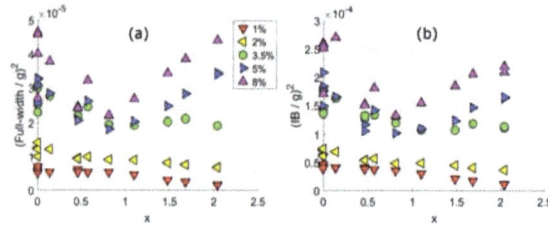

Figure 16. Uni-axial tensile tested Ti-6Al-4V samples measured at 90° between tensile and diffraction vector (the transverse direction), adapted from Simm et al. [42]. In (**a**) are plots of $(FW/g)^2$ against x at different applied strains, where FW is the full-width. In (**b**) are plots of $(IB/g)^2$ against x at different applied strains, where IB is the integral breadth.

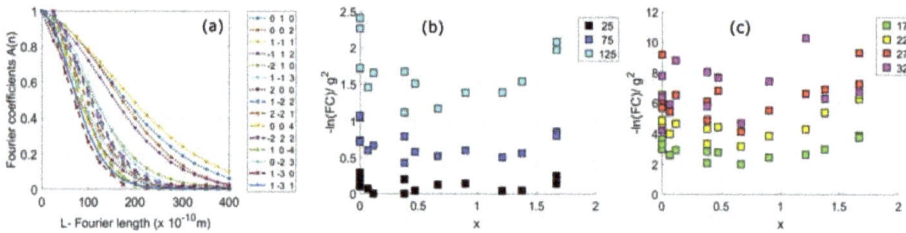

Figure 17. Uni-axial tensile tested Ti-6Al-4V sample measured at 90° between tensile and diffraction vector (the transverse direction) after 5% applied strain, this is the same data shown in Figure 16. In (**a**) the physically broadened Fourier coefficients (A_L) plotted against the Fourier length. In (**b**,**c**) are plots of $-ln(A_L)/g^2$ against x for different Fourier lengths (given in legend in units of 10^{-10} m) for the first 15 HCP diffraction peaks.

From the changes in full-width, or integral breadth, shown in Figure 16 the following trends are observed. Up to a strain of 3.5% the $(FW/g)^2$ falls almost linearly with x. Whereas for higher strains there is a bigger fall in $(FW/g)^2$ at low x, followed by an increase after $x > 1$. From Equation (9) this would be interpreted as $q_1 < 0$ and $q_2 \sim 0$ for samples up to 3.5%, and at higher strains $q_1 < 0$, but lower than q_1 at lower strains, and $q_2 > 0$. Which is what is found from fits to a modified Williamson-Hall equation, as shown in Figure 18b.

In Figure 18c,d these q values have been used to determine the slip activity, using two slightly different approaches. The first approach (Figure 18c) uses a similar approach to Dragomir and Ungar [30]. The second approach (Figure 18d) is a slight modification to the previous one, whereby the available slip system types are reduced to those expected to be the most important types in titanium [83]. This type of approach, involving reducing the available slip systems has been used by different authors [31,85]. These approaches are generally only used to predict the quantity of different Burgers vector types, <a>, <c+a> and <c>; although Mathis et al. [85] also used the approach to find the relative quantity of two different slip system types, prismatic and basal <a>. The reason for this is that using two variables (q_1 and q_2) to find details of any more than three parameters is impractical because there are too many possible combinations. However, looking at the contributions to calculated values is useful to understand the contributions to the calculated <a>, <c+a> and <c> values. Figure 18 shows that <a> type dislocations dominate for both approaches and across all applied strains. Furthermore, when the individual slip system types are considered prismatic <a> is the most active slip system type (this is because prismatic <a> and pyramidal <a> are the only slip system types with $q_1<0$ and $q_2>0$, with prismatic <a> having q values with larger magnitude). This may be expected given that <a> dislocations are much easier to activate than those with a c-component; furthermore <c> dislocations which are the hardest to activate have the lowest percentage. However, given that <a> slip does not satisfy the von Mises criterion of needing at least five independent slip systems for homogenous plastic deformation of a polycrystal [62], more dislocations with a c-component would be expected. This could be accounted for by some grains deforming more than others in a systematic way, but would go against the assumption in the Homogeneous approach that all grains are the same. There is another problem with the approaches prediction of almost 100% <a> dislocations to explain the shape of FW against x. This is the increase in broadening at high x which cannot be explained by <a> dislocations that have vanishing contrast at high x values (Figure 3).

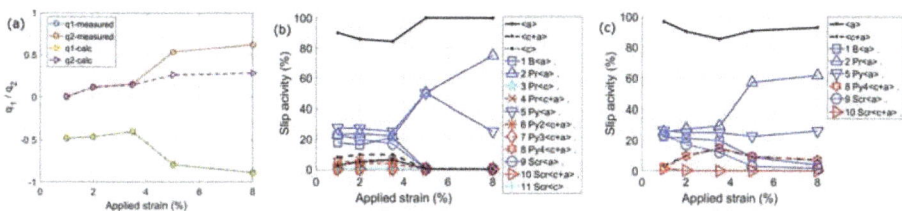

Figure 18. Uni-axial tensile tested Ti-6Al-4V samples measured at 90° between tensile and diffraction vector (the transverse direction), adapted from Simm et al [42]. In (**a**) are the fitted q values using a modified Williamson-Hall equation and the full-width values from Figure 16a. The calculated q values from the slip systems in (**c**) are also shown. In (**b,c**) are calculated activities of different slip system types, see Simm [3]. The method in (**b**) is similar to that of Dragomir and Ungar [30], whereas in (**c**) the available slip system types have been reduced to those expected in Ti-6Al-4V.

The choice of these slip systems from the q values also influences the dislocation density. This is not the case in cubic crystals since the value of b^2C_{h00} varies very little between edge and screw dislocations. However, for HCP alloys the value of $b^2C_{hk.0}$ can vary significantly between different slip system types as shown in Figure 3 and Table 1. Hence, because of the relationship in DPPA methods between the dislocation density and the contrast factor (as shown in Equation (1)), the choice of slip systems can have a large effect on the dislocation density obtained. To obtain the dislocation density it is first necessary to obtain a value for the average $b^2C_{hk.0}$ based on the expected slip systems present. $\overline{b^2C_{hk.0}}$ can be found in several ways including by averaging over the dislocation types (<a>, <c+a> and <c>) (Equation (11)), or the slip system types (the 11 slip system types in Figure 3) (Equation (12)). Where $\overline{C_i^{disloc.}}$ is the average contrast factor for the <a>, <c+a> and <a> dislocations, using the averaged contrast factor values of the different slip system types $\overline{C_i^{slipsys}}$ shown in Figure 3.

$$\overline{b^2C_{hk.0}} = \sum_i^3 b_i^2 \overline{C_i^{disloc.}}, \tag{11}$$

$$\overline{b^2C_{hk.0}} = \sum_i^{11} b_i^2 \overline{C_i^{slipsys}}, \tag{12}$$

Using the slip system activities shown in Figure 18c,d and Equations (11) and (12), the values of $\overline{b^2C_{hk.0}}$ shown in Figure 19 were calculated. These different values vary by as much as 30% with applied deformation for a particular methodology, and the values from the different methodologies can vary by up to 80%. In general, the 'All disloc. avg.' method is used in research papers, which uses all slip systems and Equation (11). But there are problems with this due to the averaging of Equation (11), because it is unlikely that all slip system types defined in Table 1 exist in the same quantity (e.g., in magnesium alloys a greater proportion of basal <a> dislocations would be expected than in a titanium alloy). However, the use of Equation (12) is also problematic as it is unlikely that the individual slip system types can be obtained from the two q values. Any method to determine $\overline{b^2 C_{hk.0}}$ is fraught with potential errors, therefore it may be advantageous to assume a constant value of $\overline{b^2 C_{hk.0}}$ for a particular set of data, unless a good justification for doing otherwise can be established.

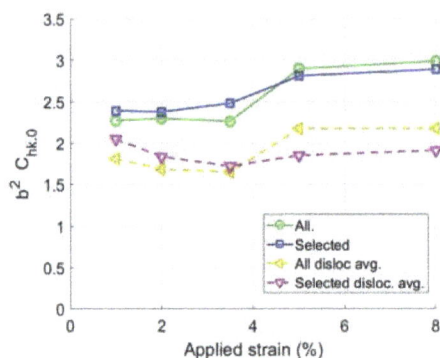

Figure 19. Changes in $\overline{b^2 C_{hk.0}}$ (since b varies with slip system) from the slip activities determined in Figure 18, for Ti-6Al-4V deformed by uni-axial tension to various strains. The results labelled 'All' use the slip activities in Figure 18c using all slip system types, and labelled 'Selected' use the slip activities in Figure 18d which use selected slip system types. Those labelled 'disloc. avg.' take the average values of the different dislocation types (<a>, <c+a> and <c>), whereas those without sum up the individual slip system types.

3.2.2. Plasticity Approach

In Figure 20 are full-width divided by g plots at different angles between the compression/tension direction and the diffraction vector, for a zirconium alloy (a) and a magnesium alloy (b) (taken from [94]). The figure shows, in the same way as seen with the FCC samples, that the full-width of a diffraction peak can vary considerably at different angles. These differences show that the assumptions used in the homogeneous approach, that all grains deform in the same way, is not valid. To examine how this would influence the use of the homogeneous approach it is worth considering the change in full-width with x.

In Figure 21a–c are plots of the change in full-width with x for three different HCP alloys for axial and transverse orientations: in (a) a titanium alloy [42], (b) a magnesium alloy [94], and (c) a zirconium alloy [94]. In the figure the angle between tensile axis and diffraction vector is $0°$ for the axial direction and $90°$ for the transverse. The changes of FW/g with x are similar for the zirconium alloy and Ti-6Al-4V. The alloys also both display a marked difference in the change in FW/g with x in the two measurement directions. Whereas, for the magnesium alloy there is very little difference in the two measurement directions, with a change in FW/g that is similar to that found for titanium and zirconium alloys in the axial direction. In Figure 21d are plots of FW/g against x for a deformed magnesium alloy (from Mathis et al. [85]) and a deformed zirconium alloy (from Long et al. [91]). The magnesium alloy is measured in the transverse direction and the zirconium alloy at an angle between transverse and axial directions. In general, the changes of FW/g with x for the zirconium alloys is similar to that of the titanium and zirconium alloys shown in Figure 21a,c and the magnesium alloy is slightly different and similar to that of the magnesium alloy shown in Figure 21b.

In general, in HCP alloys with a lower c/a ratio, such as titanium and zirconium alloys, will deform preferentially by prismatic <a> slip. Whereas, those with a higher c/a ratio, such as magnesium and cobalt alloys, will deform preferentially by basal <a> slip. Hence, the similarities in broadening anisotropy between zirconium and titanium alloys, and their difference with the broadening anisotropy of magnesium alloys may be expected. To examine these differences, it is worthwhile to consider how the activity of different slip system types can vary in the different alloys and in different orientations.

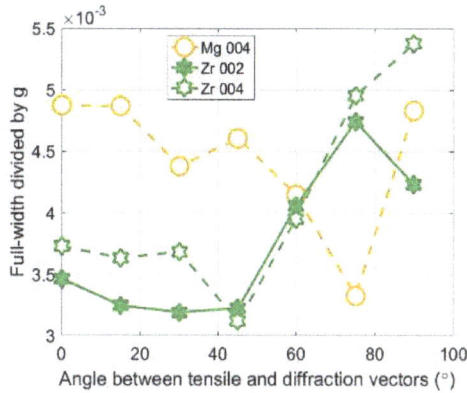

Figure 20. The change in 0002/0004 full-width with angle between the tensile/compression direction and the diffraction vector, for a magnesium alloy and a zirconium alloy. Adapted from Simm et al. [94].

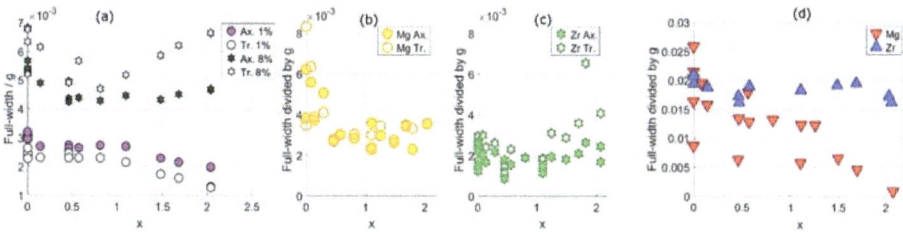

Figure 21. The change in full-width divided by *g* with *x* for different HCP alloys. In (**a**) a titanium alloy (Ti-6Al-4V) deformed in tension to strains 1% and 8% in the axial and transverse directions. In (**b**, **c**) measurements at HRPD on a magnesium alloy compressed to 7% (**b**) and a zirconium alloy deformed in tension to 5% (**c**) (data in b and c taken from Simm et al. [94]). In (**d**) a zirconium alloy deformed to 5% by compression and measured at an angle of ±32° between the transverse direction and the diffraction vector (data taken from Long et al. [91]), and a magnesium alloy deformed to 6% by compression measured in the transverse direction (data taken from Mathis et al. [85]).

There is an increased difficulty in predicting the deformation and activity of different slip systems in HCP alloys compared to cubic alloys [52,95]. This is due the greater number of slip systems that can be active and the variations in their ease of activity. The shear stress needed to activate different slip system types varies considerably, known as the critically resolved shear stress (CRSS), for different slip systems [57,84,96]. Those systems with a <c> component are much harder to activate than the <a> slip system types, and certain <a> slip system types are preferred in different alloys. In addition to this, deformation twinning is an important mode of deformation in HCP alloys [95,97]. The result of this is that deformation is more heterogeneous in HCP alloys [98–100], where different grains or orientations can have markedly different deformation microstructures. The most successful models to predict the slip systems and changes during deformation are crystal plasticity finite element models [51] and viscoplastic self-consistent models [52]. The simpler use of the Schmid factor to quantify slip activity has however shown some success in understanding the deformation of HCP alloys [101–104]. The Schmid factor can help to quantify what slip systems may be likely to be active in different orientations. The higher the value of the Schmid factor multiplied by the CRSS of a slip system, the more likely it is to be active.

In Figure 22a and b are plots of the change in Schmid factor with x for the most important slip system types in HCP alloys [84]. The figure shows that there are large changes in the Schmid factor with x, and the changes are different for the different slip system types and measurement directions. For example, prismatic <a> slip is more likely to be activated when the normal of the prismatic plane is parallel to the tensile direction ($x = 0$ in axial plot, or $x\sim2$ in the transverse plot), and less likely when the normal of the basal plane is parallel to the tensile direction ($x\sim2$ in axial plot, or $x = 0$ in transverse plot).

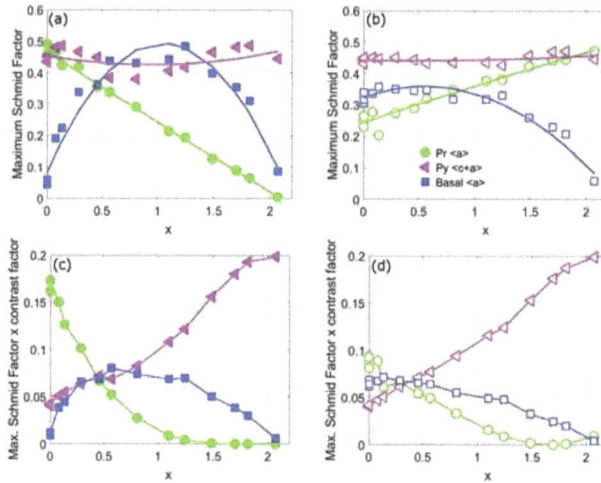

Figure 22. The maximum Schmid factor of different slip system types at different angles of x, for the axial (**a**) and transverse directions (**b**). And the average contrast factor multiplied by the maximum Schmid factor, for axial (**c**) and transverse directions (**d**). The slip systems are those expected to be most important in HCP alloys, prismatic <a> (Pr <a> in green), pyramidal <c+a> (Py <c+a> in pink) and basal <a> (Basa l<a> in blue).

To get an indication of how these predicted changes in slip activity would influence the broadening, the Schmid factors are multiplied by the contrast factors values at different values of x (Figure 22c,d). This is done for each of the different slip system types using the maximum Schmid factor and the average edge contrast factors values at different values of x (from [30]). Both of these approximations, the use of the maximum Schmid factor and the average contrast factor values, doesn't significantly affect the overall trends as shown in Simm et al. [42]. There are two observations from Figure 22c,d that are of particular interest regarding the broadening anisotropy of the alloys shown in Figure 21. Firstly, in these plots the difference in the two measurement directions is much smaller than that seen for the contrast factors. Secondly, basal <a> has a higher value at $x > 1$ than prismatic <a>, and hence should contribute more to broadening at $x > 1$. These changes go against the broadening differences observed: (a) between alloys that deform principally by basal (magnesium alloys) and prismatic (titanium and zirconium alloys) slip, and (b) between broadening in the axial and transverse directions observed in alloys that deform principally by prismatic slip. With regards to (a), from Figure 21 the full-width of magnesium alloy tends to fall more with x than seen for zirconium and titanium alloys. But this is the opposite of what is expected from the predictions in Figure 22c,d. Whereas, with regards to (b), the expected broadening from prismatic <a> slip falls to zero with increasing x in both directions. Hence, the dominant slip system is unable to explain the difference in broadening anisotropy in the two directions and the increased broadening seen in the transverse direction at increasing x. Both observations point to an unusually high broadening for alloys that preferentially deform by prismatic <a> slip in the transverse direction at high values of x. From Figure 22b it can be seen that for these orientations there is a high expected amount of prismatic <a> slip relative to the axial direction, in contrast to basal <a> or pyramidal <c+a> slip which have similar behaviour in the transverse and axial directions with x. However, because these prismatic <a> dislocations have a diminishing contrast factor (see Figure 3), they are not expected to contribute to broadening. If these dislocations were to contribute to broadening in a way that is not related to the contrast factor, a 'non-contrast factor' broadening, this could explain the discrepancies observed. The presence of these dislocations may break up the grains into domains that don't diffract coherently, providing an additional crystal size broadening. Although, the activity of different slip systems will be more complicated than given by the Schmid factors, the discussion using the Schmid factors is able to provide an explanation for the unusual broadening anisotropy observed in the HCP alloys. This 'non-contrast factor' broadening has important implications on how broadening anisotropy is dealt with in HCP alloys, and on the underlying assumption used that variations in contrast factor alone can explain broadening anisotropy.

The analysis presented in this section highlights problems in using the homogeneous approach to explain broadening anisotropy in HCP alloys. These issues place doubt on the use of the homogeneous approach for deformed HCP alloys, particularly to quantify slip activity. If broadening anisotropy is not properly accounted for there are significant potential errors that can result from DPPA methods, which are much larger than would be for cubic alloys. However, the solution to this problem is not clear and hence it may therefore be best practice to assume a constant value of $\overline{b^2 C_{hk.0}}$ when comparing different samples. Further investigation is needed as the implications of the analysis of HCP alloys, and the significance of 'non-contrast factor' broadening, has important implications for DPPA methods in HCP alloys.

3.2.3. Twinning

To examine the effect of twinning on peak profiles two different titanium alloys are compared: Ti-6Al-4V and a commercially pure titanium (Ti-CP). Both alloys were deformed by uni-axial tension to a range of applied strains and measured by synchrotron x-ray diffraction, at ID31, ESF, Grenoble (more details in [3,42]). Whereas, Ti-CP is known to display marked deformation twinning [65,103,105,106], in Ti-6Al-4V twinning is rarely found [63,64,107]. Although, there will be more differences between the alloys than just twinning, because even the ability of an alloy to twin will influence the activity of the different slip systems, it is worth considering the broadening of the two alloys.

In Figure 23 are the changes in the shape of diffraction peaks of the two alloys with x; including the change in FW/g, the shape of the tails of the peaks (given by the pseudo-Voigt mixing parameter [108]) and broadening anisotropy (the difference in FW on the left and right side of the peak). The change in the shape of the peaks with x shares similarities for the two alloys. Both have a transverse FW/g that falls then increases with x, and have higher value at high x in the transverse direction than the corresponding axial FW/g. The FW/g values in the axial direction are different for both alloys, with FW/g being more constant with x for Ti-CP than for Ti-6Al-4V, although the differences are less evident at the other measured strains of Ti-6Al-4V [3]. In addition, the shape of the tails of the peaks and the peak's asymmetry has similarities for the two alloys. The pseudo-Voigt mixing parameter of both alloys increases with x for the axial direction, and falls with x for the transverse direction. The asymmetry parameter is approximately constant with x for both alloys. The work of Balogh, Tichy and Ungar [40] on how twinning would contribute to diffraction profiles in HCP alloys, shows that twin boundaries cause changes in the broadening and asymmetry of a peak. However, the results suggest the broadening and asymmetry of peaks to be similar in both alloys. This suggests that the contribution of twinning to broadening anisotropy in HCP alloys is minimal. However, twinning will also affect the activity of different slip systems and in turn broadening anisotropy, so further work is needed to understand how twins contribute to broadening. Twins in HCP alloys are most often of the order of several micrometres in thickness [62,109], rather than the deformation twins observed in FCC alloys which tend to be fractions of micrometres in thickness [66,110,111] (as shown in Figure 24). Therefore, in HCP alloys the number of twin boundaries can be low even though the volume of twinned grains can be large. In addition, there can be many other reasons that a twin could causes broadening. These include variations in the intergranular strains in twins compared to un-twinned regions, that contribute to a diffraction peak [95,112]. Along with the dislocations present in the twin before the twinning has occurred. Hence, these factors, are probably more important considerations to quantify twinning in HCP alloys, than the traditional methods (e.g., Balogh et al. [40]) that are based on the ordering of atoms at the twin boundary.

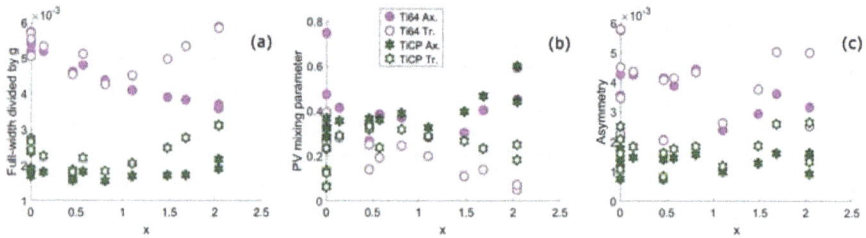

Figure 23. The change in peak-shape of Ti-6Al-4V and Ti-CP with x, for the axial and transverse directions. Before alloys were deformed by uni-axial tension, Ti-6Al-4V is deformed to 5% applied strain and Ti-CP to 3.5% applied strain. In (**a**) the changes in full-width, in (**b**) the changes of the pseudo-Voigt mixing parameter, and in (**c**) changes in the asymmetry of the peaks. Adapted from Simm et al. [42].

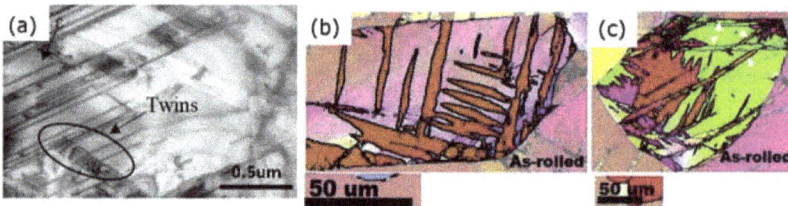

Figure 24. Deformation twins. Left (**a**) from a stainless steel alloy (adapted from [110]) and right (**b**,**c**) from a magnesium alloy (adapted from [109]).

4. Conclusions

Diffraction peak profile analysis (DPPA) is a valuable technique for science, engineering and applied physics communities to characterise the nano-structure of crystalline materials that is not possible or not easy by other techniques.

Formulism to describe broadening anisotropy, such as those given for the contrast factor and planar faults, are valuable for quantifying details of the microstructure such as the relative amount of different dislocation types, or the quantity of planar faults. In addition, their use can improve the statistics for analysis. This is achieved by allowing the use of many more diffraction peaks which represent a higher fraction of the sample, than possible by using multiple orders of the same peak.

However, there are problems with using these formalisms, which are a result of the heterogeneous nature of plastic deformation, and the practical and mathematical limitations of peak profile methods. They add errors to the results from DPPA methods. This is most evident for slip system and planar fault predictions which are used to describe broadening anisotropy. But other results such as dislocation density and crystal size will also be affected. These errors will be larger for HCP alloys due to the greater number of slip systems and differences in their ease of activity.

It is suggested that a way to retain the benefits of using the formulae to describe planar faults is by using the broadening and peak shifts of different orders of the same peak. However, a way to be able to quantify the dislocation slip system types present (i.e., edge and screw, or <a> and <c+a>) in a sample is more complicated. This is particularly the case in HCP alloys were it is shown that significant broadening anisotropy can result from 'non contrast factor' broadening. Which goes against current formalisms that assume that all broadening anisotropy is due to the contrast factor.

There are several implications of broadening anisotropy on the results of diffraction peak profile analysis methods:

(1) They add errors to the results obtained by DPPA methods.
(2) They limit the ability of DPPA methods to be used with other characterisation techniques. But,
(3) They can be used to provide additional information about the materials deformation microstructure.
(4) They offer a means to develop and verify models of plastic deformation that incorporate crystal plasticity formulations with work-hardening models that predict changes that occur at the scale of dislocations.

Acknowledgments: This work was part funded by the European Regional Development Fund as part of the Ser Cymru II program. The author is grateful to the STFC for the beam time granted on the HRPD beamline and to ESRF for beam time granted at ID31. The author is appreciative for J. Quinta da Fonseca and P. J. Withers for devising and supervising his PhD in this field, along with some of the experiments and analysis discussed in this paper. The author thankful for MATLAB and MTEX which have played an important part in this work.

Conflicts of Interest: The authors declare no conflict of interest.

References

1. Warren, B.E. *X-ray Diffraction*; Addison-Wesley Publishing Co.: Reading, UK, 1969.
2. Kuzel, R. Dislocation line broadening. *Zeitschrift Fur Kristallographie* **2006**, *1*, 75–80. [CrossRef]
3. Simm, T.H. The Use of Diffraction Peak Profile Analysis in Studying the Plastic Deformation of Metals. Ph.D. Thesis, University of Manchester, Manchester, UK, 2012.
4. Simm, T.H.H.; Withers, P.J.; da Fonseca, J.Q. An evaluation of diffraction peak profile analysis (DPPA) methods to study plastically deformed metals. *Mater. Des.* **2016**, *111*, 331–343. [CrossRef]
5. Van Berkum, J.G.M.; Vermeulen, A.C.; Delhez, R.; de Keijser, T.H.; Mittemeijer, E.J. Applicabilities of the Warren-Averbach analysis and an alternative analysis for separation of size and strain broadening. *J. Appl. Crystallogr.* **1994**, *27*, 345–357. [CrossRef]
6. Rietveld, H.M. A profile refinement method for nuclear and magnetic structures. *J. Appl. Crystallogr.* **1969**, *2*, 65–71. [CrossRef]
7. Le Bail, A. Modelling anisotropic crystallite size/microstrain in Rietveld analysis. *NIST Spec. Publ.* **1992**, *846*, 142–153. [CrossRef]
8. Stokes, A.R.; Wilson, A.J.C. The diffraction of X rays by distorted crystal aggregates—I. *Proc. Phys. Soc.* **1944**, *56*, 174–181. [CrossRef]
9. Stephens, P.W. Phenomenological model of anisotropic peak broadening in powder diffraction. *J. Appl. Crystallogr.* **1999**, *32*, 281–289. [CrossRef]
10. Popa, N.C. The (hkl) Dependence of Diffraction-Line Broadening Caused by Strain and Size for all Laue Groups in Rietveld Refinement. *J. Appl. Crystallogr.* **1998**, *31*, 176–180. [CrossRef]
11. Krivoglaz, M.A. *X-ray and Neutron Diffraction in Nonideal Crystals*; Springer-Verlag: Berlin, Germany, 1996.
12. Simm, T.H. Supplementary data and analysis for "Peak broadening anisotropy and the contrast factor in metal alloys". *OSF* **2018**. [CrossRef]
13. Simm, T.H. DPPA BIGdippa and dippaFC. *OSF* **2018**. [CrossRef]
14. Liu, Q.; Zhao, C.; Su, S.; Li, J.; Xing, Y.; Cheng, B. Strain Field Mapping of Dislocations in a Ge/Si Heterostructure. *PLoS ONE* **2013**, *8*, 4–9. [CrossRef] [PubMed]

15. Dupraz, M.; Beutier, G.; Rodney, D.; Mordehai, D.; Verdier, M. Signature of dislocations and stacking faults of face-centred cubic nanocrystals in coherent X-ray diffraction patterns: A numerical study. *J. Appl. Crystallogr.* **2015**, *48*, 621–644. [CrossRef] [PubMed]

16. Borbély, A.; Dragomir-Cernatescu, J.; Ribárik, G.; Ungár, T.; Borbely, A.; Dragomir-Cernatescu, J.; Ribarik, G.; Ungar, T. Computer program ANIZC for the calculation of diffraction contrast factors of dislocations in elastically anisotropic cubic, hexagonal and trigonal crystals. *J. Appl. Crystallogr.* **2003**, *36*, 160–162. [CrossRef]

17. Armstrong, R.W. Crystal Dislocations. *Crystals* **2016**, *6*, 9. [CrossRef]

18. Williamson, G.K.; Hall, W.H. X-ray line broadening from filed aluminium and wolfram. *Acta Metall.* **1953**, *1*, 22–31. [CrossRef]

19. Ungár, T.; Borbély, A. The effect of dislocation contrast on x-ray line broadening: A new approach to line profile analysis. *Appl. Phys. Lett.* **1996**, *69*, 3173. [CrossRef]

20. Scardi, P.; Leoni, M.; Delhez, R. Line broadening analysis using integral breadth methods: A critical review. *J. Appl. Crystallogr.* **2004**, *37*, 381–390. [CrossRef]

21. Wilkens, M. The determination of density and distribution of dislocations in deformed single crystals from broadened X-ray diffraction profiles. *Phys. Status Solidi* **1970**, *2*, 359–370. [CrossRef]

22. Ungár, T.; Tichy, G. The Effect of Dislocation Contrast on X-Ray Line Profiles in Untextured Polycrystals. *Phys. Status Solidi* **1999**, *171*, 425–434. [CrossRef]

23. Warren, B.E. X-ray studies of deformed metals. *Prog. Met. Phys.* **1959**, *8*, 147–202. [CrossRef]

24. Ungár, T.; Gubicza, J.; Hanak, P.; Alexandrov, I. Densities and character of dislocations and size-distribution of subgrains in deformed metals by X-ray diffraction profile analysis. *Mater. Sci. Eng. A* **2001**, *319–321*, 274–278. [CrossRef]

25. Ungár, T.; Dragomir, I.; Révész, Á.; Borbély, A. The contrast factors of dislocations in cubic crystals: The dislocation model of strain anisotropy in practice. *J. Appl. Crystallogr.* **1999**, *32*, 992–1002. [CrossRef]

26. Abouhilou, F.; Khereddine, A.; Alili, B.; Bradai, D. X-ray peak profile analysis of dislocation type, density and crystallite size distribution in cold deformed Pb-Ca-Sn alloys. *Trans. Nonferr. Met. Soc. China* **2012**, *22*, 604–607. [CrossRef]

27. Wang, X.L.; Wang, Y.D.; Stoica, A.D.; Horton, D.J.; Tian, H.; Liaw, P.K.; Choo, H.; Richardson, J.W.; Maxey, E. Inter- and intragranular stresses in cyclically-deformed 316 stainless steel. *Mater. Sci. Eng. A* **2005**, *399*, 114–119. [CrossRef]

28. Ungár, T.; Ott, S.; Sanders, P.; Borbély, A.; Weertman, J. Dislocations, grain size and planar faults in nanostructured copper determined by high resolution X-ray diffraction and a new procedure of peak profile analysis. *Acta Mater.* **1998**, *46*, 3693–3699. [CrossRef]

29. Ungár, T.; Ribárik, G.; Zilahi, G.; Mulay, R.; Lienert, U.; Balogh, L.; Agnew, S.R. Slip systems and dislocation densities in individual grains of polycrystalline aggregates of plastically deformed CoTi and CoZr alloys. *Acta Mater.* **2014**, *71*, 264–282. [CrossRef]

30. Dragomir, I.C.; Ungár, T. Contrast factors of dislocations in the hexagonal crystal system. *J. Appl. Crystallogr.* **2002**, *35*, 556–564. [CrossRef]

31. Seymour, T.; Frankel, P.; Balogh, L.; Ungár, T.; Thompson, S.P.; Jädernäs, D.; Romero, J.; Hallstadius, L.; Daymond, M.R.; Ribárik, G.; et al. Evolution of dislocation structure in neutron irradiated Zircaloy-2 studied by synchrotron x-ray diffraction peak profile analysis. *Acta Mater.* **2017**, *126*, 102–113. [CrossRef]

32. Treacy, M.M.J.; Newsam, J.M.; Deem, M.W. A general recursion method for calculating diffracted intensities from crystals containing planar faults. *Proc. R. Soc. Math. Phys. Sci.* **1991**, *433*, 499–520. [CrossRef]

33. Estevez-Rams, E.; Penton-Madrigal, A.; Lora-Serrano, R.; Martinez-Garcia, J. Direct determination of microstructural parameters from the x-ray diffraction profile of a crystal with stacking faults. *J. Appl. Crystallogr.* **2001**, *34*, 730–736. [CrossRef]

34. Balogh, L.; Ribarik, G.; Ungár, T. Stacking faults and twin boundaries in fcc crystals determined by x-ray diffraction profile analysis. *J. Appl. Phys.* **2006**, *100*, 1–10. [CrossRef]

35. Stokes, A.R.; Wilson, A.J.C.; Bragg, W.L. A method of calculating the integral breadths of Debye-Scherrer lines. *Math. Proc. Camb. Philos. Soc.* **1942**, *38*, 313. [CrossRef]

36. Unga, T.; Gubicza, J.; Riba, G. MWP-fit: A program for multiple whole-profile fitting of diffraction peak profiles by ab initio theoretical functions. *J. Appl. Crystallogr.* **2001**, *34*, 669–676.

37. Scardi, P.; Leoni, M. Fourier modelling of the anisotropic line broadening of X-ray diffraction profiles due to line and plane lattice defects. *J. Appl. Crystallogr.* **1999**, *32*, 671–682. [CrossRef]

38. Ungár, T.; Stoica, A.D.; Tichy, G.; Wang, X.L. Orientation-dependent evolution of the dislocation density in grain populations with different crystallographic orientations relative to the tensile axis in a polycrystalline aggregate of stainless steel. *Acta Mater.* **2014**, *66*, 251–261. [CrossRef]
39. Csiszár, G.; Balogh, L.; Misra, A.; Zhang, X.; Ungár, T. The dislocation density and twin-boundary frequency determined by X-ray peak profile analysis in cold rolled magnetron-sputter deposited nanotwinned copper. *J. Appl. Phys.* **2011**, *110*. [CrossRef]
40. Balogh, L.; Tichy, G.; Ungár, T. Twinning on pyramidal planes in hexagonal close packed crystals determined along with other defects by X-ray line profile analysis. *J. Appl. Crystallogr.* **2009**, *42*, 580–591. [CrossRef]
41. Balogh, L.; Figueiredo, R.B.; Ungár, T.; Langdon, T.G. The contributions of grain size, dislocation density and twinning to the strength of a magnesium alloy processed by ECAP. *Mater. Sci. Eng. A* **2010**, *528*, 533–538. [CrossRef]
42. Simm, T.H.; Withers, P.J.; da Fonseca, J.Q. Peak broadening anisotropy in deformed face-centred cubic and hexagonal close-packed alloys. *J. Appl. Crystallogr.* **2014**, *47*, 1535–1551. [CrossRef]
43. Borbely, A.; Driver, J.H.; Ungár, T. X-ray method for the determination of stored energies in texture components of deformed metals; application to cold worked ultra high purity iron. *Acta Mater.* **2000**, *48*, 2005–2016. [CrossRef]
44. Guiglionda, G.; Borbely, A.; Driver, J.H. Orientation-dependent stored energies in hot deformed Al-2.5%Mg and their influence on recrystallization. *Acta Mater.* **2004**, *52*, 3413–3423. [CrossRef]
45. Bate, P.S.; da Fonseca, J.Q. Texture development in the cold rolling of IF steel. *Mater. Sci. Eng. A* **2004**, *380*, 365–377. [CrossRef]
46. Sevillano, J.G.; van Houtte, P.; Aernoudt, E. Large strain work hardening and textures. *Prog. Mater. Sci.* **1980**, *25*, 69–134. [CrossRef]
47. Honniball, P.D.; Preuss, M.; Rugg, D.; da Fonseca, J.Q. Grain Breakup During Elevated Temperature Deformation of an HCP Metal. *Metall. Mater. Trans. A Phys. Metall. Mater. Sci.* **2015**, *46*, 2143–2156. [CrossRef]
48. Da Fonseca, J.Q.; Oliver, E.C.; Bate, P.S.; Withers, P.J. Evolution of intergranular stresses during in situ straining of IF steel with different grain sizes. *Mater. Sci. Eng. A* **2006**, *437*, 26–32. [CrossRef]
49. Taylor, G.I. The Mechanism of Plastic Deformation of Crystals. Part I. Theoretical. *Proc. R. Soc. A Math. Phys. Eng. Sci.* **1934**, *145*, 362–387. [CrossRef]
50. Kurdjumov, G.; Sachs, G. Over the mechanisms of steel hardening. *Z. Phys.* **1930**, *64*, 325–343.
51. Bate, P. Modelling deformation microstructure with the crystal plasticity finite-element method. *Philos. Trans. R. Soc. A Math. Phys. Eng. Sci.* **1999**, *357*, 1589–1601. [CrossRef]
52. Lebensohn, R.A.; Tomé, C.N. A self-consistent anisotropic approach for the simulation of plastic deformation and texture development of polycrystals: Application to zirconium alloys. *Acta Metall. Mater.* **1993**, *41*, 2611–2624. [CrossRef]
53. Cottrell, A.H. *Dislocations and Plastic Flow in Crystals*; Clarendon Press: London, UK, 1964.
54. Cottrell, A.H. Commentary. A brief view of work hardening. *Dislocations Solids* **2002**. [CrossRef]
55. Bertin, N.; Cai, W. Computation of virtual X-ray diffraction patterns from discrete dislocation structures. *Comput. Mater. Sci.* **2018**, *146*, 268–277. [CrossRef]
56. Bulatov, W.; Cai, V.V. *Computer Simulations of Dislocations*; Oxford University Press: Oxford, UK, 2006. [CrossRef]
57. Hull, D.; Bacon, D.J. *Introduction to Dislocations*; Elsevier Science: New York, NY, USA, 2001.
58. Zehetbauer, M. Cold work hardening in stages IV and V of F.C.C. metals-II. Model fits and physical results. *Acta Metall. Mater.* **1993**, *41*, 589–599. [CrossRef]
59. Keller, C.; Hug, E.; Retoux, R.; Feaugas, X. TEM study of dislocation patterns in near-surface and core regions of deformed nickel polycrystals with few grains across the cross section. *Mech. Mater.* **2010**, *42*, 44–54. [CrossRef]
60. Balogh, L.; Gubicza, J.; Hellmig, R.J.; Estrin, Y.; Ungár, T. Thermal stability of the microstructure of severely deformed copper. *Zeitschrift fur Kristallographie* **2006**, *2*, 381–386. [CrossRef]
61. Gubicza, J.; Nam, N.H.; Balogh, L.; Hellmig, R.J.; Stolyarov, V.V.; Estrin, Y.; Ungár, T. Microstructure of severely deformed metals determined by X-ray peak profile analysis. *J. Alloys Compd.* **2004**, *378*, 248–252. [CrossRef]
62. Christian, J.W.; Mahajan, S. Deformation twinning. *Prog. Mater. Sci.* **1995**, *39*, 1–157. [CrossRef]

63. Karaman, I.; Yapici, G.G.; Chumlyakov, Y.I.; Kireeva, I.V. Deformation twinning in difficult-to-work alloys during severe plastic deformation. *Mater. Sci. Eng. A* **2005**, *410–411*, 243–247. [CrossRef]
64. Yapici, G.G.; Karaman, I.; Luo, Z.P. Mechanical twinning and texture evolution in severely deformed Ti-6Al-4V at high temperatures. *Acta Mater.* **2006**, *54*, 3755–3771. [CrossRef]
65. Glavicic, M.G.; Salem, A.A.; Semiatin, S.L. X-ray line-broadening analysis of deformation mechanisms during rolling of commercial-purity titanium. *Acta Mater.* **2004**, *52*, 647–655. [CrossRef]
66. Byun, T.S.; Lee, E.H.; Hunn, J.D. Plastic deformation in 316LN stainless steel—Characterization of deformation microstructures. *J. Nucl. Mater.* **2003**, *321*, 29–39. [CrossRef]
67. Lee, W.-S.; Lin, C.-F. Impact properties and microstructure evolution of 304L stainless steel. *Mater. Sci. Eng. A* **2001**, *308*, 124–135. [CrossRef]
68. Zhang, P.; Chen, Y.; Xiao, W.; Ping, D.; Zhao, X. Twin structure of the lath martensite in low carbon steel. *Prog. Nat. Sci. Mater. Int.* **2016**, *26*, 169–172. [CrossRef]
69. Tian, Y.; Gorbatov, O.I.; Borgenstam, A.; Ruban, A.V.; Hedström, P. Deformation Microstructure and Deformation-Induced Martensite in Austenitic Fe-Cr-Ni Alloys Depending on Stacking Fault Energy. *Metall. Mater. Trans. A Phys. Metall. Mater. Sci.* **2017**, *48*, 1–7. [CrossRef]
70. Hansen, N.; Huang, X.; Pantleon, W.; Winther, G. Grain orientation and dislocation patterns. *Philos. Mag.* **2006**, *86*, 3981–3994. [CrossRef]
71. Dillamore, I.L.; Katoh, H. The Mechanisms of Recrystallization in Cubic Metals with Particular Reference to Their Orientation-Dependence. *Met. Sci.* **1974**, *8*, 73–83. [CrossRef]
72. Hughes, D.A.; Hansen, N. Microstructure and Stength of Nickel at Large Strains. *Acta Mater.* **2000**, *48*, 2985–3004. [CrossRef]
73. Kocks, U.F.; Mecking, H. Physics and phenomenology of strain hardening: The FCC case. *Prog. Mater. Sci.* **2003**, *48*, 171–273. [CrossRef]
74. Čížek, J.; Janeček, M.; Krajňák, T.; Stráská, J.; Hruška, P.; Gubicza, J.; Kim, H.S. Structural characterization of ultrafine-grained interstitial-free steel prepared by severe plastic deformation. *Acta Mater.* **2016**, *105*, 258–272. [CrossRef]
75. Gubicza, J.; Chinh, N.Q.; Krallics, G.; Schiller, I.; Ungár, T. Microstructure of ultrafine-grained fcc metals produced by severe plastic deformation. *Curr. Appl. Phys.* **2006**, *6*, 194–199. [CrossRef]
76. Humphreys, F.J.; Hatherly, M. *Recrystallization and Related Annealing Phenomena*; Elsevier: New York, NY, USA, 2004. [CrossRef]
77. Driver, J.H.; Theyssier, M.-C.; Maurice, C.I. Electron backscattered diffraction microtexturestudies on hot deformed aluminium crystals. *Mater. Sci. Technol.* **1996**, *12*, 851–858. [CrossRef]
78. Popova, E.; Staraselski, Y.; Brahme, A.; Mishra, R.K.; Inal, K. Coupled crystal plasticity—Probabilistic cellular automata approach to model dynamic recrystallization in magnesium alloys. *Int. J. Plast.* **2015**, *66*, 85–102. [CrossRef]
79. Murty, K.L.; Charit, I. Texture development and anisotropic deformation of zircaloys. *Prog. Nucl. Energy* **2006**, *48*, 325–359. [CrossRef]
80. Arafin, M.A.; Szpunar, J.A. A new understanding of intergranular stress corrosion cracking resistance of pipeline steel through grain boundary character and crystallographic texture studies. *Corros. Sci.* **2009**, *51*, 119–128. [CrossRef]
81. Das, Y.B.; Simm, T.H.; Gungor, S.; Fitzpatrick, M.E.; Forsey, A.N.; Moat, R.J. An experimental study of plastic deformation and transformation in austenitic stainless steel. Unpubl. Work. (n.d.). 1–35.
82. Bachmann, F.; Hielscher, R.; Schaeben, H. Grain detection from 2d and 3d EBSD data—Specification of the MTEX algorithm. *Ultramicroscopy* **2011**, *111*, 1720–1733. [CrossRef] [PubMed]
83. Lütjering, G.; Williams, J. *Titanium*, 2nd ed.; Springer: Berlin, Germany, 2007.
84. Honeycombe, R.W.K. *The Plastic Deformation of Metals*; Edward Arnold: Maidenhead, UK, 1984.
85. Máthis, K.; Csiszár, G.; Čapek, J.; Gubicza, J.; Clausen, B.; Lukáš, P.; Vinogradov, A.; Agnew, S.R. Effect of the loading mode on the evolution of the deformation mechanisms in randomly textured magnesium polycrystals—Comparison of experimental and modeling results. *Int. J. Plast.* **2015**, *72*, 127–150. [CrossRef]
86. Máthis, K.; Nyilas, K.; Axt, A.; Dragomir-Cernatescu, I.; Ungár, T.; Lukáč, P. The evolution of non-basal dislocations as a function of deformation temperature in pure magnesium determined by X-ray diffraction. *Acta Mater.* **2004**, *52*, 2889–2894. [CrossRef]

87. Dragomir, I.C.; Li, D.S.; Castello-Branco, G.A.; Garmestani, H.; Snyder, R.L.; Ribarik, G.; Ungár, T. Evolution of dislocation density and character in hot rolled titanium determined by X-ray diffraction. *Mater. Charact.* **2005**, *55*, 66–74. [CrossRef]

88. Dragomir, I.C.; Castello-Branco, G.A.; Ribarik, G.; Garmestani, H.; Ungár, T.; Snyder, R.L. Burgers vector populations in hot rolled titanium determined by X-ray peak profile analysis. *Zeitschrift fur Kristallographie* **2006**, *1*, 99–104. [CrossRef]

89. Schafler, E.; Nyilas, K.; Bernstorff, S.; Zeipper, L.; Zehetbauer, M.; Ungár, T. Microstructure of post-deformed ECAP- Ti investigated by Multiple X-Ray Line Profile Analysis. *Zeitschrift fur Kristallographie* **2006**, *23*, 129–134. [CrossRef]

90. Ungár, T.; Castelnau, O.; Ribarik, G.; Drakopoulos, M.; Bachade, J.L.; Chauveau, T.; Snigirev, A.; Snigireva, I.; Schroer, C.; Bacroix, B. Grain to grain slip activity in plastically deformed Zr determined by X-ray micro-diffraction line profile analysis. *Acta Mater.* **2007**, *55*, 1117–1127. [CrossRef]

91. Long, F.; Balogh, L.; Daymond, M.R. Evolution of dislocation density in a hot rolled Zr–2.5Nb alloy with plastic deformation studied by neutron diffraction and transmission electron microscopy. *Philos. Mag.* **2017**, *97*, 2888–2914. [CrossRef]

92. Ungár, T.; Holden, T.M.; Jóni, B.; Clausen, B.; Balogh, L.; Csiszár, G.; Brown, D.W. Dislocation structure in different texture components determined by neutron diffraction line profile analysis in a highly textured Zircaloy-2 rolled plate. *J. Appl. Crystallogr.* **2015**, *48*. [CrossRef]

93. Wilkens, M. *Fundamental Aspects of Dislocation Theory*; Natl. Bur. Stand. (US)Spec. Publ. No. 317; U.S. Department of Commerce National Bureau of Standards: Washington, DC, USA, 1970.

94. Simm, T.H.; Das, Y.; Dunlop, T.; Perkins, K.M.; Birosca, S.; Prakash, L.; da Fonseca, J.Q. In situ and ex situ measurements and modelling of the plastic deformation of HCP alloys measured by neutron diffraction. Unpubl. Work. N/A (2018).

95. Timár, G.; da Fonseca, J.Q. Modeling Twin Clustering and Strain Localization in Hexagonal Close-Packed Metals. *Metall. Mater. Trans. A Phys. Metall. Mater. Sci.* **2014**, *45*, 5883–5890. [CrossRef]

96. Lutjering, G.; Williams, J.C. Commercially Pure (CP) Titanium and Alpha Alloys. *Titanium* **2007**, 175–201. [CrossRef]

97. Staroselsky, A.; Anand, L. A constitutive model for hcp materials deforming by slip and twinning: Application to magnesium alloy AZ31B. *Int. J. Plast.* **2003**, *19*, 1843–1864. [CrossRef]

98. Zaefferer, S. A study of active deformation systems in titanium alloys: Dependence on. *Mater. Sci. Eng. A* **2003**, *344*, 20–30. [CrossRef]

99. Evans, C. Micromechanisms and Micromechanics of Zircaloy-4. Ph.D. Thesis, Imperial College London, London, UK, 2014.

100. Britton, T.B.; Dunne, F.P.E.; Wilkinson, A.J. On the mechanistic basis of deformation at the microscale in hexagonal close-packed metals. *Proc. R. Soc. A Math. Phys. Eng. Sci.* **2015**, *471*, 20140881. [CrossRef]

101. Zaefferer, S. Investigation of the Correlation between Texture and Microstrucure on a Submicrometer Scale in the TEM. *Adv. Eng. Mater.* **2003**, *5*, 745–752. [CrossRef]

102. Bridier, F.; Villechaise, P.; Mendez, J. Analysis of the different slip systems activated by tension in a α/β titanium alloy in relation with local crystallographic orientation. *Acta Mater.* **2005**, *53*, 555–567. [CrossRef]

103. Battaini, M.; Pereloma, E.V.; Davies, C.H.J. Orientation effect on mechanical properties of commercially pure titanium at room temperature. *Metall. Mater. Trans. A Phys. Metall. Mater. Sci.* **2007**, *38*, 276–285. [CrossRef]

104. Liu, Q.; Hansen, N. Deformation microstructure and orientation of F.C.C. crystals. *Phys. Status Solidi Appl. Res.* **1995**, *149*, 187–199. [CrossRef]

105. Bozzolo, N.; Chan, L.; Rollett, A.D. Misorientations induced by deformation twinning in titanium. *J. Appl. Crystallogr.* **2010**, *43*, 596–602. [CrossRef]

106. Mullins, S.; Patchett, B.M. Deformation microstructures in titanium sheet metal. *Metall. Trans. A* **1981**, *12*, 853–863. [CrossRef]

107. Liu, Y.; Yang, D.Z.; He, S.Y.; Wu, W.L. Microstructure developed in the surface layer of Ti-6Al-4V alloy after sliding wear in vacuum. *Mater. Charact.* **2003**, *50*, 275–279. [CrossRef]

108. Wertheim, G.K.; Butler, M.A.; West, K.W.; Buchanan, D.N.E. Determination of the Gaussian and Lorentzian content of experimental line shapes. *Rev. Sci. Instrum.* **1974**, *45*, 1369. [CrossRef]

109. Guan, D.; Rainforth, W.M.; Ma, L.; Wynne, B.; Gao, J. Twin recrystallization mechanisms and exceptional contribution to texture evolution during annealing in a magnesium alloy. *Acta Mater.* **2017**, *126*, 132–144. [CrossRef]

110. Gong, N.; Wu, H.-B.; Yu, Z.-C.; Niu, G.; Zhang, D. Studying Mechanical Properties and Micro Deformation of Ultrafine-Grained Structures in Austenitic Stainless Steel. *Metals* **2017**, *7*, 188. [CrossRef]

111. Lee, E.H.; Byun, T.S.; Hunn, J.D.; Yoo, M.H.; Farrell, K.; Mansur, L.K. On the origin of deformation microstructures in austenitic stainless steel: Part I—Microstructures. *Acta Mater.* **2001**, *49*, 3269–3276. [CrossRef]

112. Balogh, L.; Niezgoda, S.R.; Kanjarla, A.K.; Brown, D.W.; Clausen, B.; Liu, W.; Tomé, C.N. Spatially resolved in situ strain measurements from an interior twinned grain in bulk polycrystalline AZ31 alloy. *Acta Mater.* **2013**, *61*, 3612–3620. [CrossRef]

Review

Dislocation Structures in Low-Angle Grain Boundaries of α-Al₂O₃

Eita Tochigi [1,*] , Atsutomo Nakamura [2] , Naoya Shibata [1,3] and Yuichi Ikuhara [1,3,4]

[1] Institute of Engineering Innovation, The University of Tokyo, 2-11-16 Yayoi, Bunkyo-ku, Tokyo 113-8656, Japan; shibata@sigma.t.u-tokyo.ac.jp (N.S.); ikuhara@sigma.t.u-tokyo.ac.jp (Y.I.)
[2] Department of Materials Physics, Nagoya University, Furo-chou, Chikusa-ku, Nagoya, Aichi 464-8603, Japan; anaka@nagoya-u.jp
[3] Nanostructures Research Laboratory, Japan Fine Ceramics Center, 2-4-1, Mutsuno, Atsuta-ku, Nagoya, Aichi 456-8587, Japan
[4] Center for Elements Strategy Initiative for Structure Materials, Kyoto University, Yoshidahonmachi, Sakyo-ku, Kyoto 606-8501, Japan
* Correspondence: tochigi@sigma.t.u-tokyo.ac.jp; Tel.: +81-3-5841-7689

Received: 14 February 2018; Accepted: 7 March 2018; Published: 12 March 2018

Abstract: Alumina (α-Al₂O₃) is one of the representative high-temperature structural materials. Dislocations in alumina play an important role in its plastic deformation, and they have attracted much attention for many years. However, little is known about their core atomic structures, with a few exceptions, because of lack of experimental observations at the atomic level. Low-angle grain boundaries are known to consist of an array of dislocations, and they are useful to compose dislocation structures. So far, we have systematically fabricated several types of alumina bicrystals with a low-angle grain boundary and characterized the dislocation structures by transmission electron microscopy (TEM). Here, we review the dislocation structures in $\{11\bar{2}0\}/[0001]$, $\{11\bar{2}0\}/\langle 1\bar{1}00\rangle$, $\{1\bar{1}00\}/\langle 11\bar{2}0\rangle$, $(0001)/\langle 1\bar{1}00\rangle$, $\{\bar{1}104\}/\langle 11\bar{2}0\rangle$, and $(0001)/[0001]$ low-angle grain boundaries of alumina. Our observations revealed the core atomic structures of $b = 1/3\langle 11\bar{2}0\rangle$ edge and screw dislocations, $\langle 1\bar{1}00\rangle$ edge dislocation, and $1/3\langle\bar{1}101\rangle$ edge and mixed dislocations. Moreover, the stacking faults on $\{11\bar{2}0\}$, $\{1\bar{1}00\}$, and (0001) planes formed due to the dissociation reaction of the dislocations are discussed, focusing on their atomic structure and formation energy.

Keywords: alumina; sapphire; dislocations; low-angle grain boundaries; stacking faults; transmission electron microscopy

1. Introduction

A dislocation is one-dimensional lattice defect within a crystal structure. Dislocations strongly influence the mechanical and functional properties of crystalline materials, and thus it is essential to investigate the atomic structures of each dislocation. A dislocation is characterized by its Burgers vector and line direction. The Burgers vector represents the direction and magnitude of lattice distortion due to a dislocation, which is a critical parameter to determining the behavior of a dislocation, such as its slip direction and self-energy. Since the Burgers vector of a perfect dislocation must coincide with a lattice translation vector, the number of possible Burgers vectors is restricted in a crystal structure. In this sense, it should be efficient to characterize dislocation structures systematically in terms of Burgers vector in order to understand the dislocation behavior in a crystal.

A low-angle grain boundary is known to consist of a periodic array of dislocations, and it is useful to design and compose dislocation structures. A low-angle grain boundary is defined as the boundary between two crystal grains with a misorientation typically less than 15° [1,2]. The misorientation of a low-angle grain boundary is accommodated by the presence of dislocations. Low-angle grain boundaries are divided into two types: tilt boundaries and twist boundaries, as shown in Figure 1a,b. Basically, a low-angle tilt boundary consists of an array of edge dislocations with the Burgers vector perpendicular to the boundary plane and the line direction parallel to the rotation axis (Figure 1c), whereas a low-angle twist boundary consists of a network of screw dislocations with the Burgers vector on the boundary plane (Figure 1d). The relationship between the interval of perfect dislocations *d*, the Burgers vector *b*, and the misorientation angle of a low-angle grain boundary θ is given by Frank's equation [1,2]:

$$d = |b|/\theta. \tag{1}$$

Note that this equation is valid for both tilt and twist grain boundaries. From these geometrical laws, we can predict the configuration of dislocations in a low-angle grain boundary. In turn, we can compose various dislocation structures by artificially fabricating low-angle grain boundaries.

In this study, we demonstrate the design and characterization of dislocation structures using fabricated low-angle grain boundaries of alumina (α-Al$_2$O$_3$). Alumina has the corundum structure (space group: $R\bar{3}c$). The lattice parameters of the hexagonal unit cell are $a = 0.476$ nm and $c = 1.30$ nm ($c/a = 2.73$). Alumina is one of the representative high-temperature structural materials. Dislocations in alumina play an important role in its plastic deformation processes at elevated temperatures [3–11]. Microstructure analyses of the dislocations using transmission electron microscopy (TEM) were actively performed in the 1970s [4–7]. These studies revealed that the dislocations in deformed alumina crystals are typically dissociated into some partial dislocations with a stacking fault. However, the dislocation core structures, which are critical for the slip behavior, had not been well understood for many years because of a lack of atomic-scale observations. In the 2000s, a high-resolution TEM (HRTEM) study successfully characterized the core structure of $b = 1/3\langle11\bar{2}0\rangle$ edge dislocation associated with the $(0001)\langle11\bar{2}0\rangle$ basal slip [11]; nevertheless, the core structures of the other types of dislocation were still unidentified. In addition, dislocations in alumina strongly interact with impurity atoms [12–15], and impurity-doped dislocations have a potential to become functional nanowires [12,16,17]. To efficiently utilize such functional nanowires, the control of dislocation configuration will be a key technique. Consequently, it is of great interest to investigate the core structures and configurations of dislocations in alumina.

In alumina, the bicrystal method, joining two pieces of single crystal at high temperature, is useful to obtain well-oriented grain boundaries [18–20]. So far, we have systematically fabricated alumina bicrystals with a low-angle grain boundary and characterized the dislocation structures formed in the grain boundaries by TEM [21–32]. The low-angle grain boundaries investigated are tilt boundaries of $\{11\bar{2}0\}/[0001]$ [21], $\{11\bar{2}0\}/\langle1\bar{1}00\rangle$ [22–26], $\{1\bar{1}00\}/\langle11\bar{2}0\rangle$ [27,31], $(0001)/[1\bar{1}00]$ [30], and $\{\bar{1}104\}/\langle11\bar{2}0\rangle$ [28,32], and a twist boundary of $(0001)/[0001]$ [29], as listed in Table 1. In each notation, the indices refer to the grain boundary plane and rotation axis. From the geometrical laws regarding low-angle grain boundaries, it is expected that these five tilt grain boundaries consist of edge dislocations with $b = 1/3\langle11\bar{2}0\rangle$, $1/3\langle11\bar{2}0\rangle$, $\langle1\bar{1}00\rangle$, $[0001]$, and $1/3\langle\bar{1}101\rangle$, respectively. In the last case, the $\{\bar{1}104\}$ plane and the $1/3\langle\bar{1}101\rangle$ vector are not normal but at an angle of 84.16° (There is no low-index plane normal to the $1/3\langle\bar{1}101\rangle$ vector.), and the grain boundary should contain not only $1/3\langle\bar{1}101\rangle$ dislocation but also other types of dislocations. In the $(0001)/[0001]$ twist grain boundary, $1/3\langle11\bar{2}0\rangle$ screw dislocations are expected to be formed. In this paper, we review the configurations and atomic structures of dislocations in these low-angle grain boundaries.

Table 1. The low-angle grain boundaries investigated in this study.

Notation	Type	Misorientation Angle	Burgers Vector Expected
$\{11\bar{2}0\}/[0001]$	tilt	2°, 6°, 8°	$1/3\langle11\bar{2}0\rangle$
$\{11\bar{2}0\}/\langle1\bar{1}00\rangle$	tilt	2°, 10°	$1/3\langle11\bar{2}0\rangle$
$\{1\bar{1}00\}/\langle11\bar{2}0\rangle$	tilt	2°	$\langle1\bar{1}00\rangle$
$(0001)/\langle1\bar{1}00\rangle$	tilt	2°	$[0001]$
$\{\bar{1}104\}/\langle11\bar{2}0\rangle$	tilt	2°	$1/3\langle\bar{1}101\rangle$
$(0001)/[0001]$	twist	~0°	$1/3\langle11\bar{2}0\rangle$

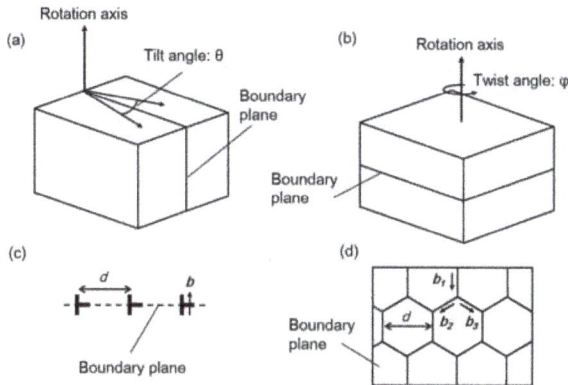

Figure 1. Schematic illustrations of low-angle grain boundaries: (**a**) tilt boundary; (**b**) twist boundary. Typical dislocation structures in (**c**) tilt boundary; (**d**) twist boundary.

2. Materials and Methods

Alumina bicrystals with a low-angle grain boundary were fabricated by joining two pieces of alumina single crystal at 1500 °C for 10 h in air. The low-angle grain boundaries examined in this study are listed in Table 1. The bicrystals were cut to small chips, and then they were thinned by mechanical grinding and Ar ion milling to obtain electron transparency. The samples were observed by TEM (JEOL JEM-2010HC, 200 kV, Tokyo, Japan), high-resolution TEM (HRTEM: Topcon 002BF, 200 kV; JEOL JEM-4010, 400 kV, Tokyo, Japan), and scanning TEM (STEM: JEOL ARM-200F, 200 kV, Tokyo, Japan).

3. Results and Discussion

3.1. $\{11\bar{2}0\}/[0001]$ Low-Angle Tilt Grain Boundary

Figure 2a shows a TEM image of the $\{11\bar{2}0\}/[0001]$ 2° low-angle tilt grain boundary. Pair contrasts are periodically arrayed with the interval of about 13.2 nm, suggesting that each dislocation is dissociated into two partial dislocations. The Burgers vector of the dislocation pairs should be $b = 1/3\langle11\bar{2}0\rangle$ in total because the translation vector of $1/3\langle11\bar{2}0\rangle$ is perpendicular to the $\{11\bar{2}0\}$ grain boundary plane. Substituting d = 13.2 nm and $|b| = |1/3\langle11\bar{2}0\rangle|$ = 0.476 nm into Equation (1), the misorientation angle θ is estimated to be 2.1°, which agrees with the designed angle of the present grain boundary.

A HRTEM image of a dislocation pair in the grain boundary is shown in Figure 2b. The two dislocation cores corresponding to partial dislocations are observed. These partial dislocations are separated along the $\{11\bar{2}0\}$ plane, suggesting that a stacking fault on the $\{11\bar{2}0\}$ plane is formed between the partial dislocations. The plane of the stacking fault does not coincide with the slip plane of the perfect dislocation. Accordingly, the partial dislocations were separated by the self-climb mechanisms. This type of dissociation is called the climb dissociation. The large Burgers circuit shows that this dislocation pair has the Burgers vector of $1/3\langle11\bar{2}0\rangle$, and the small Burgers circuits show that the upper and the lower partial dislocations have the Burgers vectors of $1/3\langle10\bar{1}0\rangle$ and $1/3\langle01\bar{1}0\rangle$, respectively. This dissociation reaction is represented as follows:

$$1/3\langle11\bar{2}0\rangle \rightarrow 1/3\langle10\bar{1}0\rangle + 1/3\langle01\bar{1}0\rangle. \tag{2}$$

The $\{11\bar{2}0\}/[0001]$ grain boundaries with the tilt angle of 6° and 8° were also found to consist of the same type of partial-dislocation pairs [21]. It is known that the slip dislocation associated with the $(0001)\langle11\bar{2}0\rangle$ basal slip (the so-called basal dislocation) has the Burgers vector of $1/3\langle11\bar{2}0\rangle$, and the

$1/3\langle11\bar{2}0\rangle$ basal edge dislocation also dissociates into two partial dislocations following the reaction of Equation (2) [4,6,10]. Note that $1/3\langle11\bar{2}0\rangle$ edge dislocation in the $\{11\bar{2}0\}/[0001]$ tilt grain boundary has the line direction of $[0001]$, whereas the $1/3\langle11\bar{2}0\rangle$ basal edge dislocation has the line direction of $[1\bar{1}00]$. Therefore, they are not equivalent to each other.

The $\{11\bar{2}0\}$ stacking faults generated by the $1/3\langle10\bar{1}0\rangle$ or $1/3\langle01\bar{1}0\rangle$ partial dislocations are known to be structurally equivalent [5,11]; that is, there is only one type for the $\{11\bar{2}0\}$ stacking fault. The formation energy of a stacking fault (stacking fault energy) γ_{SF} formed between a partial-dislocation pair can be estimated by calculating the repulsive force acting between the partial dislocations based on an elastic theory, which is the so-called Peach–Kohler equation [2]. For a partial-dislocation pair in a low-angle boundary, contributions from other dislocations also need to take into account the force calculations, and a detailed derivation is given elsewhere [21,33]. For the present case, the repulsive force f (= γ_{SF}) is represented as follows:

$$\gamma_{SF} = f = \frac{\mu b_p^2}{4\pi(1-\nu)} \cdot \frac{1}{d} \sum_{n=0}^{\infty} \left(\frac{1}{n+\alpha} - \frac{1}{n+1-\alpha} \right), \tag{3}$$

where μ is the shear modulus (~150 GPa [34]), ν is Poisson's ratio (~0.24 [35]), b_p is the magnitude of the Burgers vector of partial dislocations, $\left|1/3\langle10\bar{1}0\rangle\right|$ = 0.275 nm, and α is d_1/d (d_1: the spacing of a partial dislocation pair, or the width of the stacking fault). Using the averaged distances measured from our experiment, d = 13.2 nm and d_1 = 4.6 nm, the stacking fault energy was estimated to be 0.32 Jm^{-2}. This value agrees well with an experimental value calculated from an isolated partial-dislocation pair in a deformed crystal, 0.28 Jm^{-2} [11]. In addition, a couple of theoretical studies have been carried out to examine the $\{11\bar{2}0\}$ stacking fault [36,37], and one using first-principles calculations within the generalized gradient approximation proposed a similar value of 0.35 Jm^{-2} [37].

From Equations (1) and (3), the relationship between d_1 and θ is given. Since the stacking fault energy does not depend on the tilt angle, d_1 can be obtained as a function of θ. This indicates that the configuration of partial dislocations dissociated by climb in a low-angle boundary can be predicted by the stacking fault energy corresponding to the grain boundary plane in addition to the orientation relationship.

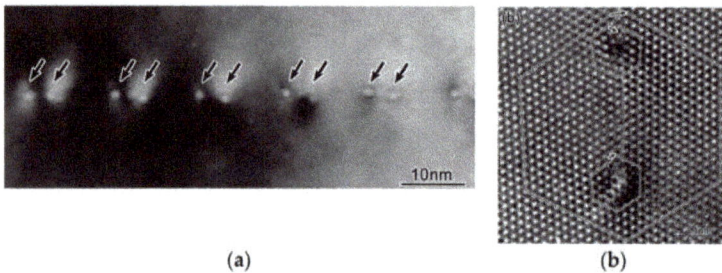

(a)   (b)

Figure 2. (a) TEM image of the $\{11\bar{2}0\}/[0001]$ 2° low-angle tilt grain boundary. Partial-dislocation pairs are formed along the grain boundary. (b) HRTEM image of a partial dislocation pair. The grain boundary is parallel to the vertical direction. The large Burgers circuit shows edge component of $1/3\langle11\bar{2}0\rangle$, and the small circuits show edge components of $1/3\langle10\bar{1}0\rangle$ and $1/3\langle01\bar{1}0\rangle$. The images shown are adapted from [21] and reprinted with the permission of the American Ceramics Society.

3.2. $\{11\bar{2}0\}/\langle1\bar{1}00\rangle$ Low-Angle Tilt Grain Boundary

Figure 3a shows a TEM image of the $\{11\bar{2}0\}/\langle1\bar{1}00\rangle$ low-angle 2° tilt grain boundary. It is seen that dislocations are periodically arrayed along the grain boundary. The dislocation structure is divided into two groups, pairs of dislocations and groups of odd-numbered dislocations. Figure 3b shows a dark-field TEM image taken at the same region in Figure 3a using the reflection of $g = \bar{3}030$, where the $\{11\bar{2}0\}$ grain boundary plane is inclined by about 30° from the observation direction. The dislocations are clearly seen as line contrasts. Figure 3c shows an HRTEM image of a dislocation pair. The dislocation is dissociated into two partial dislocations on the $\{11\bar{2}0\}$ plane. The large Burgers circuit shows that the dislocation pair has the Burgers vector of $1/3\langle 11\bar{2}0\rangle$ in total. The small Burgers circuits show that each partial dislocation has an edge component of $1/6\langle 11\bar{2}0\rangle$. This component corresponds to the $\{1\bar{1}00\}$ projection of the vectors of $1/3\langle 10\bar{1}0\rangle$ and $1/3\langle 01\bar{1}0\rangle$. Therefore, it is considered that the observed structure corresponds to the dissociation of the $1/3\langle 11\bar{2}0\rangle$ edge dislocation into the $1/3\langle 10\bar{1}0\rangle$ and $1/3\langle 01\bar{1}0\rangle$ mixed partial dislocations according to the reaction of Equation (2). The $\{11\bar{2}0\}/\langle 1\bar{1}00\rangle$ low-angle 10° tilt grain boundary was also investigated and found to consist of the same type of partial-dislocation pairs [25]. The $1/3\langle 11\bar{2}0\rangle$ edge dislocation formed in the $\{11\bar{2}0\}/\langle 1\bar{1}00\rangle$ tilt grain boundary has the dislocation line direction of $\langle 1\bar{1}00\rangle$, and thus this dislocation structure is considered to be equivalent to that of the $1/3\langle 11\bar{2}0\rangle$ basal edge dislocation associated with the $(0001)\langle 11\bar{2}0\rangle$ basal slip [10].

The distances between dislocations d and d_1 were measured to be 15 nm and 3.6 nm. Using Equation (3) (where the coefficient attributes $\frac{\mu b_p^2(2+\nu)}{8\pi(1-\nu)}$), the formation energy of the $\{11\bar{2}0\}$ stacking fault was estimated to be 0.30 Jm^{-2} [24]. This value is consistent with that estimated using the $\{11\bar{2}0\}/\langle 0001\rangle$ low-angle tilt grain boundary, as discussed in Section 3.1.

For the imaging condition of $g = \bar{3}030$ in Figure 3b, the partial dislocations with $b_1 = 1/3\langle 10\bar{1}0\rangle$ and with $b_2 = 1/3\langle 01\bar{1}0\rangle$ have strong and weak contrasts, respectively. As seen in Figure 3b, the partial-dislocation pairs (b_1–b_2 pairs) are imaged as the strong and weak line contrasts. From the contrast features, the configurations of the group of five and thirteen partials are found to be $b_1 b_1 b_2 b_1 b_1$ and $b_1 b_1 b_2 b_1 b_2 b_1 b_2 b_1 b_2 b_1 b_1$. The edge components of the partial dislocations with b_1 and b_2 are both $1/6\langle 11\bar{2}0\rangle$, whereas their screw components are $1/6\langle 1\bar{1}00\rangle$ and $1/6\langle \bar{1}100\rangle$, respectively. Therefore, it is found that the 5-partial structure has an edge component of $5/6\langle 11\bar{2}0\rangle$ and a screw component of $1/2\langle 1\bar{1}00\rangle$ in total, and the 13-partial structure has an edge component of $13/6\langle 11\bar{2}0\rangle$ and a screw component of $1/2\langle 1\bar{1}00\rangle$. Note that both of the structures have the same screw component. This suggests that the odd numbered dislocation structures are generated by the twist component of the grain boundary. The averaged interval of the odd-numbered dislocation structures was about 230 nm. Substituting $d = 230$ nm and $|b| = |1/2\langle 1\bar{1}00\rangle| = 0.407$ nm into Equation (1), the twist angle is estimated to be 0.10°. This value is possible considering the accuracy of the bicrystal fabrication processes.

The odd-numbered dislocation structures can be written by the general expression as:

$$n \times \frac{1}{3}\langle 11\bar{2}0\rangle + \langle 10\bar{1}0\rangle \rightarrow (n+3) \times \frac{1}{3}\langle 10\bar{1}0\rangle + n \times \frac{1}{3}\langle 01\bar{1}0\rangle, \tag{4}$$

where n is an integer. The odd-numbered dislocation structures with the n-value of 0–7 (corresponding to 3 to 17-partial structures) have been found so far [22,26]. The characteristic dislocation configuration in the present grain boundary is explained as follows. An additional twist component to the $\{11\bar{2}0\}/\langle 1\bar{1}00\rangle$ low-angle tilt grain boundary generates the $\langle 10\bar{1}0\rangle$ mixed dislocations into the $1/3\langle 11\bar{2}0\rangle$ edge dislocation array. The $\langle 10\bar{1}0\rangle$ dislocation dissociates into three $1/3\langle 10\bar{1}0\rangle$ partial dislocations with the $\{11\bar{2}0\}$ stacking faults, which are equivalent to that formed between the partial-dislocation pair associated with the $1/3\langle 11\bar{2}0\rangle$ dislocation. As a result, the partial dislocations generated from the $\langle 10\bar{1}0\rangle$ and $1/3\langle 11\bar{2}0\rangle$ dislocations are combined into the odd-numbered partial structures. In addition, there is a variation of the n-values for the odd-numbered partial structures. This would be because the competition between the strain energy and the excess energy of stacking faults. The net screw component of the partial structures with smaller n-values is more localized than that with larger n values, namely that the strain energy decreases with the n-value. In contrast, the area of stacking fault increases with the n-value.

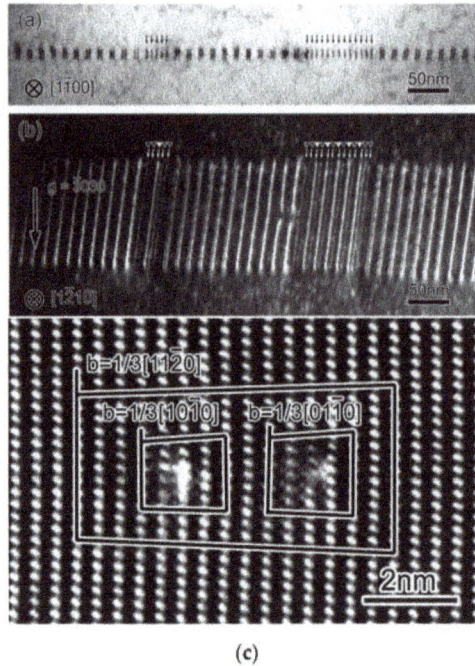

Figure 3. (**a**) TEM image of the $\{11\bar{2}0\}/\langle1\bar{1}00\rangle$ 2° low-angle tilt grain boundary. The grain boundary consists of partial-dislocation pairs and groups of 5 and 13 partial dislocations. (**b**) dark-field TEM image taken at the same region in (**a**) using $g = \bar{3}030$, where the grain boundary plane is inclined by about 30°. Open and filled triangles indicate $1/3\langle10\bar{1}0\rangle$ and $1/3\langle01\bar{1}0\rangle$ partial dislocations, respectively. (**c**) HRTEM image of a partial-dislocation pair viewed along the $[1\bar{1}00]$ zone axis. From the Burgers circuits, it is found that the dislocation pair has the Burgers vector of $1/3\langle11\bar{2}0\rangle$ and the partial dislocations have $1/3\langle10\bar{1}0\rangle$ and $1/3\langle01\bar{1}0\rangle$. The images (**a**) and (**b**) are adapted from [26] and reprinted with the permission of Elsevier B.V. (Amsterdam, Netherlands).

3.3. $\{1\bar{1}00\}/\langle11\bar{2}0\rangle$ Low-Angle Tilt Grain Boundary

Figure 4a shows a TEM image of the $\{1\bar{1}00\}/\langle11\bar{2}0\rangle$ 2° tilt grain boundary. Dislocation triplets are arrayed along the grain boundary, suggesting that the dislocations dissociated into three partial dislocations with $\{1\bar{1}00\}$ stacking faults. An HRTEM image of a dislocation triplet is shown in Figure 4b. The three partial dislocations connected with the two stacking faults are clearly observed. The Burgers circuit shows that this dislocation triplet has the Burgers vector of $\langle1\bar{1}00\rangle$ in total. Therefore, this dissociation reaction is written as:

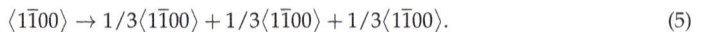

$$\langle1\bar{1}00\rangle \rightarrow 1/3\langle1\bar{1}00\rangle + 1/3\langle1\bar{1}00\rangle + 1/3\langle1\bar{1}00\rangle. \tag{5}$$

This dissociation reaction is known to occur for the $\langle1\bar{1}00\rangle$ slip dislocation associated with the $\{11\bar{2}0\}\langle1\bar{1}00\rangle$ prism-plane slip [5].

The two stacking faults on the $\{1\bar{1}00\}$ plane are formed between three partial dislocations. The stacking sequence of the $\{1\bar{1}00\}$ plane is represented as ...ABCABC.... Thus, the stacking faults generated by the fault vector of $1/3\langle1\bar{1}00\rangle$ have structural variations: ...ABC/B/ABC..., ...ABC/C/ABC..., and ...ABC//BCA..., where the position of stacking disorder is indicated by '//'. These stacking faults are called interstitial fault type-I (I_1), interstitial fault type-II (I_2), and vacancy fault (V), respectively [8,27]. Due to the geometric constraint, the combination of the stacking faults formed between the $1/3\langle1\bar{1}00\rangle$ triplet is known to be either of I_1-V or I_2-V [8,27]. To identify the structure of the stacking faults formed between the partial dislocations, we observed the stacking faults by atomic-resolution STEM.

Figure 4c,d show annular bright-field (ABF) STEM images of the stacking faults, corresponding to the left and the right ones in Figure 4b, respectively. ABF STEM is capable of visualizing atomic columns and even light elements [38,39]. In these images, strong dark contrasts correspond to aluminum columns and weak dark contrasts to oxygen columns, as shown in the atomic structure model in the figure. The stacking sequences of the stacking faults can be directly interpreted as ...ABC/C/ABC...(I_2) for the left stacking fault and ...ABC//BCAB...: (V) for the right stacking fault.

Their stacking fault energies can be estimated by the similar way as discussed in Section 3.1. For the present case, the equations become the following forms:

$$\gamma_{I2} = \frac{\mu b_p^2}{2\pi(1-\nu)} \cdot \frac{1}{d} \sum_{n=0}^{\infty} \left(\frac{1}{n+\alpha_1} + \frac{1}{n+\alpha_1+\alpha_2} - \frac{1}{n+1-\alpha_1} - \frac{1}{n+1-\alpha_1-\alpha_2} \right), \quad (6)$$

$$\gamma_{V} = \frac{\mu b_p^2}{2\pi(1-\nu)} \cdot \frac{1}{d} \sum_{n=0}^{\infty} \left(\frac{1}{n+\alpha_2} + \frac{1}{n+\alpha_1+\alpha_2} - \frac{1}{n+1-\alpha_2} - \frac{1}{n+1-\alpha_1-\alpha_2} \right), \quad (7)$$

where α_1 is d_1/d (d_1: the width of staking fault I_2) and α_2 is d_2/d (d_2: the width of staking fault V). The experimental distances of d, d_1 and d_2 were measured to be 22 nm, 4.7–5.1 nm, and 5.5–5.9 nm, respectively. Using these values, the stacking fault energies were estimated to be I_2: $\gamma = 0.41$–0.46 Jm^{-2} and V: $\gamma = 0.33$–0.37 Jm^{-2}. Theoretical calculations suggested that the stacking fault energies of I_1, I_2, and V are 0.62–0.63 Jm^{-2}, 0.46 Jm^{-2}, and 0.41 Jm^{-2}, respectively [27,37]. These results agree well with the experimental results.

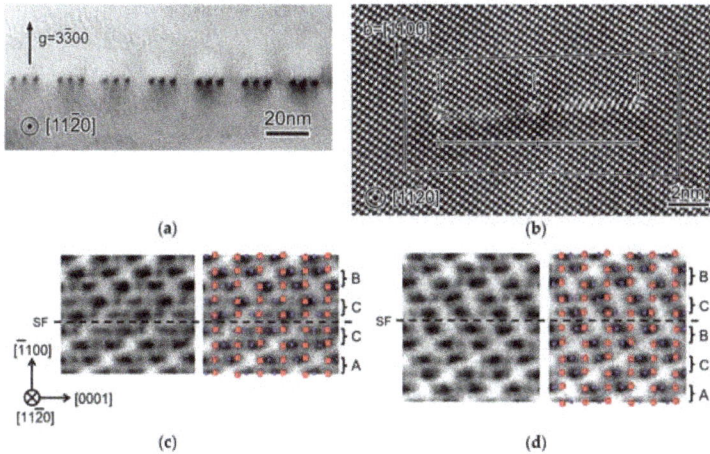

Figure 4. (a) TEM image of the $\{1\bar{1}00\}/\langle11\bar{2}0\rangle$ 2° low-angle tilt grain boundary. The grain boundary consists of dislocation triplets. (b) HRTEM image showing one of the dislocation triplets. The Burgers circuit indicates the Burgers vector of $\langle1\bar{1}00\rangle$, suggesting that the $\langle1\bar{1}00\rangle$ dislocation is dissociated into $1/3\langle1\bar{1}00\rangle$ partial dislocations with two stacking faults in between. (c) ABF STEM image of the left stacking fault in (b). The atomic structure model overlapped with the image is shown at the right panel. The stacking sequence is ...ABC/C/AB...: I_2. (d) The right stacking fault in (b). The stacking sequence is ...ABC//BCAB...: V. The images (a,b) are adapted from [27] and reprinted with the permission of Elsevier B.V. The images (c,d) are adapted from [31], a proceedings paper published by AIP Publishing LLC (Melville, New York, NY, US).

3.4. $(0001)/\langle 1\bar{1}00 \rangle$ Low-Angle Tilt Grain Boundary [30]

Figure 5a shows an HRTEM image of the $(0001)/\langle 1\bar{1}00 \rangle$ 2° low-angle tilt grain boundary. Five dislocation structures are observed and each dislocation structure appears to consist of two partial dislocations with a stacking fault on the (0001) plane. From the Burgers circuits, the edge component of these dislocations is either of $1/6[\bar{1}\bar{1}22]$, $1/6[11\bar{2}2]$, or $1/3[0001]$. These components do not correspond to a translation vector, and thus these dislocations should be a mixed dislocation with a screw component along the $[1\bar{1}00]$ direction. The screw components can be uniquely determined so as to match a possible translation vector as follows:

$$\text{(edge)} \qquad \text{(screw)} \qquad \text{(total)}$$

$$1/6[\bar{1}\bar{1}22] + 1/6[1\bar{1}00] = 1/3[0\bar{1}11], \tag{8}$$

$$1/6[11\bar{2}2] + 1/6[1\bar{1}00] = 1/3[10\bar{1}1], \tag{9}$$

$$1/3[0001] + 1/3[\bar{1}100] = 1/3[\bar{1}101]. \tag{10}$$

The sum of these three vectors is $[0001]$, and their screw components are cancelled out in total. Since the $[0001]$ vector is the translation vector perpendicular to the (0001) boundary plane, it can be said that the tilt component of the grain boundary is effectively accommodated by groups of these three equivalent dislocations of $1/3[0\bar{1}11]$, $1/3[10\bar{1}1]$, and $1/3[\bar{1}101]$. This characteristic dislocation configuration is reasonable in terms of dislocation self-energy, which is proportional to the square of Burgers vector. The magnitude of the $[0001]$ vector is 1.30 nm and that of $1/3\langle \bar{1}101 \rangle$ vector is 0.513 nm, leading the relationship of $| \boldsymbol{b}_{[0001]} |^2 > 3 \times | \boldsymbol{b}_{1/3<-1101>} |^2$. Therefore, the triplet of $1/3\langle \bar{1}101 \rangle$ dislocation is considered to be energetically favorable than the single $[0001]$ dislocation.

Figure 5b shows an ABF STEM image of the $1/3[\bar{1}101]$ dislocation. The zigzag contrasts along the $[0001]$ direction correspond to the configurations of oxygen and aluminum columns as illustrated in the atomic structure at the right. It is clearly seen that the dislocation is dissociated into two partial dislocations with the (0001) stacking fault. The Burgers circuits drawn around the partial dislocations indicate that the left and right ones have an edge component of $1/18[11\bar{2}3]$ and $1/18[\bar{1}\bar{1}23]$, respectively. Their screw components, which are necessary to determine the dissociation reaction, are not given by the STEM observations. To identify the Burgers vector of the partial dislocations, instead, we analyzed the fault vector of the stacking fault.

An ABF STEM image of the (0001) stacking fault formed between the partial dislocations is shown in Figure 5c. The position of the stacking fault is indicated by the dashed line. As shown in the figure, the stacking sequence along the $[0001]$ direction on the $\{1\bar{1}00\}$ projection is ...1 2A 3 2B 1 2C 3//1A 2 1B 3 1C..., where the single numbers refer to oxygen layers and the numbers with a letter to aluminum layers. Theoretical calculations revealed that there is the displacement of $1/6[\bar{1}100]$ or $1/6[1\bar{1}00]$ across the (0001) stacking fault shown in Figure 5c [30], namely that the partial dislocations have a screw component of $1/6[\bar{1}100]$ or $1/6[1\bar{1}00]$. Here, the screw components of the partial dislocations are uniquely determined to be $1/6[1\bar{1}00]$ because the partial-dislocation pair has the screw component of $1/3[\bar{1}100]$ in total (see Equation (10)). As a result, the Burgers vectors for the two partial dislocations are identified to be $1/18[11\bar{2}3] + 1/6[\bar{1}100] = 1/18[\bar{2}4\bar{2}3]$ and $1/18[\bar{1}\bar{1}23] + 1/6[1\bar{1}00] = 1/18[\bar{4}223]$. Consequently, the dissociation reaction of the $1/3\langle \bar{1}101 \rangle$ mixed dislocations is represented by the following equation:

$$1/3\langle \bar{1}101 \rangle \rightarrow 1/18\langle \bar{4}223 \rangle + 1/18\langle \bar{2}4\bar{2}3 \rangle. \tag{11}$$

The $1/3\langle \bar{1}101 \rangle$ dislocation is known to be associated with the $\{1\bar{1}02\}\langle \bar{1}101 \rangle$ pyramidal slip (The $\{10\bar{1}1\}$ and $\{2\bar{1}\bar{1}3\}$ planes are also possible.) [7]. The dissociation reaction of the $1/3\langle \bar{1}101 \rangle$ slip dislocation has not been reported.

The formation energy of the (0001) stacking fault can be calculated by a theoretical equation similar to Equation (3). In the present case, however, it should be too complicated to derive such theoretical equations because the grain boundary consists of six kinds of partial dislocations. Instead, the (0001) stacking fault energy was numerically calculated using the actual dislocation configurations identified from our experimental observations, and the calculated value was $0.58\,Jm^{-2}$ [30]. In addition, first-principles calculations suggested the stacking fault energy to be $0.72\,Jm^{-2}$ [30].

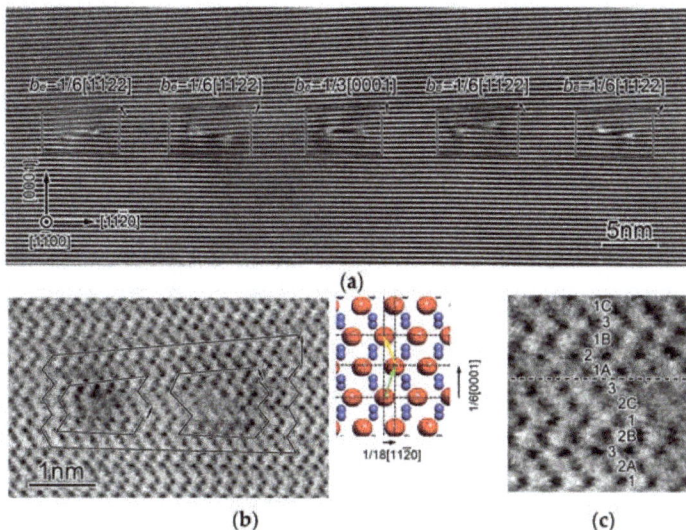

Figure 5. (a) HRTEM image of the $(0001)/\langle 1\bar{1}00 \rangle$ $2°$ low-angle tilt grain boundary. The grain boundary consists of dislocations having an edge component of $1/6\,[\bar{1}\bar{1}22]$, $1/6\,[11\bar{2}2]$, or $1/3[0001]$. Each dislocation is dissociated into two partial dislocations with a stacking fault on the (0001) plane; (b) ABF STEM image of the dislocation having an edge component of $1/3[0001]$. The Burgers circuits drawn around the partial dislocations indicate edge components of $1/18\,[11\bar{2}3]$ and $1/18\,[\bar{1}\bar{1}23]$, as seen in the atomic structure on the right-hand side. (c) enlarged image of the (0001) stacking fault formed between the partial dislocations. The stacking sequence is represented as $...1\ 2A\ 3\ 2B\ 1\ 2C\ 3//1A\ 2\ 1B\ 3\ 1C...$ in the $\{1\bar{1}00\}$ projection. The images (**a**,**b**) are adapted from [30] and reprinted with the permission of Elsevier B.V.

3.5. $\{\bar{1}104\}/\langle 11\bar{2}0 \rangle$ *Low-Angle Tilt Grain Boundary*

Figure 6a shows a dark-field TEM image of the $\{\bar{1}104\}/\langle 11\bar{2}0 \rangle$ low-angle $2°$ tilt grain boundary taken using $g = 1\bar{1}0\bar{4}$. The grain boundary appears to be wavy, and relatively broad contrasts and pair contrasts are observed, suggesting that the grain boundary consists of multiple kinds of dislocation structures.

Figure 6b shows an HRTEM image of the grain boundary. A single dislocation and two dislocation pairs are observed, as indicated by the arrows. Figure 6c shows an enlarged image of one of the single dislocations. The single dislocation has an edge component of $1/2\,[\bar{1}100]$. Since this component does not correspond to a translation vector, the single dislocation should have a screw component along the $[11\bar{2}0]$ direction. The screw component is considered to be $1/6\,[11\bar{2}0]$ (or $1/6\,[\bar{1}\bar{1}20]$), which makes the smallest translation vector of $1/3\,[\bar{1}2\bar{1}0]$ (or $1/3\,[\bar{2}110]$). Therefore, the single dislocation should be the $1/3\,[\bar{1}2\bar{1}0]$ mixed dislocation. An HRTEM image of a dislocation pair dislocation is shown in Figure 6d. This dislocation structure consists of two partial dislocations with a stacking fault on the (0001) plane. The Burgers circuits show that the total edge component is $1/3\,[\bar{1}101]$, and each partial dislocation has an edge component of $1/6\,[\bar{1}101]$. Since $1/3\,[\bar{1}101]$ vector corresponds to a translation vector, this dislocation structure is considered to be the $1/3\,[\bar{1}101]$ edge dislocation.

It was found that the $\{\bar{1}104\}/\langle 11\bar{2}0\rangle$ low-angle tilt grain boundary consists of not only one kind of dislocation structure. This is because there is no translation vector perpendicular to the $\{\bar{1}104\}$ grain boundary plane. The $1/3[\bar{1}101]$ vector corresponding to the dislocation pair is at an angle of 84.16° to the $\{\bar{1}104\}$ grain boundary plane. This vector introduces a component parallel to the grain boundary of $|1/3[\bar{1}101]| \times cos84.16° = 0.052$ nm (along the $[2\bar{2}01]$ direction) in addition to the component normal to the grain boundary. This additional component should be cancelled out in total over the grain boundary by another vector. The $1/2[\bar{1}100]$ vector corresponding to the edge component of the $1/3[\bar{1}2\bar{1}0]$ dislocation is at an angle of 141.76° to the $\{\bar{1}104\}$ plane. This vector has a component along the $[2\bar{2}01]$ direction of $|1/2[\bar{1}100]| \times cos141.76° = -0.32$ nm, which can compensate the additional components due to the $1/3[\bar{1}101]$ vector. Therefore, it is considered that the $1/3[\bar{1}101]$ dislocation pairs and the $1/3[\bar{1}2\bar{1}0]$ dislocations are mixed in the grain boundary so as to cancel out a component along the $[2\bar{2}01]$ direction. The ideal ratio of the numers of the $1/3[\bar{1}101]$ dislocation pairs to the $1/3[\bar{1}2\bar{1}0]$ dislocations is estimated to be $0.32/0.052 = \sim 6.2$.

Here, we further discuss the structure of the $1/3[\bar{1}101]$ dislocation pair. Theoretical calculations revealed that the (0001) stacking fault formed between the partial dislocations has a displacement of $1/18[\bar{1}120]$ and its structure is identical to the (0001) stacking fault shown in Figure 5c [32]. The dislocation pair has no screw component in total, and thus the Burgers vectors of the partial dislocations are $1/6[\bar{1}101]+1/18[\bar{1}120]= 1/18[\bar{4}223]$ and $1/6[\bar{1}101]-1/18[\bar{1}120]= 1/18[2\bar{4}23]$. Consequently, it is found that the dissociation reaction of the $1/3[\bar{1}101]$ edge dislocation also corresponds to Equation (11).

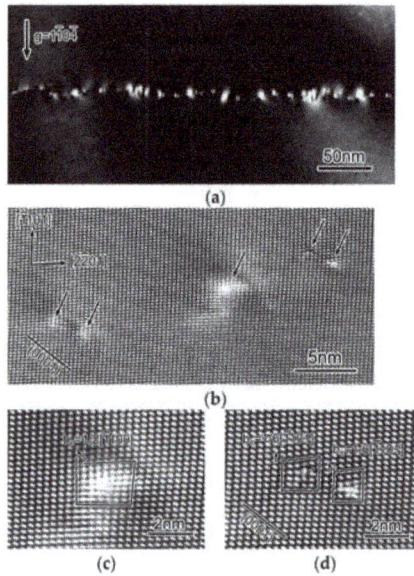

Figure 6. (**a**) dark-field TEM image of the $\{\bar{1}104\}/\langle 11\bar{2}0\rangle$ low-angle 2° tilt grain boundary taken using $g = 1\bar{1}0\bar{4}$. In the grain boundary, relatively broad contrasts and pair contrasts are seen. (**b**) HRTEM image of the grain boundary. A single dislocation and two dislocation pairs are observed. The arrows indicate the dislocation core positions. (**c**) enlarged image of the single dislocation. This dislocation has an edge component of $1/2[\bar{1}100]$. (**d**) enlarged image of the dislocation pair. Dislocation is dissociated into partial dislocations with a stacking fault on the (0001) plane. The partial dislocations have an edge component of $1/6[\bar{1}101]$. The images shown are adapted from [28] and reprinted with the permission of Springer Nature (Berlin, Germany).

3.6. (0001)/[0001] Low-Angle Twist Grain Boundary

A $(0001)/[0001]$ low-angle twist grain boundary was fabricated by joining two pieces of (0001) alumina substrate. The twist angle around the $[0001]$ axis was expected to be as much as the cutting accuracy of the substrate sides $(<\sim0.1°)$. Two TEM samples were prepared: ones for plan-view and for edge-on observations.

Figure 7a shows a plan-view TEM image of the $(0001)/[0001]$ low-angle twist grain boundary. The observation direction is along the $[0001]$ zone axis. A hexagonal dislocation network is formed on the grain boundary. Since these dislocations are parallel to either of the three equivalent directions of $\langle11\bar{2}0\rangle$, they should correspond to the $1/3\langle11\bar{2}0\rangle$ screw dislocations. The interval of the equivalent dislocations is about 60 nm. Using Equation (1), the twist angle of this grain boundary is estimated to be 0.45°.

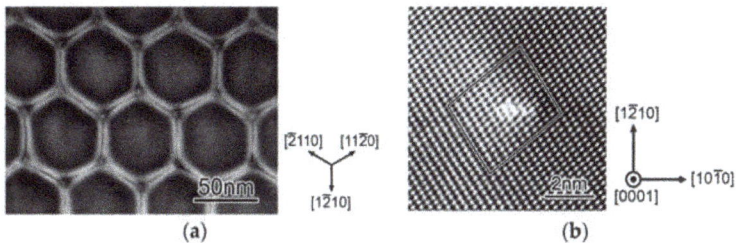

(a) (b)

Figure 7. (**a**) plan-view TEM image of the $(0001)/[0001]$ low-angle twist grain boundary. A hexagonal dislocation network is clearly seen. The dislocation lines are parallel to either of $[\bar{2}110]$, $[11\bar{2}0]$, and $[1\bar{2}10]$ directions, indicating that they are $1/3\langle11\bar{2}0\rangle$ screw dislocations. (**b**) HRTEM image showing the end-on view of the $1/3\langle11\bar{2}0\rangle$ screw dislocation. The screw dislocation is not dissociated. The images shown are adapted from [29] and reprinted with the permission of Elsevier B.V.

An HRTEM image of a $1/3\langle11\bar{2}0\rangle$ screw dislocation in the grain boundary viewed end-on is shown in Figure 7b. The image contrasts are slightly disordered over a few atomic columns, corresponding to the dislocation core region. The circuit drawn around the dislocation core is closed, indicating that this dislocation is a pure screw dislocation. From the image contrasts, the dislocation core is localized within 1 nm or less, and thus the $1/3\langle11\bar{2}0\rangle$ screw dislocation is not dissociated, in contrast to the $1/3\langle11\bar{2}0\rangle$ edge dislocation, as discussed in Section 3.2. The $1/3\langle11\bar{2}0\rangle$ screw dislocation in the present grain boundary corresponds to the $1/3\langle11\bar{2}0\rangle$ basal screw dislocation associated with the $(0001)\langle11\bar{2}0\rangle$ basal slip. The basal screw dislocation is likely to have the perfect type core structure, although its core structure has not been reported. If this is the case, it is considered that the basal screw dislocation slips easily in comparison with the basal edge dislocation. This is because the $\{11\bar{2}0\}$ stacking fault generated by the dissociation of the basal edge dislocation is not on the (0001) slip plane and does not move without atomic diffusion.

4. Findings and Future Subjects

Our investigations found the dissociation reactions and the core structures of $1/3\langle11\bar{2}0\rangle$ (edge, screw), $\langle1\bar{1}00\rangle$ (edge) and $1/3\langle\bar{1}101\rangle$ (edge, mixed) dislocations. The partial dislocations generated by the dissociation reactions of Equations (2), (5), and (11) have the Burgers vector of $1/3\langle10\bar{1}0\rangle$ or $1/18\langle\bar{4}223\rangle$. The formation of these two partial dislocations can be understood in terms of the stability of stacking faults. The stacking faults on the $\{11\bar{2}0\}$ or $\{1\bar{1}00\}$ planes formed by the $1/3\langle10\bar{1}0\rangle$ partial dislocation are stable (low energy) (formation energy: 0.3–0.5 Jm^{-2}) [11,21,24,27,37], whereas ones on the (0001) plane were calculated to be unstable (>1 Jm^{-2}) [36,40]. The stacking fault on the (0001) plane formed by the $1/18\langle\bar{4}223\rangle$ partial dislocation is relatively stable (0.6 Jm^{-2}) [30,32]. The point is that the stability of stacking faults determines the dissociation reaction of dislocations. To examine whether another dissociation reaction is possible, it would be helpful to calculate the relationship between the formation energies of stacking faults and fault vectors for every lattice planes, i.e., generalized stacking fault energies (also known as γ-surfaces) [41].

Screw dislocations except for the $1/3\langle11\bar{2}0\rangle$ dislocation have not been investigated yet. Since the translation vectors of $\langle1\bar{1}00\rangle$, $[0001]$, $1/3\langle\bar{1}101\rangle$, and $1/3\langle2\bar{2}01\rangle$ are on the $\{11\bar{2}0\}$ plane, screw dislocations with either of these vectors are expected to be formed in the $\{11\bar{2}0\}/\langle11\bar{2}0\rangle$ low-angle twist grain boundary. Structural analysis of this type of grain boundary will give us further knowledge on screw dislocations in alumina.

5. Conclusions

Alumina bicrystals with a low-angle grain boundary were systematically fabricated and the grain boundaries were observed by TEM. The dislocation structures formed in the low-angle grain boundaries are summarized below:

(1) $\{11\bar{2}0\}/[0001]$ tilt grain boundary

The $1/3\langle11\bar{2}0\rangle$ perfect edge dislocations dissociate into the $1/3\langle10\bar{1}0\rangle$ and $1/3\langle01\bar{1}0\rangle$ partial dislocations with the $\{11\bar{2}0\}$ stacking fault. The stacking fault energy was estimated to be 0.32 Jm^{-2}.

(2) $\{11\bar{2}0\}/\langle1\bar{1}00\rangle$ tilt grain boundary

The $1/3\langle11\bar{2}0\rangle$ perfect edge dislocations dissociate into the $1/3\langle10\bar{1}0\rangle$ and $1/3\langle01\bar{1}0\rangle$ partial dislocations with the $\{11\bar{2}0\}$ stacking fault. The stacking fault energy was estimated to be 0.35 Jm^{-2}. It was also found that an additional screw component in the grain boundary forms the odd-numbered partial dislocation structures represented by $(n+3)\times1/3\langle10\bar{1}0\rangle + n\times1/3\langle01\bar{1}0\rangle$.

(3) $\{1\bar{1}00\}/\langle11\bar{2}0\rangle$ tilt grain boundary

The $\langle1\bar{1}00\rangle$ perfect edge dislocations dissociate into three $1/3\langle1\bar{1}00\rangle$ partial dislocations with the $\{1\bar{1}00\}$ stacking faults of I$_2$ and V. The stacking fault energies were estimated to be I$_2$: 0.41–0.46 Jm^{-2} and V: 0.33–0.37 Jm^{-2}.

(4) $(0001)/[1\bar{1}00]$ tilt grain boundary

The misorientation of the grain boundary is accommodated by the groups of the $1/3[0\bar{1}11]$, $1/3[10\bar{1}1]$, and $1/3[\bar{1}101]$ perfect mixed dislocations. Each perfect dislocation dissociates into $1/18\langle\bar{4}223\rangle$ and $1/18\langle\bar{2}4\bar{2}3\rangle$ partial dislocations with the (0001) stacking fault. The stacking fault energy was estimated to be 0.58 Jm^{-2}.

(5) $\{\bar{1}104\}/\langle11\bar{2}0\rangle$ tilt grain boundary

The misorientation of the grain boundary is accommodated by $1/3\langle1\bar{2}10\rangle$ perfect mixed dislocations and $1/3\langle\bar{1}101\rangle$ perfect edge dislocations. The $1/3\langle1\bar{2}10\rangle$ dislocations are not dissociated, whereas $1/3\langle\bar{1}101\rangle$ edge dislocations dissociate into $1/18\langle\bar{4}223\rangle$ and $1/18\langle\bar{2}4\bar{2}3\rangle$ with the (0001) stacking fault.

(6) $(0001)/[0001]$ twist grain boundary

The hexagonal network of the $1/3\langle11\bar{2}0\rangle$ prefect screw dislocations is formed. The $1/3\langle11\bar{2}0\rangle$ screw dislocation is not dissociated into partial dislocations.

Acknowledgments: The authors gratefully thank K. Peter D. Lagerlöf, Takahisa Yamamoto, Katsuyuki Matsunaga, Teruyasu Mizoguchi, Matthew F. Chisholm, Hitoshi Nishimura, and Yuki Kezuka for collaborative works on dislocations in alumina. A part of this study was supported by Grant-in-Aid for Specially promoted Research (Grant No. JP17H06094) from the Japan Society for the Promotion of Science (JSPS), the Elements Strategy Initiative for Structural Materials (ESISM) from the Ministry of Education, Culture, Sports, Science, and Technology in Japan (MEXT), Grant-in-Aid for Scientific Research on Innovative Areas "Nano Informatics" (Grant No. JP25106002 and JP25106003) from JSPS, "Nanotechnology Platform" (Project No. 12024046) of MEXT, and JSPS KAKENHI (Grant Nos. JP15H04145 and JP15K20959, and JP17K18983).

Author Contributions: Eita Tochigi drafted the manuscript. All the authors discussed and amended the manuscript.

Conflicts of Interest: The authors declare no conflict of interest.

References

1. Sutton, A.P.; Balluffi, R.W. *Interfaces in Crystalline Materials*; Clarendon Press: Oxford, UK, 1995.
2. Hirth, J.P.; Lothe, J. *Theory of Dislocations*, 2nd ed.; Krieger Publishing Company: Malabar, India, 1982.
3. Snow, J.D.; Heuer, A.H. Slip systems in Al$_2$O$_3$. *J. Am. Ceram. Soc.* **1973**, *56*, 153–157. [CrossRef]
4. Pletka, B.J.; Heuer, A.H.; Mitchell, T.E. Dislocation structures in sapphire deformed basal slip. *J. Am. Ceram. Soc.* **1974**, *56*, 136–139. [CrossRef]
5. Bilde-Sørensen, J.B.; Thölen, A.R.; Gooch, D.J.; Groves, G.W. Structure of ⟨01$\bar{1}$0⟩ dislocation in sapphire. *Philos. Mag.* **1976**, *33*, 877–889. [CrossRef]
6. Mitchell, T.E.; Pletka, B.J.; Phillips, D.S.; Heuer, A.H. Climb dissociation in sapphire (α-Al$_2$O$_3$). *Philos. Mag.* **1976**, *34*, 441–451. [CrossRef]
7. Firestone, R.F.; Heuer, A.H. Creep deformation of 0° sapphire. *J. Am. Ceram. Soc.* **1976**, *59*, 13–19. [CrossRef]
8. Lagerlöf, K.P.D.; Mitchell, T.E.; Heuer, A.H.; Rivière, J.P.; Cadoz, J.; Castaing, J.; Phillips, D.S. Stacking fault energy in sapphire (α-Al$_2$O$_3$). *Acta Met.* **1984**, *32*, 97–105. [CrossRef]
9. Lagerlöf, K.P.D.; Heuer, A.H.; Castaing, J.; Rivière, J.P.; Mitchell, T.E. Slip and twinning in sapphire. *J. Am. Ceram. Soc.* **1994**, *77*, 385–397.
10. Bilde-Sørensen, J.B.; Lawlor, B.F.; Geipel, T.; Pirouz, P.; Heuer, A.H.; Lagerlöf, K.P.D. On basal slip and basal twinning in sapphire (α-Al$_2$O$_3$)—I. Basal slip revisited. *Acta Mater.* **1996**, *44*, 2145–2152. [CrossRef]
11. Nakamura, A.; Yamamoto, T.; Ikuhara, Y. Direct observation of basal dislocation in sapphire by HRTEM. *Acta Mater.* **2002**, *50*, 101–108. [CrossRef]
12. Nakamura, A.; Matsunaga, K.; Tohma, J.; Yamamoto, T.; Ikuhara, Y. Conducting nanowires in insulating ceramics. *Nat. Mater.* **2003**, *2*, 453–456. [CrossRef] [PubMed]
13. Tochigi, E.; Kezuka, Y.; Nakamura, A.; Nakamura, A.; Shibata, N.; Ikuhara, Y. Direct observation of impurity segregation at dislocation cores in an ionic crystal. *Nano Lett.* **2017**, *17*, 2908–2912. [CrossRef] [PubMed]
14. Bouchet, D.; Lartigue-Korinek, S.; Molis, R.; Thibault, J. Yttrium segregation and intergranular defects in alumina. *Philos. Mag.* **2006**, *86*, 1401–1413. [CrossRef]
15. Lartigue-Korinek, S.; Bouchet, D.; Bleloch, A.; Colliex, C. HAADF study of relationship between intergranular defect structure and yttrium segregation in an alumina grain boundary. *Acta Mater.* **2011**, *59*, 3519–3527. [CrossRef]
16. Ikuhara, Y. Nanowire design by dislocation technology. *Prog. Mater. Sci.* **2009**, *54*, 770–791. [CrossRef]
17. Nakamura, A.; Mizoguchi, T.; Matsunaga, K.; Yamamoto, T.; Shibata, N.; Ikuhara, Y. Periodic Nanowire Array at the Crystal Interface. *ACS Nano* **2013**, *7*, 6297–6302. [CrossRef] [PubMed]
18. Matsunaga, K.; Nishimura, H.; Saito, T.; Yamamoto, T.; Ikuhara, Y. High-resolution transmission electron microscopy and computational analyses of atomic structures of [0001] symmetric tilt grain boundaries of Al$_2$O$_3$ with equivalent grain-boundary planes. *Philos. Mag.* **2003**, *83*, 4071–4082. [CrossRef]
19. Gemming, T.; Nufer, S.; Kurtz, W.; Rühle, M. Structure and chemistry of symmetrical tilt grain boundaries in α-Al$_2$O$_3$: I, Bicrystals with "clean" interface. *J. Am. Ceram. Soc.* **2003**, *86*, 581–589. [CrossRef]
20. Lartigue-Korinek, S.; Liagege, S.; Kisielowski, C.; Serra, A. Disconnection arrays in a rhombohedral twin in α-alumina. *Philos. Mag.* **2008**, *88*, 1569–1579. [CrossRef]
21. Ikuhara, Y.; Nishimura, H.; Nakamura, A.; Matsunaga, K.; Yamamoto, T. Dislocation structures of low-angle and near-Σ3 grain boundaries in alumina bicrystals. *J. Am. Ceram. Soc.* **2003**, *86*, 595–602. [CrossRef]
22. Nakamura, A.; Matsunaga, K.; Yamamoto, T.; Ikuhara, Y. Multiple dissociation of grain boundary dislocations in alumina ceramic. *Philos. Mag.* **2006**, *86*, 4657–4666. [CrossRef]
23. Shibata, N.; Chisholm, M.F.; Nakamura, A.; Pennycook, S.J.; Yamamoto, T.; Ikuhara, Y. Nonstoichiometric dislocation cores in α-alumina. *Science* **2007**, *316*, 82–85. [CrossRef] [PubMed]

24. Tochigi, E.; Shibata, N.; Nakamura, A.; Yamamoto, T.; Ikuhara, Y. TEM characterization of 2° tilt grain boundary in alumina. *Mater. Sci. Forum* **2007**, *561–565*, 2427–2430. [CrossRef]
25. Tochigi, E.; Shibata, N.; Nakamura, A.; Yamamoto, T.; Lagerlöf, K.P.D.; Ikuhara, Y. Dislocation structure of 10° low-angle tilt grain boundary in α-Al$_2$O$_3$. *Mater. Sci. Forum* **2007**, *558–559*, 979–982. [CrossRef]
26. Tochigi, E.; Shibata, N.; Nakamura, A.; Yamamoto, T.; Ikuhara, Y. Partial dislocation configurations in a low-angle boundary in α-Al$_2$O$_3$. *Acta Mater.* **2008**, *56*, 2015–2021. [CrossRef]
27. Tochigi, E.; Shibata, N.; Nakamura, A.; Mizoguchi, T.; Yamamoto, T.; Ikuhara, Y. Structures of dissociated $\langle 1\bar{1}00 \rangle$ dislocations and $\{1\bar{1}00\}$ stacking faults of alumina (α-Al$_2$O$_3$). *Acta Mater.* **2010**, *58*, 208–215. [CrossRef]
28. Tochigi, E.; Shibata, N.; Nakamura, A.; Yamamoto, T.; Ikuhara, Y. Dislocation structures in a $\{\bar{1}104\}/\langle 11\bar{2}0 \rangle$ low-angle tilt grain boundary of alumina (α-Al$_2$O$_3$). *J. Mater. Sci.* **2011**, *46*, 4428–4433. [CrossRef]
29. Tochigi, E.; Kezuka, Y.; Shibata, N.; Nakamura, A.; Ikuhara, Y. Structure of screw dislocations in a (0001)/[0001] low-angle twist grain boundary of alumina (α-Al$_2$O$_3$). *Acta Mater.* **2012**, *60*, 1293–1299. [CrossRef]
30. Tochigi, E.; Nakamura, A.; Mizoguchi, T.; Shibata, N.; Ikuhara, Y. Dissociation of the $1/3\langle \bar{1}101 \rangle$ dislocation and formation of the anion stacking fault on the basal plane in α-Al$_2$O$_3$. *Acta Mater.* **2015**, *91*, 152–161. [CrossRef]
31. Tochigi, E.; Findlay, S.D.; Okunishi, E.; Mizoguchi, T.; Nakamura, A.; Shibata, N.; Ikuhara, Y. Atomic structure characterization of stacking faults on the $\{1\bar{1}00\}$ plane in α-alumina by scanning transmission electron microscopy. *AIP Conf. Proc.* **2016**, *1763*, 050003.
32. Tochigi, E.; Mizoguchi, T.; Okunishi, E.; Nakamura, A.; Shibata, N.; Ikuhara, Y. Dissociation reaction of the $1/3\langle \bar{1}101 \rangle$ edge dislocation in α-Al$_2$O$_3$. *J. Mater. Sci.* **2018**, in press. [CrossRef]
33. Nakamura, A.; Tochigi, E.; Shibata, N.; Yamamoto, T.; Ikuhara, Y. Structure and configuration of boundary dislocations on low angle tilt grain boundaries in alumina. *Mater. Trans.* **2009**, *50*, 1008–1014. [CrossRef]
34. Chung, D.H.; Simmons, G. Pressure and temperature dependences of isotropic elastic moduli of polycrystalline alumina. *J. Appl. Phys.* **1968**, *39*, 5316–5326. [CrossRef]
35. Gieske, J.H.; Barsch, G.R. Pressure dependence of elastic constants of single crystalline aluminum oxide. *Phys. Stat. Sol.* **1968**, *29*, 121–131. [CrossRef]
36. Kenway, P.R. Calculated stacking-fault energies in α-Al$_2$O$_3$. *Philos. Mag. B* **1993**, *68*, 171–183. [CrossRef]
37. Jhon, M.H.; Glaeser, A.M.; Chrzan, D.C. Computational study of stacking faults in sapphire using total energy methods. *Phys. Rev. B* **2005**, *71*, 214101. [CrossRef]
38. Okunishi, E.; Ishikawa, I.; Sawada, H.; Hosokawa, F.; Hori, M.; Kondo, Y. Visualization of light elements at ultrahigh resolution by STEM annular bright field microscopy. *Microsc. Microanal.* **2009**, *15*, 164–165. [CrossRef]
39. Findlay, S.D.; Shibata, N.; Sawada, H.; Okunishi, E.; Kondo, Y.; Yamamoto, T.; Ikuhara, Y. Robust atomic resolution imaging of light elements using scanning transmission electron microscopy. *Appl. Phys. Lett.* **2009**, *95*, 191913. [CrossRef]
40. Marinopoulos, A.G.; Elsässer, C. Density-functional and shell-model calculations of the energies of basal-plane stacking faults in sapphire. *Philos. Mag. Lett.* **2001**, *81*, 329–338. [CrossRef]
41. Duesberya, M.S.; Vitek, V. Plastic anisotropy in b.c.c. transition metals. *Acta Mater.* **1998**, *46*, 1481–1492. [CrossRef]

crystals

MDPI

Review

Study on Dislocation-Dopant Ions Interaction in Ionic Crystals by the Strain-Rate Cycling Test during the Blaha Effect

Yohichi Kohzuki

Department of Mechanical Engineering, Saitama Institute of Technology, Saitama 369-0293, Japan;
kohzuki@sit.ac.jp

Received: 23 October 2017; Accepted: 8 January 2018; Published: 12 January 2018

Abstract: The interaction between a dislocation and impurities has been investigated by measurements of yield stress and proof stress, micro-hardness tests, direct observations of dislocation, internal friction measurements, or stress relaxation tests so far. A large number of investigations has been carried out by the separation of the flow stress into effective and internal stresses on the basis of the temperature dependence of yield stress, the strain rate dependence of flow stress, and stress relaxation. Nevertheless, it is difficult to investigate the interaction between a dislocation and obstacles during plastic deformation by the mentioned methods. As for the original method which combines strain-rate cycling tests with the Blaha effect measurement, the original method is different from above-mentioned ones and would be possible to clear it up. The strain-rate cycling test during the Blaha effect measurement has successively provided the information on the dislocation motion breaking away from the strain fields around dopant ions with the help of thermal activation, and seems to separate the contributions arising from the interaction between dislocation and the point defects and those from dislocations themselves during plastic deformation of ionic crystals. Such information on dislocation motion in bulk material cannot be obtained by the widely known methods so far. Furthermore, the various deformation characteristics derived from the original method are sensitive to a change in the state of dopant ions in a specimen by heat treatment, e.g., the Gibbs free energy (G_0) for overcoming of the strain field around the dopant by a dislocation at absolute zero becomes small for the annealed KCl:Sr^{2+} single crystal (G_0 = 0.26 eV) in comparison with that for the quenched one (G_0 = 0.39 eV).

Keywords: ultrasonic oscillatory stress; strain-rate sensitivity of flow stress; activation energy for dislocation motion; forest dislocation density; heat treatment

---

1. Introduction

The interaction between a dislocation and point defects has been investigated by measurements of yield stress (e.g., [1–7]) and proof stress (e.g., [8,9]), micro-hardness tests (e.g., [10–14]), direct observations of dislocation (e.g., [15–21]), internal friction measurements (e.g., [22–27]), or stress relaxation tests (e.g., [28,29]) so far. A large number of investigations have been carried out by the separation of the flow stress into effective and internal stresses on the basis of the temperature dependence of yield stress, the strain rate dependence of flow stress, and the stress relaxation. Nevertheless, it is difficult to investigate the interaction between a dislocation and obstacles during plastic deformation by the mentioned methods. This is because yield stress depends on dislocation velocity, dislocation density, and multiplication of dislocations [30] and proof stress is used as a means of assessing the yield stress. On the other hand, the effect of heat treatment on the micro-hardeness is almost insensitive to the change of atomic order of point defects in a specimen. As for direct observations, electron microscopy provides the information on the interaction between a dislocation and obstacles for a thin specimen but not for bulk, and also light scattering method is useful only for a transparent specimen. X-ray topography is the lack of resolution in the photograph, so that the specimen is limited to the low dislocation density below 10^4 cm^{-2}. Internal friction measurements cannot provide the information on the motion of the dislocation which moves by overcoming the forest dislocations and the weak obstacles such as impurities during plastic deformation, because the measurements concern the motion of the dislocation which breaks away from the weak obstacles between two forest dislocations by vibration [31]. In stress relaxation tests it is generally assumed that internal structure of crystals does not change, i.e., dislocation density and internal stress are constant. In this paper, the study on interaction between a dislocation and dopant ions is made by the strain-rate cycling tests during the Blaha effect measurement.

The Blaha effect is the phenomenon that static flow stress decreases when an ultrasonic oscillatory stress is superimposed [32]. Blaha and Langenecker found it when the ultrasonic oscillatory stress of 800 kHz was superimposed during plastic deformation of Zn single crystals. Figure 1 shows the variation of the stress-strain curve by superimposed ultrasonic vibrations as an example of their experimental results [20]. The curve represents the intermittent addition of ultrasonic vibrations in a solid line and the continued addition in a broken line. This phenomenon is the so-called the Blaha effect. The same phenomena have been confirmed in many metals (e.g., [33–35]) and have been widely made to apply to the plastic working technique for industrial purpose such as wire drawing, deep drawing, rolling, and another metal forming techniques, since this phenomenon has an industrial significance (e.g., [36–44]). The Blaha effect has been explained by a temperature rise of materials [45], an abrupt increase in mobile dislocation density [46], a reduction in internal stress [47–49], or a stress superposition mechanism [50–55] so far. The temperature rise is not so high as to influence the flow stress in this test (the strain-rate cycling test during the Blaha effect measurement). It is also improbable that the mobile dislocation density markedly increases immediately after the application of oscillations. If the Blaha effect is explained by itself, the mobile dislocation density has to increase by a factor of tens or hundreds. The reduction in internal stress should be valid only for the particular case that the amplitude of oscillatory stress is so large as to change the internal structure, where a redistribution of the dislocations may occur. The stress superposition mechanism, where the internal friction due to application of oscillations is neglected and the amplitude dependent region (Q_H^{-1}) of internal friction is not considered, has been regarded as the most important one to interpret the Blaha effect. However, Lebedev et al. [49,56] reported that the stress decrement ($\Delta\tau$) due to the application of ultrasonic oscillatory stress appears at the moment when measurements of the amplitude dependence of internal friction were made. Ohgaku and Takeuchi [57] also concluded that $\Delta\tau$ is proportional to Q_H^{-1} at small amplitude for NaCl and KCl single crystals. The relation between $\ln\Delta\tau$ and $\ln Q_H^{-1}$ is shown in Figure 2, which is drawn from the experimental results for single crystals of KCl, NaCl, NaBr, and KBr. The relation is divided into three regions denoted by I, II, and III. It was furthermore found that the average length of dislocation segments become long when the internal friction measurement was carried out [58]. Therefore, it could be considered that $\Delta\tau$ of alkali halide is attributed to the increase in activation volume for dislocation motion when dislocations move with the aid of thermal activation [57,59]. Assuming that the mobile dislocation density and the enthalpy at zero effective stress keep constant before and after application of oscillatory stress, this can be interpreted on the basis of the following discussion. The activation volume (V_0) and the effective stress (τ_e) will change into V and τ, respectively, when an oscillation is applied. Then, we can obtain

$$V_0\tau_e = V\tau \tag{1}$$

because the strain rate keeps constant before and after application of oscillations. Equation (1) directly gives

$$\Delta\tau = \tau_e - \tau = \left(1 - \frac{V_0}{V}\right)\tau_e \tag{2}$$

Accordingly, the increment in activation volume, namely the increment in the average length of dislocation segments (ΔL) as given by Equation (3), causes $\Delta \tau$ of alkali halide.

$$\Delta V = b \Delta L d \tag{3}$$

where b is the magnitude of Burgers vector and d is the activation distance.

The $\Delta \tau$ increases with increasing stress amplitude and slightly increases with strain rate ($\dot{\varepsilon}$) [59]. The strain rate dependence of the $\Delta \tau$ may suggest that the stress change ($\Delta \tau'$) due to the strain-rate cycling depend on the stress amplitude and on the $\Delta \tau$. Then, Ohgaku and Takeuchi [59] carried out the strain-rate cycling tests under superposition of an oscillatory stress.

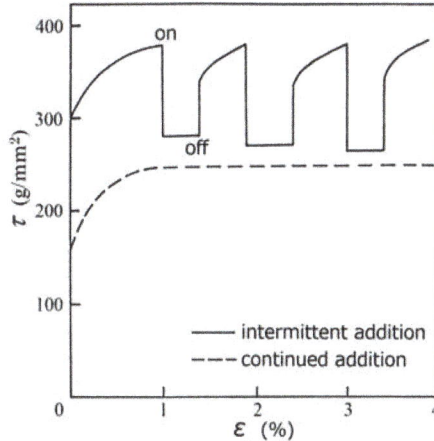

Figure 1. Variation of the stress-strain curve for Zn single crystals by superimposed ultrasonic vibrations (reproduced from reference [32] with permission from the publisher).

The strain-rate cycling test during the Blaha effect measurement is illustrated in Figure 3. The stress drop $\Delta \tau$ is caused by superposition of oscillatory stress (τ_v) during plastic deformation. When strain-rate cycling between the strain rates of $\dot{\varepsilon}_1$ and $\dot{\varepsilon}_2$ is carried out keeping the stress amplitude of τ_v constant, the stress change due to the strain-rate cycling is $\Delta \tau'$. Figure 4 shows the result at room temperature for single crystals of nominally pure alkali halide such as NaCl, KCl, NaBr, and KBr. The relative curve between $\Delta \tau'$ and $\Delta \tau$ shifts upward with strain and bends at a critical $\Delta \tau$. The critical $\Delta \tau$ corresponds to the point which separates region I and region II in Figure 2. They identified the small $\Delta \tau$ region in Figure 4 with region I in Figure 2 and the larger $\Delta \tau$ region with regions II and III. The relations between $\Delta \tau$ and Q_H^{-1} as well as $\Delta \tau$ and $\Delta \tau'$ are essentially divided into two regions, where different mechanisms are considered to dominate. In region III, it concerns the dislocation density, because Q_H^{-1} linearly decreases as applied shear stress, which is generally proportional to the square root of the dislocation density [60–62], becomes large in region III [63]. When the tests are carried out, $\Delta \tau'$ can be expressed by

$$\Delta \tau' = \Delta \tau'_i + \Delta \tau'_p + \Delta \tau'_s \tag{4}$$

where the suffixes of stress variation, i, p, and s, mean the internal stress, the part of the effective stress due to impurities, and the strain-dependent part of the effective stress, respectively. They considered Figure 4 on the basis of Equation (4) as follows. $\Delta \tau'_p$ corresponds to the linearly decreasing part of $\Delta \tau'$ at small stress amplitude in Figure 4 and is given by the difference between $\Delta \tau'$ at $\Delta \tau$ of zero and $\Delta \tau'$ at the critical point which distinguishes regions I and II. $\Delta \tau'_s$ dominates in regions II and III, where $\Delta\tau'_p$ is zero. $\Delta\tau'_s$ increases with increasing strain and dislocation density. Li reported that the effective internal stress depends on the flow stress [64]. When a dislocation velocity-stress exponent is nearly 30 and the maximum internal stress is close to the flow stress, the effective internal stress approaches the maximum internal stress [64]. With regard to $\Delta\tau'_i$, it was considered by the results in the paper [64] that the influence of internal stress during the strain-rate cycling under the oscillation is negligible. This can be examined also from the experimental result in Section 3.4 (i.e., λ_p is almost constant independently of the strain). Therefore, the strain-rate cycling tests during the Blaha effect measurement seems to separate the contributions arising from the interaction between a dislocation and dopant ions and those from dislocations themselves during plastic deformation.

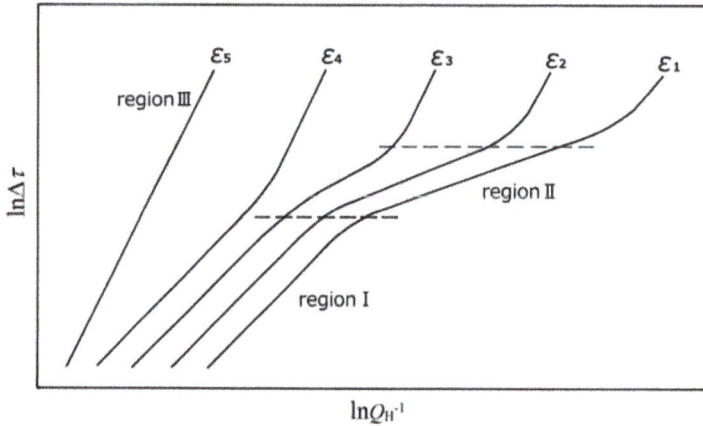

Figure 2. Schematic relation between $\ln\Delta\tau$ and $\ln Q_H{}^{-1}$ (reproduced from reference [63] with permission from the publisher). The strain, ε, increases with its suffix.

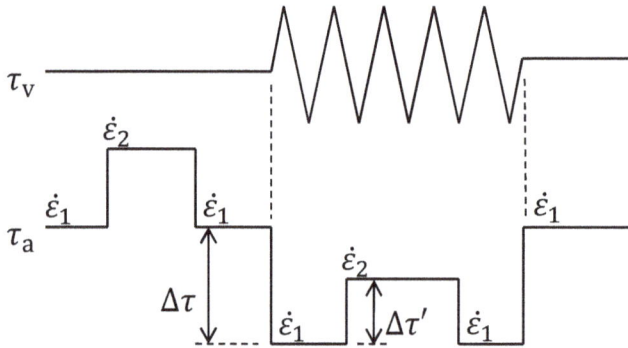

Figure 3. Variation of applied shear stress, τ_a, when the strain-rate cycling between the strain rates, $\dot{\varepsilon}_1$ (2.2×10^{-5} s^{-1}) and $\dot{\varepsilon}_2$ (1.1×10^{-4} s^{-1}), is carried out under superposition of ultrasonic oscillatory shear stress, τ_v.

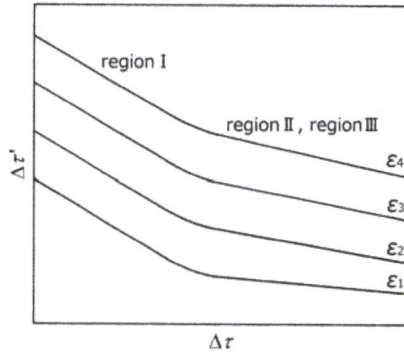

Figure 4. Schematic relation between $\Delta\tau'$ and $\Delta\tau$ at room temperature (reproduced from reference [63] with permission from the publisher). The strain increases with its suffix.

Ohgaku and Takeuchi [57,63] reported that the strain-rate cycling tests associated with the ultrasonic oscillation can separate the contributions arising from the interaction between a dislocation and dopant ions and from the dislocations themselves during plastic deformation at room temperature. Then, they discussed the temperature dependence of the effective stress due to monovalent dopants (i.e., Br^- or I^-) in KCl:Br^- (0.5, 1.0, and 2.0 mol.% in the melt) or I^- (0.2, 0.5, and 1.0 mol.% in the melt) single crystals at 77 to 420 K [65] and NaCl:Br^- (0.1, 0.5, and 1.0 mol.% in the melt) single crystals at 77 to 253 K [66]. They found that the measurement of strain-rate sensitivity under the application of ultrasonic oscillatory stress provides useful information on the interaction between a mobile dislocation and the dopant ions [65,66]. It has been reported so far through the combination method that the information on the dislocation motion can be obtained in terms of various ionic crystals. The details are reviewed mainly for KCl:Sr^{2+} single crystals.

As compared with metals, the ionic crystals used here have some advantages: A small number of glide systems, low dislocation density in grown crystal (e.g., 10^4 cm^{-2} order for NaCl single crystal [67]) as against that of annealed metals (e.g., 1 to 5×10^6 cm^{-2} in pure Cu single crystal [68] and 10^7 cm^{-2} order for pure Fe single crystal [69]), transparency and readiness for single crystal preparation, etc. Pure or impurity-doped alkali halide crystals, therefore, are excellent materials for an investigation on mechanical properties of crystals. A large number of investigations on strength of materials have been made with alkali halide crystals (e.g., [19,70,71]).

2. Experimental Procedure

2.1. Preparation of Specimens

Two kinds of single crystals used in this work, nominally pure KCl, KCl doped with SrCl$_2$ (0.035, 0.050, 0.065 mol.% in melt) were grown from the melt of reagent-grade powders by the Kyropoulos method in air. The specimens, which were prepared by cleaving out of single crystalline ingots to the size of $5 \times 5 \times 15$ mm^3, were kept immediately below the melting point for 24 h in order to reduce dislocation density as much as possible, followed by cooling to room temperature at the rate of 40 Kh^{-1}. This is termed the specimen A. The additive Sr^{2+} ions are supposed to aggregate owing to the gradual cooling in the specimen A. Accordingly, the specimens A were held at 673 K for 30 min, followed by quenching to room temperature immediately before the test, in order to disperse the additive Sr^{2+} ions into them. The temperature (i.e., 673 K) of this heat treatment was determined on the basis of the below experimental results.

Figure 5 shows the dependence of initial dislocation density (ρ), the dielectric loss peak due to the I-V pairs ($\tan\delta$), and yield stress (τ_y) on the temperature from which KCl:Sr^{2+} (0.3 and 1.0 mol.% in the melt) single crystals were quenched [72]. Etch pits technique was used to detect the density of dislocations in KCl:Sr^{2+} (0.3 mol.% in the melt) with a corrosive liquid (saturated solution of PbCl$_2$ + ethyl alcohol added two drops of water). The etching was carried out at room temperature for 30 min. The dielectric loss factor tanδ as a function of frequency was measured for KCl:Sr^{2+} (1.0 mol.% in the melt) in thermostatic bath within 300 to 873 K. The height of the loss peak is related to the concentration of isolated I-V dipole (see Equation (7)). The τ_y values were obtained at room temperature for KCl:Sr^{2+} (1.0 mol.% in the melt) compressed along the <100> axis at the crosshead speed of 20 μm min^{-1}. As a result, it was found that the dislocation density and the yield stress remarkably increase for the crystals quenched from the temperature (T) above 673 K. While the value of tanδ does not vary by quenching from the temperature below 573 K or above 673 K. In Figure 5, the values of ρ, tanδ, and τ_y at about 300 K correspond to the specimen A and those at 673 K the specimen A quenched from 673 K. When alkali halide crystals are doped with divalent cations, the dopant cations induce positive ion vacancies in order to conserve the electrical neutrality and are expected to be paired with the vacancies. They are often at the nearest neighbor sites forming a divalent impurity-vacancy (I-V) dipole, which attract them strongly [73], for crystals quenched from a high temperature. Then asymmetrical distortions (tetragonal lattice distortions) are produced around the I-V dipoles. Mobile dislocations on a slip plane interact strongly only with these defects lying within one atom spacing of the glide plane [74]. The concentration of isolated I-V dipoles affects the yield stress, as reported in the papers [75–78]. The difference in dislocation density is slight and the tanδ obviously becomes larger with higher quenching temperature between 573 and 673 K as shown in Figure 5. Therefore, the specimens are determined to be quenched from 673 K to room temperature immediately before the compression tests mentioned below.

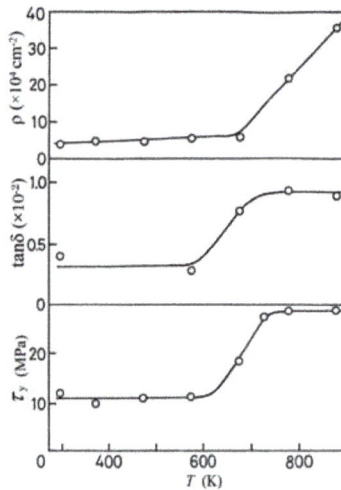

Figure 5. Dependence of initial dislocation density (ρ), the dielectric loss peaks due to the I-V pairs (tanδ), and yield stress (τ_y) on the quenching temperature (reproduced from reference [72]).

2.2. Strain-Rate Cycling Test during the Blaha Effect

The schematic illustration of apparatus is shown in Figure 6a. A resonator composed of a vibrator and a horn was attached to an Instron-type testing machine DSS-500 (Shimadzu Corp., Kyoto, Japan), in Figure 6b. The specimens, of which the upper and bottom sides were coated with molybdenum disulfide as a lubricant to prevent from barrel shape deformation during the test, were lightly fixed on a piezoelectric transducer and then cooled down to a test temperature. The specimens were deformed by compression along the <100> axis, and the ultrasonic oscillatory stress with the signal of 20 kHz from a multifunction synthesizer was intermittently superimposed for one or two minutes by the resonator in the same direction as the compression. The amplitude of the oscillatory stress was monitored by the output voltage from the piezoelectric transducer set on the bottom side of specimen (i.e., between a specimen and the support rod), which was observed by an a.c. voltmeter or an oscilloscope. The strain of the specimen is considered to be homogeneous, because the wavelength, which is 196 mm by calculating from the data of reference [79], is 13 times as long as the length of the specimen. Strain-rate cycling test during the Blaha effect measurement was already illustrated in Figure 3. Superposition of oscillatory stress (τ_v) during plastic deformation causes a stress drop (as $\Delta\tau$, in MPa). Keeping the stress amplitude of τ_v constant, strain-rate cycling between the strain rates of $\dot{\varepsilon}_1$ (2.2×10^{-5} s^{-1}) and $\dot{\varepsilon}_2$ (1.1×10^{-4} s^{-1}) was carried out. Then, the stress change due to the strain-rate cycling is $\Delta\tau'$. Slip system for rock-salt structure such as KCl crystal is {110} $< 1\bar{1}0 >$ so that shear stress (τ) and shear strain (ε) calculated for the slip system were used in this study. The strain-rate cycling tests made between the crosshead speeds of 20 and 100 μm min^{-1} were performed at 77 K up to the room temperature as shown in Figure 6c. For the tests at 77 K, the specimen was immersed in the liquid nitrogen. The other temperature measurements were made by heater controlled using thermocouples of Ni-55%Cu vs. Cu, in Figure 6d. Each specimen was held at the test temperature for 30 min prior to the compression test, and the stability of temperature during the test was kept within 2 K. The strain-rate sensitivity ($\Delta\tau'/\Delta\ln\dot{\varepsilon}$) of the flow stress, which is given by $\Delta\tau'/1.609$, was used as a measurement of the strain-rate sensitivity (as $\lambda = \Delta\tau'/\Delta\ln\dot{\varepsilon}$, in MPa).

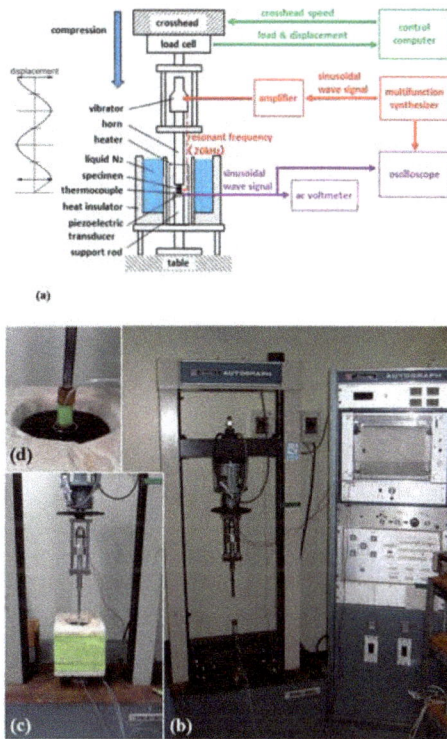

Figure 6. Experimental apparatus: (**a**) Schematic block diagram of apparatus system; (**b**) main testing machine; (**c**) resonator connected with the test machine and heat-insulator box; (**d**) heater covered specimen in the heat-insulator box.

3. Results and Discussion

3.1. Relations between $\Delta\tau$ and Strain-Rate Sensitivity (λ) for KCl:Sr^{2+} and KCl

The variation of λ with shear strain is shown in Figure 7a for KCl:Sr^{2+} (0.050 mol.%) single crystal at 200 K. Figure 7b concerns $\Delta\tau$ for the same specimen. $\Delta\tau$ does not largely change with shear strain and is almost constant independently of the strain. This means that the process of plastic deformation does not affect the decrease of flow stress due to superposition of oscillatory stress. However, $\Delta\tau$ increases with increasing stress amplitude at a given temperature and shear strain. λ tends to increase with shear strain at all stress amplitude (except for $\tau_v = 45$ below the strain of about 9%) and decrease with stress amplitude at a given shear strain. The increase of λ with shear strain is caused by the increase of the forest dislocation density. The value of $(\Delta\lambda/\Delta\tau_v)_{T,\varepsilon}$ seems to be small at low and high stress amplitude as can be seen in Figure 7a. The values of $\Delta\tau$ and λ, which is obtained from Figure 7 at the shear strains of 10%, 14% and 18%, are plotted in Figure 8. The figure provides the relationship between $\Delta\tau$ and λ for a fixed internal structure of the KCl:Sr^{2+} (0.050 mol.%) single crystal at 200 K. As can be seen from Figure 8, the variation of λ with $\Delta\tau$ has stair-like shape. That is to say, there are two bending points on the each curve, and there are two plateau regions: The first plateau region ranges below the first bending point at low $\Delta\tau$ and the second one extends from the second bending point at high $\Delta\tau$. The second plateau place is considered to correspond to the regions II and III in Figure 4, as reported by Ohgaku and Takeuchi [63]. The λ gradually decreases with $\Delta\tau$ between the two bending points. λ_p denoted in Figure 8 is introduced later (Section 3.4). The values of $\Delta\tau$ at the first and the second bending points are referred to as τ_{p1} and τ_{p2} in MPa, respectively. However, no first bending point (i.e., τ_{p1}) is on the relative curve of $\Delta\tau$ and λ for nominally pure KCl single crystal. This is revealed in Figure 9 represented the relationship of $\Delta\tau$ and λ for the KCl at several temperatures. As for KCl, the first plateau region does not appear on each curve and λ decreases with increasing $\Delta\tau$ at low stress decrement. In contrast to KCl:Sr^{2+} single crystals, there is only one bending point on each curve, which is considered to correspond to τ_{p2}. Therefore, τ_{p1} is due to the additive ions (Sr^{2+}) in the crystals. Figure 8 shows also the influence of shear strain on the relation between $\Delta\tau$ and λ. The curve shifts upwards with increasing shear strain. This phenomenon is caused by the part of the strain-rate sensitivity which depends on dislocation cuttings as well as KCl. Since the dislocation cuttings increase with increasing strain, the λ increases [63]. Stair-like relation curve of $\Delta\tau$ vs. λ is obtained for KCl:Sr^{2+} at any other temperature. Figure 10 shows the influence of temperature on the $\Delta\tau$ vs. λ curve for the crystal at strain 10%, which is obtained by the above-mentioned method. Similar result as Figure 8 is also obtained at low temperature. As the temperature is higher, τ_{p1} and τ_{p2} tend to be lower, and τ_{p1} disappears at 225 K. If τ_{p1} is the effective stress due to the weak obstacles (i.e., the additive Sr^{2+} ions) on the mobile dislocation during plastic deformation of KCl:Sr^{2+} single crystals, it should depend on temperature and Sr^{2+} concentration. Figure 11a–c shows the dependence of τ_{p1}, τ_{p2}, and yield stress τ_y on temperature for KCl:Sr^{2+} (0.065, 0.050, and 0.035 mol.%, respectively) single crystals, where the curves for these stresses are to guide the reader's eye. It is clear from the figure that τ_{p1} and τ_{p2} tend to increase with decreasing temperature as the variation of τ_y in the temperature range applied for the three specimens. Both τ_{p1} and τ_{p2} seem to depend on Sr^{2+} concentration at a given temperature. τ_{p1} may increase with the solute concentration (c_0) as below

$$\tau_{p1} = \sqrt{\frac{F^3 c_0}{2\mu b^6}} \tag{5}$$

where F is the force acted on the dislocation and μ is the shear modulus. Equation (5) is derived from the Friedel relation [80] between the effective stress and the average length of dislocation segments. Figure 12 shows the dependence of τ_{p1} and τ_{p2} on the yield stress for KCl:Sr^{2+} at 150 K. The plots correspond to the case when the Sr^{2+} concentration is 0.035, 0.050, and 0.065 mol.% from the bottom. It can be seen from this figure that both τ_{p1} and τ_{p2} are approximately proportional to the yield stress.

This means that τ_{p1} and τ_{p2} increase, depending on the Sr^{2+} concentration, because the yield stress generally increases with increasing impurity concentration [1–3,75,81].

The critical temperature (T_c), at which the T-τ_{p1} curves intersect the abscissa and τ_{p1} is zero, may be determined from Figure 11. Then, T_c appears to be constant independently of the Sr^{2+} concentration and τ_y seems to approach a constant value above the T_c.

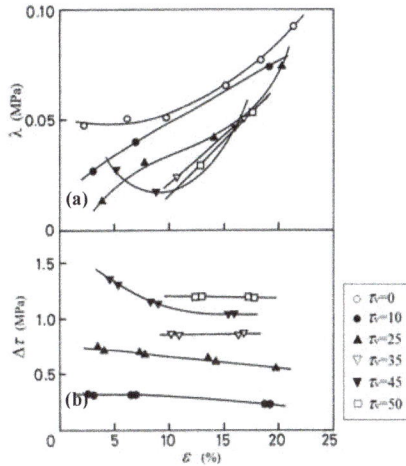

Figure 7. Variation of (**a**) the strain-rate sensitivity, λ (= $\Delta\tau'/\Delta\ln\dot\varepsilon$) of flow stress, and (**b**) the stress decrement, $\Delta\tau$, due to the superimposition of ultrasonic oscillation with shear strain at 200 K and various stress amplitude for $KCl:Sr^{2+}$ (0.050 mol.%) single crystal (reproduced from reference [82] with permission from the publisher). τ_v (arb. units): ($\bigcirc$) 0, ($\bullet$) 10, ($\blacktriangle$) 25, ($\triangledown$) 35, ($\blacktriangledown$) 45, and ($\square$) 50. The ultrasonic oscillatory stress (τ_v) increases with its numbers besides each symbol.

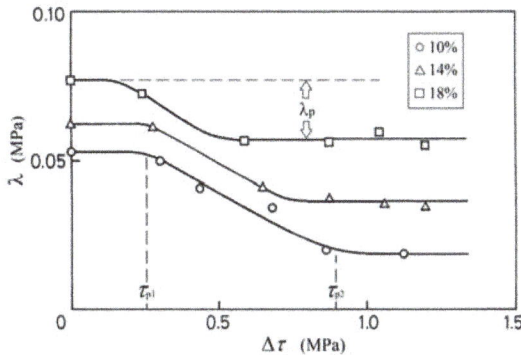

Figure 8. Relation between the strain-rate sensitivity and the stress decrement for $KCl:Sr^{2+}$ (0.050 mol.%) at 200 K (reproduced from reference [82] with permission from the publisher). ε: ($\bigcirc$) 10%, ($\triangle$) 14%, and ($\square$) 18%. The plotted points are obtained from Figure 7a,b.

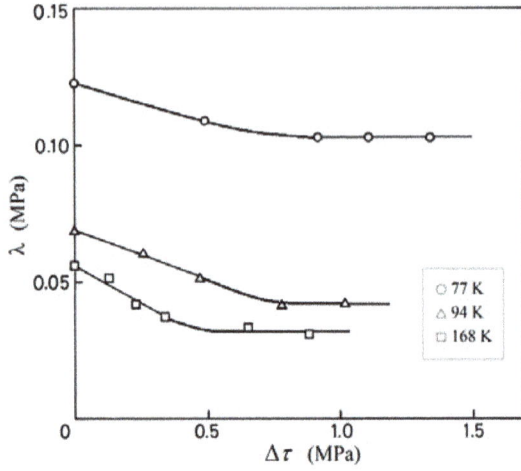

Figure 9. Relation between the strain-rate sensitivity and the stress decrement for KCl at various temperatures: (○) 77 K, (△) 94 K, (□) 168 K (reproduced from reference [82] with permission from the publisher).

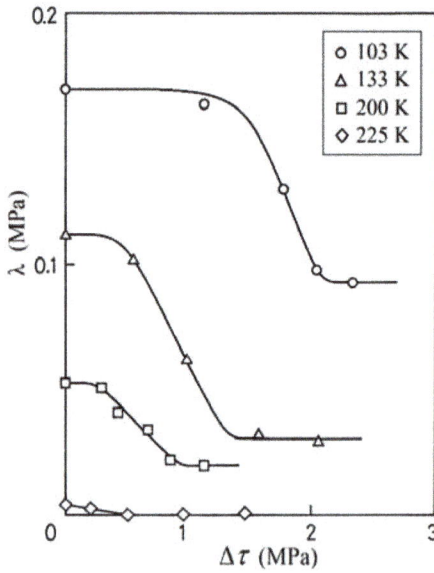

Figure 10. Relation between the strain-rate sensitivity and the stress decrement at the shear strain of 10 % for KCl:Sr^{2+} (0.050 mol.%) at various temperatures: (○) 103 K, (△) 133 K, (□) 200 K, (◇) 225 K (reproduced from reference [83] with permission from the publisher).

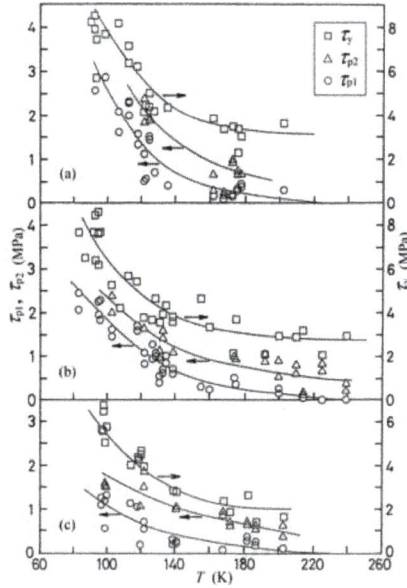

Figure 11. Dependence of ($\bigcirc$) τ_{p1}, ($\triangle$) τ_{p2}, and ($\square$) τ_y on temperature for KCl:Sr^{2+}. Sr^{2+} concentration is (**a**) 0.065, (**b**) 0.050, and (**c**) 0.035 mol.%.

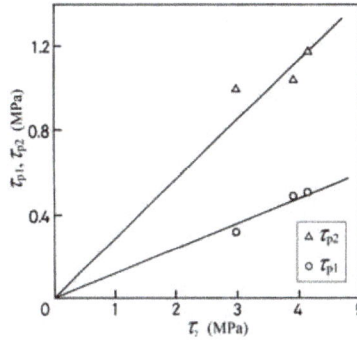

Figure 12. Dependence of ($\bigcirc$) τ_{p1} and ($\triangle$) τ_{p2} on the yield stress, τ_y, for KCl:Sr^{2+} at 150 K (reproduced from reference [82] with permission from the publisher). Concentration of Sr^{2+} is 0.035, 0.050, 0.065 mol.% from the left.

3.2. Discussion for the $\Delta\tau$ vs. λ Curve

The preceding experimental facts that the $\Delta\tau$ vs. λ curve for KCl:Sr^{2+} has two bending points and two plateau regions, and both τ_{p1} and τ_{p2} depend on Sr^{2+} concentration, suggest that the phenomena shown in Figures 8 and 10 are attributable to the interaction between a dislocation and only one type of obstacle. The reason for this is discussed below.

The strain-rate sensitivity of effective stress (τ^*) is inversely proportional to the activation volume, i.e., the average length of dislocation segment, as given by

$$\left(\frac{\Delta\tau^*}{\Delta\ln\dot{\varepsilon}}\right)_T = \frac{kT}{V}, \ (V = bLd) \tag{6}$$

where k is the Boltzmann constant. In addition, it is reported that the length of the dislocation segment increases and the strain-rate sensitivity decreases when the ultrasonic oscillatory stress is applied at room temperature during plastic deformation and that the plateau region is due to the dislocation cuttings when oscillations cause the dislocation to break away from all weak obstacles [84]. Therefore, the first plateau region, as well as the second one, indicates that the average length of the dislocation segments remains constant there. In other words, application of oscillations with low stress amplitude cannot influence the average length of the dislocation segments at low temperature, but it can do so at high temperature, such as room temperature. Therefore, the plateau region appears at low stress decrement in Figures 8 and 10. Now it is imaged that a dislocation is pinned by many weak obstacles and is bowing by applied stress between a few strong obstacles such as forest dislocations during stationary plastic deformation. When the stress amplitude increases, the dislocation begins to break away from weak obstacles by oscillation between the strong ones and the average length of the dislocation begins to increase. The λ starts to decrease at the stress decrement of τ_{p1}. This τ_{p1} should depend on temperature and on density of the obstacle. τ_{p1} is considered to represent the effective stress due to the weak obstacles which lie on the dislocation when the dislocation begins to break away from these weak obstacles with the help of oscillation. Observation of τ_{p1}, therefore, provides information on the interaction between a dislocation and weak obstacles. Consequently, the phenomena shown in Figures 8–10 seem to reflect the influence of ultrasonic oscillation on the dislocation motion on the slip plane containing many weak obstacles and a few strong ones, during plastic deformation. The weak obstacles are considered to be the dopant ions and not to be vacancies here, because the vacancies in the specimen have low density in contrast to the dopants. For example, the concentration of vacancies at thermal equilibrium state is about 1.2×10^{-7} % for KCl single crystal quenched from 673 K to room temperature. As the stress amplitude increases further, oscillating dislocations overcome all Sr^{2+} ions (i.e., Sr^{2+} ions do not act as obstacles to the translational motion of dislocations) and the λ becomes constant. The $\Delta\tau$ value at which the ultrasonic oscillatory stress helps the dislocation break away from all weak obstacles is τ_{p2}. At a stress decrement more than τ_{p2}, forest dislocations remain as the obstacles to the mobile dislocation.

KCl crystal (nominal pure KCl crystal) may contain a small amount of various impurities (e.g., Cu max.2ppm, Pb max.2ppm, Fe max.2ppm, etc. in the reagent-grade powders), although none were added. As a result, the first bending point does not appear. The second bending point, however, appears because the amount of impurities is small.

3.3. Influence of the State of Sr^{2+} Ions on $\Delta\tau$ vs. λ Curve by Heat Treatment

Here is clarified the state of impurities in the specimen and is referred to the influence of the state of dopant ions on the interaction between a dislocation and the dopants, especially on the relation between temperature and the effective stress. The specimens used here are two kinds of single crystals. The first is KCl:Sr^{2+} (0.050 mol.% in the melt) in the preceding Section 2.1 and is hereafter named the quenched specimen. The second was prepared by keeping the quenched specimen at 370 K for 500 h and gradually cooling in a furnace for the purpose of aggregating dopants in it [85]. This is termed the annealed specimen here.

Dielectric absorption of an I-V dipole causes a peak on the tanδ-frequency relation. The relative formula which gives the proportionality between the concentration of I-V dipoles and a Debye peak height is expressed by [86]

$$\tan \delta = \frac{2\pi e^2 c}{3\varepsilon' akT}, \text{ (maximum)} \tag{7}$$

where e is the elementary electric charge, c is the concentration of the I-V dipole, ε' is the dielectric constant in the matrix, and a is the lattice constant. Figure 13 shows the influence of this heat treatment on the tanδ-frequency curves for KCl:Sr^{2+} at 393 K. The upper solid and dotted curves correspond to the quenched specimen and the lower curves the annealed specimen. Dotted lines show Debye peaks obtained by subtracting the d.c. part which is obtained by extrapolating the linear part of the solid curves in the low-frequency region to the high-frequency region. Introducing the peak heights of the dotted curves into Equation (7), the concentration of the isolated dipole was determined to be 98.3 ppm for the quenched specimen and 21.8 ppm for the annealed specimen by dielectric loss measurement using an Andoh electricity TR-10C model. On the other hand, the atomic absorption gave 121.7 ppm for the Sr^{2+} concentration in the quenched specimen and 96.2 ppm for the annealed specimen. Therefore, it should be considered that 71.9% of the I-V dipoles turn into the aggregates in KCl:Sr^{2+} single crystal and form at least trimers [85] by the heat treatment. The trimer has a structure in which three dipoles (Sr^{2+}-vacancy-Sr^{2+}-vacancy-Sr^{2+}-vacancy) are arranged hexagonally head to tail in a (111) plane, as suggested by Cook and Dryden [85].

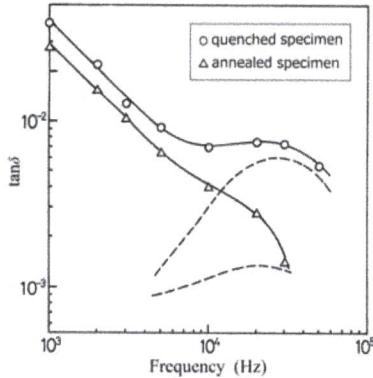

Figure 13. Dielectric loss in KCl containing Sr^{2+} ions at 393 K: (○) for the quenched specimen, (△) for the annealed specimen (reproduced from reference [87] with permission from the publisher). (- - - -) Losses coming from the dipoles.

The $\Delta\tau$ vs. λ curves for the quenched specimen are already shown in Figures 8 and 10. The same phenomena as these figures are observed for the annealed specimens. That is to say, the curve has two bending points and two plateau regions and λ decreases with the $\Delta\tau$ between the two bending points at a given shear strain and temperature. The dependence of τ_{p1} and τ_{p2} on temperature for the quenched specimen is shown in Figure 14a, while that for the annealed specimen is shown in Figure 14b (the curves for τ_{p1} and τ_{p2} are to guide the reader's eye). The value of τ_{p1} becomes small by forming into the aggregates in the crystal and this result is clearer at lower temperature. This may be caused by the result that the separation between the weak obstacles lying on the mobile dislocation becomes wider as the I-V dipoles turn into aggregates. In addition, it is supposed that the decrease in τ_{p1} due to agglomerate of the I-V dipoles, i.e., softening, would be attributable to the loss of tetragonality (lattice distortion around the dopants) in terms of the Fleischer's model [74], as suggested by Chin et al. [1]. The decrease in the effective stress due to agglomerates of I-V dipoles has been reported for alkali halide crystals doped with divalent cations so far [1,13,75,88,89]. As for τ_{p2}, no great difference is seen between the quenched specimen and the annealed one. Accordingly, as the I-V dipoles turn into the aggregates by the heat treatment, the difference between τ_{p1} and τ_{p2} obviously becomes wider at lower temperature. This suggests that the distribution of Sr^{2+} obstacles on mobile dislocation, i.e., the average length of dislocation segments, for the annealed specimen is wider than that for the quenched one.

The critical temperature T_c at which τ_{p1} becomes zero and the dislocation breaks away from the dopants only with the help of thermal activation is around 210 K for the annealed specimen. This T_c value is small in contrast to $T_c \approx 240$ K for the quenched specimen as can be seen from Figure 14a,b.

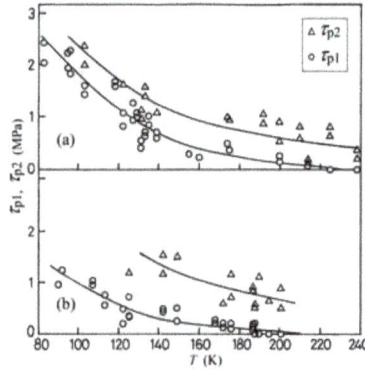

Figure 14. Dependence of (○) τ_{p1} and (△) τ_{p2} on temperature for KCl:Sr^{2+} (reproduced from reference [87] with permission from the publisher): (**a**) for the quenched specimen; (**b**) for the annealed specimen.

3.4. Activation Energy for the Break-Away of a Dislocation from the Dopant Ion

The thermally activated deformation rate is expressed by an Arrhenius-type equation:

$$\dot{\varepsilon} = \dot{\varepsilon}_0 \exp\left(\frac{-G}{kT}\right) \tag{8}$$

where $\dot{\varepsilon}_0$ is a frequency factor and is unique for a particular dislocation mechanism. G is the activation energy for overcoming of local barrier (i.e., the dopant in here) by a dislocation at temperature T. It is assumed that the interaction between a dislocation and the dopant ion (i.e., I-V dipole) for the quenched specimen can be approximated to the Fleischer's model [74] taking account of the Friedel relation [80]. The model is named F-F here. Then, the relative formula of τ_{p1} and temperature, which will reveal the force–distance relation between a dislocation and a weak obstacle, is given by

$$\left(\frac{\tau_{p1}}{\tau_{p0}}\right)^{1/3} = 1 - \left(\frac{T}{T_c}\right)^{1/2} \tag{9}$$

where τ_{p0} is the value of τ_{p1} at absolute zero.

The activation energy for the interaction between a dislocation and the obstacle is expressed from the Arrhenius-type equation for the thermally activated deformation rate at low temperature as the Equation (8), namely

$$G = \alpha kT, \left(\alpha = \ln\left(\frac{\dot{\varepsilon}_0}{\dot{\varepsilon}}\right)\right) \tag{10}$$

and the G for the F-F is also expressed by [90]

$$G = G_0 \left\{ 1 - \left(\frac{\tau^*}{\tau_0^*}\right)^{1/3} \right\}^2, (G_0 = F_0 b) \tag{11}$$

where G_0 is the Gibbs free energy for the break-away of a dislocation from the dopant in the absence of an applied stress, τ^*_0 is the effective shear stress (τ^*) at absolute zero, and F_0 is the force acted on the dislocation at 0 K. Differentiating the combination of Equations (10) and (11) with respect to the shear stress gives

$$\frac{\partial \ln \dot{\varepsilon}}{\partial \tau^*} = \left(\frac{2G_0}{3kT\tau_0^*}\right)\left(\frac{\tau_0^*}{\tau^*}\right)^{2/3}\left\{1 - \left(\frac{\tau^*}{\tau_0^*}\right)^{1/3}\right\} + \frac{\partial \ln \dot{\varepsilon}_0}{\partial \tau^*} \tag{12}$$

Further, substituting Equation (9) in Equation (12) gives

$$\frac{\partial \ln \dot{\varepsilon}}{\partial \tau^*} = \left(\frac{2G_0}{3kT\tau_{p0}}\right)\left\{1 - \left(\frac{T}{T_c}\right)^{1/2}\right\}^{-2}\left(\frac{T}{T_c}\right)^{1/2} + \frac{\partial \ln \dot{\varepsilon}_0}{\partial \tau^*} \tag{13}$$

where τ^*_0 is replaced by τ_{p0}. The result of calculations of Equation (13) for the F-F is shown in Figure 15 for the quenched specimen and the solid line is determined using the least squares method so as to fit to the experimental data (open symbols). The open symbols correspond to the λ_p^{-1} for the quenched specimen. On the basis of the slope of straight line, the G_0 is 0.39 eV. The strain-rate sensitivity due to weak obstacles, λ_p, is defined in Figure 8. It is assumed from the following two more experimental results in addition to the description in the Section 1 (introduction) that λ_p is the strain-rate sensitivity due to Sr^{2+} obstacles. Firstly, the $\Delta\tau$ vs. λ curve shifts upward with increasing shear strain, while λ_p is almost constant independently of the strain as shown in Figure 8 for $KCl:Sr^{2+}$ (0.050 mol.%) at several strains and 200 K. Secondly, λ_p is proportional to the square root of Sr^{2+} concentration ($c^{1/2}$), which is related to the inverse of the average length L of the dislocation segments (i.e., $\lambda_p = kT/bLd$), at a given temperature. This is shown in Figure 16. The concentration of the I-V dipole was estimated by the dielectric loss measurement.

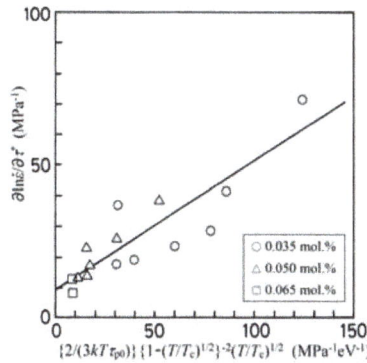

Figure 15. Linear plots of Equation (13) for the quenched specimen: $KCl:Sr^{2+}$ ((O) 0.035 mol.%, ($\triangle$) 0.050 mol.%, ($\square$) 0.065 mol.% in the melt) (reproduced from reference [90] with permission from the publisher).

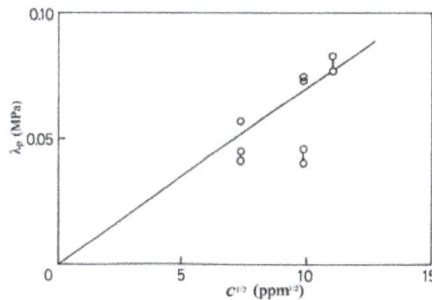

Figure 16. Relationship between λ_p and $c^{1/2}$ for $KCl:Sr^{2+}$ at the temperatures of 120 to 130 K (reproduced from reference [83] with permission from the publisher).

As for the annealed specimen, the G_0 is obtained as follows. The G is expressed for square force–distance relation between a dislocation and a weak obstacle as follows

$$G = G_0 - \tau^* Lbd \tag{14}$$

From the Fridel relation [80], the average spacing of weak obstacles along the dislocation is

$$L = \left(\frac{2L_0^2 E}{\tau^* b} \right)^{1/3} \tag{15}$$

where L_0 is the average spacing of weak obstacles on the slip plane and E is the line tension of the dislocations. Substituting Equation (15) into Equation (14), the Gibbs free energy is given by

$$G = G_0 - \beta \tau^{*2/3}, \left(\beta = \left(2\mu b^4 d^3 L_0^2 \right)^{1/3} \right) \tag{16}$$

Differentiating the substitutional equation of Equation (16) in Equation (8) with respect to the shear stress, we find

$$\tau_{p1} \left(\frac{\partial \ln \dot{\varepsilon}}{\partial \tau^*} \right) = \left(\frac{2G_0}{3kT} \right) + \frac{2}{3} \ln \left(\frac{\dot{\varepsilon}}{\dot{\varepsilon}_0} \right) \tag{17}$$

where τ^* is replaced by τ_{p1}. The result of calculations of Equation (17) is shown in Figure 17 for the annealed specimen, where the solid line is given using the least squares method and the open circles are obtained from τ_{p1}/λ_p. The interaction energy G_0 between a dislocation and the aggregate of I-V dipoles in the specimen, which is obtained from the slope of straight line in Figure 17, is 0.26 eV. The Gibbs free energy for the annealed specimen is smaller than that for the quenched one as tabulated in Table 1.

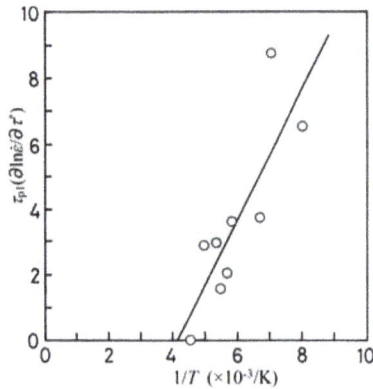

Figure 17. Linear plots of Equation (17) for the annealed specimen (reproduced from reference [91] with permission from the publisher).

Table 1. Values of G_0 for the quenched and the annealed specimens.

Specimen	G_0 (eV)
Quenched specimen	0.39
Annealed specimen	0.26

3.5. Strain-Rate Sensitivity at the Second Plateau Place on the $\Delta\tau$ vs. λ Curve

Three-stage strain hardening has been clearly established as the characteristic behavior of single crystals of rock salt structure, deforming predominantly by single glide [92]. The three-stage strain hardening is obtained for KCl [93,94]. It is also observed for the quenched and the annealed specimens. Each of the three-stages for the specimens is illustrated in Figure 18, which is generally denoted by stage I, II, and III as shown in the figure. The variation of strain-rate sensitivity at the second plateau place on the $\Delta\tau$ vs. λ curve with shear strain (see the marked part with grey circle in Figure 19 illustrated the relative curve between $\Delta\tau$ and λ), where the obstacles to a dislocation motion are only forest dislocations and the dopant ions no longer act as obstacles, is investigated in the different plastic deformation regions of stress-strain curve for the two kinds of specimens, namely, the quenched specimen and the annealed one. The size of specimens used here is within $(5 \pm 1) \times (5 \pm 1) \times (15 \pm 1)$ mm^3. Because, it has been reported that the height–width of specimens determines the shape of the work-hardening curve in the compression at a constant strain rate and temperature for CaF$_2$ single crystals [95] and besides glide geometry is directly related to the shape and size of KCl crystals [2]. The variation of λ at the second plateau place on the $\Delta\tau$ vs. λ curve with shear strain is shown in Figure 20. The $\Delta\lambda/\Delta\varepsilon$ seems to be the variation of the strain-rate sensitivity due to dislocation cuttings with shear strain. The relations of temperature and $\Delta\lambda/\Delta\varepsilon$ in stage I and in stage II of stress-strain curve are represented by a circle and a triangle, respectively. The open symbols correspond to that for the quenched specimen and the solid ones that for the annealed specimen. The full lines in this figure are to guide the reader's eye. Unfortunately, the $\Delta\lambda/\Delta\varepsilon$ could not be obtained at low temperature. The $\Delta\lambda/\Delta\varepsilon$ in stage II is obviously larger than that in stage I at a given temperature. And the values of $\Delta\lambda/\Delta\varepsilon$ in stage I and in stage II increase with decreasing temperature. Figure 20 also shows that the $\Delta\lambda/\Delta\varepsilon$ for the annealed specimens is considerably large in contrast to that for the quenched specimens in the two stages at the temperature. This may result from a rapid increase in forest dislocation density with shear strain in the annealed specimen. Accordingly, the increase in it in the annealed specimen seems to be remarkable in the two stages under the compression test, compared with it in the quenched specimen. According to Michalak [96] some dislocation cutting process becomes an important hardening mechanism and the strain-hardening increases. If the consideration about forest dislocation density for the two kinds of specimens is accurate, it is expected that a strain-hardening rate $(d\tau/d\varepsilon)$ for the annealed specimen is larger than that for the quenched specimen at a given temperature and shear strain.

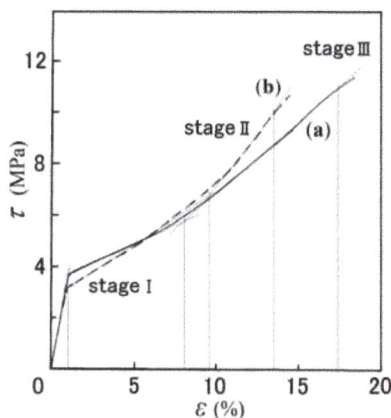

Figure 18. Shear stress-shear strain curves for (**a**) (—) the quenched specimen at 175 K and (**b**) (- - -) the annealed specimen at 176 K.

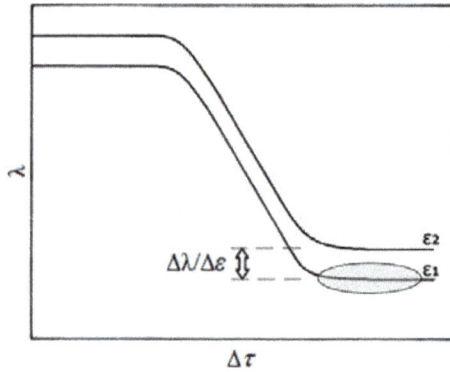

Figure 19. Illustration of relationship between the strain-rate sensitivity of flow stress and the stress decrement due to the superimposition of ultrasonic oscillation at a given strain, ε ($\varepsilon_1 < \varepsilon_2$). $\Delta\lambda/\Delta\varepsilon$ represents the variation of the strain-rate sensitivity due to dislocation cuttings with shear strain.

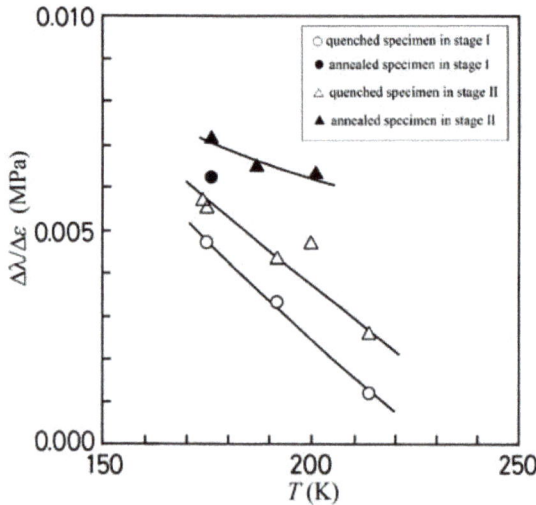

Figure 20. Dependence of $\Delta\lambda/\Delta\varepsilon$ on the temperature in the different plastic deformation regions: (○) for the quenched specimen and (●) for the annealed specimen in stage I; (△) for the quenched specimen and (▲) for the annealed specimen in stage II (reproduced from reference [97] with permission from the publisher).

When the whole crystal is filled with primary dislocation dipoles and multipoles, stage I of stress-strain curve ends and the stage II, which is characterized by the onset of glide on oblique systems, follows. Afterwards, the stage III, which is probably associated with the cross glide of screw dislocation around obstacles, ensues in the process of plastic deformation [92]. Accordingly, it will be also expected that the extent of stage I and stage II for the annealed specimen is shorter than that for the quenched specimen at a given temperature.

The relation between $d\tau/d\varepsilon$ and temperature for the two kinds of specimens is shown in Figure 21a for stage I and b for stage II of the plastic deformation regions. The value of $d\tau/d\varepsilon$ in stage II is about two times larger than that in stage I at a given temperature for both the specimens. Furthermore, the $d\tau/d\varepsilon$ for the annealed specimen is obviously large as against that for the quenched specimen in the two stages as can be seen from Figure 21a,b. Although the extent of stage I is further investigated for the two kinds of specimens, the difference between the extents of it could not be found out within the temperature. While the extent of stage II could not also be made a comparison between them, because most of the specimens were fractured within stage II by the compression test at low temperature. Then, we attempted to investigate the extent between shear strains of yield point and of fractured point, i.e., plastic deformation region, at various temperatures. The result of the extent of plastic deformation region is shown in Figure 22. The open circles and the solid ones correspond to those for the quenched specimen and for the annealed specimen, respectively. The extent of it tends to be large with increasing temperature for both the specimens. Figure 22 also shows that the extent of it for the annealed specimen seems to be shorter than it for the quenched specimen at a given temperature. Consequently, it is clear that the increase in forest dislocation density in the annealed specimen is remarkable in contrast to it in the quenched specimen under the compression test. This may be attributed to the phenomenon that a small amount of the much larger aggregates than trimers, which are contained in the annealed specimen, contribute to the increase in forest dislocation density.

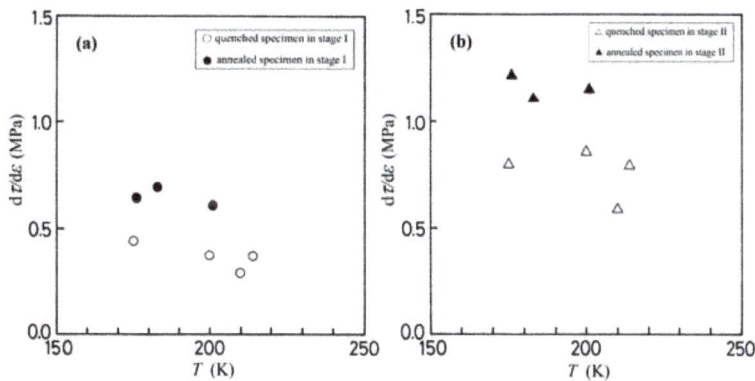

Figure 21. Dependence of the strain-hardening rate on the temperature in the different plastic deformation regions (reproduced from reference [97] with permission from the publisher): (**a**) (○) for the quenched specimen and (●) for the annealed specimen in stage I; (**b**) (△) for the quenched specimen and (▲) for the annealed specimen in stage II.

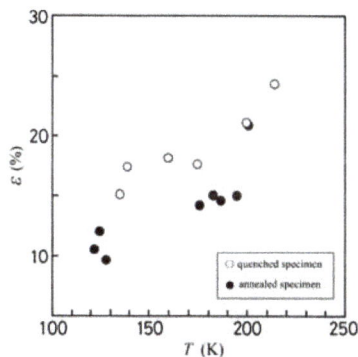

Figure 22. Dependence of the extent of plastic deformation region on the temperature for both the specimens: (○) for the quenched specimen and (●) for the annealed specimen (reproduced from reference [97] with permission from the publisher).

4. Summary

By the original method which combines strain-rate cycling tests with the Blaha effect measurement, the dislocation-Sr^{2+} ions (I-V dipoles) interaction has been investigated during the plastic deformation of bulk material ($KCl:Sr^{2+}$ single crystal). The data obtained by the original method is sensitive to a change in the state of dopant ions in the specimen by heat treatment, as reviewed in the each Sections 3.3–3.5. The following results are reviewed mainly on the basis of $\Delta\tau$ vs. λ curves derived from the original method.

In Sections 3.1 and 3.2, we reviewed the process of the measurement of strain-rate sensitivity under the application of ultrasonic oscillatory stress in detail. The plots of the strain-rate sensitivity and the stress decrement due to oscillation has two bending points and two plateau regions for $KCl:Sr^{2+}$ and seem to reflect the influence of ultrasonic oscillation on the dislocation motion on the slip plane. The first bending point, τ_{p1}, at low stress decrement corresponds to the effective stress due to Sr^{2+} ions on the mobile dislocation.

In Section 3.3, we reviewed the influence of the state of dopant ions on the effective stress due to the Sr^{2+} ions in $KCl:Sr^{2+}$ single crystals. If the dopant ions in $KCl:Sr^{2+}$ single crystals diffuse into the matrix, or they aggregate by the heat treatment, various deformation characteristics (τ_{p1} and T_c) become changed. This is because the state of a small amount of dopants in the crystals strongly influences the resistance to movement of a dislocation.

In Section 3.4, the change in obstacle size due to the agglomeration of I-V dipoles was reviewed as an influential factor of the Gibbs free energy for overcoming the local barrier by a dislocation after the heat treatment. The specimens used here are two kinds of heat-treated ones, i.e., the quenched and the annealed $KCl:Sr^{2+}$ (0.050 mol.% in the melt) single crystals. The G_0 value (0.26 eV) for the annealed specimen is smaller than that (0.39 eV) for the quenched one.

In Section 3.5, we reviewed the remarkable increase in density of the forest dislocation under the compression test by annealing the quenched specimen. This is examined from the variation of strain-rate sensitivity at the second plateau place on the $\Delta\tau$ vs. λ curve with shear strain in both stage I and stage II of stress-strain curve at the temperatures of 170 to 220 K. As a proof, the experimental results on the strain-hardening rate and on the extent of plastic deformation region are obtained for the two kinds of the specimens (i.e., the quenched specimen and the annealed one).

The strain-rate cycling test during the Blaha effect measurement has been successively provided the information on the dislocation motion breaking away from the strain fields around dopant ions with the help of thermal activation, and seems to separate the contributions arising from the interaction between dislocation and the point defects and those from dislocations themselves during plastic deformation of ionic crystals. Such information on dislocation motion in bulk material cannot be obtained by the widely known methods, e.g., the measurements of yield stress and proof stress, micro-hardness tests, direct observations of dislocation, internal friction measurements, or stress relaxation tests. However, the original method (the strain-rate cycling test during the Blaha effect measurement) is different from above-mentioned ones and would be possible to clear it up.

Conflicts of Interest: The author declares no conflict of interest.

References

1. Chin, G.Y.; Van Uitert, L.G.; Green, M.L.; Zydzik, G.J.; Kometani, T.Y. Strengthening of Alkali Halides by Divalent-ion Additions. *J. Am. Ceram. Soc.* **1973**, *56*, 369–372. [CrossRef]
2. Suszyńska, M. Effect of Impurity Concentration and Plastic Deformation on Dislocation Density of KCl Crystals. *Kristall. Technik.* **1974**, *9*, 1199–1207. [CrossRef]
3. Kataoka, T.; Yamada, T. Yield Strength and Dislocation Mobility of KCl-KBr Solid Solution Single Crystals. *Jpn. J. Appl. Phys.* **1977**, *16*, 1119–1126. [CrossRef]
4. Boyarskaya, Y.S.; Zhitaru, R.P.; Palistrant, N.A. The anomalous behaviour of the doped NaCl crystals compressed at low temperatures. *Cryst. Res. Technol.* **1990**, *25*, 1469–1473. [CrossRef]

5. Boyarskaya, Y.S.; Zhitaru, R.P.; Palistrant, N.A. Influence of the state of the impurity on the deformation-rate dependence of the yield stress of NaCl:Ca single crystals. *Sov. Phys. Solid State* **1990**, *32*, 1989–1990.

6. Okazaki, K. Solid-solution hardening and softening in binary iron alloys. *J. Mater. Sci.* **1996**, *31*, 1087–1099. [CrossRef]

7. Tabachnikova, E.D.; Podolskiy, A.V.; Smirnov, S.N.; Psaruk, I.A.; Liao, P.K. Temperature dependent mechanical properties and thermal activation plasticity of nanocrystalline and coarse grained Ni-18.75 at.% Fe alloy. *IOP Conf. Ser. Mater. Sci. Eng.* **2014**, *63*, 012105. [CrossRef]

8. Pratt, P.L.; Harrison, R.P.; Newey, C.W.A. Dislocation mobility in ionic crystals. *Disc. Faraday Soc.* **1964**, *38*, 211–217. [CrossRef]

9. Newey, C.W.A.; Harrison, R.P.; Pratt, P.L. Precipitation Hardening and Dislocation Locking in Doped NaCl. *Proc. Brit. Ceram. Soc.* **1966**, *6*, 305–316.

10. Chin, G.Y.; Van Uitert, L.G.; Green, M.L.; Zydzik, G. Hardness, Yield Strength and Young's Modulus in Halide Crystals. *Scripta Metall.* **1972**, *6*, 475–480. [CrossRef]

11. Green, M.L.; Zydzik, G. Effect of heat treatment on the microhardness of some mixed and doped alkali halides. *Scripta Metall.* **1972**, *6*, 991–994. [CrossRef]

12. Andreev, G.A.; Klimov, V.A. Influence of the State of an Impurity on the Microhardness of NaCl:Sr Single Crystals. *Sov. Phys. Solid State* **1980**, *22*, 2042–2043.

13. Buravleva, M.G.; Rozenberg, G.K.; Soifer, L.M.; Chaikovskii, E.F. Changes in the flow stress of LiF:Mg^{2+} and LiF:Co^{2+} crystals during precipitation of solid solutions. *Sov. Phys. Solid State* **1980**, *22*, 150–152.

14. Narasimha Reddy, K.; Subba Rao, U.V. Influence of Gadolonium Impurity on Microhardness of Host Alkali Halide Crystal. *Cryst. Res. Technol.* **1984**, *19*, K73–K76. [CrossRef]

15. Strunk, H. Investigation of Cross-slip Events in NaCl Crystals by Transmission Electron Microscopy. *Phys. Status Solidi (a)* **1975**, *28*, 119–126. [CrossRef]

16. Appel, F.; Messerschmidt, U. The interaction between dislocations and point obstacles: A comparison of the interaction parameter distributions obtained from computer simulation and from in situ high voltage electron microscopy straining experiments. *Mater. Sci. Eng.* **1982**, *52*, 69–74. [CrossRef]

17. Messerschmidt, U.; Appel, F.; Schmid, H. The radius of curvature of dislocation segments in MgO crystals stressed in the high-voltage electron microscope. *Philos. Mag. A* **1985**, *51*, 781–796. [CrossRef]

18. Kataoka, T.; Ohji, H.; Morishita, H.; Kishida, K.; Azuma, K.; Yamada, T. In-situ observation of moving dislocations in KCl crystal by laser-light topography. *Jpn. J. Appl. Phys.* **1989**, *28*, L697–L700. [CrossRef]

19. Kataoka, T.; Ohji, H.; Kishida, K.; Azuma, K.; Yamada, T. Direct observation of glide dislocations in a KCl crystal by the light scattering method. *Appl. Phys. Lett.* **1990**, *56*, 1317–1319. [CrossRef]

20. Kataoka, T. The light scattering topography method: Direct observation of moving dislocations. *Butsuri* **1992**, *47*, 713–716. (In Japanese)

21. Messerschmidt, U. *Dislocation Dynamics during Plastic Deformation*; Springer: Berlin, Heidelberg, 2010.

22. Indenbom, V.L.; Chernov, V.M. Determination of characteristics for the interaction between point defects and dislocations from internal friction experiments. *Phys. Status Solidi (a)* **1972**, *14*, 347–354. [CrossRef]

23. Schwarz, R.B.; Granato, A.V. Measurement of the force-distance profile for the interaction between a dislocation and a point defect. *Phys. Rev. Lett.* **1975**, *34*, 1174–1177. [CrossRef]

24. Ivanov, V.I.; Lebedev, A.B.; Kardashev, B.K.; Nikanorov, S.P. Interaction of dislocations with pinning centers in magnesium at temperatures 295-4.2K. *Sov. Phys. Solid State* **1986**, *28*, 867–868.

25. Kosugi, T.; Kino, T. Experimental Determination of the Force-Distance Relation for the Interaction between a Dislocation and a Solute Atom. *J. Phys. Soc. Jpn.* **1987**, *56*, 999–1009. [CrossRef]

26. Kosugi, T. Temperature Dependence of Amplitude-dependent Internal Friction due to Simultaneous Breakaway of a Dislocation from Several Pinning Points. *Mater. Sci. Eng. A* **2001**, *309–310*, 203–206. [CrossRef]

27. Gremaud, G. Dislocation-Point Defect Interactions. *Mater. Sci. Forum* **2001**, *366–368*, 178–246. [CrossRef]

28. Dotsenko, V.I. Stress Relaxation in Crystals. *Phys. Status Solidi (b)* **1979**, *93*, 11–43. [CrossRef]

29. Urusovskaya, A.A.; Petchenko, A.M.; Mozgovoi, V.I. The influence of strain rate on stress relaxation. *Phys. Status Solidi (a)* **1991**, *125*, 155–160. [CrossRef]

30. Johnston, W.G.; Gilman, J.J. Dislocation velocities, dislocation densities, and plastic flow in lithium fluoride crystals. *J. Appl. Phys.* **1959**, *30*, 129–144. [CrossRef]

31. Granato, A.V.; Lücke, K. Theory of mechanical damping due to dislocations. *J. Appl. Phys.* **1956**, *27*, 583–593. [CrossRef]

32. Blaha, F.; Langenecker, B. Dehnung von Zink-Kristallen unter Ultraschalleinwirkung. *Naturwiss.* **1955**, *42*, 556. [CrossRef]

33. Nevill, G.E.; Brotzen, F.R. The effect of vibrations on the static yield strength of low-carbon steel. *Proc. ASTM* **1957**, *57*, 751–758.

34. Langenecker, B. Effects of ultrasound on deformation characteristics of metals. *IEEE Trans. Sonic Ultrasonic* **1966**, *13*, 1–8. [CrossRef]

35. Izumi, O.; Oyama, K.; Suzuki, Y. Effects of superimposed ultrasonic vibration on compressive deformation of metals. *Trans. JIM* **1966**, *7*, 162–167. [CrossRef]

36. Evans, A.E.; Smith, A.W.; Waterhouse, W.J.; Sansome, D.H. Review of the application of ultrasonic vibrations to deforming metals. *Ultrasonics* **1975**, *13*, 162–170.

37. Jimma, T.; Kasuga, Y.; Iwaki, N.; Miyazawa, O.; Mori, E.; Ito, K.; Hatano, H. An application of ultrasonic vibration to the deep drawing process. *J. Mater. Process. Technol.* **1998**, *80–81*, 406–412. [CrossRef]

38. Susan, M.; Bujoreanu, L.G. The metal-tool contact friction at the ultrasonic vibration drawing of ball-bearing steel wires. *Rev. Metal. Madrid* **1999**, *35*, 379–383. [CrossRef]

39. Murakawa, M.; Jin, M. The utility of radially and ultrasonically vibrated dies in the wire drawing process. *J. Mater. Process. Technol.* **2001**, *113*, 81–86. [CrossRef]

40. Susan, M.; Bujoreanu, L.G.; Gǎluşcǎ, D.G.; Munteanu, C.; Mantu, M. On the drawing in ultrasonic field of metallic wires with high mechanical resistance. *J. Optoelectron. Adv. Mater.* **2005**, *7*, 637–645.

41. Lucas, M.; Gachagan, A.; Cardoni, A. Research applications and opportunities in power ultrasonics. *Proc. IMechE Part C J. Mech. Eng. Sci.* **2009**, *223*, 2949–2965. [CrossRef]

42. Siddiq, A.; El Sayed, T. Ultrasonic-assisted manufacturing processes: Variational model and numerical simulations. *Ultrasonics* **2012**, *52*, 521–529. [CrossRef] [PubMed]

43. Makhdum, F.; Phadnis, V.A.; Roy, A.; Silberschmidt, V.V. Effect of ultrasonically-assisted drilling on carbon-fibre-reinforced plastics. *J. Sound Vib.* **2014**, *333*, 5939–5952. [CrossRef]

44. Graff, K.F. Ultrasonic Metal Forming: Processing. In *Power Ultrasonics: Applications of High-intensity Ultrasound*; Gallego-Juarez, J.A., Graff, K.F., Eds.; Elsevier: Cambridge, UK, 2015; pp. 377–438.

45. Puškár, A. Effect of grain size, cold work, loading frequency and temperature on fatigue limit of mild steel. *Metallurg. Trans.* **1976**, *7A*, 1529–1533. [CrossRef]

46. Jon, M.C.; Mason, W.P.; Beshers, D.N. Internal friction during ultrasonic deformation of alpha-brass. *J. Appl. Phys.* **1976**, *47*, 2337–2349. [CrossRef]

47. Shetty, D.K.; Meshii, M. Plastic deformation of aluminum under repeated loading. *Metallurg. Trans.* **1975**, *6A*, 349–366. [CrossRef]

48. Bradley, W.L.; Nam, S.W.; Matlock, D.K. Fatigue perturbed creep of pure aluminum at ambient temperatures. *Metallurg. Trans.* **1976**, *7A*, 425–430. [CrossRef]

49. Lebedev, A.B.; Kustov, S.B.; Kardashev, B.K. Amplitude-dependent ultrasound absorption and acoustoplastic effect during active deformation of sodium chloride crystals. *Sov. Phys. Solid State* **1982**, *24*, 1798–1799.

50. Baker, G.S.; Carpenter, S.H. Dislocation mobility and motion under combined stresses. *J. Appl. Phys.* **1967**, *38*, 1586–1591. [CrossRef]

51. Kaiser, G.; Pechhold, W. Dynamic-mechanical investigations for the study of dislocation motion during plastic flow. *Acta Metall.* **1967**, *17*, 527–537. [CrossRef]

52. Suzuki, T. Effects of vibrational stress on static flow stress (I). *Seisan-Kenkyu* **1970**, *22*, 194–197. (In Japanese)

53. Suzuki, T. Effects of vibrational stress on static flow stress (II). *Seisan-Kenkyu* **1970**, *22*, 344–347. (In Japanese)

54. Endo, T.; Suzuki, K.; Ishikawa, M. Effects of superimposed ultrasonic oscillatory stress on the deformation of Fe and Fe-3%Si alloy. *Trans. Japan Inst. Metals* **1979**, *20*, 706–712. [CrossRef]

55. Endo, T.; Tasaki, M.; Kubo, M.; Shimada, T. High temperature deformation of an Al-5at%Mg alloy under combined high frequency stresses. *J. Japan Inst. Met. Mater.* **1982**, *46*, 773–779. (In Japanese) [CrossRef]

56. Lebedev, A.B.; Kustov, S.B.; Kardashev, B.K. Investigation of the amplitude-dependent internal friction during plastic deformation of sodium chloride single crystals. *Sov. Phys. Solid State* **1983**, *25*, 511–512.

57. Ohgaku, T.; Takeuchi, N. The relation of the Blaha effect with internal friction for alkali halide crystals. *Phys. Status Solidi (a)* **1988**, *105*, 153–159. [CrossRef]

58. Hikata, A.; Johnson, R.A.; Elbaum, C. Interaction of dislocations with electrons and with phonons. *Phys. Rev.* **1970**, *B2*, 4856–4863. [CrossRef]

59. Ohgaku, T.; Takeuchi, N. The Blaha effect of alkali halide crystals. *Phys. Status Solidi (a)* **1987**, *102*, 293–299. [CrossRef]

60. Hesse, J.; Hobbs, L.W. Dislocation density in highly deformed NaCl single crystals. *Phys. Status Solidi (a)* **1972**, *14*, 599–604. [CrossRef]

61. Frank, W. Theorie der Verfestigung von Alkalihalogenid-Einkristallen I. Qualitative betrachtungen. *Mater. Sci. Eng.* **1970**, *6*, 121–131. [CrossRef]

62. Frank, W. Theorie der Verfestigung von Alkalihalogenid-Einkristallen II. Quantitative formulierung. *Mater. Sci. Eng.* **1970**, *6*, 132–148. [CrossRef]

63. Ohgaku, T.; Takeuchi, N. Relation between plastic deformation and the Blaha effect for alkali halide crystals. *Phys. Status Solidi (a)* **1989**, *111*, 165–172. [CrossRef]

64. Li, J.C.M. Kinetics and Dynamics in Dislocation Plasticity. In *Dislocation Dynamics*; Rosenfield, A.R., Hahn, G.T., Bemont, A.L., Jaffee, R.I., Eds.; Mc Graw-Hill Publ. Co.: New York, NY, USA, 1968; pp. 87–116.

65. Ohgaku, T.; Takeuchi, N. Interaction between a dislocation and monovalent impurities in KCl single crystals. *Phys. Status Solidi (a)* **1992**, *134*, 397–404. [CrossRef]

66. Ohgaku, T.; Teraji, H. Investigation of interaction between a dislocation and a Br$^-$ ion in NaCl:Br$^-$ single crystals. *Phys. Status Solidi (a)* **2001**, *187*, 407–413. [CrossRef]

67. Argon, A.S.; Nigam, A.K.; Padawer, G.E. Plastic deformation and strain hardening in pure NaCl at low temperatures. *Philos. Mag.* **1972**, *25*, 1095–1118. [CrossRef]

68. Young, F.W., Jr. Etch pits at dislocations in copper. *J. Appl. Phys.* **1961**, *32*, 192–201. [CrossRef]

69. Takeuchi, S. Solid-Solution strengthening in single crystals of iron alloys. *J. Phys. Soc. Jap.* **1969**, *27*, 929–940. [CrossRef]

70. Sprackling, M.T. *The Plastic Deformation of Simple Ionic Crystals*; Alper, A.M., Margrave, J.L., Nowick, A.S., Eds.; Academic Press: London, UK, 1976.

71. Zakrevskii, V.A.; Shul'diner, A.V. Dislocation interaction with radiation defects in alkali-halide crystals. *Phys. Sol. Stat.* **2000**, *42*, 270–273. [CrossRef]

72. Katoh, S. Influence of impurity concentration on the Blaha effect. Bachelor's Thesis, Kanazawa University, Kanazawa, Japan, 1987; pp. 16–21.

73. Pick, H.; Weber, H. Dichteänderung von KCl-Kristallen durch Einbau zweiwertiger Ionen. *Z. Phys.* **1950**, *128*, 409–413. [CrossRef]

74. Fleischer, R.L. Rapid solution hardening, dislocation mobility, and the flow stress of crystals. *J. Appl. Phys.* **1962**, *33*, 3504–3508. [CrossRef]

75. Dryden, J.S.; Morimoto, S.; Cook, J.S. The hardness of alkali halide crystals containing divalent ion impurities. *Philos. Mag.* **1965**, *12*, 379–391. [CrossRef]

76. Orozco, M.E.; Mendoza, A.A.; Soullard, J.; Rubio, O.J. Changes in yield stress of NaCl:Pb^{2+} crystals during dissolution and precipitation of solid solutions. *Jpn. J. Appl. Phys.* **1982**, *21*, 249–254. [CrossRef]

77. Zaldo, C.; Solé, J.G.; Agulló-López, F. Mechanical strengthening and impurity precipitation behaviour for divalent cation-doped alkali halides. *J. Mater. Sci.* **1982**, *17*, 1465–1473. [CrossRef]

78. Reddy, B.K. Annealing and ageing studies in quenched KBr:Ba^{2+} single crystals. *Phys. Status Solidi (a)* **1987**, *99*, K7–K10. [CrossRef]

79. Sirdeshmukh, D.B.; Sirdeshmukh, L.; Subhadra, K.G. *Alkali Halides*; Hull, R., Osgood, R.M., Jr., Sakaki, H., Zunger, A., Eds.; Springer-Verlag: Berlin, Germany, 2001; p. 41.

80. Friedel, J. *Dislocations*; Pergamon Press: Oxford, UK, 1964; p. 224.

81. Stoloff, N.S.; Lezius, D.K.; Johnston, T.L. Effect of temperature on the deformation of KCl-KBr alloys. *J. Appl. Phys.* **1963**, *34*, 3315–3322. [CrossRef]

82. Kohzuki, Y.; Ohgaku, T.; Takeuchi, N. Interaction between a dislocation and impurities in KCl single crystals. *J. Mater. Sci.* **1993**, *28*, 3612–3616. [CrossRef]

83. Kohzuki, Y.; Ohgaku, T.; Takeuchi, N. Interaction between a dislocation and various divalent impurities in KCl single crystals. *J. Mater. Sci.* **1995**, *30*, 101–104. [CrossRef]

84. Ohgaku, T.; Takeuchi, N. Study on dislocation-impurity interaction by the Blaha effect. *Phys. Status Solidi (a)* **1990**, *118*, 153–159. [CrossRef]

85. Cook, J.S.; Dryden, J.S. An investigation of the aggregation of divalent cationic impurities in alkali halides by dielectric absorption. *Proc. Phys. Soc.* **1962**, *80*, 479–488. [CrossRef]

86. Lidiard, A.B. Ionic Conductivity. In *Handbuch der Physik*; Springer: Berlin, Germany, 1957; Volume 20, pp. 246–349.

87. Kohzuki, Y.; Ohgaku, T.; Takeuchi, N. Influence of a state of impurities on the interaction between a dislocation and impurities in KCl single crystals. *J. Mater. Sci.* **1993**, *28*, 6329–6332. [CrossRef]

88. Johnston, W.G. Effect of impurities on the flow stress of LiF crystals. *J. Appl. Phys.* **1962**, *33*, 2050–2058. [CrossRef]

89. Gaiduchenya, V.F.; Blistanov, A.A.; Shaskol'skaya, M.P. Thermally activated slip in LiF crystals. *Sov. Phys. Solid State* **1970**, *12*, 27–31.

90. Kohzuki, Y. Study on the interaction between a dislocation and impurities in KCl:Sr^{2+} single crystals by the Blaha effect part I Interaction between a dislocation and an impurity for the Fleischer's model taking account of the Friedel relation. *J. Mater. Sci.* **2000**, *35*, 3397–3401. [CrossRef]

91. Kohzuki, Y.; Ohgaku, T. Study on the interaction between a dislocation and impurities in KCl:Sr^{2+} single crystals by the Blaha effect Part II Interaction between a dislocation and aggregates for various force-distance relations between a dislocation and an impurity. *J. Mater. Sci.* **2001**, *36*, 923–928. [CrossRef]

92. Sprackling, M.T. Three-stage Hardening. In *The Plastic Deformation of Simple Ionic Crystals*; Alper, A.M., Margrave, J.L., Nowick, A.S., Eds.; Academic Press: London, UK, 1976; pp. 203–206.

93. Alden, T.H. Latent hardening and the role of oblique slip in the strain hardening of rock-salt structure crystals. *Trans. Met. Soc. AIME* **1964**, *230*, 649–656.

94. Davis, L.A.; Gordon, R.B. Plastic deformation of alkali halide crystals at high pressure: Work-hardening effects. *J. Appl. Phys.* **1969**, *40*, 4507–4513. [CrossRef]

95. Evans, A.G.; Pratt, P.L. Work hardening in ionic crystals. *Philos. Mag.* **1970**, *21*, 951–970. [CrossRef]

96. Michalak, J.T. The influence of temperature on the development of long-range internal stress during the plastic deformation of high-purity iron. *Acta Metall.* **1965**, *13*, 213–222. [CrossRef]

97. Kohzuki, Y. Study on the interaction between a dislocation and impurities in KCl:Sr^{2+} single crystals by the blaha effect Part IV influence of heat treatment on dislocation density. *J. Mater. Sci.* **2009**, *44*, 379–384. [CrossRef]

MDPI
St. Alban-Anlage 66
4052 Basel
Switzerland
Tel. +41 61 683 77 34
Fax +41 61 302 89 18
www.mdpi.com

Crystals Editorial Office
E-mail: crystals@mdpi.com
www.mdpi.com/journal/crystals

www.ingramcontent.com/pod-product-compliance
Lightning Source LLC
Chambersburg PA
CBHW051715210326
41597CB00032B/5491